Water Resources Development and Management

More information about this series at http://www.springer.com/series/7009

Jilili Abuduwaili · Gulnura Issanova
Galymzhan Saparov

Hydrology and Limnology of Central Asia

Springer

Jilili Abuduwaili
State Key Laboratory of Desert and Oasis Ecology, Xinjiang Institute of Ecology and Geography, Chinese Academy of Sciences
Urumqi
China

and

Research Centre of Ecology and Environment of Central Asia (Almaty)
Almaty
Kazakhstan

Gulnura Issanova
State Key Laboratory of Desert and Oasis Ecology, Xinjiang Institute of Ecology and Geography, Chinese Academy of Sciences
Urumqi
China

and

Faculty of Geography and Environmental Sciences
Al-Farabi Kazakh National University
Almaty
Kazakhstan

and

Research Centre of Ecology and Environment of Central Asia (Almaty)
Almaty
Kazakhstan

Galymzhan Saparov
Research Centre of Ecology and Environment of Central Asia (Almaty)
U.U. Uspanov Kazakh Research Institute of Soil Science and Agrochemistry
Almaty
Kazakhstan

ISSN 1614-810X ISSN 2198-316X (electronic)
Water Resources Development and Management
ISBN 978-981-13-0928-1 ISBN 978-981-13-0929-8 (eBook)
https://doi.org/10.1007/978-981-13-0929-8

Library of Congress Control Number: 2018943710

Printed on acid-free paper

This Springer imprint is published by the registered company Springer Nature Singapore Pte Ltd.
The registered company address is: 152 Beach Road, #21-01/04 Gateway East, Singapore 189721, Singapore

Preface

Water resources of Central Asia are formed mainly in mountainous areas and occupied more than 20% of the Aral Sea basin area (3,50,000 km^2). About 90% of the surface runoff is formed in this area that is 115 billion m^3 of water (in the average water year). Mainly, they are located within two large river systems: Amudarya and Syrdarya rivers.

Lakes are a source of significant water resources, organic and mineral raw materials. Lakes are used for water supply, irrigation, watering of livestock, and for the development of fishing. Some salt lakes are deposits of valuable mineral raw materials for the chemical, food, and other industries (sodium and magnesium chlorides, sodium sulfates, etc.). In addition, most of the highly mineralized lakes contain significant reserves of medicinally valuable mud (mineral mud), and they have a great balneological significance. In addition, some lakes have a great recreational value.

Lakes are original water complexes and one of the most important elements of natural landscapes. They differ from the surrounding natural complexes of land. There are thousands of large and small lakes spread on the territory of Central Asia and Northwest China. They distributed on the plains and in the mountains as well. There are more than 51,000 lakes with a total area of 14,571 km^2 on the territory of Central Asia, including in the basins of the Amudarya, Syrdarya, Shu, Talas rivers, Issyk-Kul Lake, Eastern Pamir, and Tien Shan, as well as the drainless regions of Turkmenistan. There are 700 lakes in the arid regions of Northwest China, including 29 lakes in size more than 10 km^2.

Limnology is an independent complex science of continental water reservoirs with slow water exchange such as lakes, water reservoirs, ponds belonging to the family of geographical disciplines. This complex science studies the interaction of physical, chemical, and biological processes taking place in water reservoirs, as well as the history and evolution of lakes.

At present, the problem of integrated study of lakes is becoming increasingly important. Its correct solution is the study of the lakes in the arid regions of Central Asia and Northwest China. Since the Lobnor lake was dried up because of increasing anthropogenic factor, the same situation was observed in the Aral Sea

Basin. The basin is characterized by crisis level and became a catastrophical region. The rapid development of the economy requires the full utilization of all natural resources.

The main aim of this book is a comprehensive study of the development of lake systems and water reservoirs and impact of climate change on water resources in Central Asian countries. And this book provides information about genesis of lake basins, physical and chemical properties of water in lakes, hydrological regime (water balance and fluctuation levels) of lakes of Central Asia and Northwest China.

Study and research on lake systems are important and required in the background of climate change and sustainable development and use of natural resources in Central Asia. The study of the possible impact of climate change on water sources is very relevant, since the role of lakes in human life is determined primarily by large reserves of freshwater. Lakes are very sensitive in terms of climate change and anthropogenic factors, since lake sediments are the main parameter or source in climate change and anthropogenic factors. In addition, lakes are natural runoff regulators and the core of specially protected areas such as national parks, reserves of various levels. In modern conditions, the role of lakes is significantly increasing, because they remain custodians of clean freshwater in conditions of permanent anthropogenic factor and impact.

This book mainly addressed to scientists and researchers whose research has been focused on lakes and use of natural resources, irrigation, hydropower, and water supply as well as students and planners. We believe that our contribution on the study of lakes in Central Asia and in the arid regions of Northwest China will be a source of information and knowledge for all readers who feel responsible for the sustainable use and development of natural resources such as lakes and other water bodies.

Urumqi, China — Jilili Abuduwaili
Urumqi, China/Almaty, Kazakhstan — Gulnura Issanova
Almaty, Kazakhstan — Galymzhan Saparov

Acknowledgements

Many scientists have contributed, supported and provided the basis for this book.

We sincerely thank all of the scientists: Dr. Ma Long and Dr. Yilinuer (Ilnur) Alifujiang from China and Dr. Salamat Abdyzhapar uulu, Burul Alymkulova and Talant Asankulov from Kyrgyzstan, who had provided help during the writing of this book by sharing their scientific publications, all of which became part of this book.

Separate thanks go to Dr. Azamat Kaldybayev from Kazakhstan and Dr. Kanat Samarkhanov, Dr. Alim Samat, and for technical support in GIS mapping for some chapters of this book.

We owe special thanks to Dr. Ge Yongxiao for his support and making comprehensive conditions to write this valuable book.

Thanks to library staffs in Kazakhstan and China, namely Wu Miao, for their help and assistance in gathering such base materials.

We thank Gulmira Nazanova, Shen Hao, Zhang Weiyan, Pu Jia, and Hua Yousheng for their contributions in the designing of tables and some illustrative materials.

We sincerely thank all of the colleagues from Xinjiang Institute of Ecology and Geography, Chinese Academy of Sciences, in particular Prof. Chen Xi and Prof. Lei Jiaqiang, and U.U. Uspanov Kazakh Research Institute of Soil Science and Agrochemistry, in particular professor and academician Saparov Abdulla.

Thanks to all the colleagues from Research Centres of Ecology and Environment of Central Asia, including the branches in Kazakhstan (Almaty), Kyrgyzstan (Bishkek), and Tajikistan (Dushanbe).

The authors thank all the contributors, colleagues, friends, and well-wishers for their support and engagement.

This research was funded and supported by the National Natural Science Foundation of China (U1603242), the Science and Technology Service Network Initiative (Y838031) and CAS "Light of West China" Program (2016-QNXZ-A-4), and the Strategic Priority Research Program of Chinese Academy of Sciences (Grant No. XDA2006030102).

Content and Structure of This Book

This book summarizes the outcomes of research results and studies on water resources and lakes and lake systems in Central Asian countries (Kazakhstan, Kyrgyzstan, Uzbekistan, and Tajikistan) and Northwest China. This book has twelve individual chapters. Studies and research results on lake systems and water resources were analyzed and overviewed within those chapters. Chapter 1, "Water Resources and Impact of Climate Change on Water Resources in Central Asia," provides an overview and analyzes the lakes in Central Asia and impact of climate change on water resources. Chapter 2, "Water Resources in Kazakhstan," contains three sections providing general information on water resources (water reservoirs, rivers, groundwater) in Kazakhstan, as well as water availability and state of water resources within water-economic basins in Kazakhstan. Chapter 3, "Lakes in Different Regions of Kazakhstan," considers a regional division of lakes within Kazakhstan. It describes regional distribution of lakes in different regions of Kazakhstan. Chapter 4, "Lakes in the Central Kazakhstan," contains three sections providing information on morphology and genetic types of the lakes in Central Kazakhstan, as well as hydrological and hydrochemical regimes of the lakes in Central Kazakhstan. Chapter 5, "Morphometry and Genesis of Lakes in Kazakhstan," provides and describes information about morphometry and genesis of lake basins in Kazakhstan. Chapter 6, "Water Balance and Physical and Chemical Properties of Water in Lakes of Kazakhstan," considers a water balance, fluctuation of levels in lakes, and physical and chemical properties of water in lakes of Kazakhstan. Chapter 7, "Lake and Sea Basins in Kazakhstan," provides an overall information (hydrography, water regime, hydrographic network, etc.) on water resources and water use and state of water resources in the Lake Basin, Lake System, and Sea basin such as Aral Sea and Balkash Lake, Alakol–Sasykkol and Tengiz-Korgalzhyn lake systems. In addition, environmental impact of the Aral Sea desiccation was discussed in the chapter. Chapter 8, "Water Resources and Lakes in Kyrgyzstan," contains five sections providing information on lakes, and water resources and hydrographic system in Kyrgyzstan, water balance of natural belts and artificial hydrological network, as well as water reservoirs, fresh groundwater, and wetlands in Kyrgyzstan. In addition, use of water resources in Kyrgyzstan was discussed and analyzed. Chapter 9, "Hydrographical and Physical–Geographical Characteristics of the Issyk-Kul Lake Basin and Use of Water Resources of the Basin, and Impact of Climate Change on it," describes and considers physical–geographical characteristics of the Lake and provides information about hydrography and hydrochemistry of the Issyk-Kul lake, as well as use of water resources of the Lake in different sectors of economy and sustainable water management, and impact of climate change on the Lake basin was analyzed and discussed. Chapter 10, "Water Resources and Lakes in Uzbekistan," offers a general information on water reservoirs as a source of manufacture, and management of lakes, as well as use of water resources in Uzbekistan, was considered. Chapter 11, "Water Resources in Tajikistan," provides an overview on glaciers and rivers in Tajikistan. Chapter 12, "Lakes in Arid Regions of Northwest China," contains three sections providing a general information on lake distribution and lake types in arid regions of China, as well as hydrographical characteristics of the Ebinur and Sayram Lakes and environmental issues in the region, and climate and environmental changes over the past 150 years in the Chaiwopu Lake were considered and discussed.

Contents

1 Water Resources and Impact of Climate Change on Water Resources in Central Asia 1
1.1 Water Resources, Lakes and Water Reservoirs, and Glaciers in Central Asia 1
1.2 Impact of Climate Change on Water Resources in Central Asia 5
1.3 Impact of Climate Warming on Glacier and Permafrost Melting in Central Asian Mountains 7
References 9

2 Water Resources in Kazakhstan 11
2.1 General Characteristics of Water Resources in Kazakhstan 11
2.2 Water Reservoirs, Glaciers, Rivers, and Groundwaters in Kazakhstan 12
2.3 Water Availability and State of Water Resources Within Water-Economic Basins in Kazakhstan 32
2.4 Hydrochemical regime of rivers in Kazakhstan 42
References 45

3 Lakes in Different Regions of Kazakhstan 47
3.1 Distribution of Lakes Within Kazakhstan 47
3.2 Lakes in the Northern, Western, Eastern, and Southern Kazakhstan 50
References 175

4 Lakes in the Central Kazakhstan 177
4.1 Morphology and Genetic Types of the Lakes in the Central Kazakhstan 177
4.2 Hydrological Characteristics of the Lakes in the Central Kazakhstan 178

4.3 Hydrological and Hydrochemical Regimes of the Lakes in the Central Kazakhstan . . . 192
References . . . 196

5 Morphometry and Genesis of Lakes in Kazakhstan . . . 199
5.1 Morphometry and Genesis of Lake Basins in Kazakhstan . . . 199
5.2 Tectonic Depressions . . . 202
5.3 Depressions of Exogenous Origin . . . 207
References . . . 211

6 Water Balance and Physical and Chemical Properties of Water in Lakes of Kazakhstan . . . 213
6.1 Water Balance and Fluctuation of Levels in Lakes of Kazakhstan . . . 213
6.2 Physical and Chemical Properties of Water in Lakes of Kazakhstan . . . 214
References . . . 220

7 Lake and Sea Basins in Kazakhstan . . . 221
7.1 Caspian Sea . . . 221
7.2 Aral Sea Basin . . . 224
7.2.1 Water Resources of the Aral Sea Basin . . . 227
7.2.2 Environmental Impact of the Aral Sea Desiccation . . . 229
7.2.3 Modern State of the Aral Sea . . . 232
7.2.4 Use of Water Resources in the Aral Sea Basin . . . 246
7.3 Balkash Lake Basin . . . 252
7.3.1 Hydrography and Water Regime of the Balkash Lake . . . 253
7.4 Alakol–Sasykkol Lake System in the Alakol Basin . . . 256
7.4.1 Formation of the Alakol Lake System and Hydrographic River Network of the Basin . . . 262
7.4.2 Modern State of Water Resources in the Alakol Lake System . . . 265
7.5 Tengiz–Korgalzhyn Lake System . . . 266
References . . . 268

8 Water Resources and Lakes in Kyrgyzstan . . . 271
8.1 Water Resources and Hydrographic System in Kyrgyzstan . . . 271
8.2 Water Balance of Natural Belts in Kyrgyzstan and Artificial Hydrological Network . . . 275
8.3 Water Reservoirs, Fresh Groundwaters, and Wetlands in Kyrgyzstan . . . 278
8.4 Use of Water Resources in Kyrgyzstan . . . 285
8.5 Lakes in Kyrgyzstan . . . 288
References . . . 294

9 Hydrographical and Physical–Geographical Characteristics of the Issyk-Kul Lake Basin and Use of Water Resources of the Basin, and Impact of Climate Change on It 297
9.1 Physical–Geographical Characteristics of the Issyk-Kul Lake Basin 297
9.2 Hydrography of the Issyk-Kul Lake 300
9.3 Hydrochemistry of the Issyk-Kul Lake Basin 303
9.4 Use of Water Resources of the Issyk-Kul Lake Basin and Sustainable Water Management 316
9.5 Impact of Climate Change on the Issyk-Kul Lake Basin 319
9.6 System Dynamics Modeling of Water-Level Variations of the Issyk-Kul Lake 329
References 349

10 Water Resources and Lakes in Uzbekistan 359
10.1 The Main River Basins in Uzbekistan 359
10.2 Lakes and Water Reservoirs as a Source of Manufacture/Production 360
10.3 Use of Water Resources in Uzbekistan 362
References 364

11 Water Resources in Tajikistan 367
11.1 Lakes and Water Reservoirs in Tajikistan 367
11.2 The Main River Basins and Rivers in Tajikistan 369
11.3 Glaciers in Tajikistan 371
References 373

12 Lakes in Arid Regions of Northwest China 375
12.1 Distribution and Types of Lakes in Arid Regions of Northwest China 375
12.2 Ebinur Is the Largest Salt Lake in the Northwest China 397
12.3 Hydrograhical Characteristics of the Sayram Lake and Environmental Issues in the Region 400
12.4 Climate and Environmental Changes Over the Past 150 Years in the Chaiwopu Lake 402
References 407

Index 411

About the Authors

Jilili Abuduwaili is a professor and deputy director at the Xinjiang Institute of Ecology and Geography, Chinese Academy of Sciences, China, and co-director of the Research Center of Ecology and Environment of Central Asia, Almaty. He holds a doctoral degree from the Lomonosov Moscow State University, Russia.

He is a member of a council of the China Natural Resources Society and a deputy chairman of the Xinjiang Geographical Society as well as a deputy chairman of the Xinjiang Natural Resources Society.

He is an associate editor of the *Journal of Arid Land* and editor in chief of the journals *Arid Land Geography* and *Arid Land Research* (both in Uyghur language). He is also an editorial board member of the journals *Arid Ecosystems* (*Russia*) and *Soil Science and Agrochemistry* (*Kazakhstan*). He is a supervisor for postgraduate students.

Since 2015, he has been a foreign academician at the Academy of Agriculture of Kazakhstan. His main fields of work and expertise are in the areas of natural resources and environments, ecological safety, lake environments, disaster, and landscape changes in arid lands.

He has built international cooperation in research between Russia and Central Asian countries. He has published more than 200 papers in international and domestic journals and 7 books: *Lakes and Salt Dust Storms in Arid Lands* (in Chinese), *Overview of Central Asian Environments* (in Chinese), *Aeolian processes as dust storms in the deserts of Central Asia and Kazakhstan* (in English) *by Springer Nature, Salt Dust*

Storms: Special Disaster in Arid Lands (in Uyghur), *Climate Change Effects to the Ecological System in Tien Shan Mountains* (in Russian), *Physical Geography of Kyrgyzstan* (in Russian), *Arid Soil in Kazakhstan: Current States and Utilization* (in Russian).

Gulnura Issanova holds a doctorate degree in Natural Sciences and is an associate professor, scientist, and researcher at U.U. Uspanov Kazakh Research Institute of Soil Science and Agrochemistry and a scientific secretary at the Research Centre of Ecology and Environment of Central Asia, Almaty, Kazakhstan.

She studied at the Al-Farabi Kazakh National University for bachelor's degree (B.Sc.) and master's (M.Sc.) degree in Physical Geography and Xinjiang Institute of Ecology and Geography, Chinese Academy of Sciences, China, for her doctoral degree. She did a postdoc under the CAS President's (Bai Chunli) International Fellowship Initiative (PIFI) for 2017 at the Xinjiang Institute of Ecology and Geography, Chinese Academy of Sciences, China. She is a holder of the *Foreign Expert Certificate of the People's Republic of China.*

Her research interest was focused on problems of soil degradation and desertification, in particular the role of dust and sand storms in the processes of land and soil degradation and desertification during Ph.D. study. Currently, she is interesting and focusing on water resources and lakes (availability, state, and consumption/use) in Central Asian countries.

She participates regularly in the International Scientific Activities (Conference, Forum, and Symposium) on Environmental Problems as well as writes articles and monographs on the subject and takes part in local and international projects.

She has published many SCI papers in international peer-reviewed journals with high level and wrote a handbook, "How to Write Scientific Papers for International Peer-Reviewed Journals."

She is an author of the monograph *Aeolian processes as dust storms in the deserts of Central Asia and Kazakhstan* by Springer Nature (2017) and co-author of *Man-Made Ecology of East Kazakhstan* by Springer Nature (2017) and *Overview of Central Asian*

Environments (in Chinese) and the handbook *Methodical Handbook on Interpretation of Saline Soils* (in four languages: Kazakh, Russian, English, and Chinese).

She became a laureate of the International Award "Springer Top Author" and was awarded in the Nomination "Springer Young Scientist Awards-2016" for high publication activity in scientific journals published by Springer Nature. She was included to the list of "Top-8 Young Scientists and Top-18 leading Scientists from Kazakhstan," 2016.

Galymzhan Saparov is a Ph.D. candidate in Agricultural Sciences. He is a head of department of Soil Ecology at U.U. Uspanov Kazakh Research Institute of Soil Science and Agrochemistry as well as head of laboratory at the Research Centre of Ecology and Environment of Central Asia, Almaty, Kazakhstan.

He graduated from Kazakh Agricultural Institute in 1995 and holds a Ph.D. degree from U.U. Uspanov Kazakh Research Institute of Soil Science and Agrochemistry in 2004.

His research interest is focused on problems of soil degradation and remediation of disturbed lands.

He has published more than 45 scientific papers, including 2 recommendations in production and 3 innovative patents.

Chapter 1
Water Resources and Impact of Climate Change on Water Resources in Central Asia

1.1 Water Resources, Lakes and Water Reservoirs, and Glaciers in Central Asia

Central Asia covers the territory of five countries such as Kazakhstan, Kyrgyzstan, Tajikistan, Turkmenistan, and Uzbekistan. It is located in the center of the Eurasian continent with an area of 3,882,000 km^2 and with a population of about 55 million people. It borders on Afghanistan and Iran in the south, China in the east, and Russia in the west and north (Grinyayev and Fomin 2009).

The territory of Central Asia, with the exception of the highlands, is inadequate in humidification, and the humidification for the most part is extremely insufficient: The ratio of the average annual precipitation to the average annual evaporation is less than unity. This is connected with the great (in comparison with other CIS countries) sparseness of the hydrographic network. The density of the river network in the desert plains of Central Asia is about 2 m per 1 km^2, while in the northern half of the Russian Plain it reaches 300–350 m per 1 km^2 (Grinyayev and Fomin 2009).

Central Asian countries are located in the arid zone, where without irrigation it is impossible to cultivate crops and to obtain sustainable crops. Therefore, there is and prevails irrigation in almost all countries of the region, which requires large amount of water resources. Central Asia has approximately 170–180 km^3 of water resources, of which today more than 90% is used.

Water resources between the countries of the region are divided unevenly. More than 90% of Central Asia's water resources sources are concentrated in Kyrgyzstan and Tajikistan. At the same time, Uzbekistan and Kazakhstan are the main water users in the region. The share of Uzbekistan accounts for more than half of the water.

Syrdarya and Amudarya river flows are the main source of water in the region. They are formed in the mountains of the Pamir and Tien Shan. The Syrdarya river flows from Kyrgyzstan through Tajikistan to Uzbekistan and through the densely

J. Abuduwaili et al., *Hydrology and Limnology of Central Asia*, Water Resources Development and Management, https://doi.org/10.1007/978-981-13-0929-8_1

populated Ferghana Valley to Kazakhstan and the Amudarya river from Tajikistan to Turkmenistan and Uzbekistan.

Lakes in Central Asia are distributed unevenly (Fig. 1). There are many lakes in the mountainous area and plains of Central Asia.

Fig. 1.1 xxx

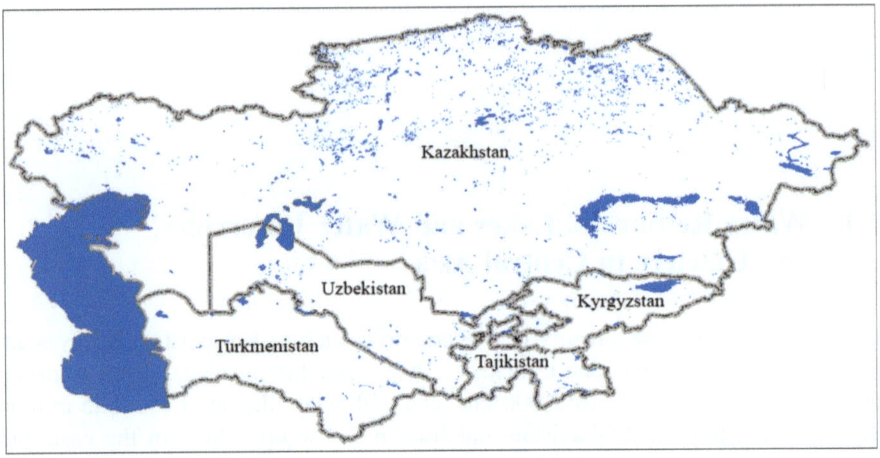

The lake is a natural pond with a slowed down water exchange. It is formed as a result of water filling with a deepening of the earth's surface. The size of the lake varies from very large, such as the Caspian Sea and the Great Lakes in North America, to tiny pools of several hundred square meters and even less. The lake water can be fresh as in the Baikal lake or salty as in the Dead Sea. Lakes are found at any altitude; from the lowest point on the Earth surface that is 408 m (Dead Sea) and high in the mountains (Lake Merzbacher on the glacier Engilchek, Kyrgyzstan). Many lakes exist permanently, while others are only occasionally filled with water. However, lakes of all types have a number of common physical, chemical, and biological characteristics and are subject to many general laws.

There are more than 51,245 lakes in Central Asia, namely in Kazakhstan more than 48,000 lakes, 1923 lakes in Kyrgyzstan, in Uzbekistan 22 lakes with size more than 1 km^2, and in Tajikistan more than 1300 lakes. The drying lake–sea Aral Sea and Lake Balkash with freshwater in western and salty eastern parts are the largest lakes and one of the deepest lakes in the world—Issyk-Kul Lake. The lakes in Kyrgyzstan (84%) and in Tajikistan (78%) are distributed in the mountains at an altitude of 3000–4000 m a.s.l and 3500 m a.s.l, respectively. Most mountain lakes are concentrated in modern glaciers and alpine zones. There are more than 3000 very small high-altitude glacial lakes, dozens of seasonal regulation reservoirs, thousands of ponds of decade-old and daily regulation (Gagloeva 2016). The plain lakes are mostly distributed in Kazakhstan and Uzbekistan. In Uzbekistan, plain lakes can be found in the Deltas of the Amudarya and Syrdarya rivers. There are many natural lakes in the mountainous areas and hollows of Central Asia. The mountainous or alpine lakes are mostly concentrated in Tajikistan and Kyrgyzstan. There are different types of the mountainous lakes. Most large lakes occupy basins of tectonic origin (Issyk-Kul, Chatyr-Kul, Karakul, Sarychelek). In addition, Sarez and Yashinkul in the Pamirs belong to the lagoon lakes.

There are numerous lakes of glacial origin. Lake Zorkul is one of the largest lakes, located at an altitude of 4125 m in the East Pamirs. Besides, there are karst lakes. The water in the lakes is usually fresh or brackish, depending on the quality of the inflow (Grinyayev and Fomin 2009).

Due to the discharge of drainage water into the drainless basins, many lakes of anthropogenic origin have arisen. Most of them are small. The largest lakes of this type are the Sarykamysh (in the lower reaches of the Amudarya river) and Arnasai (in the middle reaches of the Syrdarya river). Due to the low capacity of the Syrdarya river bed which is below the Shardara water reservoirs (on the border between Kazakhstan and Uzbekistan), in excessively large water years, excess water is discharged into Lake Arnasai. The volume of water in lakes of anthropogenic origin is estimated at 40 km^3. However, pumping is required to use this water. In addition, the water in the lakes is highly mineralized. In the future, this water can be best used for purposes in fisheries and biodiversity conservation (Grinyayev and Fomin 2009).

The total capacity of the water reservoirs in Central Asia is 176.9 km^3, of which 54% are in Kazakhstan (Table 1.1).

There are fifteen water reservoirs in the region, and each reservoir has a capacity of more than 1 km^3, of which six are in Uzbekistan, four in Kazakhstan, two in Turkmenistan, and two in Tajikistan. Most of them are multi-purpose, and they are used for the production of hydropower, irrigation, water supply, and flood control. The total capacity of these fifteen water reservoirs is 129.4 km^3 or 72% of the total capacity of all water reservoirs in Central Asia. The Bukhtyrma water reservoir in Kazakhstan has the largest capacity (50 km^3). The capacity of Toktogul water reservoir in Kyrgyzstan, Kapshagai water reservoir in Kazakhstan and Nurek water reservoir in Tajikistan is 20, 19 and 11 km^3, respectively. In Uzbekistan, a large water reservoir is the Tuyamuyun (8 km^3), the Zeid water reservoir (2 km^3) in Turkmenistan (FAO 2012) (Table 1.2).

Table 1.1 Water reservoirs in Central Asia (FAO 2012)

Country	Water reservoir capacity (m^3)	% from the region
Kazakhstan	95.5	54
Tajikistan	29.5	17
Kyrgyzstan	23.5	14
Uzbekistan	22.2	12
Turkmenistan	6.2	3
Central Asia	176.9	100

Table 1.2 Water reservoirs with capacity of more than 1 km^3 (FAO 2012)

Water reservoir	River	Basin	Exploitation year	Volume (m^3)	Surface area (km^2)	Reservoir function	Country
Bukhtyrma	Ertis	Ob	1960	49.6	5490	I, H, W, F	Kazakhstan
Toktogul	Naryn	Naryn	1974	19.5	–	H	Kyrgyzstan
Kapshagai	Ile	Balkash–Alakol	1970	18.6	1847	I, H, W	Kazakhstan
Nurek	Vakhsh	Amudarya	1980	10.5	98	I, H, W, F	Tajikistan
Tuyamuyun	Amudarya	Amudarya	1980	7.8	790	I, H, F	Uzbekistan
Shardara	Syrdarya	Syrdarya	1968	5.2	783	I, H, W, F	Kazakhstan
Kairakkum	Syrdarya	Syrdarya	1959	4.2	5450	I, H	Tajikistan
Shulba	Ertis	Ob	1988	2.4	255	I, H, W, F	Kazakhstan
Zeid	Karakum canal	Amudarya	1986	2.2	465	I, W	Turkmenistan
Charvak	Chirchik	Syrdarya	1977	2	22	I, H	Uzbekistan
Andizhan	Karadarya	Syrdarya	1978	1.9	55	I	Uzbekistan
Talimardzhan	Amudarya	Amudarya	1985	1.5	–	I	Uzbekistan
Pachkamar	Guzor	Amudarya	1961	1.5	–	I	Uzbekistan
Dostluk	Tedzhen	Tedzhen	2004	1.3	48	I, H, W, F	Turkmenistan
Tudakul	Tudakul	Amudarya	1953	1.2	–	I, H	Uzbekistan
Central Asia				129.4			

I irrigation, *H* hydropower, *W* water supply, *F* flood protection
Tuyamuyun water reservoirs consist of four reservoirs such as Ruslovoi, Sultansandzhar, Kapparas, and Koshbulak

1.2 Impact of Climate Change on Water Resources in Central Asia

Global climate change is one of the main issues among the major environmental problems facing the world community in recent decades. The issue of the impact of climate change on water resources in the Central Asian countries requires more detailed study from a regional perspective.

The arid and semi-arid climates are common for most part of Central Asia with minor precipitation during the year. Climate change created problems related to the ability of countries to meet the growing demand for water. As a result of a decrease in precipitation and a rise in temperature, the problem of water scarcity may worsen, especially in those parts of Asia where water resources are already stressed due to the growing water demand and inefficient water use.

There has been an inter-seasonal and spatial variability in rainfall throughout Asia over the past several decades. In general, the frequency of more intense precipitation events has increased in many parts of Asia, causing severe floods, landslides, and debris flows and mudflows, while the number of rainy days and total annual rainfall decreased. Nevertheless, there are reports that the frequency of extreme rainfall in some countries shows a decreasing tendency (UNEP 2006).

Global climate change affects processes occurring in the environment and changing the existing mechanism functions of the entire environmental system. Global climate warming will cause an increase in the melting intensity of glacial cover of the planet. Reducing the areas of arctic ice sheets can lead to significant changes in the water cycle in nature. The ice melting in the Arctic Ocean and the subsequent warming in the Arctic region can cause a change in the existing interaction between the underlying surface and the atmosphere of the entire Northern Hemisphere, and consequently the entire planet (Gagloeva 2016).

In general, the mechanism of moisture circulation between oceans and continents can change. The intensity of the zonal atmospheric circulation can decrease because of the decrease in the temperature gradient between the poles and the equator (WERCA 2008).

The role of the temperature difference between the sea and land will increase, and will enhance the monsoon circulation, especially on the coasts of Eurasia. The decrease in the rates of zonal transport of air masses and the intensity of cyclonic activity due to the decrease in the temperature gradient in the "pole–equator" system will cause an increase in atmospheric precipitation near the seas and it decreases within the continent, especially in winter.

The total water resources of Central Asia consist of river runoff formed from the water of precipitation, melted glacial water and underground feeding, and underground waters pumped by wells. The wide-scale impact of climate change on the globe, first of all, affects the state of glaciers, the seas of the World Ocean, and the snow cover of the Earth. In the period of cooling, the world water balance changes in the direction of increasing moistening of the continents and the mass of glaciers is increased. The water balance of the oceans is becoming negative, and their level

The wide-scale impact of climate change on the globe affects the state of its glaciers, the seas of the World Ocean, and the snow cover of the Earth. As a result of climate warming, the glaciers started to melt. Asian glaciers are melting since the 1960s and in a constant rate. However, individual glaciers may fall out of this list, and some of them actually come in, their thickness increases and it is possibly due to increased precipitation. As a result of the continuing melting of glaciers, the glacial runoff and the frequency of breakthrough of glacial lakes that cause mudflows and avalanches increases. So, the current climate warming leads to a steady reduction of the glaciers in the Tien Shan and a reduction in their glacial coefficients which show the ratio of the areas of glacial accumulation to the entire area of glaciers (Gagloeva 2016).

As is known, small glacial coefficients are characteristic for degrading glaciers, where the arrival of ice matter does not compensate for its consumption. Thus, there is a deepening process of the decay of glaciation in the basins of the Kichi Naryn, Talas, and Assa rivers and on the southern slope of Kungei Alatau, where the glacial coefficients remained 0.45 (Gagloeva 2016).

Hydrometeorological observations showed that the asynchronous nature of atmospheric precipitation and air temperature in the high-mountain zone of the Tien Shan negatively affects the balance of glaciers and affects the total water content of rivers with a significant glacier catchment.

Past and future changes in water resources of Tajikistan are also associated with climate change: a decrease in precipitation and an increase in air temperature. At that time, according to the most conservative estimates, the glaciers of Tajikistan lost more than 20 km^3 of ice only in the twentieth century. Intensively degrading small glaciers with areas less than 1 km^2, which account for 80% of all glaciers. The average annual river runoff of Tajikistan over the last 30 years has decreased annually by 110 million m^3 per year (Gagloeva 2016). The forecasts of Tajik specialists and scientists show that thousands of small glaciers will disappear in Tajikistan until 2050. Therefore, the area of its glaciation will be reduced by 20%, and the volume of ice will decrease by 25%. Consequently, this will lead to a reduction in glacial feeding of rivers by 20–40%. At the same time, there is a prediction that on increasing in the amount of atmospheric precipitation by 14–18% will not have a significant effect on the runoff, since most of the precipitation will be spent for evaporation from watershed surfaces.

In the mountain systems of Kazakhstan, there is also a reduction in the number and size of glaciers. According to forecasts, due to global warming the water resources of the main rivers of Kazakhstan can decrease by 20–40% in the next decades.

According to studies and researches, since the middle of the twentieth century, the glaciation in Central Asia has decreased so much that by the middle of the twenty-first century the revealed trend may lead to the disappearance of many glaciers (IPCC 2007).

Changes in the quantity and regime of water resources in Central Asia are associated with global climate warming, and with intensive development of irrigation in the river basins of the region. Studies also show that intensive melting of glaciers continues in the formation zone of the Syrdarya and Amudarya river flows. For 50 years, the volume of glaciers has decreased by 20–40%, and in recent years, the rate of reduction is about 1% per year (UN 2002).

References

Alamanov SK, Lelevkin VM, Podrezov OA (2006) Climate change and water problems in Central Asia: A training course for students of natural and humanitarian specialties. Bishkek (in Russian)

FAO (2012) Irrigation in Central Asia in numbers. In: Frenken K (ed) Review AQUASTAT, water reports (in Russian)

Gagloeva AE (2016) The climate change impact on water resources of Central Asia. J Water Econ Kaz 3(71):49–54 (in Russian)

Grinyayev SN, Fomin AN (2009) Actual issues of application of the mechanism of exchange trade for solving water and energy problems of the Central Asia countries, Analytical report. Moscow, 52 p (in Russian)

IPCC (2007) Climate change. The fourth assessment report of the intergovernmental panel on climate change. Summary for Policymakers. The physical science basis, vol 1, p 18. Impacts, Adaptation and Vulnerability, Mitigation, vol 2, pp 24–26

UN (2002) A diagnostic report for the preparation of a regional strategy for the rational and efficient use of water resources. Project working group on energy and water resources, 83 p (in Russian)

UNEP (2006) Climate change and water problems in Central Asia. In: World wildlife fund of Russia, Moscow-Bishkek (in Russian)

WERCA (2008) Water-energy resources of Central Asia: problems in use and development. Branch survey of the Eurasian Development Bank, Almaty (in Russian)

Chapter 2
Water Resources in Kazakhstan

2.1 General Characteristics of Water Resources in Kazakhstan

The territory of Kazakhstan is 2717.3 thousand km^2 and ninth biggest country in the world. It extends from west to east by 2995 km and from north to south by 1600 km. Kazakhstan borders with Russia in the north and west; in the south with Turkmenistan, Uzbekistan, and Kyrgyz Republics; and with the People's Republic of China in the southeast (Dzhanalieva et al. 1998).

Intra-continental location in the center of the Eurasian continent is a characteristic feature of Kazakhstan, which affects the entire physical–geographical conditions: climate, hydrography, soil, and vegetation covers and fauna. There are four main hydrological regions in Kazakhstan: the Basin of Ob river, which flows into the Arctic Ocean; the Caspian Sea basin; the Aral Sea basin; and basin of inland lakes, depressions, or deserts. Surface water resources are distributed over the country is extremely uneven and characterized by significant long-term and seasonal dynamics. Only 3% of the total water resources is formed in Central Kazakhstan. The western and southwestern regions (Atyrau, Kyzylorda and, in particular, Mangystau) suffer from a significant deficit of water, and there is almost no freshwater. Almost 75% of the surface water resources generated within the country belong to the Balkash–Alakol and Ertis water-economic basins in the east and northeast parts of Kazakhstan. About 90% of the surface runoff is formed in spring, exceeding the capacity of existing water reservoirs (FAO 2012).

Kazakhstan is located in four natural climate zones: forest steppe, steppe, semidesert, and desert. In Kazakhstan, severe Siberia and sultry Central Asia, mountainous taiga and desert, vast plain steppes and mountains covered with eternal snows come together. The diversity of the terrain or relief and climatic conditions of Kazakhstan causes the uneven distribution of surface waters over its area.

J. Abuduwaili et al., *Hydrology and Limnology of Central Asia*, Water Resources Development and Management, https://doi.org/10.1007/978-981-13-0929-8_2

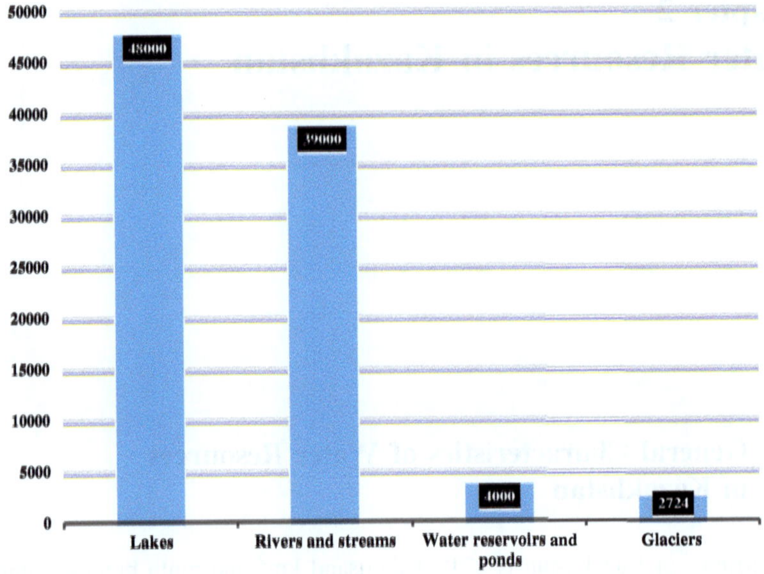

Fig. 2.1 Water resources of Kazakhstan

There are 48,000 lakes, 39,000 rivers and streams, 4000 water reservoirs and 2724 glaciers in Kazakhstan (Fig. 2.1) (FAO 2012). There are very few lakes and rivers in desert regions; it is much larger in the north part of Kazakhstan and in the mountains (Philonets and Omarov 1974).

Lakes play a significant role in providing the national economy with certain types of industrial raw materials. They have such important resources as water, fish, various salts, building materials (gravel, gravel, sand, and mud), reed beds, and other aquatic vegetation, sapropels, peat deposits, therapeutic mud, and brine. Some lakes can be breed a waterfowl and muskrat and create hunting farms and nature reserves. In addition, lakes can serve as rest or tourism places.

2.2 Water Reservoirs, Glaciers, Rivers, and Groundwaters in Kazakhstan

Water reservoirs There are more than 200 water reservoirs within Kazakhstan with a total capacity of more than 95.5 km^3 (excluding ponds and small reservoirs designed to contain spring runoff) (UNDP 2003; Riabtsev et al. 2004).

The water reservoirs are distributed according to their volume as follows: More than 50% of the reservoirs have a volume of 1–5 million m^3 of water (Fig. 2.2). Most of the reservoirs are designed for seasonal regulation of wastewater. The volume of annual runoff is affected by water reservoirs with a regime of long-term

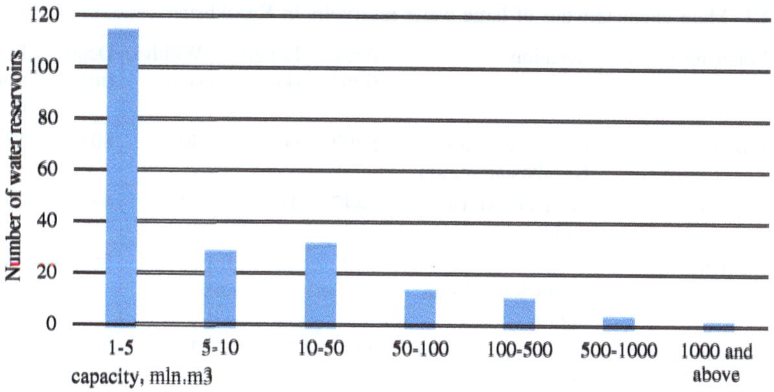

Fig. 2.2 Water reservoirs in Kazakhstan

regulation of sewage, of which about 20. Reservoirs with a capacity of more than 1 km^3 are considered as the largest ones. The largest water reservoirs are Bukhtyrma (on the Ertis River) with a total volume of 49.6 km^3, Kapshagay (on the Ile River) with a volume of 18.6 km^3, Shardara (on the Syrdarya river)—5.2 km^3 (UNDP 2003), the Verkhne–Tobol and the Karatomar (on the Tobyl River)—0.82 and 0.59 km^3, respectively, Vyacheslav and Sergeev water reservoirs (on the Esil River)—0.4 and 0.7 km^3, respectively (Philonets 1981).

The water reservoirs can be classified by their area. Small water reservoirs with area of up to 50 km^2; up to 250 km^2 is medium; up to 1000 km^2 is large, and over 1000 km^2 are the largest ones (Philonets 1981). According to the economic use, they are divided into integrated (complex) and special purpose. Most water reservoirs are multi-purpose that providing hydropower generation, irrigation, and flood control. The largest and large water reservoirs and many of the medium reservoirs belong to complex reservoirs. They are designed to solve the problems of energy, transport, and irrigation. Small ones have a designated purpose. They are used mainly for irrigation in arid regions and in less arid region for small-scale energy, water supply, and fish farming (depends on the water needs). The water reservoirs in the eastern and southeastern regions of Kazakhstan are mainly used for agriculture, and in the central, northern, and western regions—for drinking and industrial needs. The Bukhtyrma, Kapshagay, Shardarya, and Shulbi water reservoirs are connected to HPPs for power generation.

The water reservoirs due to their nature of formation are divided into lake, river, watershed, and liquid types (Vikulina 1979).

The acute need of the national economy of Kazakhstan in water is partially ensured by the creation of water reservoirs on its territory. There are more than four thousand ponds and water reservoirs with different size in Kazakhstan. Their total area is about 10 thousand km^2, and with a freshwater volume is about 90 km^3 (Table 2.1). A big number of the water reservoirs concentrated in central, southern, and eastern parts of Kazakhstan. Such large water reservoirs Bukhtyrma,

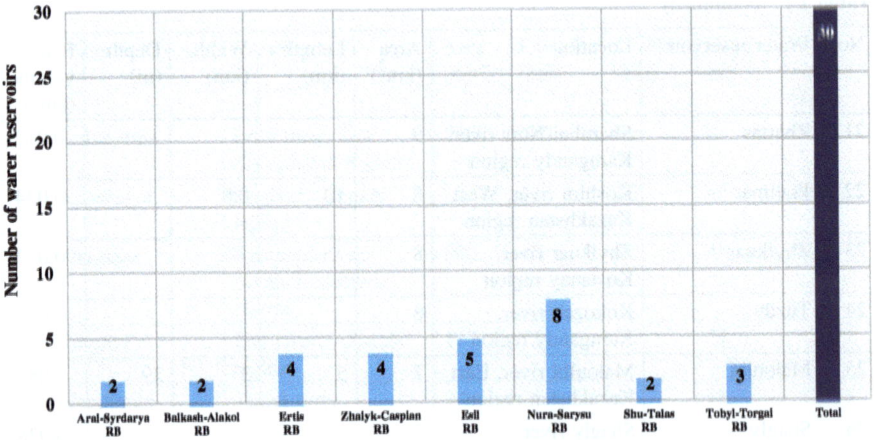

Fig. 2.3 Large water reservoirs in the main water-economic basins (WEB) of Kazakhstan

A total capacity of the Bukhtyrma, Kapshagay, Oskemen, and Shardara hydroelectric power stations is 1.54 million kW. More than 0.2 million hectares of land are irrigated with water from the reservoirs, and 0.5 million hectares are flooded. In addition, reservoir water is used to provide cities and settlements, for technical needs and cattle watering. Some water reservoirs are important for water transport (Philonets 1981).

Bukhtyrma water reservoir is located on the Ertis river in the southeast of the Oskemen City. The capacity of the Bukhtyrma HPP is 675 thousand kW. The area of Bukhtyrma water reservoir is 5490 km^2, the width is 40 km, and the greatest depth is 80 m. The volume of water with full filling will be 49.6 km^3. The water reservoir is the 11th largest reservoir in the world and the 4th in the CIS countries and by the water volume is 24th in the world and 7th in the CIS countries (Philonets 1981). It has a great importance for the energy sector, the fishing industry, intensive farming in floodplain of the Ertis river and water transport.

The structure of the Bukhtyrma reservoir includes the delta of the Black Ertis, the lake Zaisan, White Ertis. According to hydrological and morphometric features, Bukhtyrma reservoir is divided into three parts: upper (lake), middle (mountain–valley), and lower (mountain). Lake is a shallow part covers the area from the delta of the Black Ertis to the mouth of Kurshim river. Soils are mainly represented by gray and dark gray silts, sands, and muddy sands. Gray silt covers the bottom of the whole open part of the lake Zaisan, and the dark silts are concentrated in bays and delta lakes. The mountain–valley part occupies the area from the mouth of the Kurshim river to the mouth of the Narym river. It is characterized by medium depths and a moderately developed littoral. The mountainous part (from the Narym expansion to the dam of the hydroelectric station) is distinguished by great depths and almost completes absence of the littoral. The sands are concentrated in river beds and coastal lakes, and muddy sands are in creeks and streams of rivers, and in Zaisan river, they form a transition zone from coastal sands to gray mud of the open part.

The water reservoir began to fill in April 1960. In 1962, after the confluence of the Ertis spur of the reservoir with Zaisan lake formed a single water reservoir that is Bukhtyrma reservoir. The highest level was reached in August 1971 that is 401.3 m, which is 0.7 m lower than the design level. The reservoir feeds on the flow of the rivers such as Ertis, Bukhtyrma, Kurshim, and a large number of small rivers and atmospheric precipitation on the water mirror.

The rise in water level in Bukhtyrma reservoir occurs from the beginning of May, reaching the highest point in August; the autumn–winter runoff begins from September and lasts until the end of April. From the beginning of April–May, water is released for irrigation of the floodplain of the Ertis river (2.3–2.9 km^3). During this time, the water level in the reservoir is reduced by 0.4–0.65 m. The amplitude of long-term fluctuations of the level reaches more than 5 m. The fluctuation of the water level in the reservoir depends on the volume of water of the falling rivers, evaporation, and also the forced water releases for irrigation of the floodplain of the Ertis river. Seasonal drainage of the reservoir is 1.3–3.8 m (Philonets, 1981).

The *mineralization* and *chemical composition* of water are associated with the hydrochemical characteristics of feeding rivers and with the arrival of thawed waters in spring and their evaporation in summer. The water is fresh and soft slightly alkaline and refers to the hydrocarbonate class in the calcium group (Tables 2.2 and 2.3). Its transparency is 0.2–6 m, and the color is greenish-yellow.

The content of dissolved oxygen is close to normal and often higher up to 181% saturation, free carbon dioxide is usually absent, the water oxidizability is low—1.76–10 mg O_2/l, the active reaction is slightly alkaline. Iodine, bromine, and boron in water are not found, and fluorine is small (Philonets 1981).

Kapshagay water reservoir is located on the Ile river in the southeast of Kazakhstan. It is one of the largest water reservoirs in Kazakhstan. The reservoir has an area of 1847 km^2, a length of 187 km, a width of 23 km, and an average depth is 15 m (the largest depth is about 45 m). The volume of water reaches 28 billion m^3, which is 1.9 times higher than the average long-term annual Ile river flow. Its useful capacity is 6.64 billion m^3 (Philonets 1981).

The Kapshagay water reservoir extends from east to west along the Ile river. It occupies the lowest part of its floodplain. The northern shore is sandy–pebbly and mostly high and steep. This shore is waterless and only in places to the reservoir is suitable dry channels of temporary streams flowing down from the mountains. The northern part of the reservoir is deep in water. The southern coast is low, gently sloping, sandy, and loamy. Numerous mountain streams flow into the reservoir from the south: Sharyn, Shelek, Esik, Turgen, Talgar, Kaskelen, and others.

A capacity of the Kapshagay HPP is 434 thousand kW. The Kapshagay reservoir replenishes due to the flow of the Ile river and its tributaries, precipitation as well. The total catchment area is 113 thousand km^2. The average long-term flow of water from Ile river to the Kapshagay HPP was 14.9 km^3 of water for the observation period from 1911 to 1970 (Philonets 1981).

Table 2.2 A chemical composition of water in the large water reservoirs of Central, Southern, and Eastern Kazakhstan (Mcg/l) (Philonets 1981)

Water reservoirs	pH	Iodine	Fluorine	Bromine	Boron	Zinc	Copper	Manganese	Plumbum	Molybdenum	Cobalt	Nickel	Iron
East Kazkhstan region													
Bukhtyrma	7.2	–	200	–	–	5	2.5	<20	–	2.5	<5	<2.5	<200
Almaty region													
Kapshagay	7.7	1	700	–	<100	>2.5	>2.5	10	<2	<2.5	5	<2.5	<100
Kurti	8.1	–	700	<1000	–	15	<2.5	<20	<5	10	2.5	<2.5	<100
Zhambyl region													
Akkol	8.4	60	300	–	–	–	–	–	–	2.5	–	–	<100
Teris-Ashybulak	7.8	15	200	–	–	15	–	–	–	–	–	–	<100
South Kazakhstan region													
Shardara	7.4	–	300	–	–	–	–	–	10	–	–	–	–
Bugun	7.2	–	500	–	–	–	–	–	–	–	–	–	<100
Burzhar	7.2	–	200	–	–	20	–	–	20	–	–	–	<100
Karagandy region													
Kengir	7.6	120	1250	400	–	20	<2.5	<10	<2.5	5	<1	<1	<100
Klych	7.2	–	1000	400	–	10	<2.5	30	<5	5	5	<2.5	>100
Samarkand	7.5	50	1250	130	–	20	<2.5	10	<2.5	10	<1	<1	<100
Sherubai-Nura (Topar)	7.3	30	10000	330	100	30	<2	<10	–	–	–	–	<100
Botakara	7.0	–	700	1500	–	60	2.5	30	5	5	5	2.5	<100
Tuzdy	7.4	100	500	1500	–	50	<2.5	<20	<5	5	5	2.5	–
Molodezhnoe	7.1	–	600	1000	–	15	2.5	<20	<5	5	20	2.5	–
Kyzylorda region													
Koltugan	8.2	10	500	<100	100	5	2	48	–	15	<2.5	<2.5	<100

Table 2.3 Chemical composition of water of the main tributaries in the Bukhtyrma water reservoir (Philonets 1981)

River	Dry residue (g/l)	pH	Chemical class of water
Black Ertis	0.073	7.45	Bicarbonate, calcium group
Kurshum, estuary	0.139	7.3	Bicarbonate, sodium group
Bukhtyrma river	0.073	6.8	Bicarbonate, sodium group
Bukhtyrma water reservoir	0.096	7.25	Bicarbonate, calcium group

The hydrochemical regime of the reservoir is determined by the Ile river with tributaries, arid climate, and biological processes taking place in the reservoir itself.

The *average mineralization* of water is about 0.5 g/l, with minimum amount in spring, the maximum in autumn. The total hardness varies from 2.8 mg-meq/l in the upper reaches to 3.8 mg-meq/l in the dam. Transparency of water increases from 0.1 m in the upper reaches to 6 m near the dam. The temperature of the upper water layer fluctuates from 21 to 28 °C in the summer.

The chemical composition of water refers to the hydrocarbonate class in the calcium group (Table 2.2). The oxygen content in it is 4.48–12.43 mg/l at 50–146% saturation, and pH is 7.59–8.72. The CO_2 content is from 0.6 to 6.6 mg/l, and its dynamics have a pronounced seasonal character: It is absent in the spring and autumn. Permanganate oxidability of water is high, with an oscillation interval of 2.2–14.29 mg O_2/l. The content of biogenic elements (mg/l) is the following: Ammonium nitrogen (NH_4) varies from 0.2 to 6.6, and it is higher in spring than in other seasons of the year, nitrites (NO_2) from 0.001 to 0.05, nitrates (NO_3) from 0.012 to 1.5, silicon from 1.25 to 12.75, phosphorus from 0.0004 to 0.015, total iron from 0.05 to 1.2 (Philonets 1981).

Shardara water reservoir was built in the middle course of the Syrdarya river, near the town of Shardara in the South Kazakhstan region. It is the largest artificial water reservoir in Southern Kazakhstan. The Shardara water reservoir includes a hydroelectric power plant (HPP) with a capacity of 100,000 kW. Its main purpose is irrigation, more than 100,000 ha of lands have been used, and 500,000 ha of pastures have been flooded. In addition, it is important for fisheries, energy, and water supply (Philonets 1981).

The area of the reservoir is 900 km^2, a length is 100 km along the main watercourse, 25 km in wide, with an average depth is 6.3 m (maximum is 25 m). Most of the water reservoir (80%) is shallow (Ereshchenko 1972).

The reservoir is located in the alluvial terraces of the Syrdarya river. Left bank of the river is flat, and the right bank is steep and hilly. A considerable part of the basin of the Shardara reservoir is composed of clayey deposits and covered with gray and dark gray silts. Black silt occupies a smaller area, and the rest part is sandy.

The filling of the Shardara reservoir began in April 1965 and ended at the end of 1968. The reservoir feeds on the runoff of the Syrdarya river and atmospheric precipitation. The water volume was 5.7 billion m^3.

The amplitude of perennial fluctuations in the water level is 8 m, and the average is 5.4 m. The fluctuation of the water level is caused by the uneven flow of the Syrdarya river and evaporation from the water surface and the generation of water for economic and technical needs.

The *hydrochemical regime* of the Shardara reservoir depends on fluctuations in the water level for the seasons of the year, and the aridity of the climate. Evaporation reaches 1300 mm/year, as a result, the mineralization of water rises from 6.8 to 10%), from return water, from irrigated areas, and from river water and the biological processes taking place in the reservoir itself.

The *mineralization* of water during the year ranges from 710 to 1650 mg/l. As a result of the drainage water from irrigated areas (the Fergana Valley and the Golodnaya Steppe) with a salinity of 4–5 g/kg, the mineralization of water increases significantly (Lopareva and Amirgaliev 1974). Its chemical composition refers to the sulfate class in the sodium group. The pH of the water fluctuated within the range of 7–8.5, the content of CO_2 was 1.5–8.5 mg/l, and the saturation with oxygen was 85–120%. The permanganate oxidizability is not high—1.7–7.7 mg O_2/l. The phosphorus content varies from 0.001 to 0.02 mg/l, silicon is 3–5 mg/l, and nitrate and nitrite nitrogen from 0.01 to 7.0 mg/l. Water hardness is 5.8–29.27 mg-meq/l (Philonets 1981).

Samarkand water reservoir was built in 1934, around the Samarkand village. It is located in the valley of the Nura river in the northwest of the Karagandy City. The filling of the reservoir began in 1939, and in 1941, after the secondary emptying of the basin chalice, it was filled and commissioned. The reservoir was named after Samarkand by the name of the settlement. It is also called Nura, Big Karagandy, and Temirtau. In 1935–1938, the Karagandy hydropower station was built on the basis of the reservoir. A railway branch was established, connecting Karagandy and Temirtau. As well as, the metallurgical combine ("Kazakh Magnet") and a number of other industrial objects were built, consuming the water of the Samarkand water reservoir.

The reservoir extends from east to west for more than 15 km. The area is 82 km^2, and the volume is 267 million m^3 of water. Its maximum width is 7 km, the average depth is 3.25 m (the maximum is 12 m), and the guaranteed water yield is 51 million m^3 per year. The catchment area is 11,500 km^2 (Philonets 1981). The reservoir is the largest one in the Karagandy region.

The Samarkand water reservoir occupies the floodplain of the Nura river. Its basin is composed of powerful sand and pebble deposits, covered with clays and loam. The bottom is flat, with an insignificant inclination to the west, where the greatest depth is.

The northern shore of the reservoir is flat, cased with a forest stratum 200–300 m wide. It is a good place for recreation.

The northwestern, western, and southwestern coasts are steep and occupied by residential and industrial buildings in Temirtau. The southeast and east coasts are gentle, and the east coast is cut and bogged. The coastline near the eastern shore is unstable; with fluctuations in the water level during the year, it then comes and then recedes (in some years up to 3 km). The banks are covered with thinned thickets of reeds, sedges, and shrubs (Philonets 1981).

The Samarkand water reservoir replenishes the waters of the Nura river, atmospheric precipitation on the water mirror, and also due to a partial discharge of water from the Ertis–Karagandy canal. The hydrological regime of the Nura river is characterized by long-term variability of runoff and extremely uneven intra-annual distribution. It has snow feeding, and more than 90% of annual runoff occurs in spring, during high water. The average long-term water discharge near the city of Temirtau is 6.5 m^3/s, and the flow volume is more than 200 million m^3. The average annual rainfall per water mirror is 260 mm (Philonets 1981).

The fluctuation in the level of the water reservoir depends on the amount of runoff of the Nura river, water discharge from the channel, evaporation (736 mm), and water consumption for technical and economic needs. It is characterized by a gradual summer–autumn–winter drop in the water level and a sharp increase in April–May. The annual amplitude of the oscillation is more than 2 m. The transparency of the water is 1.5 m, and the color is greenish-yellow. Mineralization is from 0.7 to 1 g/kg of water. According to the chemical composition, water belongs to the chloride class, the sodium group, moderately hard, and slightly alkaline. The nitrate content does not exceed 0.01 mg/l, and the amount of phosphates is 0.01–0.02 mg/l. The water contains from 6.11 to 6.7 mg/l of oxygen and the oxidizability of 8.5–9.0 mg O_2/l. The content of microelements: zinc, copper, manganese, and iron is low (Table 2.2).

Bugun water reservoir was built in the lower reaches of the Bugun river and of the south part in South Kazakhstan region. It was established in 1967, which made it possible to dramatically improve the water availability of irrigated agriculture in the Turkestan, Bugun, and Kyzylkum districts. The increase in irrigated lands amounted to 90 thousand hectares for those periods.

The Bugun water reservoir has two dams: a southern one with length more than 1 km, bounding the water spill, and the western retaining one with length about 4 km. The water reservoir area is 65 km^2, the length is 13 km, the width is 6 km, the average depth is 6 m (the largest is 15 m). It contains 377 million m^3 of water.

The bottom has a slight slope to the northwest. The soil is gray silt with an admixture of clay, and the its thickness sometimes reaches more than 1 m. The reservoir feeds on the waters of the Arys and Bugun rivers. The average annual flow is 618 million m^3, which exceeds the reservoir capacity by more than 1.5 times (Philonets 1981).

Since the end of October, there is a gradual accumulation of water. In winter and early spring, before high floods occur, there is a short increase in the water discharge of feeding systems caused by thaws and rains. The mainspring flood is used both for filling the reservoir and for the pre-irrigation of the fields. By mid-March–early April, the level reaches a normal retaining horizon (259.5 m) and persists with small fluctuations within a half to two months. From the middle of May, the development of water reserves begins and accompanied by an intensive drop in the level. Within three to four months, the level is reduced to a minimum. The maximum difference in the horizons of water is more than 11 m, the area is reduced from

65 to 5.1 km^2, and the volume of water is from 377 to 7.3 million m^3. In September–October, the water discharge in the headwater intake decreases and the level drop slows down. After the end of the irrigation season, there is a gradual accumulation of water (Kuznetsova 1974).

The *hydrochemical regime* of the Bugun water reservoir depends on the runoff of the Arys and Bugun rivers, evaporation from the water area and fluctuations in the level. The water in the reservoir is fresh and does not exceed 600 mg/l. The chemical composition refers to the hydrocarbonate class and the magnesium group (Table 2.2). Oxygen content is 10–11 mg/l, pH is 6.6–6.8, and oxidizability is 8.1–8.9 mg/l. The water is grayish in color, the transparency is 0.65-0.7 m, the hardness is 3.80 mEq/l, and the NO_3 content is 0.85–1.8 mg/l (Philonets 1981).

Enrichment of water with biogenic elements is facilitated by an annual drainage of the greater part of the reservoir basin. As the shoals dry, the amphibian vegetation is replaced by terrestrial grass and shrub vegetation, which grows during the summer season. These plots are used as pastures for cattle, hayfields, and partly for vegetable gardens.

Sherubai-Nura water reservoir (Topar) was built in 1960 on the Sherubai-Nura river, near the Topar village. It is used for water supply in the Karagandy industrial region, as well as for irrigation. Its area is 38.8 km^2 and the volume of 274 million m^3 of water. It extends from the southeast to the northwest by 11 km, the maximum width is 4.2 km, and the depth is 24 m (average is 7 m). Guaranteed water yield is 60 million m^3 per year (Philonets 1981).

The banks are shallow, sometimes overgrown with a willow. In the southwest along the northern shore stretches the forest belt. The western part of the reservoir is the deepest. The coastal areas grow reed, cattail, sedge, rdest, urr, pemphigus.

The Sherubai-Nura water reservoir is replenished due to the Sherubai-Nura river, as well as atmospheric precipitation on a mirror of the water surface. The fluctuation of the water level depends on the flow of the river, evaporation from the water surface, and the flow of water for household needs. A gradual summer–autumn–winter fall in the level common for the reservoir and a sharp increase in April–May are typical. The annual amplitude of the water level fluctuation is several meters.

In the spring, a large amount of snow water enters the reservoir, and in summer, there is a strong evaporation. Water transparency is 1.5 m, the color is greenish, the oxygen content is 9.92–7.97 mg/l, and oxidizability is 6.8–5.5 mg O_2/l. The water is fresh, moderately hard, and slightly alkaline. Chemical composition refers to the hydrocarbonate class and the sodium group. It contains a low content of trace elements: zinc, copper, manganese (Table 2.2).

Kengir water reservoir is located on the Kengir river, near the Zhezkazgan City. It is an important source of water supply to the enterprises of the copper industry in the region. Its waters irrigate the land of a suburban vegetable farm. At the same time, the dam part of it is used as a cooling pond for thermal power plants.

The construction of the Kengir water reservoir began in 1940 and resumed only in 1946. The area of the mirror is 37.5 km^2, and the volume of water is 319 million m^3 (Philonets 1981).

The body of the dam is composed of silty loam and silty clay of alluvial origin. It is protected from freezing by a layer of sand in 1.5 m. Along the entire dam is made anti-filtration bitumen curtain.

The water reservoir has a long elongated shape, and most of it is enclosed in a narrow natural canyon. The length of the reservoir is 33 km, the average depth is 8.6 m, the width is 1.6 km, and the guaranteed water loss is 50 million m^3 per year (Philonets 1981). The catchment area is 12,500 km^2.

Hydrological regime of the Kengir river is characterized by high long-term variability of runoff and extremely uneven intra-annual distribution. Feeding of the Kengir river is snow (95–98% of annual runoff). A sanitary protection zone has been established to prevent the possibility of contaminating the reservoir.

The shores of the reservoir are mainly composed by dense marly sandstones. The left bank in the upper part is steep and cut by many small valleys and temporary streams; the right bank is sloping. In the middle and head parts of the reservoir, the banks are elevated, sometimes abrupt, steeply entering the water. The bottom is hard and stony with a slight siltation.

The water reservoir is fed by the Kengir river and precipitation on a water mirror. The fluctuation of the water level depends on the volume of the Kengir river flow and economic use. The reservoir is characterized by a gradual summer–autumn–winter fall in the level and a sharp increase in April—during the spring high water period. The annual amplitude of water level fluctuations is several meters.

The Kengir reservoir is characterized by considerable *mineralization of water*, and its highest values are recorded in winter; up to 0.97 g/l.

The water is fresh, hard, and slightly alkaline. The chemical composition refers to the sulfate class and the sodium group. It has a low content of zinc and manganese (Table 2.2).

The *oxygen regime* of the reservoir is favorable for the life of plant and animal organisms. Oxidizability of water in summer is small that is 4–12 mg O_2/l, and in winter it doubles.

Oskemen water reservoir is located in the southeast of the Oskemen City, where the Ertis river breaks through the Albaket Range. Preparatory work was started as early as 1939, and it was completed only in 1953. The hydropower station on the Ertis river is designed to supply energy to the mining industry and the Oskemen City. A hydroelectric power plant has a capacity of 331,000 kW.

The area of the Oskemen water reservoir is 37 km^2, the width is 0.2–1.2 km, the maximum depth is 46 m, and the water volume is 659 million m^3 (useful volume is 68 million m^3).

The reservoir is deepwater, and the average depth is more than 17 m. The depths increase toward the dam. The annual water production starts in March, and in the second decade of April, the level drops to 4 m, in May by 1.8 m, and in August–September by 1.7–2.2 m. Water exchange in the reservoir takes place in about 15 days (Philonets 1981).

The water warms up in July at maximum. The temperature of the surface layer ranges from 8.2 to 23.5 °C, the bottom from 8 to 13.5 °C. Its gradual cooling begins in the end of September. Autumn ice drifting occurs in late October. The difference between surface and bottom temperatures is 6–10 °C. In general, the reservoir is a cold water body (Kiseleva 1964).

The thickness of the sediment in the reservoir is from 0.2 to 0.8 m. The mud (silt) is gray, sometimes scabrous and brown.

The content of total organic matter in the mud varies from 1.76 to 4.96% in the channel part and 2.8–6.1% in the bays. The total number of bacteria in the surface layers of soils averages 1.0–2.7 billion/g and in some cases reaches 4.5 billion cells/g (Gulaya and Kanachinova 1961).

Transparency of water depends on the time of year and the areas of the reservoir: In the upper part, it varies from 0.73 to 5.3 m; in the central part, it is 1.1–4.0 m; and in dam, it is 1.8–3.3 m. The content of dissolved oxygen in water within the norm: from 9.8 to 11.31 mg/l, free carbon dioxide from 2.2 to 6.5 mg/l, and the reaction of water is slightly alkaline. The water is ultra-fresh and refers to the hydrocarbonate class with the calcium group.

Botakara water reservoir is located in the Kazakh hilly plain of the Karagandy region. On its eastern shore is the district center of Ulyanovsk and on the north-western branch of the state farm Petrovsky. The water reservoir was formed in 1953 along the eastern, southeastern and western shores of the Botakara lake. The area of the lake was 7.3 km^2. The area of the reservoir is 25 km^2, a length of 8 km, a width of about 4 km, and a depth of 8 m (Philonets 1981).

The banks of the reservoir are shallow and sandy loamy. From the south, west, and northwest, water is close to the hills (hills have a height of 100 m above the water level). The northern and northeastern coasts are overgrown with reeds, cat-tails, reeds, and ponds.

An unnamed stream flows into the reservoir in the north. It often dries up during the dry season. During the spring snowmelt, up to 53 million m^3 of water accu-mulate in the reservoir. Its high level is observed in April, low one in October. The annual amplitude of the oscillation is more than 3 m, the transparency is 2 m, the color is brownish-yellow (Philonets 1981).

The water in the water reservoir is fresh. But in some arid years, during the period of low water, the *mineralization* can reach 2.5 g/kg. The water is moderately hard and slightly alkaline. The chemical composition refers to the chloride class and the sodium group. It contains a low content of trace elements: zinc, copper, iron, and manganese (Table 2.2). Oxygen is 11.14 mg/l, and oxidizability of water is 21.7 mg O_2/l (Philonets 1981).

Klych water reservoir was created in 1960 in the middle reaches of the Atasu river (tributary of the Sarysu river), in the Karagandy region. The water from the water reservoir is used for technical needs and watering cattle and vegetable gar-dens. The reservoir has a dam in a B-shaped form (its length is more than 7 km). It delays the flood waters of the Atasu river. The area of the reservoir is about 4 km^2 with an average depth of up to 2 m. Water is pumped through the dam into a second water reservoir of more than 1 km^2 with a maximum depth of 7 m.

The banks of the reservoir are shallow and ungrown. The bottom is even and sandy. Water vegetation is more developed in the upper part of the reservoir, where there are thickets of cattail, cane, rdest, and urti.

The highest water level is set in late April—early May, low one in February. The amplitude of its oscillation is several meters. Transparency of water is 1–2.5 m.

The water is brackish, stiff, and slightly alkaline. The chemical composition refers to the chloride class with the sodium group. There is little zinc, copper, manganese, but excess fluorine (Table 2.2). The oxygen content is 7.46–7.93 mg/l; oxidizability of water is 8.5–9.0 mg O_2/l (Philonets 1981).

Rivers There are about 39,000 rivers and temporary streams on the territory of Kazakhstan. More than eight rivers have a length over 10 km. Density of the river network is uneven. It is in the range 0.03–0.05 km/km^2 in the north, and then it decreases significantly in the desert zone, while it is 0.4–1.8 km/km^2 in the well-moistened mountain regions such as Altai, Zhetysu (Dzhungar), and Ile Alatau.

Most of rivers in Kazakhstan belong to the internal closed basins of the Caspian Sea and Aral Sea, Balkash and Tengiz lakes, and only the rivers of the Ob, Esil, and Tobyl basins bring their waters to the Kara Sea. In Kazakhstan, there are six rivers with water flows from 100 to 1000 m^3/s; 7 rivers—from 50 to 100 m^3/s, and 40 rivers—from 5 to 50 m^3/s (Riabtsev et al. 2004).

There are rivers of almost all types of water regime in Kazakhstan.

Large rivers such as Ertis, Zhaiyk, Ili, Syrdarya flow only along the outskirts of Kazakhstan with the exception of Ertis and Ile rivers. They are of little use for shipping.

The rivers of Central and Western Kazakhstan, as well as the tributaries of the Ertis, Esil, and Tobol rivers, are low in water. They are ladle out in some areas during spring. The overwhelming part of them either completely dries up in the summer or breaks up into a chain of salting tufts. Some of the rivers end blindly reaching the water only to the temporal lakes that turn into solonchaks when the snow melts in the summer. Sharp runoff fluctuations in such rivers limit the possibilities of water use for irrigation and meet the needs of industry and transport.

The rivers in plain mainly have snow feeding. Rainfall and groundwater make up between 5 and 15% of the annual runoff. Most of the rivers dry after the spring flood for a period of several days to eleven months. Underground feeding of rivers reaches 40–60% in the mountain rivers. The least water expenditure in rivers and their drying are observed more often at the end of winter and less frequently in summer during lowest-water level of rivers. The poverty of river network in Kazakhstan indicated by the extent of the runoff. Surface water resources with taking into account the flow coming from contiguous areas amount to 110–115 km^3 per year; local runoff is 59 km^3. The average water supply of the entire territory of Kazakhstan is about 20 thousand m^3 of water per 1 km^2 per year.

Glaciers are powerful accumulators of pure freshwater. They occupy about 16 million km^2 of land and have accumulated 30 million km^3 of water. This is almost two-thirds of freshwater on the Earth (Philonets 1981).

Highland chains prevail in the east Kazakhstan, which create characteristic features in the usual cycle of water, where glaciers play an important role, being the only freshwater reservoirs.

The bulk of Kazakhstan's glaciers in the form of a huge ice belt is located in the south and east of Kazakhstan, where the height of more than four thousand meters above sea level. These are the mountain ranges of the Tian Shan, Ile, Kungei and Terskey Alatau and ridges of Zhetysu (Dzhungar) Alatau and Altai Mountains, Sauyr–Tarbagatai range, and Kyrgyz and Talas Alatau (Table 2.4). On the territory of Kazakhstan, there were 2724 glaciers in size from 0.01 km^2 and above with a total glaciation area was 2033 km^2, including the 1975 glaciers with an area of 0.6 km^2 and more and 70 km^2 of firn fields at the end of 1980 s of twentieth century (Table 2.5). The total amount of saved water reserves in glaciers is 95 km^3, which is close to the value of the annual flow of all rivers in Kazakhstan (Riabtsev et al. 2004).

Almost half of the area of glaciation within Kazakhstan falls on the mountains of Zhetysu (Dzungar) Alatau (1000 km^2), the second place is occupied by Ile and Kungei Alatau (610.7 km^2), the third is Terskey Alatau (141.9 km^2), then followed by Kazakhstan Altai with Saur–Tarbagatai (106.2 km^2) and ridges of the Kyrgyz and Talas Alatau (101.5 km^2) (Fig. 2.4).

About 100 km^3 of freshwater in the form of ice was canned in glaciers, which was equal to the volume of water in Lake Balkash (Cherkasov and Vilesov 1972).

Glaciation in the mountains of Kazakhstan is represented by three main groups: valley glaciers, which account for 19.2% of the total number of glaciers and 65.7% of the glaciation area; glaciers of mountain slopes, respectively, 79.1 and 33.4% and glaciers on denudation surfaces (flat tops) is 1.7 and 0.9% (Pal'gov 1969).

There is a regularity in the change of the lower boundary of the glaciation area and firn line. The absolute heights of which increase from north to south and from west to east. In Altai, glaciers descend to an altitude of 2600 m, in the Tian Shan to 3600 m, respectively, and the height of the firn line varies from 2800 to 3800 m. The total range of glaciation in the mountains of Kazakhstan is 4000 m (Pal'gov 1969).

Glaciers have a low-temperature regime. In winter, a layer of glaciers deep down to 12–15 m is strongly damped. On the surface, the temperature of the ice drops to −10 °C. In the summer during the thaw, the surface layers of ice 1 m in depth are heated to 0 °C. In the bottom layers of the powerful valley glaciers, zero temperatures are maintained throughout the year.

Due to the harsh climate of the highlands, there is a zone of permafrost. The area of permafrost was 16,760 km^2 in the territory of the mountains of Kazakhstan, including in Altai is 5670 km^2, in Sauyr–Tarbagatai is 920 km^2, in Zhetysu (Dzhungar) Alatau is 4890 km^2, in Terskey Alatau and Ketmen is 1990 km^2, in Ile and Kungei Alatau is 2250 km^2, in the Kyrgyz Alatau is 460 km^2, and in Talas Alatau is 580 km^2 (Gorbunov 1974). The lower boundary of the permafrost distribution in the Altai is 2000 m, in Sauyr is 2300 m, in Zhetysu (Dzhungar) Alatau is 2800 m, and on the ridges of the Tian Shan (within Kazakhstan) is 3000 m.

The glaciers of Kazakhstan are moving at a speed approaching from 0 to 70 m/year, and the Shokalsky Glacier in the Ile Alatau is 300 m/year and more.

Table 2.4 Largest glaciers in Kazakhstan (Philonets 1981)

Glacier	Location, river basin	Height of the end of tongue, a.s.l. (m)	Area (km^2)	Length (m)	Ice volume (km^3)
Ile Alatau					
Korzhenevskogo	Shelek	3270	38.0	11.7	6.32
Bogatyr'	Talgarlgar	3420	30.3	9.1	4.50
Zhangyryk	Shelek	3370	17.7	8.9	2.01
Dmitrieva	Talgar	3400	17.4	5.7	1.90
Shokal'skogo	Talgar	3370	10.8	4.7	0.96
Gornogo instituta	Turgen	3450	9.8	4.5	0.83
Grigor'eva	Esik	3400	8.9	4.7	0.72
Pal'gova	Esik	3250	7.5	5.1	0.55
Kassina	Esik	3400	6.6	4.5	0.45
Kalesnika	Talgar	3400	6.6	4.3	0.24
Turistov	Talgar	3420	6.2	3.4	0.14
K. Makarevicha	Kaskelen	3450	6.0	4.9	0.40
Konstitutsii	Talgar	3270	6.0	6.0	0.40
Toguzak (left)	Talgar	3470	5.6	4.2	–
Zharsai (right)	Esik	3380	5.0	4.6	0.30
Central Tuiyksu	Kishi Almaty	3200	3.8	4.7	0.18
V. Shnitnikova	Aksai	3390	3.2	3.5	0.15
Teriskei Alatau					
Simonova (geographer)	Tekes	3490	28.1	9.2	4.02
Mramornoi steny	Tekes	3350	22.5	7.8	2.88
Karasai 1	Tekes	3430	9.7	4.8	0.81
Baiankol	Tekes	3400	6.9	4.9	0.49
Zhetysu (Dzhungar) Alatau					
Berg	Lepsi	2850	16.7	8.0	1.83
Kalesnika	Lepsi	2940	15.3	8.1	1.52
Voyeikova	Chyzhyn	2720	13.6	8.6	1.38
Abay	Baskan	2830	13.3	10.9	1.10
Bezsonova	Karatal	2903	12.6	6.0	1.21
Tronova	Karatal	2890	12.4	7.9	1.18
Zhambula	Baskan	3120	11.2	6.0	1.01
Nekrasova	Tentek	3120	10.9	5.8	0.80
Korolenko	Tentek	3000	9.5	6.2	0.60
Abolina	Bien	3140	8.1	4.5	0.62
Gerasimova	Baskan	2920	8.0	6.8	0.55
Shukina	Baskan	3110	7.5	5.5	0.59
Satpaev	Lepsi	2950	7.5	5.5	0.50

(continued)

Table 2.4 (continued)

Glacier	Location, river basin	Height of the end of tongue, a.s.l. (m)	Area (km^2)	Length (m)	Ice volume (km^3)
Lepsinskii	Lepsi	2980	6.2	5.3	0.37
Kheierdala	Bien	3130	6.2	3.8	0.30
Akchaganaksai	Aksu	3110	6.2	4.7	–
Arkhipovoi (west)	Lepsi	3160	5.6	2.5	–
Dal'nii	Aksu	3420	5.6	2.0	0.35
Gagarina	Aksu	3110	5.6	4.8	–
Markova	Baskan	2920	5.3	5.5	0.23
Karelina	Tentek	3180	5.0	2.7	–
Kungei Alatau					
Novyi	Shelek	3370	13.2	6.4	1.29
Iuzhnyi Zhangyryk	Shelek	3370	9.2	8.0	0.75
Kensai	Shelek	3460	6.7	3.9	0.47
Sutbulak	Sutbulak	3480	6.4	4.0	0.44
Korimdik	Korimdik	3350	6.0	4.9	0.40
Altai					
Bol'shoi Berel	Berel	1920	10.3	10.4	0.89
Malyi Berel	Berel	2100	8.9	8.3	0.72
Bol'shoi Bukhtyrma	Bukhtyrma	2520	8.1	5.3	0.62

Table 2.5 Distribution of glaciation within Kazakhstan (Philonets 1981)

Glaciation area	Number of glaciers	Area of modern glaciation (km^2)	Area under moraine (km^2)	Total area of glaciation (km^2)	Ice volume (km^3)
Altai (Kazakh part)	328	72.3	17.3	89.6	3.5
Sauyr–Tarbagatai	18	14.8	1.8	16.6	0.5
Zhetysu (Dzhungar) Alatau	1369	813.9	186.1	1000	45.9
Terskey Alatau	169	137.8	7.1	141.9	11.2
Kungei Alatau	163	126.4	15.0	141.4	6.6
Ile Alatau	393	422.7	46.6	469.3	27.7
Kyrgyz Alatau	34	9.5	5.6	15.1	0.7
Talas Alatau	250	76.5	9.9	86.4	2.3
Total	2724	1673.9	289.4	1963.3	98.4

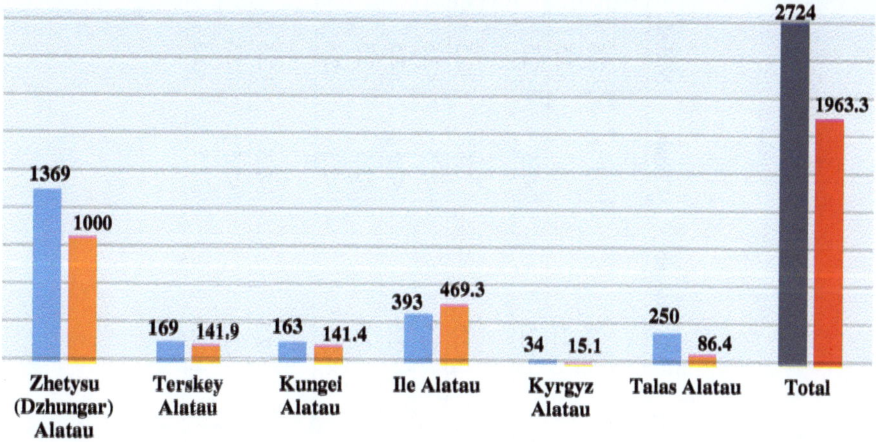

Fig. 2.4 Area of glaciation within Kazakhstan

The period of ablation on glaciers usually lasts 2–2.5 months: from July to mid-September. At glaciers with low descending tongues, during the summer, an ice layer melts about 3 m, and in exceptional cases up to 5–6 m (Makarevich 1969).

The waters formed from the ice melting participate in the feeding of mountain rivers and lakes. In summer, the rivers in the sources are 83% composed of thawed glacial waters, and at the outlet of the mountains, their share is reduced to 35%. The melting of glaciers annually gives a runoff of about 2 km^3 of water.

During the ablation of glaciers, lakes are formed at the ends of their tongue, mostly where there are moraines. The area of lakes from several hundred to several tens of thousands of square meters. Their depth sometimes reaches several tens of meters. The number of lakes and the volume of water in them vary from year to year. By the end of the ablation period, in most lakes, the basins are devoid of water. Most moraine lakes are mudflowy.

According to the *chemical composition*, the water of glaciers refers to a hydrocarbonate class with the predominance of Ca and Na ions and it has a neutral reaction and very soft. The content of microelements in water of glaciers is very low (Table 2.6). The *mineralization of glacial* waters varies from 20 to 60 mg/l, but mainly about 40 mg/l. Approximately 80 thousand tons of salts are annually taken out with thawed glacial water (Philonets 1981).

At present, most of Kazakhstan's glaciers are in the process of reducing (retreating).

The groundwater in Kazakhstan Groundwater is distributed throughout the country extremely unevenly, and the quality of this water prevents the use of part of the groundwater resources. Groundwater is available in almost all mountainous areas.

Table 2.6 A chemical composition of some glacial water in Kazakhstan (mcg-l) (Philonets 1981)

Glacier, a heigh of sampling, m	pH	Fluorine	Zinc	Copper	Manganese	Plumbum	Molybdenum	Cobalt	Nickel
Ile Alatau									
Central Tuiyksu, 3380	6.95	200	<2.5	2.5	40	<5	<2.5	2.5	<2.5
Igly Tuiyksu, 3450	6.85	200	2.5	<2.5	20	<5	<2.5	7.5	<2.5
Molodezhnyi, 3450	6.80	200	<2.5	<	5	<5	<2.5	<2.5	<2.5
Zhetysu (Dzhungar) Alatau									
Obrucheva, 390	6.75	200	2.5	<2.5	<5	<5	<2.5	<2.5	<2.5
Gagarina, 3110	6.70	200	2.5		10	<5	<2.5	<2.5	<2.5
Shumskogo, 3020	6.45	200	7.5	<2.5	< 5	<5	<2.5	<2.5	<2.5
Abay, 2830	6.60	200	5	<2.5	20	<5	<2.5	<2.5	<2.5
Gerasimova, 2920	6.70	200	15	<2.5	10	<5	<2.5	<2.5	<2.5
Altai									
Malyi Berelskii, 2326	6.45	200	200	<2.5	20	<5	<2.5	<2.5	<2.5
Malyi Berelskii (end of tongue)	7.2	200	7.5	<2.5	80	<5	<2.5	<2.5	<2.5
Malyi Berelskii (lake), 984	6.90	200	10	<2.5	10	<5	<2.5	<2.5	<2.5

Kazakhstan has considerable reserves of groundwater. About half the groundwater resources (about 50%) are concentrated in Southern Kazakhstan. A significantly smaller amount of these resources (up to 20%) is formed in Western Kazakhstan. About 30% of all groundwater resources are located in Central, Northern, and Eastern Kazakhstan (FAO 2012).

The reserves of artesian waters are great in the southern regions of Kazakhstan, under the sands of Kyzylkum, Moiyunkum, Betpakdala. The largest underground seas are located in other arid regions. There are large reserves of fractured groundwaters in mountain areas of Southeast and East Kazakhstan.

There are Tobyl and Ertis artesian basins on the plains of Northern Kazakhstan. The regions that adjacent to the eastern slope of the Southern Urals, and the Pre-Altai plains of Ertis is most rich in fresh layer (formation) water.

The karstic limestones in Zhezkazgan–Ulytau, Atasu, Bozshakol–Ekibastuz, Maikaiyn, and other areas of Central Kazakhstan are rich in undergroundwaters. Their total volume reaches 55–60 km^3. The total reserves of the fissured waters in Central Kazakhstan and in the low-mountainous regions of Bayanaul, Karkaraly, Ulytau, and other regions exceed 60 km^3.

There are artesian waters in the Northern Aral Sea region. Their total reserves are 180 km^3.

Fresh and slightly salty fractured and fissured-layer groundwater are found in the mountainous part of Mangyshlak. Their resources are 20 km^3.

Thermal waters: Warm and hot ones are widespread in most of the artesian basins of Caspian and Mangyshlak basins, Kyzylkum desert, Northern Aral Sea region, Turgai and Ertis Rivers, and Ili River Valley.

As a whole, about 70 artesian basins and powerful underground groundwater flows have been found on the territory of Kazakhstan. They occupy almost 2 million km^2 with a volume of about 7.000 km^3. The total resources of the under-groundwaters in Kazakhstan correspond to the 25 Azov seas, 140 times exceeds the volume of the Kuibyshev reservoir and 600 times the Bukhtyrma reservoir (Akhmedsafin and Sydykov 1964, 1970). Underground seas are updated annually. Their inflow is up to 45 km^3 which makes it possible to use underground reserves for a long time without fear of their depletion. There is a real opportunity to extract a water up to 3900 m^3 per second. This would allow irrigation of up to 4 million hectares of land and flooding 150 million hectares of pastures (Philonets and Omarov 1973). In Kazakhstan, 750 million m^3 of water is extracted from bowels of the Earth annually. It receives a number of mining, metallurgical, chemical enterprises, oil fields, 35 cities, including the former capital of Kazakhstan is Almaty, where there are 150 artesian wells were drilled in the region. Over 70 million hectares of desert and semidesert pastures are flooded. More than five thousand wells are supplied with groundwater by livestock collective farms.

2.3 Water Availability and State of Water Resources Within Water-Economic Basins in Kazakhstan

Water is the most important natural resource and without it no human activity is possible as it cannot be replaced by anything. Water resources are integral parts of the natural environment, and they play a decisive role in the economic development, especially in the agricultural–industrial complexes.

The water resources of Kazakhstan depend more on river and lake runoff. Kazakhstan is relatively poor in water resources (6.0 thousand m^3/capita/year) in comparison with Northern Europe (43 thousand m^3/capita/year) and the Siberian parts (30 thousand m^3/capita/year) of the CIS, but it is richer than the other Central Asian countries, namely Uzbekistan and Turkmenistan (1.8 and 4.0 thousand m^3/capita/year, respectively) (Mal'kovskii 2012; Zaurbek et al. 2012). The scarcity of freshwater is the acutest environmental problem that hinders the sustainable development of the Central Asian countries, including Kazakhstan (Espolov et al. 2006). The reasons for the lack of water resources are natural conditions (90% of spring runoff), and that about half of the river runoff (44.0 km^3) is formed in the territory of neighboring countries: China (18.9 km^3), Uzbekistan (14.6 km^3), Russia (7.5 km^3), and Kyrgyzstan (3.0 km^3) (Riabtsev et al. 2004). Only 56.5 km^3 of the runoff are formed on the territory of Kazakhstan, while the total average of 100.5 km^3/year of surface water resources in Kazakhstan. So, the geographical location of the country is one of the most important factors in water shortage. The surface water resources are extremely unevenly distributed across Kazakhstan, and they are subject to significant temporary fluctuations (Mal'kovskii 2012; Espolov et al. 2006).

The flow of river systems provides the main volume of water consumption in the world. It determines the water availability within a territory, the degree of water resources usable by the population, and through the natural discharge variability it also determines periods of water excess and water shortage. The shortage of water resources can lead to a weakening of the agro-industrial complex of countries like Kazakhstan in which the farming sector is an important economic factor. The threat of water shortage and inefficient water resources management may become the main obstacle to sustainable economic growth and social development of Kazakhstan. Furthermore, low water periods can also lead to environmental problems as the concentration of pollutants is higher while the oxygen saturation is lower, putting the aquatic ecosystems under considerable stress. Further, important parameters are the extensive water use and excessive irretrievable water consumption for irrigation. In addition, irrational development of irrigated agriculture, as well as regulation of runoff in arid climate conditions, led to a water deficit in the basins of small and large rivers. The use of available water resources in Kazakhstan is very wasteful and uneconomical. Inefficient irrigation methods, destroyed irrigation structure, and the lack of a functional drainage system contribute to water losses and soil salinization in irrigated areas as well as water demand is growing due to a strong population growth (Mal'kovskii 2008, 2012). The population of

Central Asian countries has increased over the past decades, and the problem of water scarcity has become even more acute respectively. Consequently, the demand for water rapidly has grown and as a result of the reduction of sustainable water supplies, a water deficit of 14 billion m^3 is expected by 2030, by 2050 the deficit could be 20 billion m^3 (70% of the demand for water) (Green Economy 2013). In addition, the climate in most parts of Central Asia is arid and semiarid. This is especially true for Southern Kazakhstan, where the limited precipitation (up to 130 mm) can easily lead to water shortages (Gagloeva 2016; Danilov-Danilian 2008; Danilov-Danilian and Khranovich 2010).

Besides many water sources are polluted with industrial wastewater as water treatment is still limited in its regional coverage and efficiency (Espolov et al. 2006; Riabtsev et al. 2004; Dostai 2012). In general, the quality of surface waters is considered good, although a number of river water basins are contaminated. Mining, metallurgical and chemical industries, and municipal services of cities are the main sources of water pollutants (Dostai 2012; Sarsembekov 2004). As a result of the deterioration of the quality of natural waters in Kazakhstan, the problem of providing the population with good quality drinking water is increasing. This leads to an increase in infectious morbidity, the emergence of environmental refugees, and the growth of social tension (Espolov et al. 2006).

The scarcity of freshwater and the deterioration of the water quality hinder the sustainable development of the country and cause environmental issues (Dostai 2012). The problems of sustainable water supply in the Central Asian countries are currently acquiring an acute socioeconomic, ecological, and political dimension, which is due to the increasing role of anthropogenic factors related to water consumption for the needs of the population, industry, and agriculture on the one hand and the impact of the climate change on the other hand (Kulmatov 2014; Medeu et al. 2012; Alamanov 2016; Muhabbatov 2016; Veisov and Khamrayev 2016; Khamrayev and Rahimov 2016). The consequences of global and regional climate changes, the inconsistency of interstate water relations, the use of water-consuming technologies and the imperfection of technical means of water regulation and water distribution are the main threats and challenges in the field of water supply in the Central Asian countries (Gagloeva 2016; Moerlins et al. 2008; Wouters et al. 2007). Assessment and forecast of changes in water resources due to the expansion of economic activities and climate change; water resources management and modeling; harmful impacts on water resources; and international cooperation in the field of water issues are modern topical hydrological aspects in Central Asia and the adjacent countries. From this review, it is clear that studying the water resources, and especially their availability, demand, and supply are important and required.

In terms of water availability, Kazakhstan ranks last among the CIS countries. The specific water availability was 37 thousand m^3 per 1 km^2 and 6.0 thousand m^3 per 1 person per year (Duskayev 2004). However, according to our calculation is estimated 15.3 thousand m^3/capita with total population is 18.1 million and water resource is 278.04 km^3 (Tables 2.7 and 2.8). The water availability in the Central Asian countries is very uneven. It is 5.7 thousand m^3 for five Central Asian countries. The average annual water availability per capita is 13.5 thousand m^3 in

Table 2.7 Main characteristics of water-economic basins in Kazakhstan

Water-economic basins	Basin area (thousand km^2)	Main river	Flow formation (%)	Population within basin (million people)		% from total population in Kazakhstan	Mean long-term runoff (km^3)
				Urban	Rural		
Aral–Syrdarya	345	Syrdarya	75.2 in Kyrgyzstan 15.2 in Uzbekistan 6.9 in Kazakhstan 2.7 in Tajikistan	1.3	1.5	16	From 11.3 to 27.8
Zhaiyk–Caspian	415	Zhaiyk	8 in Kazakhstan 92 in Russia	2.2		13	
Balkash–Alakol	353	Ile	56.6 in Kazakhstan 43.4 in China	1.8	1.5	19	From 27.8 to 33.8
Ertis	316.5	Ertis	74 in Kazakhstan 26 in China	3.7		23.4	
Esil	245	Esil	100 in Kazakhstan	1.09	0.81	11	From 2.2 to 11.3
Shu–Talas	64.3	Shu, Talas	24 in Kazakhstan 76 in Kyrgyzstan	1.0		5.8	
Nura–Sarysu	139.7	Nura and Sarysu	100 in Kazakhstan	1.0		5.8	From 1.3 to 2.2
Tobyl–Torgai	214	Tobyl	78 in Kazakhstan 22 in Russia	1.05		6	

Tajikistan and 8.4 thousand m^3 in Kyrgyzstan. In Russia, it is about 30.0 thousand m^3/capita (Boiarkina 2011).

The research was conducted within WEBs in Kazakhstan (Fig. 2.5). The WEB is a natural–anthropogenic complexes of interconnected natural objects and engineering and technical facilities that jointly function to meet the various water-related socio-ecological and economic needs of people, as well as the rational management of which provides a safe and sustainable ecologic–economic development. The basin is a managed system of social, economic, technical, legal, and ecological relationships regarding rational water use (Abdukhalikov 2015).

Table 2.8 A share of water resources in the river basins and large water reservoirs in Kazakhstan

Water-economic basins	Water resources (km^3)	Share of rivers (%)	Share of lakes (%)	Share of water reservoirs (%)	Share of groundwater (%)	Large water reservoir	
						Name	Volume (km^3)
Aral–Syrdarya	37.9	70	–	21–23	7–9	Shardara	5.2
Zhaiyk–Caspian	28	94	3	3	3	–	–
Balkash–Alakol*	149.4	14	77	5	4	Kapshagai	14.0
Ertis	43.8	59	16	18	7	Bukhtyrma	49.0
Esil	5.34	34	55	7	4	Viacheslavskoye and Sergeevskoye	0.4 and 0.7
Shu–Talas	6.11	59	6	8	27	–	–
Nura–Sarysu	4.59	33	20	4	43	–	–
Tobyl–Torgai	2.9	35	33	17	15	Verhne-Tobolskoye and Karatomar	0.82 and 0.59

**Notice* the main volume (77%) of water is concentrated mainly in Balkash lake

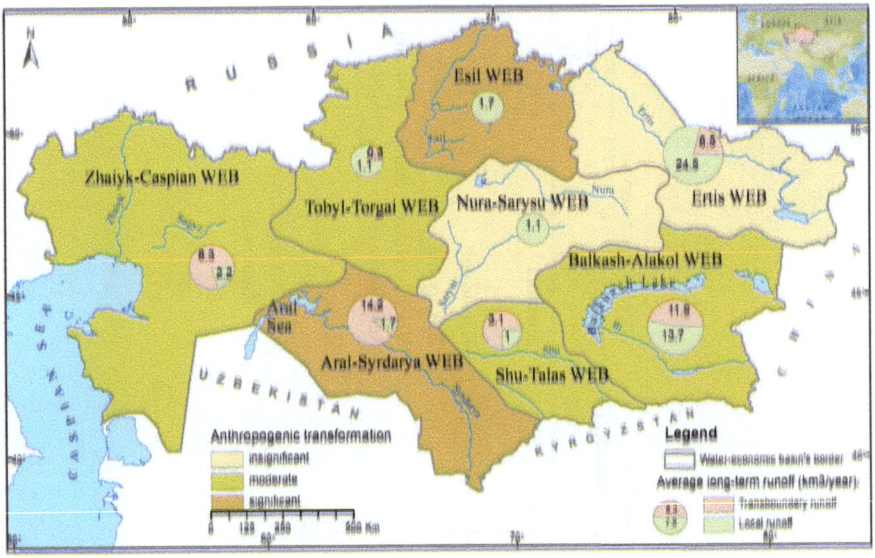

Fig. 2.5 Main water-economic basins (WEB) in Kazakhstan

There are eight water-economic basins in Kazakhstan such as Aral–Syrdarya, Balkash–Alakol, Ertis, Zhaiyk–Caspian, Esil, Nura–Sarysu, Shu–Talas, and Tobyl–Torgai basins (Table 2.7; Fig. 2.5). The Ertis WEB has with 43.8 km^3 the biggest share (%) of the water resources. The Tobyl–Torgai WEB is the basin with the smallest share of the water resources (2.9 km^3) (Duskayev 2004). The largest WEB in Kazakhstan is the Zhaiyk–Caspian basin with an area of 415 thousand km^2 (Table 2.7). River water is 94%, the share of water reservoirs is 3%, and groundwater is 3% in the basin (Table 2.8) (Duskayev 2004).

The water availability in Kazakhstan differs significantly. The East Kazakhstan region is a water-rich region within the Ertis WEB, and there are regions where water is scarce such as Mangystau region. The water availability and state of water resources within WEB in Kazakhstan were considered and analyzed.

A long-term dynamics of river runoff in Kazakhstan according to the data (annual average) on river runoff (total, transboundary and local runoffs) for the period of 1965–2016 was analyzed. In addition, numerous cartographic materials (four maps from NAKZ 2010) related to the anthropogenic transformation of the WEBs and water demand were analyzed in order to classify the category of anthropogenic transformation and water demand within the WEBs in Kazakhstan.

The Arc map software was used as the main tool to analyze the anthropogenic impact to the WEBs as well as for preparing the map of anthropogenic transformation and water demand within WEBs in Kazakhstan.

The original map with the border of WEB was scanned and registered/geo-referenced to specify its location by using vector layer of country's border. Thereafter, the collected data on average long-term runoff, anthropogenic transformation, and water demand were given as input parameters for the generation of maps using ArcGIS applications.

A share of water resources and water availability in the WEBs. In dry years, the level of water availability is 60% of the water demand, and for some regions such as Central Kazakhstan is only 5–10% due to mainly irrigated agriculture.

The Ertis WEB is the most water-rich basin. The water resource is 43.8 km^3. A river flow forms the main water reserves with a volume of 26.04 km^3 that is 59% from basin water (Table 2.8). The volume of water reservoirs is 7.7 km^3 (18%) and is the largest basin in Kazakhstan where there are powerful water reservoirs are located (Table 2.8).

The water resource in the Balkash–Alakol WEB is significant and amounts to 149.4 km^3; however, the main volume (77%) of water is in lakes, mainly in Lake Balkash. The share of river water is 14%, and water reservoirs are 5% in the basin. The water resources of the Aral–Syrdarya WEB are estimated 37.9 km^3. The main volume of the runoff which is 70% is formed in the upper part of the basin until the outlet from the Fergana Valley. The flow of right-bank tributaries above the Shardara water reservoir is 21–23% of the total water resources flowing into Kazakhstan (Table 2.8). The share of the Arys river runoff and other rivers flowing from the Karatau Range in Kazakhstan is 9–7% (Riabtsev et al. 2004). Currently, there are about 100 water reservoirs and 24,000 km of irrigation canals in the Aral Sea basin (Bogomolov et al. 2007). A share of river water accounts for 94% in the Zhaiyk–Caspian WEB (Table 2.8). This WEB has the highest share of river waters among the rest WEBs.

The water availability of Kazakhstan varies significantly. There are high water availability regions such as East Kazakhstan region (Ertis WEB). The Ertis WEB is rich in water availability as half of its renewable water resources are generated in the Kazakh part of the catchment (Abishev et al. 2016), and it is used as a donating water basin during times of water shortage (Table 2.8). Significant water resources are also concentrated in the Aral–Syrdarya and Balkash–Alakol WEBs (Table 2.8).

Dynamics of river runoff in Kazakhstan. River runoff is an objective indicator for determining the water amount of a river. It is an important element of the natural water cycle, through which water moves from land to sea and also within the catchment.

The total surface water resources of Kazakhstan were 91.3 km^3/year in 2010; 44.3 km^3 of these resources come from neighboring countries, while the remaining 47.0 km^3 are the local runoff. Since 1965 total runoff decreased by 39.4 km^3/year. The transboundary part of the runoff decreased to 23 km^3/year and local runoff to 16.5 km^3/year (Fig. 2.6). This reduction might be explained by global and regional climate change and, as well as, is due to the acceleration economic and social human activity on catchment areas and river valleys in the neighboring countries such as China and Russia, Uzbekistan and Kyrgyzstan, where 44% of surface water resources are generated.

Unfavorable implementation of climatic and transboundary hydrological threats in the future will contribute to a real decrease in river flow resources in Kazakhstan as a whole, and they will decrease to 81.6 km^3/year by 2020 (Abishev et al. 2016),

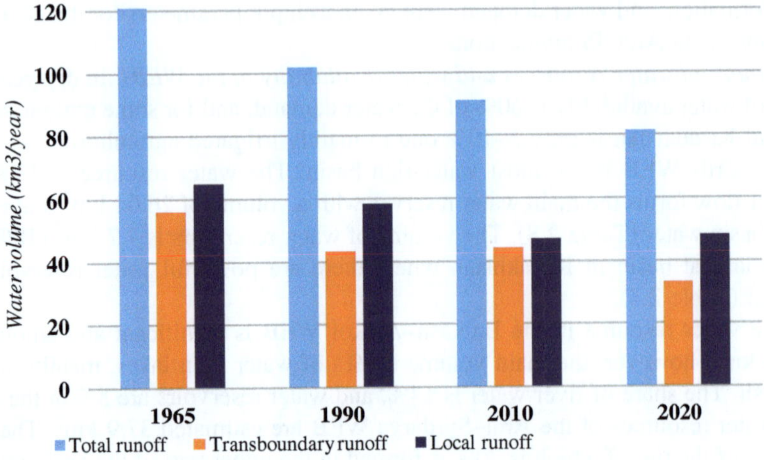

Fig. 2.6 Dynamics of river runoff in Kazakhstan (Abishev et al. 2016)

including transboundary resources up to 33.2 km^3/year, local up to 48.3 km^3/year (Fig. 2.6). The Aral–Syrdarya (89%), Zhaiyk–Caspian (79%), Shu–Talas (76%) water basins are the most dependent on the transboundary flow.

Global and regional climate changes and also economic and human activity in catchments and river valleys including in the territories of neighboring states are the main reasons for the instability of river flow.

Quality of surface water and Anthropogenic impact to the WEBs in Kazakhstan. Water quality is one of the limiting factors for the sustainable development of Kazakhstan. The water quality of almost all water basins in Kazakhstan remains unsatisfactory, despite the decline in production and the reduction in the volume of wastewater discharged. The main pollutants enter water bodies with wastewaters from chemical, oil-refining, machine-building, and non-ferrous metallurgy sectors. Contaminated sewage waters discharged without any purification are the greatest environmental threats to the water bodies.

Kazakhstan is located in zones of transit and dispersion of flow, as well as on delta zones of large water basins such as Syrdarya, Ile, Zhaiyk, and Ertis. Since the runoff of almost all rivers is regulated, the regime of runoff formation in river parts located below the water reservoirs is strongly transformed. As a result of the mutual impact of rivers and catchment areas and intensive water use, the hydrological regime and the water quality have changed in the zone of flow transit and dispersion. This mutual impact is characterized by intensive water abstraction from rivers for industry and irrigated agriculture and the discharge of return flow with salts, chemicals, and other contaminants into rivers (Espolov et al. 2006; Riabtsev et al. 2004; Dostai 2012).

To assess the quality of water in rivers and reservoirs, water is divided by contamination into several classes. The classes are based on the intervals of the water pollution index (WPI), which is an aggregate indicator based on several factors, such as the concentration of pollutants (nitrates, nitrites, ammonium nitrogen, heavy metals, oil products, etc.), hydrobiont characteristics, trophicity, and saprobity of water reservoirs (Shitikov et al. 2005). The water quality in the Aral–Syrdarya WEB can be classified as the fourth to the sixth class of water pollution, and overall it is considered to be a highly polluted WEB (Table 2.9). The main pollutants in this basin are nitrogen, sulfate, copper, and phenol. The maximum pollution was observed in spring, when nitrite and copper exceed three maximum allowable ratios (MAR) (with mean concentration of 0.08 and 0.001 mg/l, respectively), while sulfate exceeds seven MAR (100 mg/l), phenol by six MAR (0.001 mg/l), and oil products by four MAR (0.05 mg/l) (Riabtsev et al. 2004). The main tributaries of the Syrdarya river and the Shardara water reservoirs are also polluted by those chemicals. The surface waters of the Ertis WEB are contaminated with heavy metals such as zinc and copper, and the basin has been sorted into the fourth class of pollution. The main pollutants in the Zhaiyk–Caspian WEB are organic matter, boron, and chromium, and the sources of this contamination are located in the Russian part of the catchment as there are no wastewater discharges into the mainstream within Kazakhstan. The resulting pollution class for the Zhaiyk–Caspian basin is 4–6 (Riabtsev et al. 2004). The main pollutants in the Lake Balkash are heavy metals, copper, zinc, phenol, fluorides, and oil products, but overall the basin belongs to the moderate class of pollution. The pollution with heavy metals is a common problem for rivers in the Nura–Sarysu basin. The main metallurgical factories and coal basin of the country are concentrated within the basin. That is why water resources of the basin were polluted by wastewaters. The water quality of the basin is estimated to the fourth class of water pollution. The pollution degree of the surface waters in the Esil and Tobyl–Torgai basins is significantly lower, and the water quality class only shows a moderate level of pollution (Espolov et al. 2006; Riabtsev et al. 2004).

The chemical pollution of waters of open water bodies has significantly increased during the last several decades. There is a high coefficient (56.8%) of pollution in the Kyzylorda region (Aral–Syrdarya WEB), Akmola region is 15% (Esil WEB), 16.7% in the West Kazakhstan region (Zhaiyk–Caspian WEB), in Karagandy region is 15.6% (Nura–Sarysu WEB) (Espolov et al. 2006). The quality

Table 2.9 Criteria of water pollution index

Class of water quality	Characteristics of water quality	Hydrochemical parameter (water pollution index)
I	Very clean	0.0–0.3
II	Clean	0.3–1.0
III	Moderately polluted	1.01–2.5
IV	Polluted	2.6–4.0
V	Dirty (filthy or muddy)	4.1–6.0
VI	Very dirty (filthy or muddy)	6.1–10.0
VII	Extremely dirty	More than 10

of the surface water largely depends on the ecological situation and the economic development of the region. The anthropogenic activities negatively affect the regional ecological situation and cause enormous changes in the environment.

According to the anthropogenic transformation of the river flow in the WEBs, water-economic basins are divided into three classes such as slight, moderate, and significant. The anthropogenic transformation of the river flow was estimated in percentage (%): Slight class is 0–10%, moderate one is 10–20%, and significant class of the anthropogenic transformation is >20%. Study results and comprehensive analysis showed that the Aral–Syrdarya and Esil WEBs belong to the significant class of the anthropogenic transformation. Consequently, the most affected by human influences are Aral–Syrdarya WEB (47%), and 23% of the river in the Esil WEB was deteriorated by human activities. The Zhaiyk–Caspian and Tobyl–Torgai WEBs were affected by human activities to 18% and belong to the moderate class of the anthropogenic transformation as well as the moderate class includes the Shu–Talas (13%) and Balkash–Alakol (12%) WEBs. The remaining two WEBs: The Nura–Sarysu and the Ertis are included in the class of slight anthropogenic transformation with 8%.

Use and state of water resources in Kazakhstan. The main amount (about 78%) of the water resources in Kazakhstan is used in the agricultural sector, followed by the water use for industry (16%), domestic water consumption (5%), and 1% for other purposes (Riabtsev et al. 2004; Duskayev 2004). 85% of the water is supplied by surface water sources.

The average annual water consumption of the branches of the economy of Kazakhstan is 20 km^3 and tends to increase. The surface water resources are the main water source in water supply; 85% of the water is provided by surface water sources, the remaining water is groundwater (6%), water from the Caspian Sea (5%) and reused sewage and other water sources (4%) (Riabtsev et al. 2004; Duskayev 2004).

The main water consumer is the agricultural sector. Although, since 1990, there has been a tendency to reduce the water use from natural sources through the reduction of water use for agricultural, industrial, domestic, and drinking needs, as a result of the economic situation after the independence of Kazakhstan, irrigation areas were reduced and water withdrawal for agricultural needs was halved, respectively. Consequently, the water consumption decreased to 57% due to the reduction of the irrigated areas and industrial activities in Kazakhstan. The area of irrigated land decreased to 52,000 ha, namely in Almaty (18,100 ha) and Zhambyl (33,600 ha) regions (Riabtsev et al. 2004). Currently, about 15 km^3 of water is used for irrigation, which is mainly based on the surface runoff of the Syrdarya, Ile, Shu, Talas, and Ertis rivers. The decrease in water consumption is typical for industries where water consumption in 2000 decreased to 2.2 km^3, while in 1992, the water consumption was 4.8 km^3. Such a reduction in water consumption is due to a decrease of industrial activity in Kazakhstan (Riabtsev et al. 2004).

In 2002, water intake was 20.07 km^3/year, including 14.67 km^3/year for agriculture, 3.97 km^3/year for an industry, 0.87 km^3/year for communal services, and 0.55 km^3/year of water for other needs (Riabtsev et al. 2004). While the economy water intake for 2010 amounted to 23.3 km^3/year, including irretrievable water consumption of 15.3 km^3/year (66%) and water disposal of 8 km^3/year (34%), the agricultural sector consumed 15.4 km^3/year (66%), the industry used 4 km^3/year (17%), while communal services demanded 2.2 km^3/year (9%) and 8% (1.84 km^3/year) of the water was used for other needs. The water consumption in 2012 was increased by 1.33 and 1.25 km^3/year in communal services and other needs, respectively (Satenbayev et al. 2012). In addition, a slight increase was observed in the agriculture. Economic water intakes should not exceed this actual volume (amount of water) in the future, and the volumes are considered as limits of domestic water withdrawal. Otherwise, there will be a water crisis in the WEBs of Kazakhstan (Abishev et al. 2016).

Demand for water resources is established based on scientific and research developments, which includes the needs of natural water bodies, including transboundary runoff, releases and unproductive water losses for industrial use. The total demand for water resources within WEBs in Kazakhstan is set at 64.2 km^3/year, including environmental needs, mandatory transboundary water releases, as well as unproductive losses as a limitation of water use (Medeu et al. 2012).

In order to find the degree of ecological demand for water resources in WEBs, the WEBs are conditionally divided into five classes: from satisfactory to catastrophic categories. These categories are obtained based on the coefficient of water demand which is 0.0–0.4—satisfactory, 0.4–0.6—tense, 0.6–0.8—critical, and the crisis is 0.8–1.0, and there is >1 that is catastrophic degree. The coefficient of water demand and the norms of ecological demand for water are established by a political decision based on the need to balance the environmental, social, and economic development goals of the country. Over time, the constants can vary in the direction of both tightening and be alleviating the threshold of permissible anthropogenic loads. According to comprehensive analysis, there is own coefficient of water demand in each WEB. It is 0.81 in Zhaiyk–Caspian WEB, in the Balkash–Alakol WEB is 0.80, 0.60 in the Tobyl–Torgai WEB, and 0.51 in the Aral–Syrdarya WEB, in the Ertis WEB is 0.56, and in the Esil WEB is 0.54, 0.46, and 0.36 in Nura–Sarysu and Shu–Talas WEBs, respectively. So, the demand for water in Zhaiyk–Caspian WEB is categorized as the highest class (crisis degree with 0.81 coefficient) (Fig. 2.7). This is due to the development of the oil industry in the region, which requires a significant amount of water. A critical degree with 0.80 coefficient was observed in the Balkash–Alakol WEB (Fig. 2.7). The greatest amount of water is also used by the oil industry, as well as by non-ferrous metallurgy and heat power engineering sectors.

Water resources are the main component of the environment and play a decisive role in the development of all sectors of country's economy. The study of water availability and the state of water resources within water basins in Kazakhstan allow the general assessment of the conditions and a current state of the water-economic basins in Kazakhstan. During the last several decades, natural factors and human activities have changed the Kazakh landscape considerably, including its water resources. Consequently, the water resources were polluted and water quality was deteriorated as well as the total runoff decreased. The water consumption in agricultural irrigation and industry has decreased to 57%. It is due to a reduction of

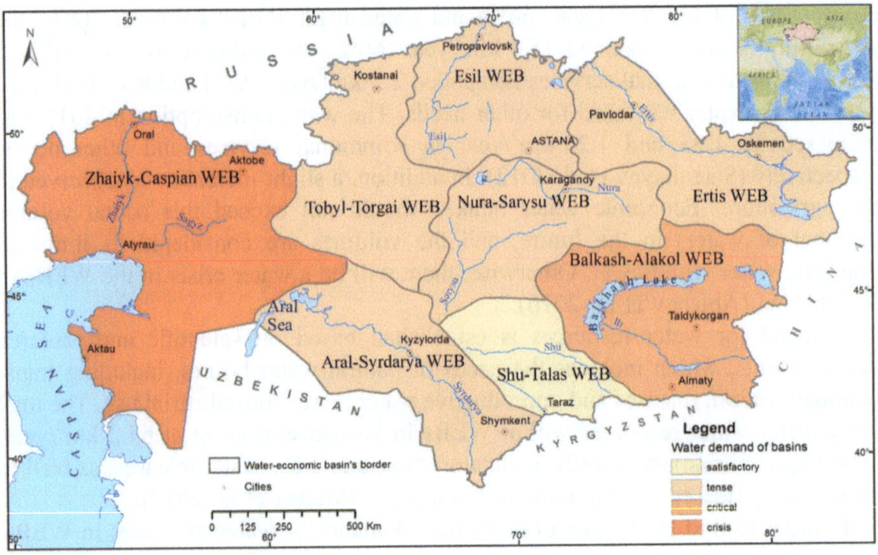

Fig. 2.7 A demand for water resources in water-economic basins of Kazakhstan (modern state)

irrigation areas and industrial activities as a result of economic situation after the independence of Kazakhstan. However, due to the population growth, the water consumption has increased in communal services and other needs.

The Aral–Syrdarya and Esil WEBs are the most affected by the direct anthropogenic factors, and consequently waters of the basins are highly polluted. The Zhaiyk–Caspian and Balkash–Alakol WEBs are characterized by a very high degree in water demand and belong to the crisis and critical classes, respectively. In perspective, the water demand creates a lack and misuse of the limited freshwater resources and consequently causes a serious threat to the sustainable development and protection of the environment. Wide application of modern water-saving technologies (drip irrigation, water treatment, reuse of irrigation drainage water, and so on) in various economic sectors; improvement of interstate water relations; inter-basin; and transboundary river flow transfers can become a real basis for ensuring a sustainable development of water resources in Kazakhstan.

2.4 Hydrochemical regime of rivers in Kazakhstan

The analysis results show that in the last 40 years, under the influence of anthropogenic factors, the hydrochemical regime of the main watercourses of Kazakhstan changes radically. Before 1960, all the main rivers of Kazakhstan belonged to the hydrocarbon class with the calcium group and in the first type, but now these indicators are characteristic only of the Ertis, Talas, Ile, Karatal and Asy (Assa) rivers, whereas in other watercourses, there are fundamental changes. Rivers such as Zhaiyk, Yesil, Tobyl, and Nura passed to the hydrocarbonate class in the first type of sodium group from the hydrocarbonate class with the calcium group in the first type. The hydrochemical regime of the Shu and Syrdarya rivers is completely changed. The regime of these basins was estimated as a hydrocarbonate class in the first type of calcium group before the beginning of extensive development; now the hydrochemical regime is estimated as the sulfate class in the second type of sodium group. The changes in the hydrochemical regimes are affected by anthropogenic factors rather than natural ones.

The deterioration of thc hydrological and hydrochemical regimes of the Syrdarya River began since 1960 associated with creation of the Shardara water reservoir, which initiated the radical reconstruction of the historically formed lower reaches of the Syrdarys and the Aral Sea.

Intra-annual distribution of the Syrdarya River flow in the long-term observation period (1912–1960) shows that this process was completely subordinated to the natural factors of flow formation, with the exception of the insignificant influence of the Farhad and Kairakum water reservoirs. At the same time, there are all kinds of basic factors of flow formation (thawed snow, rain, glacial, and groundwater, depending on the location of the components of the hydrographic network) in the Syrdarya river basin. The bulk of the transported runoff is mainly due to floods: the beginning and the end of which depends on the altitude of the catchment, the nature of the distribution of the snow cover, the climatic conditions of the rainfall, the presence of glaciers and snowfields, and the hydrological conditions of the basin.

The changes in total mineralization and the main ions (Ca^{2+}, Mg^{2+}, Na^{++} K^{+}, HCO_{3-}, SO_4^{2-} , Cl^{-}) were analyzed in a rigid relationship with the hydrological regime to obtain specific results of the dependence of the hydrochemical from hydrological regimes. The dependence of the hydrochemical regime from the hydrological regime has been analyzed for both periods with a conditionally natural and with disturbed regimes.

The results of the intra-annual distribution of total mineralization in the stations of Tomenaryk, Kyzylorda, and Kazaly for the conditionally natural period of the hydrological regime (25% of availability) show that a gradual decrease in the actual concentration depending on the water content and the period of the year. At the same time, maximum concentration from 800 to 950 mg/l were observed in the Tomenaryk station in autumn–winter periods and with minimum of up to 400 mg/l in the summer months, i.e., coinciding in time with the high water period. A sharp increase in those three stations was observed in summer season that associated with an increase in the discharge of return water from irrigated areas (Medeu, 2010). For instance, the previously maximum was observed in the autumn–winter months and minimum in spring–summer time. Almost all conditions of the conditionally natural period showed maxima of the mineralization concentration up to 1000 mg/l, whereas in modern conditions, they everywhere exceed this level and reach 2000 mg/l irrespective of the water content of the year and the period in the intra-annual distribution.

Analysis of changes in the intra-annual distribution of calcium shows that in all those three stations there is also an ambiguous process; i.e., in comparison with the conditionally natural period, there is a decrease in the winter months, and in the summer months—an increase in their concentration. For instance, if before 1960 winter concentrations were noted at 120 mg/l, then after creation of the Shardara water reservoir, the same values were recorded at 80 mg/l, regardless of availability. Summer calcium minima from 60 to 100 mg/l increased to 150 mg/l simultaneously with the smoothing of winter maximum and summer minimum (Medeu, 2010).

Compared to calcium, the analysis of the change in the intra-annual distribution of magnesium shows that regardless of the period of the year, this ingredient is constantly growing, with the maximum growth being recorded in the Kyzylorda station. Maxima of magnesium concentration from 7.0 to 60 mg/l are now stably at the level of 100 mg/l and more. The minimum concentrations of 10 to 20 mg/l noted during the flood under current conditions are fixed from 60 to 120 mg/l, differing depending on the observation points. Magnesium is also characterized by a loss of winter highs and summer lows with a simultaneous increase in concentration throughout the year.

Changes in the intra-annual distribution of the sum of sodium and calcium are identical to changes in magnesium; i.e., there is a steady increase in this indicator throughout the year, regardless of the availability. In the Tomenaryk station, during the winter period, the actual concentrations of the sum of sodium and potassium in the range of 100 mg/l (25% of the provision) in the conditionally natural period; they are already noted at 175 mg/l in the disturbed period of the same availability. The same indicator for Kyzylorda station is equal to 75 and 175 mg/l, respectively. In the Tomenaryk, the summer minimums increased from 25 mg/l to 150 mg/l, while in the Kazaly station was 98 and 325 mg/l, respectively. In some periods, in the intra-annual distribution, the amounts of sodium and potassium in the Kyzylorda and Kazaly stations reach 408 and 440 mg/l, respectively (Medeu, 2010).

Studies of changes in the intra-annual distribution of sulfates show an unambiguous tendency to ubiquitous growth of this ingredient. The maximum sulfate concentrations for the Tomenaryk station were from 200 to 300 mg/l in the winter months and the minimum of 180 mg/l in summer. However, at the current level, they are observed 500-600 mg/l in the winter months and 200 mg/l in summer. For the Kyzylorda station, the maxima were 200-320 mg/l and minima 150-180 mg/l in winter under the conditionally natural period (before 1960th). Current maximums of the actual concentration of sulfates are 620-670 mg/l with minimums of summer months from 400 to 480 mg/l. The same indices for the Kazaly station were observed 300 mg/l and with the summer minimums of 140-145 mg/l at the conditional-natural period. The current winter maximum concentration is at 650-670 mg/l, while the summer minimum is 500 mg/l.

The trend of chloride growth is also evident for all three stations considered above. At present, for instance, there are no pronounced winter and summer minimums in the intra-annual distribution of chlorides, which are characteristic of the conditionally natural period of the hydrological regime. The natural maxima of 50-60 mg/l at the current level are equal to 100-140 mg/l, while the summer minimums range from 20-30 to 250 mg/l. The current maxima reached 400 mg/l (75 % and 95% of the water availability), whereas these peaks in the conditionally natural period of the hydrological regime never exceeded 100 mg/l (Medeu, 2010). The current minimum concentration of chlorides was within 200 mg/l and exceeds natural minima by 4 times. The general analysis of changes in the intra-annual distribution of total mineralization and the group of main ions summarizes, and it must be emphasized that with the exception of calcium and hydrocarbonates; radical changes occurred both in the intra-annual distribution and in their actual concentrations associated only with growth. On the contrary, the reverse process is observed with respect to calcium and hydrocarbonates, i.e., their ubiquitous decrease. It is common for everyone that in the intra-annual distribution of all the ingredients considered, at the current level, there are no winter maxima and summer minima (Medeu, 2010).

References

Abishev IA, Medeu AR, Mal'kovskii IM, Toleubayeva LS (2016) Water resources of Kazakhstan and their use. J Water Econ Kaz 3(71):6–15 (in Russian)

Abdukhalikov RS (2015) Water resource potential of the Nura-Sarysu and Tobyl-Torgai water-economic basins in Kazakhstan. Bull Satpaev Kazakh Natl Tech Univ 6:11–16

Akhmedsafin UM, Sydykov ZhS (1964) The flow of underground waters of Kazakhstan, the zone of intensive water exchange. Alma-Ata: Nauka, 88 p (in Russian)

Akhmedsafin UM, Sydykov ZhS (1970) The formation of groundwater runoff on the territory of Kazakhstan. AlmaAta: Nauka, 148 p (in Russian)

Alamanov SK (2016) Water resources of Kyrgyzstan and their use. Water resources of Kazakhstan and their use. J Water Econ Kaz 3(71):17–25 (in Russian)

Boiarkina OA (2011) Analysis of the situation of water safety in Central Asia and options in water development. In: Towards the 6th world water forum joint action in the direction of water security, Tashkent (in Russian)

Bogomolov Yu G, Griniaev SN, Nebrenchin SM, Fomin AN (2007) Water resources of Central Asia in market relations, Moscow, 76 p (in Russian)

Cherkasov PA Vilesov EN (1972) Ice reserve of Kazakhstan. Nature (11) (in Russian)

Concept on "Green Economy" in Kazakhstan (2013) Astana, 52 p (in Russian)

Danilov-Danilian VI (2008) Global issue on water shortage. J Globalization Century 1:45–56

Danilov-Danilian VI, Khranovich IL (2010) Water management: Harmonization of the water use strategy, Moscow, Nauchnyi mir, 248 p (in Russian)

Dzhanalieva KM, Budnikova TI, Vilesov EN (1998) Physical geography of Kazakhstan. Kazak Universiteti, Almaty, p 266 (in Russian)

Duskayev K (2004) Relevant water issues in Kazakhstan. In: Water resources of Kazakhstan in the new millennium. Review. UNDP report, Almaty, pp 35–44 (in Russian)

Dostai ZhD (2012) Natural waters in Kazakhstan: resources, regime, quality and prognosis. In: Water resources of Kazakhstan: assessment, prognosis, management. Almaty, vol 2, 330 p (in Russian)

Espolov TI, Sagyndykova SZ, Kanayev AT, Kanayeva ZK (2006) Ecology of water. Almaty, 231 p (in Russian)

Ereshenko VI (1972) Carp of the Shardara water reservoir and ways to increase his herd. In: Fish resources of water bodies in Kazakshtan and their use. Alma-Ata, Kainar (in Russian)

FAO (2012) Irrigation in Central Asia in numbers. In: Frenken K (ed) Review AQUASTAT, Water reports (in Russian)

Gagloeva AE (2016) The climate change impact on water resources of Central Asia. J Water Econ Kaz 3(71):49–54 (in Russian)

Gorbunov AP (1974) Belt of the permafrost of the Tien Shan. Dissertation for Doctoral degree in Geography, Moscow, 32 p (in Russian)

Gulaya NK and Kanatchinova MK (1961) Microflora and chemical composition of soils of the Oskemen water reservoir. Trudy of Institute of microbiology and virusology Academy of Sciences of KazUSSR, vol 5 (in Russian)

Khamrayev ShR, Shkh Rahimov (2016) Water management in Uzbekistan in order to increase their productivity. J Water Econ Kaz 3(71):40–48 (in Russian)
Kulmatov R (2014) Problems of sustainable use and management of water and land resources in Uzbekistan. J Water Res Prot 6:35–42
Kiseleva VA (1964) Benthos of the Oskemen water reservoir. Izvestia of the Academy of Sciences of the USSR, Biological series, vol 4 (in Russian)
Kuznetsova VM (1974) Hydrological features of the Bugun water reservoir and their impact on fish productivity. In: Fish resources of water bodies in Kazakshtan and their use, vol 8 (In Russian)
Lopareva TYa and Amirgaliev NA (1974) Ion-salt composition and some technical properties of soil of the Shardara water reservoir. In: Fish resources of water bodies in Kazakshtan and their use. Alma-Ata, Kainar (in Russian)
Makarevich KG (1969) The glaciation of the Ile Alatau. Moscow, Nauka (in Russian)
Mal'kovskii IM (eds) (2012) Territorial distribution of water resources in Kazakhstan: possibility and appropriateness. Almaty, 414 p (in Russian)
Mal'kovskii IM (2008) Geographical basis on water supply of water basins in Kazakhstan, 248 p (in Russian)
Medeu AR, Mal'kovskii IM, Toleubayeva LS (2012) Water resources of Kazakhstan: assessment, prognosis, management (concepts). In: Water resources of Kazakhstan: assessment, prognosis, management. Almaty, vol 1, 94 p (in Russian)
Muhabbatov KhM (2016) The problems of formation and use of water resources in Tajikistan. Water resources of Kazakhstan and their use. J Water Econ Kaz 3(71):26–30 (in Russian)
Moerlins JE, Khankhasayev MK, Leitman SF, Makhmudov EJ (eds) (2008) Transboundary water resources: A foundation for regional stability in Central Asia. Springer, Netherlands, p 313
National Atlas of Kazakhstan (NAKZ) (2010) Environment and ecology, vol 3 (in Russian)
Pal'gov NN (1969) Glaciers. In: Kazakhstan, Moscow, Nauka (In Russian)
Philonets PP (1981) Essays on the geography of inland waters of the Central, Southern and Eastern Kazakhstan (lakes, water reservoirs and glaciers). Nauka, KazUSSR, Alma-Ata, 283 p (in Russian)
Philonets PP,Omarov TR (1973) Lakes in the Central and Southern Kazakhstan (reference book). Nauka, KazUSSR, Alma-Ata, 192 p (in Russian)
Philonets PP, Omarov TR (1974) Lakes in the Northern, Western and Eastern Kazakhstan (reference book). Leningrad, Hydrometeoizdat. 137 p (in Russian)
Riabtsev A, Akhmetov S, Kudaibergenuly K (2004) Water use and Protection of water resources. In: Water resources of Kazakhstan in the new millennium. Review. UNDP report, Almaty, pp 12–26 (in Russian)
Sarsembekov TT (2004) River basin management plans in Central Asian countries. Almaty, Atamura, 208 p (in Russian)
Satenbayev EN, Ibatullin SR, Balgabayev NN (2012) Water consumption in economy sectors of Kazakhstan: Assessment and prediction. In: Water resources of Kazakhstan: assessment, prognosis, management, vol 3. Almaty, 262 p (in Russian)
Shitikov VK, Rosenberg GS, Zinchenko TD (2005) Quantitative hydroecology: methods, criteria, solutions: Part 1. Nauka, Moscow, 281 p (in Russian)
UNDP (2003) Review of water situation in Kazakhstan. National Human Development Report, Kazakhstan Chapter 3
Vikulina ZA (1979) Water balance of lakes and water reservoirs of the Soviet Union. Leningrad, Hydrometeoizdat (In Russian)
Veisov SK, Khamrayev GO (2016) Rational use of water resources of Turkmenistan in the context of climate change. J. Water Econ Kaz 3(71):31–39 (in Russian)
Wouters P, Dukhovny V, Allan A (eds) (2007) Implementing integrated water resources management in Central Asia. Springer, Netherlands, p 177
Zaurbek AK, Espolov TI, Kalybekova EM, Zaurbekova ZhA (2012) Regulation and distribution of water resources in Kazakhstan. In: Water resources of Kazakhstan: assessment, prognosis, management, vol 17. Almaty, 282 p (in Russian)

Chapter 3
Lakes in Different Regions of Kazakhstan

3.1 Distribution of Lakes Within Kazakhstan

Lakes of Kazakhstan are very numerous. There are 48,262 lakes in size of one hectare and above with a total surface area of 45,002 km^2. This list is not included in Caspian and Aral Seas, ponds, reservoirs, and plexus lakes. Small lakes in size <1 km^2 are 94, and 10% in by area (Table 3.1). There are 3014 large lakes >1 km^2 with an area of 40,769 km^2 (90%) (Fig. 3.1). There are 20 lakes larger than 100 km^2 and are with an area of 26,886 km^2 (Fig. 3.1), which is 59% of the water surface of all lakes. The total volume of water in these natural bodies is 190 km^3 (Fig. 3.2).

Lakes are distributed very unevenly on the territory of Kazakhstan. They are removed for hundreds of kilometers from each other. They are distributed so dense that they form lake areas. Most of all lakes are concentrated in the forest-steppe and the northern part of the steppe zone, and also there are many in floodplains of large drainage rivers that lost in the sands. Thus, there are 21,580 lakes (45%) in Northern Kazakhstan with a total area of 15.623 km^2 (35%), while in Central and Southern Kazakhstan, combined 17.554 lakes (36%) with a total area of 4658 km^2 (10%) (Figs. 3.3 and 3.4). The Balkash Lake is not included to the total number of lakes in the region. There are 2.61 km^2 of water area of lakes in Northern Kazakhstan for 100 km^2 of territory and in Central and Southern Kazakhstan are 0.23 and 0.53 km^2, respectively (Table 3.2) (Philonets and Omarov, 1973). The largest lakes are located in the southwest and southeast. These are the Caspian and Aral Seas, Balkash Lake, Tengiz in Central Kazakhstan, Alakol and Sasykkol lakes near the Dzungar Gate and Markakol in the Altai Mountain.

Small lakes are located mainly in river floodplains; medium and large ones are mainly confined to ancient river valleys and tectonic depressions. The bulk of water bodies is at absolute altitudes from 100 to 350 m, mainly in loose Cenozoic deposits. The catchment areas of the lakes usually have an area from 0.2 to 4.4%

J. Abuduwaili et al., *Hydrology and Limnology of Central Asia*, Water Resources Development and Management, https://doi.org/10.1007/978-981-13-0929-8_3

Table 3.1 Distribution of lakes in Kazakhstan by lake size and area (Philonets and Omarov 1974)

Size of lakes (km^2)	Number of lakes	Total area	
		km^2	%
0.01–0.1	35,763	1025	2.3
0.11–0.2	3435	467	1
0.21–0.3	1764	438	1
0.31–0.4	1196	420	1
0.41–0.5	879	396	0.9
0.51–0.6	656	395	0.9
0.61–0.7	477	310	0.7
0.71–0.8	448	338	0.7
0.81–0.9	362	308	0.7
0.91–1	268	254	0.6
1.01–5	2357	4900	10.9
5.01–10	361	2424	5.4
10.01–50	252	4869	10.8
50.1–100	23	1572	3.5
Above 100	21	26,886	59.6
Total	48,262	45,002	100

Notice Caspian and Aral Seas, plexus lakes, ponds, and water reservoirs are not included to the total number of lakes in Kazakhstan. There are 6941 plexus lakes with a total area of 235 km^2

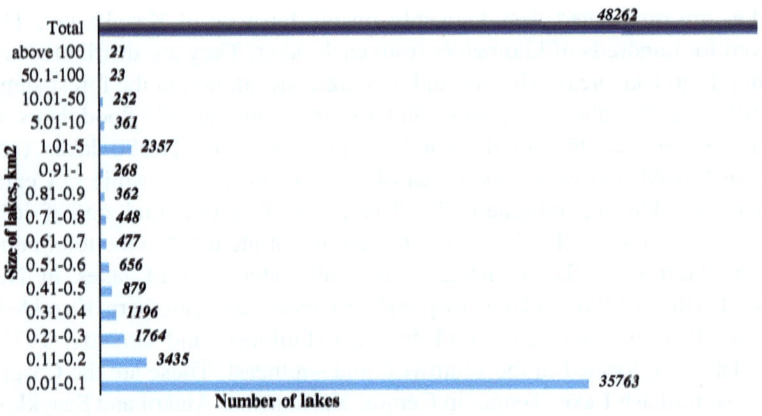

Fig. 3.1 Distribution of lakes in Kazakhstan by lake size

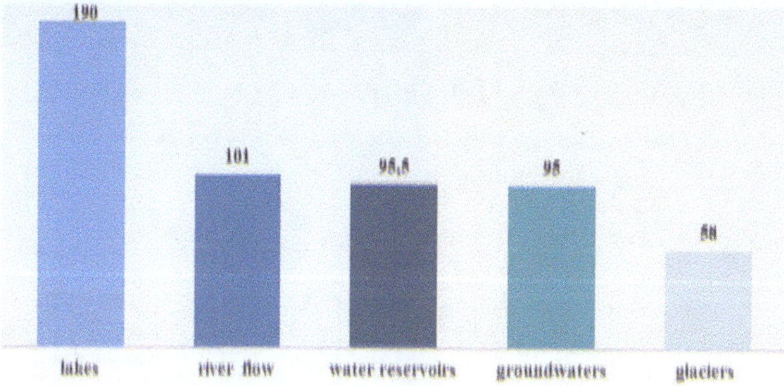

Fig. 3.2 Water volume (km^3) in different water bodies within Kazakhstan

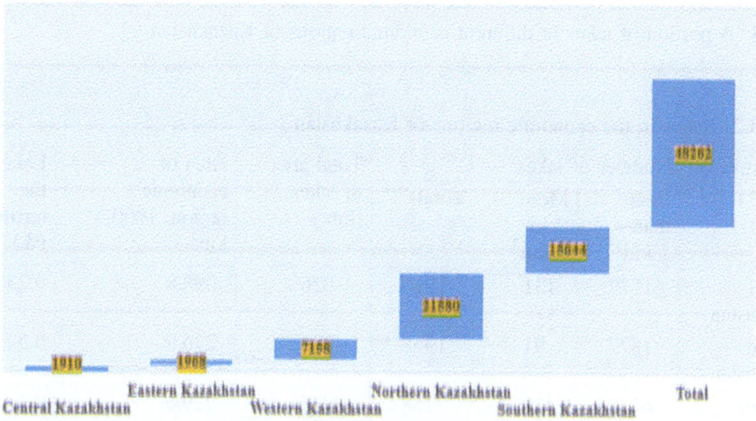

Fig. 3.3 Lakes in the economic regions of Kazakhstan

(Tables 3.3 and 3.4). The lake area is 1.65% in whole Kazakhstan. The total volume of water in the lakes exceeds 190 km^3 (Tables 3.13 and 3.14).

Lakes are unevenly distributed within the regions of Kazakhstan. Maximum number of the lakes is observed in Almaty region (10,842 lakes), in Kostanay region (8604), and in Akmola region (7322) (Fig. 3.5).

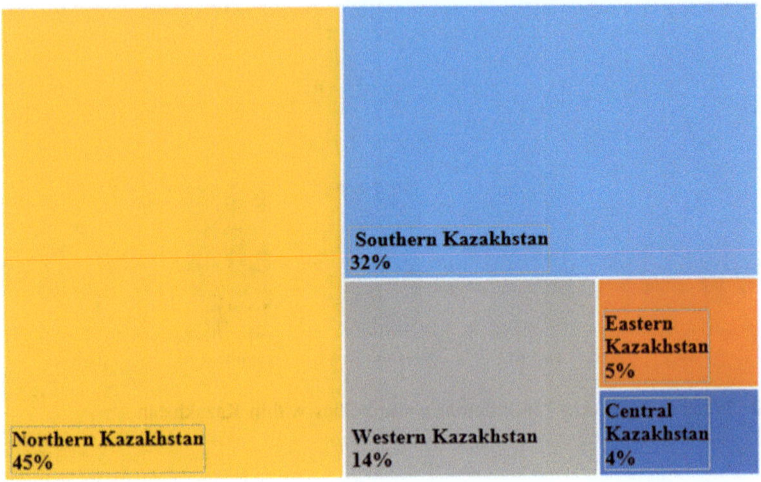

Fig. 3.4 A portion of lakes in different economic regions of Kazakhstan

Table 3.2 Lakes in the economic regions of Kazakhstan

Economic regions	Number of lakes			Total area of lakes (km^2)	Area of economic region, 1000 km^2	Lakes in the territory (%)
	Less than 1 km^2	More than 1 km^2	Total			
Central Kazakhstan	1779	131	1910	926	398.8	0.23
Eastern Kazakhstan	1877	91	1968	896	276.9	0.32
Western Kazakhstan	6738	420	7158	2975	729.6	0.4
Northern Kazakhstan	19,518	2062	21,580	15,623	600.9	2.61
Southern Kazakhstan	15,336	308	15,644	3732	708.8	0.53
Total	45,248	3014	48,262	45,002	2717.3	1.65

Notice Balkash and Alakol lakes are included to the total number of lakes; however, they do not enter any economic region of Kazakhstan

3.2 Lakes in the Northern, Western, Eastern, and Southern Kazakhstan

Lakes of the Northern Kazakhstan. There are 21,580 lakes in Northern Kazakhstan with a total area of 15,623 km^2 (Table 3.5; Figs. 3.6 and 3.7), including 19,518 lakes with size less than 1 km^2 and 2062 lakes in size more than

Table 3.3 Lakes in the administrative regions of Kazakhstan

	Administrative regions	Number of lakes			Total area of lakes (km^2)	Lakes in the territory (%)
		Less than 1 km^2	More than 1 km^2	Total		
1	Almaty	10,775	67	10,842	1598	1.41
2	Aktobe	1521	227	1748	1211	0.40
3	Atyrau	2041	109	2150	856	0.29
4	Akmola	6775	547	7322	6538	6.42
5	Karagandy	1779	131	1910	926	0.23
6	Kostanay	7750	854	8604	5068	2.59
7	Kyzylorda	2453	129	2582	1164	0.53
8	East Kazakhstan	1877	91	1968	896	0.74
9	North Kazakhstan	2252	238	2490	1458	3.24
10	Pavlodar	2741	423	3164	2559	2.05
11	South Kazakhstan	975	74	1049	502	0.41
12	West Kazakhstan	3176	84	3260	908	0.59
13	Zhambyl	1133	38	1171	468	0.31
14	Mangystau	–	–	–	–	–
	Total	45,248	3014	48,262	45,002	1.65

1 km^2 (Table 3.6). In addition, there are 2491 plexus lakes in the region with a total area of 49.79 km^2. Lakes are concentrated unevenly in the different regions of Northern Kazakhstan.

The Northern Kazakhstan includes four regions such as Kostanay, Akmola, North Kazakhstan, and Pavlodar regions. There are 8604 lakes in Kostanay region, 7322 lakes in Akmola region, and 2490 and 3164 lakes in North Kazakhstan and Pavlodar regions, respectively, (Fig. 3.7).

There are 2490 lakes in North Kazakhstan region with a total area of 1458 km^2 (Fig. 3.8), including 2252 lakes with size less than 1 km^2 and 238 lakes in size more than 1 km^2 (Tables 3.7 and 3.8). In addition, there are seven plexus lakes in the region with a total area of 0.19 km^2.

There are 8604 lakes in Kostanay region with a total area of 5068 km^2 (Fig. 3.9), including 7750 lakes with size less than 1 km^2 and 854 lakes in size more than 1 km^2 (Tables 3.9 and 3.10). In addition, there are 877 plexus lakes in the region with a total area of 15.81 km^2.

There are 7322 lakes in Akmola region with a total area of 6538 km^2 (Fig. 3.10), including 6543 lakes with size less than 1 km^2 and 779 lakes in size more than 1 km^2 (Tables 3.11 and 3.12). In addition, there are 1607 plexus lakes in the region with a total area of 33.79 km^2.

Table 3.4 Largest lakes in Kazakhstan and their morphological characteristics

No.	Lake name and its location	Inflow and flow	Catchment area (km^2)	Altitude (m a.s.l.)	Lake area (km^2)	Length (km)	Max width (km)	Length of shoreline (km)	Max depth (km)	Volume of water mass, billion (km^3)
1	Balkash	Ile, Karatal, Aksu, Lepsi, Ayagoz, Bakanas, and Tokrau rivers	501,000	312	18,200	614	74	2383	26.5	106.000
2	[a]Alakol, Almaty and East Kazakhstan region	Urzhar, Katynsu, Emel, Zhamanty rivers and Zhamaputkol	68,700	347.3	2650	104	52	381	54	58.560
3	Teniz Akmola region	Nura river and 15 streams which most of them dries up	94,900	301.4	1162	74.4	32.2	488	–	–
4	Seletiteniz, Akmola region	Sholaksai, Seleti rivers and 19 streams without names	23,400	64.7	750	64.7	22.2	274.6	23.2	1.540
5	Sasykkol, Almaty region	Tentek and Karakol rivers; Zhenishke river is outflow	–	350.5	736	49.6	19.8	182	4.7	2.661
6	Kusmuryn, Kostanay region	Ubagan, Kaib, Shalaterek, Karagan, Duzbai, Ashysai rivers; Ashysai is outflow	10,480	102.9	460	60.7	12.5	226	3.5	–
7	Markakol, East Kazakhstan region	27 small rivers and streams, and Kalzhir river is outflow	1180	1449.3	455	38	19	106	25	6.300
8	Ulken-Karoi, Akmola region	Karasu river and 10 streams without name	7490	56.8	306	67.7	11.9	126.2	–	–
9	Shaglyteniz, Akmola region	Shagly river	10,900	135.6	267	42.9	12.5	95.8	2.0	0.600
10	Teke, Akmola region	37 dried streams without name	4240	28	257	33.3	20.1	155.2	1.0	0.130

(continued)

Table 3.4 (continued)

No.	Lake name and its location	Inflow and flow	Catchment area (km^2)	Altitude (m a.s.l.)	Lake area (km^2)	Length (km)	Max width (km)	Length of shoreline (km)	Max depth (km)	Volume of water mass, billion (km^3)
11	Shalkar, West Kazakhstan region	Skalakankaty and Isenankaty rivers; Solianka river is outflow in high water years	–	16.7	206	18.4	14.7	57.1	13.0	1.00
12	Sarykopa, Kostanay region	Saryozen river	985	101.2	184	27.6	13.2	179.4	–	–
13	Kamyshlybash, Kyzylorda region	Connected with channel in Syrdarya river	–	58.1	178	27.6	9.5	115.6	9.5	0.995
14	Kyzylkak, Pavlodar region	16 dried streams without name	2280	42.6	175	18.8	14	106.5	0.5	0.180
15	Zhalauly, Pavlodar region	Karasu river	–	71.9	171	21	15	81	–	–
16	Karasor, Karagandy region	14 rivers and streams. The main rivers are Taldy and Karasu	8740	621.5	135	43	7	103.8	2.0	0.126
17	Aralsor, West Kazakhstan region	Ashysu river during high-water years	6850	−1.6	124	34.3	13	117.4	–	–
18	Koshkarkol, Almaty region	Zhinishkesu river	–	349.8	120	18.3	9.6	57.3	5.8	0.501
19	Inder, Atyrau region	Streams without names	425	23.8	110	13.5	11	40	–	–

(continued)

(continued)

Table 3.4 (continued)

No.	Lake name and its location	Inflow and flow	Catchment area (km^2)	Altitude (m a.s.l.)	Lake area (km^2)	Length (km)	Max width (km)	Length of shoreline (km)	Max depth (km)	Volume of water mass, billion (km^3)
20	Kishi-Karoi, Akmola region	19 dried streams without names	2090	53.6	101	12.7	12.1	73.4	–	–
21	[b]Aral Sea	Amudarya and Syrdarya rivers	About 94,0 000	53.3	66,458	428	235	6617	66.0	1022.600
22	[c]Caspian Sea	Edil (80% of river flow to the sea), Zhaiyk, Kura, Terek, Samur, and Sulak rivers	3.6 mln	−28.0	About 400,00	1200.0	About 300.0	About 7000.0	980.0	76,000 km^3

Notice [a]The catchment area is shown for the whole Alakol depression; [b]The catchment area of the Aral Sea is 28.5000 km^2 within Kazakhstan. The total area of the Aral is indicated with the islands, and islands are 2345 km^2; [c]The length of the Caspian Sea is given along the meridian; the width is indicated on average; the volume of the water mass is from the average multiyear level

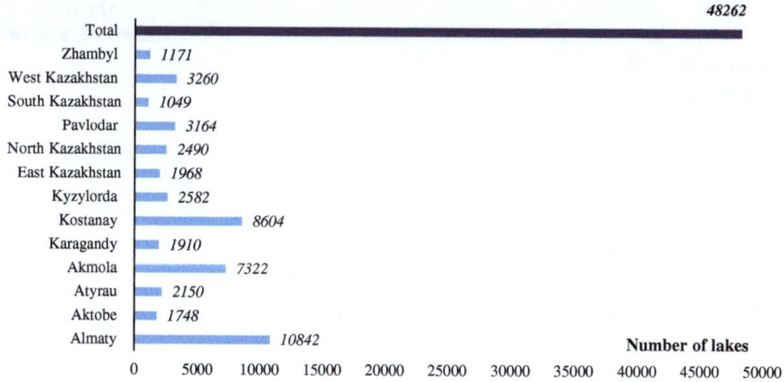

Fig. 3.5 Distribution of lakes in different administrative regions of Kazakhstan

Table 3.5 Distribution of lakes in Northern Kazakhstan by lake size and area (Philonets and Omarov 1974)

Area (km^2)	Number of lakes	Total area	
		km^2	%
0.01–0.1	13,689	381.2	2.4
0.11–0.2	1836	266.4	1.7
0.21–0.3	1091	270.3	1.7
0.31–0.4	779	273.7	1.8
0.41–0.5	576	259.7	1.7
0.51–0.6	461	255.8	1.6
0.61–0.7	336	219	1.4
0.71–0.8	307	230.6	1.5
0.81–0.9	245	207.2	1.3
0.91–1	198	183.3	1.2
1.01–5	1629	3320	21.3
5.01–10	242	1616.5	10.4
10.01–50	166	3283	21
50.01–100	15	1020	6.5
Total	21,580	15,623	100

Notice Ponds, water reservoirs, and plexus lakes are not included to the total number of lakes
There are 2491 plexus lakes in the region with a total area of 49.79 km^2

There are 3164 lakes in Pavlodar region with a total area of 2559 km^2 (Fig. 3.11), including 2741 lakes with size less than 1 km^2 and 423 lakes in size more than 1 km^2 (Tables 3.13 and 3.14). In addition, there are 1607 plexus lakes in the region with a total area of 33.79 km^2.

Lakes of the Western Kazakhstan. There are 7158 lakes in Western Kazakhstan with a total area of 2975 km^2 (Fig. 3.12), including 6738 lakes with size less than 1 km^2 and 420 lakes in size more than 1 km^2 (Tables 3.15 and 3.16).

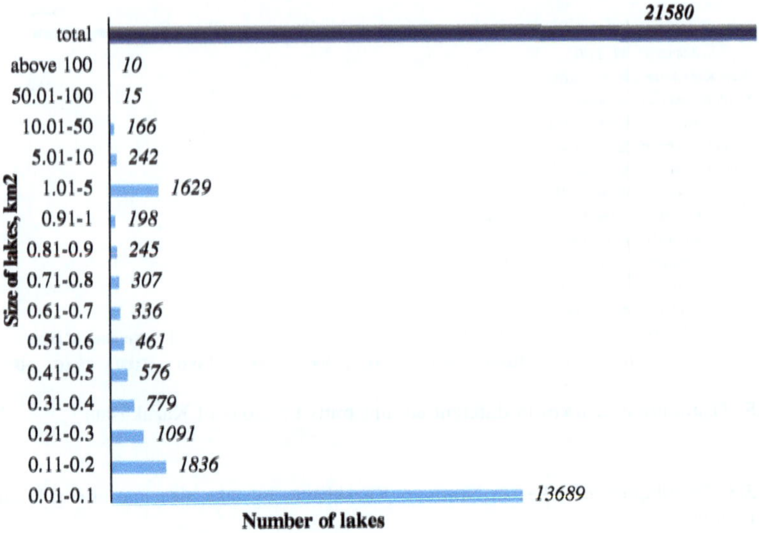

Fig. 3.6 Distribution of lakes in the Northern Kazakhstan by lake size

In addition, there are 579 plexus lakes in the region with a total area of 9.77 Km^2. Lakes are concentrated unevenly in the different regions of the Western Kazakhstan.

The Western Kazakhstan includes four regions such as Aktobe, West Kazakhstan region, Atyrau, and Mangystau regions. There are 1748 lakes in Aktobe region, 3260 lakes in West Kazakhstan region, and 2150 lakes in Atyrau region. However, there is no lake in Mangystau region (Fig. 3.13).

There are 1748 lakes in Aktobe region with a total area of 1211 km^2 (Fig. 3.14), including 1521 lakes with size less than 1 km^2 and 227 lakes in size more than 1 km^2 (Tables 3.17 and 3.18). In addition, there are eight plexus lakes in the region with a total area of 7 ha.

There are 2150 lakes in Atyrau region with a total area of 856 km^2 (Fig. 3.15), including 2041 lakes with size less than 1 km^2 and 109 lakes in size more than 1 km^2 (Tables 3.19 and 3.20). In addition, there are 19 plexus lakes in the region with a total area of 0.78 km^2.

There are 3260 lakes in West Kazakhstan region with a total area of 908 km^2 (Fig. 3.16), including 3176 lakes with size less than 1 km^2 and 84 lakes in size more than 1 km^2 (Tables 3.21 and 3.22). In addition, there are 552 plexus lakes in the region with a total area of 8.92 km^2.

Lakes of the Eastern Kazakhstan. There are 1967 lakes in the Eastern Kazakhstan with a total area of 896 km^2 (Fig. 3.17), including 1876 lakes with size less than 1 km^2 and 91 lakes in size more than 1 km^2 (Tables 3.23 and 3.24).

In addition, there are ten plexus lakes in the region with a total area of 0.41 km^2. The Eastern Kazakhstan includes only East Kazakhstan region.

Table 3.6 Distribution of lakes in different regions of Northern Kazakhstan (Philonets and Omarov 1974)

Region name	Number of lakes			Total area of lakes (km²)	Area of region (km²)	Lakes in the territory (%)
	Less than 1 km²	More than 1 km²	Total			
North Kazakhstan region	2252	238	2490	1458	44,948	3.24
Kostanay region	7750	854	8604	5068	195,257	2.6
Pavlodar region	2741	423	3164	2559	124,756	2
Akmola region	6775	547	7322	6538	233,226	6.39
Total	19,518	2062	21,580	15,623	598,187	14.23

Notice Ponds, water reservoirs, and plexus lakes are not included to the total number of lakes. There are 2491 plexus lakes in the region with a total area of 49.79 km²

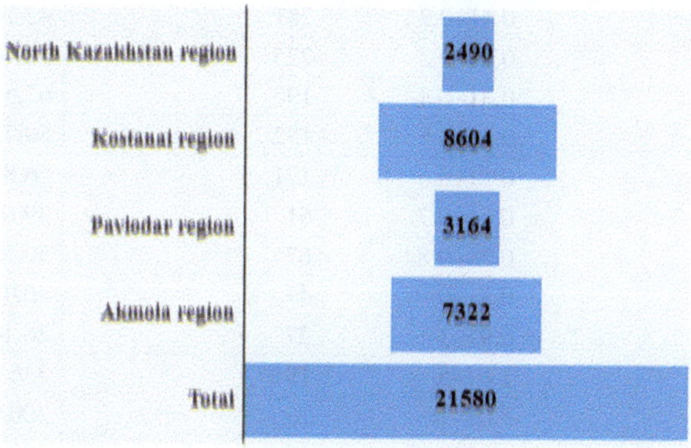

Fig. 3.7 Lakes in Northern Kazakhstan by different regions

The Lake Dubygaly is located in the East Kazakhstan region at an altitude of 343.7 m a.s.l. It is a drainage lake with area of 1.84 km², the length is 2.4 km, the maximum width is 1.2 km, and the length of coastline is 6.8 km (Fig. 3.18).

The lake is located at the foot of the northern slopes of Dungaly Mountain (749.3 m). The catchment area is 22 km². Most part of its is mountainous, with frequent outcrops on the surface of bedrock (gray sandstone) with traces of intensive weathering, and the rest part is sloping and plowed for cereals. The catchment area is widely used for pasturing cattle.

Shores of the lake are mostly low and composed of gray sandstone with an admixture of loam. But some coastal cliffs reach heights of up to 10 m. The coasts

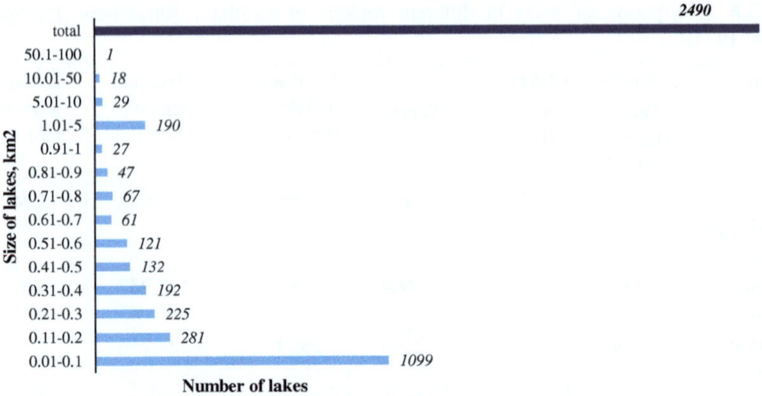

Fig. 3.8 Distribution of lakes in North Kazakhstan region by lake size

Table 3.7 Distribution of lakes in North Kazakhstan region by lake size and area (Philonets and Omarov 1974)

Area (km^2)	Number of lakes	Total area	
		km^2	%
0.01–0.1	1099	36.17	2.8
0.11–0.2	281	42.74	2.9
0.21–0.3	225	56.46	3.9
0.31–0.4	192	67.6	1.6
0.41–0.5	132	59.73	4.5
0.51–0.6	121	66.87	4.6
0.61–0.7	61	39.66	2.8
0.71–0.8	67	50.33	3.6
0.81–0.9	47	40.41	2.8
0.91–1	27	26.14	1.8
1.01–5	190	336.08	23.6
5.01–10	29	200.96	13.8
10.01–50	18	382.6	26.8
50.1–100	1	52.24	3.8
Total	2490	1458	100

are overgrown with fescue, feather grass, white wormwood, bushes of strawberry and dogrose.

The lake occupies a shallow basin of deflationary origin that extends from the southwest to the northeast. There are rare thickets of reed ordinary with an admixture of water and nettles along the coast. The lake bottom is solid, and the lake sediment is composed of fine-grained sand in gray color.

The lake is replenished by melting water and groundwater, as well as atmospheric precipitation. There are two springs on the western and southern shores of the lake. A high water level is observed in April, a low level in October. The annual

Table 3.8 Characteristics of lakes with a size more than 1 km^2 in North Kazakhstan region (Philonets and Omarov 1974)

No.	Lake and its location	Altitude (m a.s.l.)	Lake area (km^2)	Length (km)	Max width (km)	Length of coastline (km)
1	Beloye, 2 km from Sumnoye town	120	2.12/ (3.7)	1.8	1.5	5.2
2	Sumnoye, 2 km from Sumnoye town	120.4	1.36	2.8	0.9	5.3
3	Iakush, 3 km from Dubrovnoye town	121.5	25.81	8.6	3.8	20.9
4	Lake without name, Gaidukovo town	130.4	1.44	1.5	1.3	4.3
5	Gusinoe, Zhiliakova town	88.3	1.53	1.9	0.8	5.9
6	Kulomzino, Kulomzino town	129	1.35	1.4	1.3	4.2
7	Lake without name, Dubrovnoye town	132	1.32	1.4	1.2	4.2
8	Nikolskoe, 1.5 km from Krasnoiarsk town	93.5	2.22	2.3	1.2	9.6
9	Glubokoe, 2 km from Nalobino town	133.9	1.13	1.2	1.2	3.8
10	Nikolskoe, 11.4 km from Krasnoiarsk town	93.5	1.42	2	1	5.2
11	Sergino, 4 km from Dubrovnoye town	130	1.29	1.4	1.2	4.1
12	Sivkovo, Sivkovo town	129	1.48/ (1.6)	1.5	1.2	4.6
13	Krivoe, 3 km Kustovoe town	95.4	2.86	2.5	1.7	7.8
14	Kamennoe 7 km from Dubrovnoye town	130	1.46	1.4	1.4	4.4
15	Kustovoe, Kustovoe town	91.9	1.61	2.2	1.2	6.0
18	Kamyshovo, 2 km Kamyshovo town	128.1	5.26	4.5	1.9	11.2
19	Ploskoe, Glubokoe town	128.6	2.12/ (1.8)	1.9	1.3	5.6
20	Lebiajie, 1 km Sokolovka town	91.5	6.04/ (5.8)	3.1	2.7	10.2
21	Palochnoe, 1 km Glubokoe town	131	1.40/ (1.0)	1.4	1.4	4.3
22	Koniukhovskoe, Koniukovskoe town	129.1	7.98/ (7.3)	4.1	3.8	13.4
23	Shelegino, Nikolaevka town	132.3	15.50/ (21.7)	6.0	3.5	17.8

(continued)

Table 3.8 (continued)

No.	Lake and its location	Altitude (m a.s.l.)	Lake area (km^2)	Length (km)	Max width (km)	Length of coastline (km)
24	Lake without name, Slavnoe town	132.6	1.47	1.4	1.3	4.4
25	Lake without name, Chistoe town	128.8	1.04	1.2	1.1	3.8
26	Kamennoe, Beloye town	129.4	2.60	1.9	1.8	5.8
27	Pravdino, Suvorovka town	129.6	2.12/ (3.2)	1.9	1.6	7.0
28	Karabne, Oktiabrskoe town	129.8	2.56	2.1	1.4	5.9
29	Polovinnoe	108.5	11.92	6.1	3.0	16.3
30	Sevriugino	133.3	1.20	1.8	1.0	4.8
31	Lake without name, town	131.6	2.90	2.1	1.7	6.2
32	Lake without name, town	97.0	2.17	2.0	1.5	5.6
33	Zarosloe	129.2	1.19	1.3	1.2	3.9
34	Mogishe	113.9	1.17	4.1	0.6	9.4
35	Baikal	89.1	1.45	1.7	1.2	4.5
36	Beloye	129.9	2.25	1.9	1.5	5.4
37	Gusinoe	129	1.03/ (1.5)	1.2	1.1	3.9
38	Lake without name, town	129.5	1.19	1.7	1.2	5.0
39	Uglovoe	91.7	1.72	2.0	1.4	7.1
40	Bolshoi Chirok	133	1.54	1.9	1.0	5.0
41	Cherepkovo	128	2.0	2.0	1.3	5.4
42	Shirokoe	128	2.68	2.4	1.7	6.8
43	Riavkino	129.9	2.57/ (3.6)	1.9	1.6	6.0
44	Penkovskoe	128.7	1.39	2.0	0.9	5.3
45	Gor'koe	95.6	8.25	4.0	2.5	11.6
46	Kishibash	125	3.29	2.7	1.1	13.3
47	Beloye	128.9	10.03	4.5	3.2	15.5
48	Solenoye	109.7	9.51	5.8	2.3	13.9
49	Mokhovichok	135.1	1.20	1.3	1.2	4.2
50	Slivnoye	135.9	1.12	1.6	1.1	4.6
51	Lake without name	135.9	1.12	1.5	1.0	4.4
52	Za Vysokoi Dubravoi	136.0	1.1	2.0	0.6	5.9
53	Za Bugrom	132.5	1.24	2.2	0.8	5.6
54	Shirokoe-Pestroe	93.5	2.22	3.0	1.3	8.4
55	Rybnoye	132.4	1.28	1.4	1.1	4.3
56	Baranovo	111.7	5.75	3.1	2.5	9.2
57	Solodka	109.3	4.14	2.8	2.1	7.9
58	Arlagul	139.3	5.17	2.6	2.5	9.0

(continued)

Table 3.8 (continued)

No.	Lake and its location	Altitude (m a.s.l.)	Lake area (km^2)	Length (km)	Max width (km)	Length of coastline (km)
59	Bashkirskoe	138.3	1.28	1.4	1.1	4.3
60	Lebiajie	138.4	1.58	1.7	1.1	5.0
61	Krivoye	132.9	3.84	4.7	1.0	13.8
62	Pestroe	132.5	1.86	3.0	1.1	8.2
63	Pestroe	96.0	1.14	1.4	1.0	4.3
64	Kaselnoye	131.8	1.37	1.7	1.2	4.6
65	Polovinnoye	121.6	7.41	3.3	2.8	10.8
66	Kamyshovo	118.7	13.95	8.3	2.1	21.4
67	Bol'shoye Solenoye	118.5	8.61	5.4	2.6	14.8
68	Solenoye	121.3	3.02	2.1	1.7	6.9
69	Polovinnoye	114.3	8.07	4.3	2.5	11.6
70	Lake without name, town	96.0	1.19	1.8	0.9	8.0
71	Poganoye	99.5	1.06	2.7	1.0	5.2
72	Glinianoye	133.1	1.50	2.0	1.0	5.8
73	Bol'shoye Dolgoe	131.8	1.83	1.8	1.3	5.0
74	Maloe Dolgoe	131.8	1.18	1.3	1.1	4.0
75	Pitnoye	120.6	9.70/ (11.0)	4.8	2.9	13.0
76	Pasynki	129.2	3.44	3.3	1.5	10.0
77	Lake without name	138.2	1.18/ (1.3)	1.2	1.2	4.2
78	Solenoye	137.5	5.69	3.1	2.4	9.1
79	Gor'ko-Solenoye	136.0	6.53	3.4	2.5	10.9
80	Utinoye	137.6	1.20	1.3	1.2	4.2
81	Gor'koye	137.4	1.89/ (1.9)	2.0	1.4	5.7
82	Semilovo	133.2	9.40	5.4	2.5	15.2
83	Kamyshnoye	136.5	1.10	1.3	1.0	4.2
84	Kaskyrly	133.3	2.30	2.1	1.5	6.6
85	Stanovoye	117.5	29.77	9.4	4.5	32.6
86	Grachinoye	133.3	1.19	1.4	1.1	4.3
87	Ploskoye	128.1	8.52	4.0	3.0	12.0
88	Krivoi Koldaman	136.6	1.27	2.4	0.8	6.2
89	Rybnoye	128.9	2.41	2.1	1.5	6.0
90	Travnoye	99.1	3.80	2.5	1.9	7.6
91	Savak	115.0	1.22	1.7	1.0	5.0
92	Maloye	128.6	1.62	1.5	1.3	4.7
93	Pitnoye	138.3	1.96/ (2.0)	2.0	1.2	6.4

(continued)

Table 3.8 (continued)

No.	Lake and its location	Altitude (m a.s.l.)	Lake area (km^2)	Length (km)	Max width (km)	Length of coastline (km)
94	Lake without name	194.2	1.45	1.5	1.2	4.6
95	Lagernoye	137.1	1.82	1.7	1.3	5.4
96	Gor'koye	135.7	1.23/ (1.6)	1.8	1.1	4.9
97	Pustoye	135.9	1.21	1.2	1.2	4.2
98	Pitnoye	138.4	1.92	1.5	1.4	5.8
99	Shitovo	136.9	1.35	1.5	1.1	4.6
100	Mokhovoye	133.4	1.25	1.8	0.8	4.8
101	Golen'koye	99.1	3.64	5.1	1.1	15.7
102	Shirokoye	99.1	2.70	4.2	0.8	12.3
103	Khar'kovskoye	130.9	2.48	2.2	1.4	9.4
104	Lake without name	124.1	2.51	2.6	1.7	8.2
105	Lake without name	137.0	1.13/ (0.8)	1.4	1.0	4.0
106	Arykbalyk	138.0	1.12	1.5	0.9	4.0
107	Gor'koye	140.2	1.36	1.8	1.0	4.4
108	Gor'koye	137.8	1.12	1.3	1.1	3.8
109	Pitnoye	141.5	2.63	1.8	1.6	5.9
110	Piterovo	137.9	1.21	1.7	1.1	4.0
111	Lake without name	123.0	1.09	1.7	0.8	4.5
112	Dolbilovo	150.9	3.62	2.7	2.1	7.8
113	Liagushie	99.2	3.20	4.1	1.5	12.9
114	Lake without name	127.0	1.36	1.5	1.0	4.6
115	Zhaltyr	133.1	3.64	2.1	1.9	8.0
116	Ulken-Zharma	126.3	10.71/ (9.6)	4.5	2.8	12.8
117	Mokhovoye	133.4	1.26	1.5	1.0	4.9
118	Sitovo	101.4	1.40	3.5	0.6	8.0
119	Lake without name	150.0	1.15/ (0.2)	1.4	1.0	4.0
120	Ialy	134.7	2.02/ (2.0)	1.8	1.5	5.3
121	Zhaltyrkol	132.4	1.79	2.2	1.1	5.6
122	Keltesor	114.2	12.10	7.1	2.6	19.2
123	Ekaterinovskoye	150.6	5.18/ (6.4)	3.9	2.0	8.8
124	Gor'koye	143.2	6.70	3.4	2.4	9.4
125	Lake without name	128.0	1.56	1.5	1.3	4.8
126	Zhaltyr	134.3	1.54	1.8	1.2	5.2

(continued)

Table 3.8 (continued)

No.	Lake and its location	Altitude (m a.s.l.)	Lake area (km²)	Length (km)	Max width (km)	Length of coastline (km)
127	Aiakkol	129.9	2.19	1.9	1.5	5.5
128	Sasykkol	147.1	5.49	3.2	2.3	9.0
129	Yagodnoye	147.8	1.24	1.4	1.2	4.4
130	Akshysai	145.1	1.51	2.3	0.9	5.1
131	Lake without name	141.4	1.87	2.4	1.6	8.2
132	Krivoye	123.4	38.04	9.1	6.1	25.1
133	Mangiser	138.4	1.48/ (1.3)	1.4	1.2	4.6
134	Uzynkol	129.1	1.03	1.3	1.1	4.5
135	Lake without name	131.4	2.61	2.1	2.0	6.8
136	Bozaral	149.1	2.23	2.3	1.4	5.6
137	Solenoye	103.8	1.55	1.7	1.1	5.0
138	Shalkovo	134.5	1.85	1.0	1.3	5.3
139	Zhaltyr	128.3	1.27/ (1.4)	2.0	1.0	5.0
140	Zhylandykol	128.0	2.71	2.3	1.6	6.6
141	Uzynkol	148.5	1.22	1.6	1.0	4.0
142	Zheldikol	148.6	1.35	1.5	1.1	4.6
143	Gor'koe	146.3	1.43	2.1	1.2	5.2
144	Lake without name	143.9	1.04	1.3	1.1	3.8
145	Voniuchee	143.1	1.11	1.5	1.0	4.6
146	Krivoye	127.5	1.04	2.0	0.7	4.8
147	Espal	128.3	20.41	9.1	3.5	28.3
148	Taldyaral	136.1	1.21	1.5	1.2	4.5
149	Issyk-Kul	148.2	1.89	1.9	1.2	5.0
150	Tulkol	144.0	1.75	1.6	1.4	4.8
151	Tukaevo	144.5	1.18/ (1.0)	2.3	0.9	5.1
152	Tulymbai	145.0	1.78	1.8	1.4	5.5
153	Bol'shoi Karashoky	105.0	1.22	2.6	0.5	7.3
154	Pestroe	148.0	5.01	3.2	2.2	10.2
155	Krasnoye	143.8	3.05	3.1	1.9	7.9
156	Ulkenkol	143.1	1.14	2.0	0.8	4.4
157	Solenoye	105.0	2.04	6.2	0.6	17.8
158	Shirokoye	125.1	1.08	1.3	1.0	4.2
159	Baisa	104.0	1.04	3.2	0.4	6.7
160	Koshkar	126.6	12.00	6.2	3.0	18.3
161	Zholdyozek	152.7	1.93	1.9	1.2	5.1
162	Izbasor	142.4	5.71	3.4	2.4	9.5

(continued)

Table 3.8 (continued)

No.	Lake and its location	Altitude (m a.s.l.)	Lake area (km^2)	Length (km)	Max width (km)	Length of coastline (km)
163	Solenoye	117.5	7.66	4.4	2.7	11.8
164	Alva	129.9	2.36	2.0	1.8	11.2
165	Terenkol	127.0	1.84	2.4	1.0	6.2
166	Lake without name	130.9	1.97/ (1.4)	2.1	1.2	5.8
167	Obalykol	120.4	1.99	1.7	1.5	5.2
168	Bol'shaya Alua	148.9	1.05	1.4	1.0	3.9
169	Konaklaisor	149.4	1.63	2.1	1.3	5.2
170	Alkasor	153.3	1.44/ (14.7)	2.2	0.9	5.0
171	Streletskoe	127.9	5.02	3.8	1.5	11.4
172	Bozaral	140.1	1.26/ (1.4)	2.0	1.1	5.1
173	Kortupkol	124.0	1.12	1.6	1.0	4.3
174	Assau	151.8	2.74	3.0	1.4	8.3
175	Korzhynkol	150.6	1.71	1.9	1.0	5.5
176	Krivoye	124.5	1.37	2.1	0.9	6.1
177	Uzynkol	128.0	2.64	2.3	1.5	7.4
178	Kindikty	152.7	1.49	2.6	0.9	5.8
179	Tanat-Uzynkol	120.4	1.17	1.2	1.1	3.9
180	Kamyshnoye	150.7	1.73	2.3	1.2	6.3
181	Bol'shoi Karakol	128.5	1.39	1.7	1.0	4.7
182	Krugloye	126.3	1.76/ (1.3)	1.8	1.3	5.0
183	Maibalyk	152.7	1.61/ (1.9)	1.7	1.3	5.0
184	Maibalyk	151.2	1.62	1.7	1.2	5.0
185	Zhaltyrsor	128.3	4.38	2.7	2.3	8.9
186	Aksuat	151.2	3.60	3.5	1.5	10.4
187	Sarybalyk	152.0	5.26	3.5	2.3	10.1
188	Truperdisor	153.7	1.10	1.6	0.9	4.1
189	Kishkenesor	153.3	1.26	1.8	1.0	4.6
190	Itbalyk	149.4	4.19	3.5	1.5	10.1
191	Taimassor	153.1	1.04	2.1	0.8	5.0
192	Naizakol	156.4	5.78	3.8	2.7	12.8
193	Taktakol	141.0	1.35	2.4	0.8	7.2
194	Sarykol	131.9	2.0	1.7	1.4	5.2
195	Sandykol	154.8	1.35	2.0	1.0	4.7
196	Lake without name	156.1	3.07	3.8	0.9	12.9

(continued)

Table 3.8 (continued)

No.	Lake and its location	Altitude (m a.s.l.)	Lake area (km^2)	Length (km)	Max width (km)	Length of coastline (km)
197	Takmola	157.0	1.30	2.1	0.8	5.2
198	Suatkol	155.5	1.62	1.8	1.3	4.8
199	Akbasty	153.2	2.21	2.1	1.4	5.5
200	Sor	158.0	1.1	2.5	0.7	5.6
201	Lake without name	157.7	1.57/ (1.0)	2.6	0.9	7.0
202	Koskol	156.9	3.83	4.1	1.0	10.8
203	Inzhimbay	159.5	8.82	3.7	3.2	10.9
204	Zhaltyrkol	160.4	1.20	2.6	0.9	5.7
205	Lake without name	153.4	42.20/ (36.8)	10.0	4.9	29.6
206	Tarangul	152.6	13.55	5.7	3.1	17.2
207	Zhaltyr	164.1	8.43	3.7	3.1	10.6
208	Koldar	159.0	1.38/ (1.0)	4.7	0.69	8.1
209	Uzynkol	164.6	2.17	2.8	1.1	6.4
210	Kashgar	154.3	3.37	2.3	1.8	7.1
211	Zhamankol	164.1	1.31/ (1.9)	1.9	0.9	4.5
212	Akzhan	163.8	20.94/ (17.4)	8.4	3.1	22.8
213	Al'pash	166.6	2.66	2.4	1.6	6.5
214	Kunshalgan	156.5	45.12	9.0	6.1	37.6
215	Kak	165.0	3.25/ (2.9)	4.3	1.0	11.9
216	Soltan	168.0	1.28	1.6	1.2	4.5
217	Lake without name	169.0	5.23/ (5.2)	4.1	1.7	11.2
218	Sulu	179.2	25.47/ (25.7)	6.4	5.0	18.4
219	Tarangul	169.8	1.26/ (1.7)	2.2	0.9	6.0
220	Malkar	185.0	1.11	1.5	0.9	6.5
221	Lake without name	165.9	18.50/ (17.4)	7.3	3.8	29.4
222	Aksuat	163.7	52.24	9.3	7.3	31.0
223	Kak	132.3	1.02	1.4	1.0	3.8
224	Glubokoye	132.9	1.01	1.3	1.0	3.9
225	Bugrovoye	136.8	1.02	1.3	1.0	4.5
226	Lake without name	136.0	1.00	1.2	1.1	3.7

(continued)

Table 3.8 (continued)

No.	Lake and its location	Altitude (m a.s.l.)	Lake area (km^2)	Length (km)	Max width (km)	Length of coastline (km)
227	Karasie	136.0	1.00	2.00	0.6	4.7
228	Surgutnoye Zaimishe	132.4	1.00	1.2	0.9	3.6
229	Artykol	135.5	1.02	1.2	1.0	3.7
230	Shagalaly	150.6	1.01	1.5	0.9	4.0
231	Satybaldy	148.8	1.01	1.4	1.0	3.9
232	Suatkol	158.8	1.00	2.4	0.6	5.5
233	Kairan	135.9	1.12	1.5	1.0	4.4

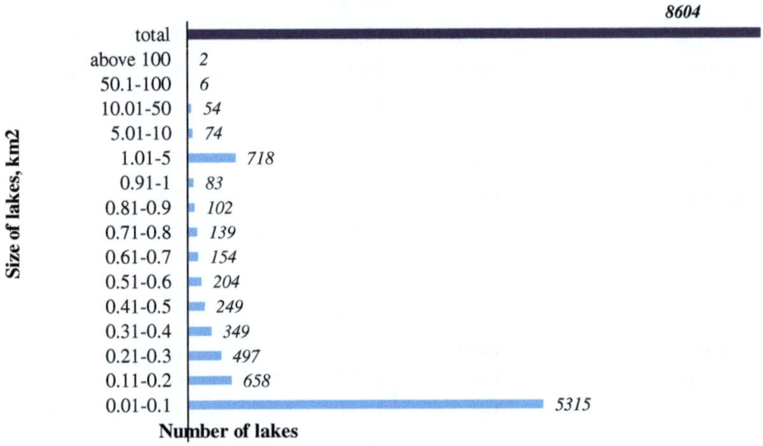

Fig. 3.9 Distribution of lakes in Kostanay region by lake size

amplitude of the oscillation reaches 85 cm (Philonets 1981). According to local residents, the low water level in the lake was observed in 1956–1957 and in 1974–1975. The maximum depth of the lake is 11.2 m, the transparency is 2.5 m, and the color is brown-yellow. It accumulates more than 10 million m^3 of water (Philonets 1981).

The water of the lake is slightly brackish, mild, and slightly alkaline. It refers to the hydrocarbonate class with the sodium group. The content of iodine, bromine, boron, zinc, copper, manganese, cobalt, molybdenum, and nickel in water is low, and fluorine is high (almost 20 times higher than the norm for drinking water) (Philonets 1981). The low content of the mobile form of microelements is characteristic for soils on the catchment area. The total mineralization of water and its chemical composition vary slightly from year to year.

Table 3.9 Distribution of lakes in Kostanay region by lake size and area (Philonets and Omarov 1974)

Area (km^2)	Number of lakes	Total area	
		km^2	%
0.01–0.1	5315	146	2.9
0.11–0.2	658	97.22	1.9
0.21–0.3	497	122.8	2.4
0.31–0.4	349	122.28	2.4
0.41–0.5	249	113.04	2.2
0.51–0.6	204	114.78	2.3
0.61–0.7	154	101.05	2
0.71–0.8	139	104.39	2
0.81–0.9	102	85.35	1.7
0.91–1	83	78.81	1.5
1.01–5	718	1451.89	28.6
5.01–10	74	468.13	9.2
10.01–50	54	997.78	19.6
50.1–100	6	420.44	8.3
Above 100	2	644.47	12.7
Total	8604	5068	100

Notice Plexus lakes, ponds, and water reservoirs are not included to the list. Plexus lakes are 877 with a total area of 15.81 km^2

Table 3.10 Characteristics of lakes with a size more than 1 km^2 in Kostanay region (Philonets and Omarov 1974)

No.	Lake and its location	Altitude (m a.s.l.)	Lake area (km^2)	Length (km)	Max width (km)	Length of coastline (km)
1	Piankovo lake, Kaban' village	143.4	4.28	2.9	2.3	8.4
2	Kamyshnoye, Kaban' village	146.3	1.05	1.5	0.9	3.9
3	Surkino, Kaban' village	151.8	1.05	1.6	0.9	4.1
4	Bolshoye Klushino, Presnogor'kovka village	154.9	1.10	2.0	0.9	4.8
5	Volgarevo, Makarievka village	158.6	1.4	1.9	1.0	4.8
6	Yastrebinoye, Makarievka village	157.1	2.28	2.4	1.4	6.2
7	Lake without name, Kaban' village	158.0	1.05	1.6	1.0	4.2
8	Lake without name, Kaban' village	154.6	1.39	1.8	1.2	4.9
9	Pitnoye, Kaban' village	153.0	1.92	1.7	1.6	5.1
10	Orlovskoye, Presnogor'kovka village	157.8	1.41	1.5	1.2	4.3

(continued)

Table 3.10 (continued)

No.	Lake and its location	Altitude (m a.s.l.)	Lake area (km^2)	Length (km)	Max width (km)	Length of coastline (km)
11	Lake without name, Presnogor'kovka village	156.4	1.46	1.9	1.1	4.8
12	Lebiajie, Presnogor'kovka village	157.0	1.04	1.3	1.2	3.8
13	Kamyshnoye, Presnogor'kovka village	155.6	1.66	2.2	1.2	5.5
14	Kolomenskoye, Presnogor'kovka village	151.0	1.21	1.8	0.8	4.4
15	Stolareno, Presnogor'kovka village	145.6	1.72	1.8	1.0	5.0
16	Bol'shoye Gor'koye, Presnogor'kovka village	148.6	2.37	3.6	1.0	8.8
17	Bol'shoye, Makar'evka village	157.4	1.32	1.8	1.2	5.5
18	Lake without name, Makar'evka village	155.4	1.13	1.4	1.1	4.0
19	Bol'shoye, Peschanka village	158.0	5.13	4.1	2.1	10.5
20	Mokhovoye, Presnogor'kovka village	158.2	1.02	1.2	1.1	3.8
21	Shortabulat, Presnogor'kovka village	150.7	1.11	1.2	1.1	3.9
22	Piazhno, Kamyshlovka village	157.9	2.37	2.4	1.8	7.0
23	Sedel'nikovo, Makar'evka village	159.0	1.15	1.7	1.1	4.7
24	Shubnoye, Makar'evka village	155.4	1.46	2.0	1.2	5.4
25	Lake without name, Chapaevka village	153.0	1.1	1.7	0.9	4.7
26	Kuji, Chapaevka village	152.4	1.41	1.5	1.4	4.5
27	Lake without name, Chapaevka village	154.0	1.23	1.6	1.0	4.5
28	Krasnoye, Chapaevka village	157.0	1.07	1.7	1.0	4.8
29	Zaimishe, Presnogor'kovka village	159.9	1.95	2.1	1.6	1.59
30	Zhukovo, Presnogor'kovka village	157.0	1.22	2.7	0.9	7.7
31	Bol'shoye Momonovo, Presnogor'kovka village	158.0	2.21	2.5	1.5	7.1
32	Dolgoye, Presnogor'kovka village	162.4	1.31	2.3	0.7	5.1

(continued)

Table 3.10 (continued)

No.	Lake and its location	Altitude (m a.s.l.)	Lake area (km^2)	Length (km)	Max width (km)	Length of coastline (km)
33	Markino, Presnogor'kovka village	157.3	2.19	2.0	1.7	5.9
34	Bol'shoye Markino, Presnogor'kovka village	157.0	5.65	6.2	1.9	14.6
35	Maloye Krivoye, Presnogor'kovka village	159.9	3.00	3.5	2.2	11.4
36	Troebratnoye, Makar'evka village	156.6	4.06	4.3	1.5	11.4
37	Mogil'noye, Makar'evka village	152.7	2.32	2.8	1.1	9.7
38	Tolybay, Makar'evka village	153.0	5.71	6.5	1.4	18.0
39	Dolgoye, Makar'evka village	153.0	1.68	2.7	0.9	6.8
40	Solenoye, Zhanazhol village	150.3	1.30	1.9	1.0	5.0
41	Bol'shoye Krivoye, Presnogor'kovka village	158.3	4.77	4.4	2.6	17.3
42	Chechnevo, Presnogor'kovka village	159.4	1.55	1.8	1.1	5.0
43	Kazennoye, Presnogor'kovka village	159.2	1.02	1.5	0.9	3.9
44	Bol'shoy Baidachek, Makar'evka village	164.4	1.06	1.6	1.0	4.2
45	Dolbilovo, Peschanka village	158.6	5.12	9.0	1.2	20.4
46	Shavrino, Peschanka village	159.9	2.18	2.1	1.7	6.4
47	Sokolovo, Presnogor'kovka village	160.6	1.04	1.5	1.1	4.1
48	Taldykol, Presnogor'kovka village	158.0	2.38	2.3	2.3	8.7
49	Ortakol, Presnogor'kovka village	158.0	1.3	1.7	1.2	4.3
50	Popovo, Presnogor'kovka village	159.9	4.14	5.8	1.2	13.4
51	Zhamankol, Presnogor'kovka village	158.0	2.33	3.2	1.9	9.5
52	Beznosovo, Presnogor'kovka village	158.1	1.84	2.7	1.8	6.4
53	Kondykol, Presnogor'kovka village	154.8	1.16	1.7	0.7	4.3

(continued)

Table 3.10 (continued)

No.	Lake and its location	Altitude (m a.s.l.)	Lake area (km^2)	Length (km)	Max width (km)	Length of coastline (km)
54	Ploskoye, Presnogor'kovka village	158.9	2.12	2.2	1.3	5.8
55	Kondykol, Presnogor'kovka village	154.8	1.03	2.2	0.7	5.2
56	Sarybarlyk, Presnogor'kovka village	157.3	1.23	1.7	1.1	4.4
57	Khokhlovatoye, Presnogor'kovka village	156.0	2.0	2.2	1.7	7.4
58	Sladkoye, Evgenievka village	157.0	10.73	5.9	3.0	18.4
59	Toksambay, Makar'evka village	153.0	1.02	1.5	1.4	4.5
60	Solenoye, Makar'evka village	150.7	3.82	3.2	1.5	9.0
61	Zhaltyrsha, Makar'evka village	155.4	1.09	1.7	1.2	4.4
62	Solenoye, Zhanazhol village	149.6	1.3	1.5	1.3	4.2
63	Lake without name, Zhanazhol village	150.2	1.81	1.7	1.4	5.0
64	Lake without name, Chapaevka village	154.0	1.16	1.2	1.2	4.2
65	Toksan, Ksenievka village	159.2	2.25	2.2	1.7	7.6
66	Lake without name, Ksenievka village	158.5	1.25	1.5	1.1	4.3
67	Torakol, Fedorovka village	156.8	3.57	4.3	1.8	10.9
68	Bol'shoy Arylbalyk, Fedorovka village	158.0	1.81	2.2	1.2	5.4
69	Ayagankol, Chapaevka village	158.0	1.18	1.4	1.1	4.0
70	Gniloye, Evgenievka village	162.3	1.65	2.0	0.9	5.3
71	Lake without name, Evgenievka village	155.9	1.4	1.5	1.3	4.4
72	Kurttykol, Ksenievka village	158.1	1.2	1.9	1.0	5.0
73	Kurinoye, Fedorovka village	156.0	4.93	3.7	2.0	10.8
74	Orlovo, Fedorovka village	155.4	1.19	1.5	1.3	4.4
75	Ogyzbalyk,	158.8	1.65/1.7	2.5	1.2	6.3
76	Lake without name,	154.8	1.1	1.4	1.0	4.0
77	Tokpankol	158.0	1.62	1.9	1.2	5.1

(continued)

Table 3.10 (continued)

No.	Lake and its location	Altitude (m a.s.l.)	Lake area (km^2)	Length (km)	Max width (km)	Length of coastline (km)
78	Akpancha	158.3	2.16	2.5	1.2	9.4
79	Koldushka, Evgenievka village	156.0	1.86	2.1	1.1	5.9
80	Goreloye, Evgenievka village	163.0	1.72	1.9	1.4	5.2
81	Koiukamys, Kara-Kamys village	163.0	1.04	1.6	0.9	4.6
82	Suatkol, Kara-Kamys village	163.0	1.22	1.5	1.3	4.5
83	Baitogay Sor, Kara-Kamys village	154.5	3.23	3.5	1.1	9.4
84	Aupildek, Oktiabr village	152.7	1.59	2.5	1.0	6.2
85	Balyktykol, Balykty village	90.7	1.42	1.4	1.3	4.8
86	Kurkopa, Balykty village	91.1	1.05	1.6	1.0	4.6
87	Elshibek, Krasnaya Presnia village	160.1	1.7	2.5	1.5	5.9
88	Gusinoye, Ksenievka village	163.5	1.34	1.4	1.4	4.4
89	Mokhovoye, Ksenievka village	165.8	2.02	2.5	1.3	6.6
90	Bol'shoy Teriskol, Fedorovka village	160.3	5.7/5.8	3.1	2.5	9.0
91	Orlovo, Fedorovka village	158.9	1.52	1.7	1.1	4.6
92	Zhylandy, Fedorovka village	154.2	2.9	2.9	1.5	7.6
93	Lake without name	157.8	1.01	1.7	1.0	4.7
94	Kairov	157.7	1.2	1.8	0.9	4.2
95	Esetkol	158.3	1.52	1.6	1.2	4.4
96	Antancha	158.3	1.76	1.8	1.4	4.8
97	Zharkaiyn	156.4	17.27	7.3	4.7	24.4
98	Ashykol, Evgenievka village	161.0	1.39	1.5	1.2	4.2
99	Solenoye, Kara-Kamys village	155.5	1.6	1.6	1.4	4.7
100	Solenoye, Kara-Kamys village	164.0	1.23	1.7	1.1	4.5
101	Ulkenkol, Kara-Kamys village	104.0	1.51	2.0	1.2	5.0
102	Kazaly, Oktiabr village	152.0	1.32	1.6	1.2	4.2
103	Solenoye, Kara-Kamys village	152.6	1.12	1.2	1.2	3.8

(continued)

Table 3.10 (continued)

No.	Lake and its location	Altitude (m a.s.l.)	Lake area (km^2)	Length (km)	Max width (km)	Length of coastline (km)
104	Lake without name, Oktiabr village	152.7	2.16	2.2	1.4	5.8
105	Sarykamys, Uzynagash village	92.3	2.44	3.0	0.8	6.8
106	Sarykamys, Uzynagash village	93.6	1.37	2.9	1.4	9.5
107	Karakamys, Krasnaya Presnia village	89.3	11.51	4.6	3.0	14.2
108	Loba, Krasnaya Presnia village	89.7	1.81	1.8	1.4	4.8
109	Tutukol, Krasnaya Presnia village	90.7	1.8	1.9	1.3	5.4
110	Uzynkol, Granichnyi village	164.2	1.83	3.3	1.1	8.9
111	Lake without name, Granichnyi village	166.0	1.08	1.6	0.9	4.2
112	Ortakol	155.5	5.54	3.9	2.3	11.2
113	Kairalykol	157.4	1.46	1.4	1.4	4.3
114	Aitpaikol	157.4	1.65	2.5	0.9	5.5
115	Shanbay, Kara-Kamys village	159.3	1.23	1.9	0.9	4.8
116	Zharkol, Kara-Kamys village	158.0	1.73	2.2	1.4	5.9
117	Alkasor, Kara-Kamys village	153.8	2.8	4.8	1.0	10.3
118	Ulkenkol, Balykty village	173.1	1.26	1.9	1.1	4.8
119	Alkakol, Krasnaya Presnia village	91.0	1.16	1.7	1.0	4.4
120	Kogaly, Krasnaya Presnia village	90.6	1.1	1.4	1.0	4.2
121	Kurkol, Ershovka village	168.0	4.26	2.7	2.2	8.0
122	Ulykol, Ershovka village	168.0	2.86	2.1	1.9	6.1
123	Lake without name, Fedorovka village	164.4	1.03	1.5	1.0	3.8
124	Aizhan	156.0	14.27	5.9	3.3	17.4
125	Kun'-Koman	157.0	1.49	1.9	1.3	5.0
126	Eshki	157.1	3.4	4.8	2.1	13.6
127	Aralkol, Kara-Kamys village	158.7	2.0	1.9	1.6	5.6
128	Karakamys, Kara-Kamys village	158.4	10.61	4.6	4.1	20.4

(continued)

Table 3.10 (continued)

No.	Lake and its location	Altitude (m a.s.l.)	Lake area (km^2)	Length (km)	Max width (km)	Length of coastline (km)
129	Lake without name, Riajskii village	160.4	1.45	1.5	1.5	5.0
130	Babie, Riajskii village	159.5	7.91	4.4	3.0	19.8
131	Kachizkol	157.0	1.67	2.0	1.1	5.7
132	Lake without name, Kara-Kamys village	159.9	2.0	2.2	1.2	6.1
133	Lake without name, Kara-Kamys village	162.0	1.24	1.5	1.0	4.6
134	Lake without name, Kara-Kamys village	154.4	1.96	2.1	1.5	6.6
135	Teniz, Krasnaya Presnia village	89.4	64.2	14.7	7.3	44.4
136	Byzhargan, Krasnaya Presnia village	92.7	1.06	1.5	1.0	4.0
137	Karakamys, Krasnaya Presnia village	89.4	2.04	2.1	1.3	5.6
138	Kutur, Demianovka village	165.7	1.39	1.8	1.1	5.0
139	Lake without name, Riajskii village	160.4	2.07	2.7	1.1	6.8
140	Rechnoye, Riajskii village	159.6	22.67	6.2	5.3	21.6
141	Ashykol	159.0	1.15	1.5	1.2	4.0
142	Lake without name	158.2	1.21	1.5	1.1	4.2
143	Lake without name, Kara-Kamys village	159.0	1.28	1.5	1.3	4.8
144	Lake without name, Kara-Kamys village	157.2	1.24	1.8	1.2	6.0
145	Medet, Krasnaya Presnia village	91.6	1.56	1.9	1.3	5.3
146	Balykty, Demianovka village	168.9	4.52/4.4	3.1	1.7	10.2
147	Bol'shoye, Demianovka village	168.7	2.8	3.3	1.3	11.0
148	Lake without name, Riajskii village	164.1	1.22	2.0	1.0	5.2
149	Lake without name, Riajskii village	168.0	1.03	1.6	1.1	4.4
150	Karangyr, Riajskii village	162.3	6.08	4.0	2.6	11.8
151	Ushkol, Riajskii village	161.7	2.51	2.4	1.4	6.2
152	Ekibas	157.1	1.02	1.7	0.9	4.1
153	Ashykol	161.0	1.05	1.6	0.9	4.0
154	Lake without name	159.7	1.03	1.5	1.2	4.2

(continued)

Table 3.10 (continued)

No.	Lake and its location	Altitude (m a.s.l.)	Lake area (km^2)	Length (km)	Max width (km)	Length of coastline (km)
155	Lake without name, Korolevka village	163.2	1.03	1.4	1.0	4.0
156	Koskol, Korolevka village	158.4	1.58	1.9	1.5	5.6
157	Lake without name, Korolevka village	158.0	1.21	1.5	1.1	4.1
158	Shoshkaly, Korolevka village	158.8	7.23	6.9	1.8	20.8
159	Karasor, It-Sary village	88.7	5.3	4.4	1.6	14.2
160	Akpas, Krasnaya Presnia village	90.6	1.43	1.7	1.3	5.1
161	Zharkol, It-Sary village	91.2	5.33	3.4	1.7	11.8
162	Kolkriuk, Magnay village	221.4	1.32	1.4	1.2	4.6
163	Karakamys, It-Sary village	93.0	5.9	4.1	1.7	11.8
164	Kamyshov, Pavlovka village	170.5	2.28	2.3	1.3	6.4
165	Lake without name, Annovka village	163.8	1.13	1.8	0.8	4.6
166	Shoshkaly, Annovka village	162.8	7.41	3.4	2.8	13.6
167	Lebedinoye, Korolevka village	162.0	4.66	2.6	2.2	9.1
168	Zhaman, Korolevka village	164.9	7.53	3.5	3.2	11.4
169	Kumdykol, Annovka village	165.4	3.31	2.1	2.0	7.0
170	Bol'shoy Koskol, Novoshumnyi village	178.0	6.02/7.8	4.6	1.6	11.5
171	Baigun, Sosna village	90.2	2.6	2.3	1.6	6.4
172	Bol'shoy Koskol, It-Sary village	94.4	2.87	2.2	1.6	6.4
173	Tuzkol, Sosna village	90.4	2.43	2.0	1.9	7.6
174	Malyi Koskol, It-Sary village	94.4	1.18	1.3	1.3	4.0
175	Itsary, It-Sary village	94.4	3.53	3.0	1.8	7.8
176	Solenoye, Novotroitsk village	228.0	1.13	1.5	1.0	4.3
177	Sor, Magnay village	237.7	4.79	2.7	2.2	9.8
178	Solenoye, Magnay village	228.4	1.05	1.3	1.1	3.8
179	Togylbay, Aralskii village	179.0	1.29	1.8	0.9	4.8
180	Mendeisor, Sosna village	96.9	1.78	1.8	1.6	8.2
181	Alakol, Aksuat village	94.9	17.63/ 19.8	8.2	4.1	36.9

(continued)

Table 3.10 (continued)

No.	Lake and its location	Altitude (m a.s.l.)	Lake area (km^2)	Length (km)	Max width (km)	Length of coastline (km)
182	Akras, Novotroitsk village	225.4	1.88	1.9	1.3	5.2
183	Sartkol, Novotroitsk village	217.8	1.9	1.8	1.4	5.1
184	Sasykkol, Novotroitsk village	209.5	20.09	5.4	5.0	19.5
185	Ashykol, It-Sary village	96.1	5.43	3.7	1.9	9.2
186	Borovskoye, Borovskoye village	167.2	1.9	2.4	1.3	6.4
187	Ashykol, It-Sary village	165.7	1.91	2.0	1.4	5.6
188	Tenizkol, It-Sary village	97.0	1.69	3.0	1.0	12.6
189	Kashykbay, Boskol	229.0	4.08	2.5	2.1	7.8
190	Ashilykol, It-Sary village	9.28	3.56	4.3	1.5	10.2
191	Chistyi-Chaidak	174.6	1.51	1.6	1.0	6.3
192	Tokbuldy, Boskol	238.6	1.65	1.6	1.3	5.0
193	Sopyi, Peshkov village	202.9	3.72/4.7	2.8	2.3	10.8
194	Zhaksyzharkol, Kopychenskii	186.0	8.66	4.0	3.	11.3
195	Zhadrasor, It-Sary village	9.39	3.	2.8	1.5	7.1
196	Bozkol, Bozkol	243.2	1.82	1.8	1.5	5.0
197	Belokamennoye, Novotroitsk village	217.2	5.08	4.4	2.1	11.4
198	Sauj, Peshkov village	202.2	4.45	3.1	1.8	8.3
199	Toktas, Tsabelevka	179.9	10.76/11.2	4.4	3.2	12.2
200	Sorkol, Kara-Kopa village	174.9	1.75	1.7	1.4	5.1
201	Zhangirkol, Kara-Kopa village	175.7	1.11	1.4	1.0	4.0
202	Zherebets, Kara-Kopa village	175.5	1.16	1.6	1.1	4.5
203	Solenoye, Boskol	243.2	1.14	1.4	1.2	4.0
204	Svetloye, Mikhailovka village	213.9	4.9	2.4	2.4	9.1
205	Sorkol, Boskol	178.7	1.75	1.8	1.4	5.1
206	Ulykankol, Kumtobe village	94.0	2.18	3.4	1.4	11.1
207	Batsalykol, Zlatoust village	96.3	2.85	2.4	1.7	6.7
208	Malye Borli, Borli village	194.0	6.32/6.6	3.0	2.7	9.6
209	Borsha-Beskol, Fedorovka village	199.9	2.93	2.0	1.8	6.3
210	Zhaman-Zharkol, Fedorovka village	194.4	2.63/2.6	2.1	1.7	6.2

(continued)

Table 3.10 (continued)

No.	Lake and its location	Altitude (m a.s.l.)	Lake area (km^2)	Length (km)	Max width (km)	Length of coastline (km)
211	Solenoye, Bikilek village	177.9	2.85	2.8	1.9	7.1
212	Bol'shoy Shishikol, Shishinckii village	175.0	1.52	1.9	1.1	4.7
213	Zholtay, Zlatoust village	96.0	3.01	2.9	1.7	7.6
214	Ulkenborli, Borli village	202.5	14.53	6.3	3.3	17.4
215	Burzhak, Smirnovskii village	209.1	1.81	1.9	1.2	5.0
216	Sharshildak, Batmanovskii village	208.4	2.74	2.1	1.8	6.4
217	Zhurzhulgan, Kalinovskii village	201.1	12.54	5.1	3.1	23.4
218	Baitemir, Uspenovka village	181.2	1.51	1.5	1.5	4.7
219	Bikilek, Bikilek village	178.2	1.38	1.5	1.2	4.3
220	Lake without name, Shishinckii village	179.0	1.05	1.9	0.8	4.7
221	Ulykol, Tatyanovka village	125.4	1.69	1.9	1.2	5.0
222	Minzhasar, Uspenovka village	183.9	3.44	2.3	2.0	7.3
223	Uzynkol, Bikilek village	176.0	4.28	4.1	1.4	11.4
224	Rybnoye, Shishinckii village	176.3	1.43	1.8	1.0	5.0
225	Bukharbaykol, Shishinckii village	176.9	1.13	1.6	1.1	4.2
226	Bol'shaya Karakopa, Timinskii village	178.6	4.82	5.0	1.1	12.1
227	Lake without name, Timinskii village	179.2	1.11	1.8	0.9	4.2
228	Syrdak, Mikhailovka village	220.4	5.07/6.9	3.3	2.4	9.6
229	Bol'shoy Kansulik, Narazovka village	209.8	1.10	1.6	0.9	4.2
230	Korov'e, Andreevka village	187.5	1.09	1.3	1.0	3.9
231	Malaya Karakopa, Timinskii village	178.6	3.18	3.2	1.4	7.8
232	Karakol, Timinskii village	176.1	1.77	2.0	1.3	5.2
233	Nazarkerei, Nazarovka village	216.2	2.16	1.8	1.5	5.3
234	Koiandy, Andreevskii village	193.0	1.49/1.5	1.5	1.3	4.6
235	Zharzhaltyrkol, Andreevskii village	187.2	5.23/6.0	2.8	2.5	8.2

(continued)

Table 3.10 (continued)

No.	Lake and its location	Altitude (m a.s.l.)	Lake area (km^2)	Length (km)	Max width (km)	Length of coastline (km)
236	Lysanovskoye, Ozernyi village	182.2	1.95	2.1	1.7	5.7
237	Korchagino, Ozernyi village	178.8	1.15	2.2	0.7	6.0
238	Lake without name, Ozernyi village	189.0	1.62	2.1	1.1	4.9
239	Ordavaevo, Ozernyi village	180.6	1.10	1.4	1.3	4.3
240	Teniz, Zlatoust village	94.8	1.81	1.8	1.4	5.4
241	Solenoye, Zlatoust village	95.4	2.79	2.3	1.5	6.9
242	Shalkol, Ozernyi village	184.0	1.96	2.6	1.3	6.7
243	Bisarkol, Narazovka village	217.7	1.38	1.5	1.2	6.5
244	Nurkopa, Andreevskii village	190.9	3.42	2.3	1.8	6.8
245	Mel'nichnoye, Ozernyi village	186.9	1.23	1.7	1.0	4.5
246	Svetloye, Kansor village	236.4	1.23	1.4	1.1	5.2
247	Karachakol	207.7	1.62	2.2	1.2	5.0
248	Barak	208.1	1.76	1.9	1.4	5.4
249	Karakol, Kostychevskii village	204.7	2.94	2.2	1.7	6.7
250	Lake without name, Svetlyi-Zharkol village	187.2	1.73	3.4	0.9	7.9
251	Perereznoye, Sheminovskii village	183.2	1.27	1.6	1.2	4.3
252	Bagay, Ermakovka village	199.9	3.97	3.1	1.9	8.7
253	Bol'shoy Kostomar, Svetlyi-Zharkol village	189.6	2.04	2.7	1.3	7.4
254	Konaykol, Svetlyi-Zharkol village	188.0	1.02	1.6	1.1	4.4
255	Solenoye, Kansor village	237.4	2.22	1.8	1.7	5.6
256	Tyzgun, Kostychevskii village	194.0	10.6	3.6	3.6	12.2
257	Sorkol, Kostychevskii village	194.2	4.29	2.6	1.9	8.3
258	Bol'shoy Koiandy, Kostychevskii village	201.6	1.30	1.3	1.2	4.2
259	Otegenkol, Kostychevskii village	193.0	5.10	2.7	2.3	8.3
260	Lelegie, Sheminovskii village	186.9	1.63	1.7	1.5	5.0

(continued)

Table 3.10 (continued)

No.	Lake and its location	Altitude (m a.s.l.)	Lake area (km^2)	Length (km)	Max width (km)	Length of coastline (km)
261	Davydova, Svetlyi-Zharkol village	189.0	1.04	1.5	1.1	4.0
262	Popova, Riazanovka village	183.4	1.14	1.5	1.2	4.5
263	Sarykol, Uritskii village	201.9	37.68/ 37.9	8.2	6.2	24.6
264	Kosoba, Karabalyk	212.6	10.06	4.0	3.1	15.4
265	Savatkol, Kostychevskii village	199.9	1.07	1.2	1.2	3.8
266	Zhakunei, Kostychevskii village	199.9	1.06	1.4	1.0	3.9
267	Sarzhav, Kostychevskii village	198.9	5.59	4.0	2.3	10.8
268	Zhersor,	234.2	1.27	1.6	1.2	4.4
269	Kyldyk, Ornek village	229.9	1.04	1.5	1.0	4.0
270	Salamat, Karabalyk	209.7	6.45	4.5	1.7	12.2
271	Shakirt, Kostychevskii village	198.3	2.30	1.2	1.3	6.5
272	Lake without name, Maikol	187.9	1.08	1.3	1.1	3.9
273	Lake without name, Kara-Oba village	99.0	1.40	2.7	0.9	6.3
274	Kirsur	233.4	1.06	1.3	1.0	4.9
275	Ornek, Ornek village	232.7	3.83	2.5	2.0	7.2
276	Karasor, Leninskoye village	213.9	6.66	3.1	2.4	9.9
277	Zhubaly, Kostychevskii village	198.8	3.36	2.5	1.8	7.2
278	Sorolinskoye, Ornek village	235.7	1.84	1.8	1.4	5.4
279	Zharkol	209.4	1.49	2.1	1.0	5.1
280	Koshkar, Zuevka village	175.7	1.34	1.8	1.0	4.8
281	Lake without name, Ornek village	174.0	1.36	1.8	0.9	4.8
282	Kunshash, Ornek village	173.0	2.37	3.0	1.3	7.3
283	Kairankol, Kara-Oba village	98.9	1.16	1.5	0.9	4.7
284	Togyzbay, Leninskoye village	218.0	4.30	2.5	2.1	8.0
285	Solenoye, Leninskoye village	213.6	2.27	2.2	1.3	8.0
286	Lake without name	209.9	2.07	2.0	1.2	6.0

(continued)

Table 3.10 (continued)

No.	Lake and its location	Altitude (m a.s.l.)	Lake area (km^2)	Length (km)	Max width (km)	Length of coastline (km)
287	Tungatar	212.4	1.69	1.4	1.4	5.8
288	Zholdybaykol	212.3	2.74	2.3	1.7	6.3
289	Bol'shoy Taldykol	213.3	4.22	2.9	2.1	8.0
290	Kemebaykol, Kostychevskii village	207.9	4.11	3.4	1.7	12.0
291	Kaiyndykol, Kostychevskii village	196.1	4.15	3.1	2.0	8.8
292	Manbay, Kostychevskii village	203.3	2.73	2.0	1.5	8.7
293	Temir, Kostychevskii village	202. 6	1.25	1.5	1.2	4.3
294	Alakol, Kostychevskii village	202.8	1.27	1.9	0.9	4.9
295	Uzynkol	189.6	4.73	4.6	1.6	11.0
296	Taly, Kara-Oba village	94.5	13.91/ 11.3	6.7	3.1	18.3
297	Bol'shoy Koskol, Pregradnyi village	199.6	4.11	2.6	2.1	7.9
298	Malyi Koskol, Pregradnyi village	200.3	1.65	2.0	1.0	5.3
299	Ulken Manas	213.8	2.50	2.3	2.0	8.4
300	Malyi Taldykol	213.3	1.56	2.2	1.2	5.4
301	Sarybaksy	208.8	2.70	2.4	1.8	7.4
302	Kumiakol, Kostychevskii village	198.3	3.21	2.2	1.9	6.7
303	Slivnoye, Kara-Oba village	95.9	5.87/5.7	5.7	1.7	14.4
304	Shengel, Leninskoye village	218.5	2.58	2.2	1.9	7.8
305	Kishkentai Manas	214.8	1.22	1.3	1.2	4.1
306	Kepebaikol	189.9	12.22	5.1	3.9	14.2
307	Baizharakkol, Biriukovskii	177.7	5.38	4.1	1.6	11.6
308	Lake without name, Kara-Oba village	97.0	1.62	2.8	0.9	5.8
309	Koskol	207.6	1.03	1.2	1.1	3.9
310	Sarkol	208.7	2.75	2.3	1.5	6.2
311	Esenkol, Kostychevskii village	204.9	1.14	2.1	0.9	5.1
312	Uzynkol, Kostychevskii village	198.1	3.13	3.4	1.5	9.1
313	Esenkol	210.8	2.81	2.1	1.8	6.4
314	Elshibek	210.9	1.44	1.8	1.0	4.7

(continued)

Table 3.10 (continued)

No.	Lake and its location	Altitude (m a.s.l.)	Lake area (km^2)	Length (km)	Max width (km)	Length of coastline (km)
315	Kishtay	206.7	1.16	1.8	1.1	4.6
316	Malyi Mamatek, Kostychevskii village	202.8	2.05	2.1	1.3	5.3
317	Bol'shoy Mamatek, Kostychevskii village	201.7	4.59	5.1	1.5	12.2
318	Kamyshinnoye, Zuevka village	176.3	1.72	1.9	1.2	5.2
319	Sor, Kostychevskii village	204.1	1.24	1.6	1.1	4.2
320	Shamatay, Biriukovskii village	186.0	1.30	1.7	1.2	4.6
321	Beloye, Leninskoye village	209.0	5.27	2.8	2.6	9.6
322	Karpyldak, Leninskoye village	216.6	3.31	2.2	2.0	6.8
323	Baiterek	213.6	1.38	1.7	1.2	4.6
324	Karasor, Leninskoye village	209.6	5.29	3.0	2.3	8.4
325	Karakipshak, Nikolaevka village	212.7	3.14	2.1	1.8	7.4
326	Obalgun	209.1	1.96	2.5	1.1	6.5
327	Eginkol, Kostychevskii village	208.0	1.14	1.6	1.1	4.1
328	Kurbay, Lomonovskii village	200.5	1.31	1.9	1.0	4.8
329	Baishalkar, Kostychevskii village	206.5	4.04	3.0	1.9	8.9
330	Uzynkol, Kostychevskii village	204.7	1.85	2.1	1.1	5.7
331	Talkol, Lomonovskii village	194.9	1.62	2.5	0.9	6.1
332	Sartovskoye, Zhdanovka village	188.5	1.45	2.1	1.0	5.2
333	Aksor, Likhachevka village	110.1	1.69	1.5	1.3	4.8
334	Bozshakol, Sevastopol'skii	201.7	24.19	6.0	5.0	18.5
335	Kukamys	210.4	1.52	1.5	1.2	4.7
336	Sartaykol, Kostychevskii village	206.2	1.38	1.4	1.2	4.5
337	Begezhan, Kostychevskii village	208.1	1.15	1.3	1.3	3.9
338	Zhalkol, Lomonovskii village	202.4	2.85	2.5	1.6	6.9
339	Konaizharkol, Pavlovskii village	190.7	4.51	4.1	1.6	10.8

(continued)

Table 3.10 (continued)

No.	Lake and its location	Altitude (m a.s.l.)	Lake area (km^2)	Length (km)	Max width (km)	Length of coastline (km)
340	Kurpildek, Pregradnyi village	193.3	4.85	2.8	2.3	8.4
341	Ospantomar, Lomonovskii village	209.2	1.23	1.4	1.2	4.2
342	Izentemir, Lomonovskii village	203.4	2.59	2.2	1.5	6.3
343	Koshelivskoye, Vasilevka village	191.0	1.17	1.3	1.2	4.1
344	Lake without name, Sverdlovka village	179.0	1.69	2.2	1.2	6.8
345	Dolgoye, Sverdlovka village	179.5	1.30	1.5	1.0	4.6
346	Teke	208.4	1.37	1.8	0.9	4.8
347	Takaykol, Lomonovskii village	207.3	3.19	2.4	2.1	7.3
348	Kumbaykol, Lomonovskii village	200.6	2.31	2.2	17	6.0
349	Botabaykol, Arystankol village	187.6	1.53	1.7	1.1	4.7
350	Sulu, Sherbinovka village	207.2	1.68	1.6	1.4	5.2
351	Zhianabaskol, Sherbinovka village	203.2	3.19	2.7	1.7	7.8
352	Shubazkol, Anastasievskii village	184.0	1.33	1.6	1.2	4.6
353	Kandykol, Sherbinovka village	209.4	1.18	1.5	1.0	4.4
354	Rynsay, Rynsay village	187.0	1.18	1.5	1.2	4.8
355	Narulchepkol, Shukalovskii	186.9	1.89	1.9	1.1	5.7
356	Alakol, Novo-Alekseevka village	102.1	1.63	2.1	1.3	5.6
357	Tenizkol, Sviatagorka village	106.8	7.29	4.6	2.1	12.8
358	Karakamys, Sviatagorka village	107.0	4.09	2.8	1.9	8.6
359	Istay, Shuvalovskii village	193.0	1.89	2.0	1.0	5.8
360	Turantaykol, Anastasievskii village	192.6	1.38	1.4	1.2	4.5
361	Sarmaykol, Shuvalovskii village	189.0	1.26	1.5	1.0	4.2
362	Sarmantopar, Shuvalovskii village	189.0	1.14	1.5	1.1	4.8

(continued)

Table 3.10 (continued)

No.	Lake and its location	Altitude (m a.s.l.)	Lake area (km^2)	Length (km)	Max width (km)	Length of coastline (km)
363	Zhaltyrkol, Arkhangel'skii village	231.0	1.23	1.7	0.9	5.1
364	Tottybaykol, Sherbinovka village	197.4	1.93	2.0	1.4	5.6
365	Uzynkol, Polovinkovki village	192.0	2.35	2.8	1.3	7.6
366	Bektaskol, Novo-Alekseevka village	190.2	1.29	1.6	1.3	5.0
367	Telenshi, Novo-Alekseevka village	191.7	1.38	1.4	1.3	4.6
368	Kirgizsoye, Arkhangel'skii village	230.6	1.58	1.5	1.3	5.2
369	Mashe, Polovinkovki village	189.0	2.13	2.5	1.2	6.6
370	Malyi uzynkol, Eriskovskii village	197.2	3.79	3.5	1.8	8.7
371	Zhanibek, Polovinkovki village	189.5	1.45	1.4	1.3	4.8
372	Uzyntomar, Eriskovskii village	194.5	1.12	1.9	0.8	4.8
373	Zhaydarkol, Kubekovka village	189.2	1.22	1.9	1.2	5.1
374	Uglovoye, Kubekovka village	195.7	1.61	1.8	1.5	5.2
375	Bol'shoy Korzhynkol, Semenovka village	200.7	2.26	2.2	1.4	6.6
376	Ulkenkol, Koskol village	197.7	1.18	1.9	0.9	4.8
377	Lake without name, Novo-Alekseevka village	192.2	1.25	1.4	1.2	4.3
378	Kulaktykol, Kubekovka village	191.8	1.75	2.1	1.5	6.1
379	Svetloye, Georgievka village	267.0	3.20	2.1	1.8	6.6
380	Karashakpay, Kubekovka village	190.1	1.75	1.6	1.4	4.8
381	Arakol, Koskol village	194.0	1.90	2.4	1.4	6.6
382	Zhaman-Alakol,	206.4	8.95	3.4	3.4	14.4
383	Tau-Bersenkol, Novo-Alekseevka village	193.5	1.24	1.9	1.0	4.8
384	Karatomar, Novo-Alekseevka village	191.9	1.09	1.4	1.0	4.0
385	Lake without name, Koskol village	194.0	1.15	1.3	1.2	3.9

(continued)

Table 3.10 (continued)

No.	Lake and its location	Altitude (m a.s.l.)	Lake area (km^2)	Length (km)	Max width (km)	Length of coastline (km)
386	Alakol, Kubekovka village	193.2	1.56	1.7	1.1	5.4
387	Kolendikol, Polovinkovki village	194.4	2.43	1.8	1.6	6.4
388	Saiyrkol, Semenovka village	203.5	3.19	3.3	1.8	11.0
389	Uglovoye, Lugovoy village	201.5	1.03	1.6	0.9	4.1
390	Tanymbike, Polovinkovki village	189.9	3.67	2.7	1.7	9.0
391	Sapula, Lugovoy village	208.2	1.32	1.5	1.1	4.3
392	Borybaykol, Kubekovka village	198.4	1.17	1.5	1.0	4.2
393	Karakempir, Kubekovka village	198.4	1.17	1.5	1.0	4.1
394	Kisykeldi, Lugovoy village	195.9	2.62	3.2	2.1	10.2
395	Zhaialgan, Semenovka village	203.0	1.84	2.1	1.2	6.3
396	Shandakty, Polovinkovki village	197.0	1.21	1.6	1.0	4.6
397	Alakol, Kubekovka village	199.1	1.38	2.2	1.1	5.4
398	Tura-Airagay, Kubekovka village	199.9	1.59	1.8	1.2	5.0
399	Shegebay, Semenovka village	199.0	1.94	2.0	1.5	6.2
400	Alabota, Sviatogorka village	199.4	36.39	8.4	5.1	27.9
401	Obalykol, Koskol village	199.6	1.58	1.6	1.3	4.8
402	Shunkyrkol, Kubekovka village	199.2	2.15	2.2	1.3	6.3
403	Tastemir, Koskol village	198.3	3.73	2.4	2.2	9.9
404	Malyi Zhaugashty, Krasnoarmeiskii village	253.6	1.11	1.5	1.1	4.6
405	Sart, Zhaysylyk village	198.0	2.12	2.3	1.3	6.4
406	Birkazan, Lavrentievka village	200.6	2.69	2.1	1.6	6.6
407	Karakol, Krymskii village	224.4	1.18	1.7	1.2	5.0
408	Lake without name, Lavrentievka village	202.4	1.35	2.2	1.1	5.9
409	Shiet, Koskol village	198.5	1.60	2.0	1.3	5.5
410	Shildebaykol, Lugovoy village	204.0	2.53	2.1	1.9	7.0
411	Eshimbay, Lugovoy village	197.9	1.37	1.7	1.1	4.8

(continued)

Table 3.10 (continued)

No.	Lake and its location	Altitude (m a.s.l.)	Lake area (km^2)	Length (km)	Max width (km)	Length of coastline (km)
412	Bol'shoy Maiyl-Tau, Lugovoy village	194.2	1.98	2.6	1.0	6.3
413	Kashartomar, Koskol village	199.9	2.17	2.0	1.3	6.1
414	Zhaukashty, Denisovka village	252.4	3.71	2.7	1.9	7.8
415	Kermenkol, Lugovoy village	193.3	4.09	2.8	2.0	8.8
416	Auynkol, Koskol village	198.0	1.67	1.5	1.5	6.2
417	Koybagar	202.6	95.95	17.8	9.1	49.2
418	Matay, Evgenievka village	191.8	1.69	1.7	1.4	4.9
419	Makbalkol, Pridorozhnyi village	213.4	1.82	1.6	1.4	5.2
420	Supulusor, Denisovka village	226.4	3.27	3.1	1.5	7.1
421	Aksukol, Pridorozhnyi village	208.4	1.57	1.8	1.3	4.8
422	Karakoga, Karakalpak	205.8	2.94	2.5	1.8	7.0
423	Prikupok, Karakalpak	201.5	2.97	2.2	1.6	7.2
424	Kenderli, Pavlovka village	189.8	8.33	3.9	2.7	12.6
425	Kazbskol, Karakalpak	209.7	2.12	1.7	1.5	5.3
426	Kopa, Karakalpak	201.5	1.15	1.5	1.0	4.5
427	Saumalkol, Karakalpak	209.4	1.48	2.0	1.3	5.7
428	Balymbay, Karakalpak	209.4	1.34	1.4	1.1	4.5
429	Abkura, Karakalpak	203.7	3.14	3.4	1.8	9.6
430	Kuset, Denisovka village	227.4	3.13	2.4	1.8	7.2
431	Sasyksor, Pridorozhnyi village	209.5	9.46	4.7	2.9	13.8
432	Obalykol, Novonezhinka village	203.1	1.42	1.9	1.1	4.5
433	Solenoye, Novoilnovka village	194.5	2.66	2.3	1.6	7.4
434	Kodak, Natalovka village	192.0	2.78	2.1	2.0	6.8
435	Tret'e, Karakalpak	202.5	1.64	1.7	1.2	4.8
436	Kuzhay, Novoilnovka village	183.1	1.53	1.9	1.1	5.3
437	Korzhynkol, Novoilnovka village	201.1	1.08	1.4	1.0	3.8
438	Astaukamys, Karakalpak	209.7	1.10	1.9	1.0	5.5
439	Solenka, Pavlovka village	197.8	1.41	2.1	1.2	7.6
440	Aksan, Karakalpak	208.6	4.12	2.8	2.4	8.0

(continued)

Table 3.10 (continued)

No.	Lake and its location	Altitude (m a.s.l.)	Lake area (km^2)	Length (km)	Max width (km)	Length of coastline (km)
441	Tomar, Novoilnovka village	202.7	1.09	1.4	1.0	3.7
442	Revul', Karakalpak	201.0	3.58	3.6	1.6	9.9
443	Kishkinekol, Karakalpak	204.9	1.53	2.2	1.7	5.4
444	Karakamys, Kara-Kamys village	206.0	3.6	3.1	1.7	8.7
445	Lake without name, Karakalpak	209.0	1.61	1.8	1.5	5.4
446	Berezovoye, Karakalpak	207.1	3.53	2.6	2.0	7.8
447	Lake without name, Karakalpak	199.1	1.41	1.5	1.2	4.8
448	Lake without name, Karakalpak	208.1	1.03	1.4	0.9	4.0
449	Balyktykol, Karakalpak	108.1	3.31	2.4	2.0	7.0
450	Sabynkol, Zhambaskol village	191.6	5.37	3.5	2.2	9.9
451	Zharkol, Kushmuryn	104.0	2.51	2.0	1.6	5.8
452	Karatkol, Kushmuryn	101.9	1.75	1.6	1.6	5.0
453	Lake without name, Kushmuryn	104.5	2.13	2.5	1.3	8.2
454	Sorkol, Denisovka village	248.5	2.86	2.3	1.7	6.4
455	Kishkentaysor, Zhambaskol village	188.1	1.04	1.7	0.9	4.4
456	Karasor (malyi) Zhambaskol village	196.9	4.07	4.1	1.8	11.4
457	Tenteksor, Zhambaskol village	189.8	14.78	8.0	3.1	26.3
458	Kokdyru, Kushmuryn	107.0	1.19	2.9	0.8	8.2
459	Karatomar, Neliubinki village	199.1	1.83	1.8	1.4	5.2
460	Arystansor, Pridorozhnyi village	214.7	7.06	3.4	2.8	10.1
461	Lake without name, Smailovskii village	199.7	1.35	1.9	1.0	5.1
462	Akkusor, Eltay village	210.8	3.26	2.6	1.6	6.9
463	Bol'shoy Kaiyndykol, Smailovskii village	202.2	1.41	1.9	1.1	5.1
464	Bayzaksor, Timofeevka village	198.7	4.19	3.7	1.5	20.1
465	Zhemedey, Smailovskoye village	181.2	2.73	3.3	1.3	10.4
466	Alasor	185.7	10.0	7.9	2.4	40.0

(continued)

Table 3.10 (continued)

No.	Lake and its location	Altitude (m a.s.l.)	Lake area (km^2)	Length (km)	Max width (km)	Length of coastline (km)
467	Lake without name	187.8	1.03	2.2	0.6	6.4
468	Karasor	184.0	25.29	11.6	3.1	38.8
469	Karatomar, Timofeevka village	204.5	3.19	3.1	1.4	7.8
470	Saz, Smailovskoye village	189.2	3.08	3.2	1.3	7.2
471	Tokalsor	203.9	1.86	2.4	1.0	10.8
472	Nogaikol, Nogaikol village	198.0	3.56/3.8	3.0	2.3	9.2
473	Taksor, Smailovskoye village	214.0	3.31	2.1	2.0	6.6
474	Aykanys, Livanovka village	227.9	7.30	3.7	2.7	11.4
475	Karakol, Eltay village	223.0	2.60	2.5	1.6	6.9
476	Zhalanash	196.8	5.73	2.9	2.3	9.0
477	Shili, Iulievka village	112.5	14.32/ 0.8	15.9	1.7	45.9
478	Merkamel'sor, Eltay village	223.5	2.30	1.8	1.7	5.5
479	Eginsor, Smailovskoye village	215.7	1.51	1.6	1.2	5.1
480	Tysygunkopa, Livanovka village	229.7	1.91	1.8	1.3	5.2
481	Shunkyr, Iulievka village	114.8	1.15	1.6	1.0	4.9
483	Lake without name	206.2	1.27	2.7	1.1	7.1
484	Aksakalkopa, Livanovka village	233.1	4.56	3.5	2.9	8.6
485	Oryntaykopa, Livanovka village	257.1	1.78	2.0	1.2	5.4
486	Korzhynkopa, Livanovka village	234.3	1.94	1.8	1.5	5.7
487	Korzhynkopa, Livanovka village	234.8	1.82	1.8	1.2	4.8
488	Bol'shoye, Smailovskoye village	227.4	1.11	1.6	1.0	4.5
489	Tuzdyborli	206.8	3.98	3.0	1.6	11.2
490	Salmakkol, Smailovskoye village	214.6	1.47	1.6	1.3	4.6
491	Kamystyburli	214.8	3.78	3.0	2.8	11.4
492	Aralsor	203.3	2.97	3.0	1.6	15.3
493	Tunkuiukty, Eltay village	239.8	3.52	3.2	1.7	11.8
494	Sunaly, Eltay village	241.3	3.39	2.2	2.0	7.2
495	Karakol, Eltay village	244.3	1.70	2.1	1.1	5.5
496	Kamyssyzburli	214.8	1.82	3.7	1.0	12.2

(continued)

Table 3.10 (continued)

No.	Lake and its location	Altitude (m a.s.l.)	Lake area (km^2)	Length (km)	Max width (km)	Length of coastline (km)
497	Tokalsor	214.9	1.14	2.3	0.9	6.5
498	Atysu, Pervomaiskii village	201.0	1.48	2.3	1.1	5.6
499	Sulukol, Pervomaiskii village	209.5	10.16	6.0	2.8	18.9
500	Lake without name, Pervomaiskii village	209.0	1.55	2.2	1.2	5.9
501	Sorkol, Livanovka village	238.9	3.08	2.3	1.8	9.6
502	Lake without name, Kamyshnyi village	257.4	1.51	2.0	1.0	4.8
503	Dosaykopa, Kamyshnyi village	256.4	8.90	4.1	3.0	11.4
504	Tulybaykopa, Mechetnyi village	250.8	4.91	2.6	2.3	9.7
505	Shubaykol, Abay village	228.0	6.11	4.6	3.2	13.2
506	Tokaskopa, Adayevka village	253.0	1.14	1.5	1.1	4.3
507	Kamystykol, Adayevka village	259.8	1.80	1.7	1.3	5.1
508	Kostomarkol, Abay village	219.0	1.65	1.6	1.3	5.3
509	Kishkene, Abay village	231.0	1.09	1.8	1.0	5.1
510	Kozhakol, Abay village	237.0	2.61	3.1	1.1	7.2
511	Oraz, Kamyshnyi village	256.4	9.13	3.8	2.9	11.1
512	Tileukobe, Philikovka village	255.0	1.04	1.3	1.1	3.7
513	Ulkenkopa, Bogdanovka village	251.0	5.27	3.0	2.8	9.3
514	Adaykol, Adayevka village	267.0	10.32	4.2	2.9	17.6
515	Sor, Sholoksay village	116.6	1.54	1.6	1.5	4.8
516	Blizkopa, Bogdanovka village	255.5	10.53	4.3	3.7	14.8
517	Erkebay, Bradlavki village	359.8	1.35	1.5	1.2	4.4
518	Konysbay, Bogdanovka village	259.9	4.21	2.8	2.0	7.6
519	Shagyrkol, Philikovka village	259.9	4.54	2.7	2.4	8.1
520	Zharkol, Nauryzym village	115.8	2.65	2.9	1.4	7.8
521	Shaptykol, Akkarga village	319.8	1.49	1.5	1.3	5.1
522	Karabala, Akkarga village	317.2	1.43	1.5	1.3	4.7
523	Mamyrkol, Urkash village	259.0	4.72	3.0	2.0	8.3
524	Karakamys, Sholakkopa village	226.2	4.31	3.7	1.8	9.6

(continued)

Table 3.10 (continued)

No.	Lake and its location	Altitude (m a.s.l.)	Lake area (km^2)	Length (km)	Max width (km)	Length of coastline (km)
525	Sholakkopa, Nauryzym village	218.9	2.17	2.7	1.1	7.0
526	Atzhargan, Sholakkopa village	228.1	2.58	2.4	1.5	6.6
527	Zhaksybay, Abay village	228.9	13.77	4.8	4.0	15.1
528	Ulendikol, Sholakkopa village	228.1	1.08	2.0	0.8	4.6
529	Shagyrkopa, Akkarga	276.2	2.67	2.6	1.2	7.2
530	Saumalkol, Abay village	222.1	1.56	1.8	1.2	5.1
531	Malyi Aksuat, Nauryzym village	120.1	1.29	2.0	1.2	5.9
532	Karasor, Abay village	217.1	10.66	9.2	2.3	30.7
533	Zharsor, Philipovka village	224.8	1.88	1.8	1.6	5.3
534	Taldykol, Zhaiylma village	249.8	11.20	4.0	3.8	13.2
535	Molokanovo, Zhaiylma village	256.6	2.53	2.2	1.7	6.5
536	Asansor, Zhaiylma village	216.3	2.78	2.3	2.0	10.2
537	Bol'shoy Sankebay, Sholakkopa village	221.0	1.47	2.6	0.9	6.3
538	Zharsor, Urkash village	217.2	6.07	4.3	2.1	11.1
539	Kulykol, Zhaiylma village	247.7	33.65	10.0	4.9	33.65
540	Konaksor, Urkash village	224.4	2.39	1.8	1.5	7.5
541	Kaiyndysor, Abay village	214.2	4.47	4.7	1.6	12.7
542	Shoshkakol, Smolokur village	124.8	5.05	4.1	1.8	14.9
543	Koskol, Abay village	226.3	2.00	2.3	1.4	7.5
544	Shukyrkol, Karasu village	229.3	2.37	2.2	1.5	6.3
545	Urkash, Urkash village	204.4	16.34	5.9	3.8	18.7
546	Syrmimbay, Urkash village	214.1	1.39	1.7	1.0	7.0
547	Teniz, Karasu village	225.1	5.44	3.8	2.5	9.2
548	Kensuat, Suly village	122.2	2.50	2.5	1.3	8.8
549	Otynkol, Karasu village	227.5	3.86	3.0	2.1	11.9
550	Malyi Suly, Suly village	122.0	1.23	1.6	1.1	4.2
551	Alakol, Karasu village	225.2	10.76	6.2	3.0	20.1
552	Kokyssor, Karasu village	228.6	5.42	3.7	2.2	10.4
553	Zholshara	227.1	5.38	2.9	2.5	9.0
554	Sorkol	235.2	2.70	2.3	1.5	6.4
555	Bol'shoy Suly, Suly village	121.2	2.63	3.0	1.5	7.5
556	Koskol	232.4	1.95	2.3	1.2	7.0
557	Lake without name	220.0	1.39	1.6	1.1	4.6

(continued)

Table 3.10 (continued)

No.	Lake and its location	Altitude (m a.s.l.)	Lake area (km^2)	Length (km)	Max width (km)	Length of coastline (km)
558	Uzynsor, Karasu village	221.6	3.57	2.8	2.0	8.4
559	Koklak, Suly village	124.1	1.55	1.6	1.9	5.2
560	Ashudastysor	219.0	1.65	2.0	1.2	5.4
561	Bolek, Karasu village	244.6	2.47	1.8	1.6	6.9
562	Kaiyndysor	214.4	5.61	3.8	2.7	10.5
563	Zharkol, Karasu village	225.6	2.39	2.0	1.6	5.7
564	Tauyksor, Karasu village	219.5	7.82	6.7	2.1	17.2
565	Aralkol	244.2	2.65	4.2	1.2	10.8
566	Lake without name	243.3	1.34	1.7	1.1	4.5
567	Shetkol	252.0	4.71	3.7	1.7	9.4
568	Lake without name	252.8	1.94	1.9	1.3	5.4
569	Zharsor	232.2	1.75	2.6	1.0	6.2
570	Kiikkol, Suly village	117.5	30.7	10.4	4.2	31.9
571	Aralkol, Batpakty village	264.0	1.11	1.6	1.1	4.3
572	Bol'shoya Koskopa, Karasu village	111.8	6.10	3.2	2.6	10.5
573	Malaya Koskopa, Karasu village	111.7	6.82	3.9	2.9	11.2
574	Zhekekol	274.9	1.22	1.6	1.1	4.2
575	Tatimek, Mukyr village	121.0	10.22	5.2	2.7	15.2
576	Skukyrkopa, Mukyr village	113.9	2.40	2.2	1.3	6.3
577	Baybishe, Mukyr village	110.5	3.61	2.5	1.9	7.6
578	Sadyrbay, Mukyr village	109.9	1.71	1.8	1.3	5.7
579	Ozhagul, Mukyr village	110.7	18.77	7.2	3.0	20.4
580	Zhalpakkuga, Mukyr village	107.2	2.85	1.9	1.7	6.6
581	Dulbay, Saga village	103.1	1.62	2.0	1.8	7.2
582	Uzynkoga, Mukyr village	109.6	1.90	3.5	1.0	8.1
583	Lake without name, Saga village	103.0	1.23	1.4	1.2	8.8
584	Sarykopa, Saga village	101.2	184.42	27.6	13.2	179.4
585	Lake without name, Kumkeshu village	108.0	3.73	3.3	1.8	8.5
586	Lake without name, Tuz village	152.6	1.13	1.6	1.1	4.5
587	Shumekty, Sholakkarasu village	105.7	2.74	2.8	1.2	8.1
588	Shalma, Sholakkarasu village	101.2	6.35	4.1	2.3	10.8
589	Oishumekty, Sholakkarasu village	102.0	4.67	4.2	1.9	16.5

(continued)

Table 3.10 (continued)

No.	Lake and its location	Altitude (m a.s.l.)	Lake area (km^2)	Length (km)	Max width (km)	Length of coastline (km)
590	Lake without name, Sholakkarasu village	98.0	1.39	2.4	0.9	6.0
591	Lake without name, Sholakkarasu village	102.0	1.18	1.5	1.0	5.2
592	Lake without name, Sholakkarasu village	99.9	3.12	4.5	1.2	10.8
593	Domalak, Sholakkarasu village	101.0	4.41	4.5	2.5	15.9
594	Saumalkol, Kumkeshu village	143.5	1.65	1.6	1.4	4.6
595	Tokabaykol, Kumkeshu village	134.1	1.32	2.8	0.8	6.9
596	Sarykopa, Sholakkarasu village	99.8	89.10	22.7	4.5	140.2
597	Kishiborly, Shoban village	109.9	2.04	2.8	1.4	8.0
598	Malyi Ushkamys, Sholakkarasu village	99.0	1.78	2.3	1.0	6.1
599	Sasykkol, Shoban village	131.0	1.66	2.0	1.3	6.6
600	Kulakkol, Shoban village	114.8	1.20	1.5	1.3	5.8
601	Bol'shoy Ushkamys, Sholakkarasu village	99.8	3.87	3.0	2.0	10.8
602	Agashtykol, Amangeldi village	138.2	3.37	2.7	1.7	7.2
603	Ayakkamys, Sholakkarasu village	99.8	11.89	6.7	2.5	22.2
604	Alakol, Shoban village	103.6	1.96	5.0	1.4	14.6
605	Zharkol, Sholakkarasu village	99.8	20.87	7.2	4.0	27.7
606	Shybyndysor, Amangeldi village	139.0	9.66	4.2	3.3	12.3
607	Zharkol	137.0	2.56	4.0	1.1	9.4
608	Tuzkudyk, Shubalan village	127.4	1.09	2.0	0.8	4.6
609	Katpagan, Zholoba village	196.5	5.32	3.3	1.8	9.9
610	Lake without name, Karakamys village	114.0	2.06	2.3	1.2	8.4
611	Nasbaykol, Karakamys village	110.8	1.84	3.0	0.9	7.5
612	Aynakol, Akshiganak village	78.3	4.29	2.8	2.0	8.4
613	Zhaman-Akkol, Zban village	–	38.61	7.4	7.0	23.1

(continued)

Table 3.10 (continued)

No.	Lake and its location	Altitude (m a.s.l.)	Lake area (km^2)	Length (km)	Max width (km)	Length of coastline (km)
614	Donyz, Zban village	103.1	1.04	1.3	1.1	4.0
615	Akkol, Zban village	105.6	62.26	11.4	8.0	43.3
616	Lake without name, Zban village	98.0	1.57	2.8	0.9	7.4
617	Karasor, Zban village	98.0	9.27	4.4	3.0	16.0
618	Gor'koye, Presnogor'kovka village	154.2	1.02	1.2	1.1	3.7
619	Ataman, Makar'evka village	156.0	1.03	1.4	1.0	3.8
620	Krugloye, Makar'evka village	158.0	1.03	1.2	1.0	3.8
621	Velikoye, Makar'evka village	158.3	1.05	1.4	1.0	4.2
622	Bol'shoy Sorkol, Mikhailovka village	158.2	1.02	1.4	1.1	4.3
623	Oguzbalyk-Sarykol, Ol'govka village	150.9	1.01	1.3	1.1	4.0
624	Lake without name, Ryjskii village	161.1	1.01	1.4	0.9	4.2
625	Lake without name, Korolevka village	161.4	1.01	1.5	0.9	3.9
626	Svetlyi Zharkol, Svetlyi Zharkol village	188.0	1.01	1.3	1.0	4.0
627	Taldytomar, Kostychevskii village	194.0	1.03	1.7	1.0	4.5
628	Sur, Kostychevskii village	197.2	1.01	1.4	1.0	4.2
629	Shandykol, Kostychevskii village	209.9	1.03	1.3	1.3	4.5
630	Ulkenkol, Maikol	189.7	1.01	1.3	1.1	4.0
631	Strunovoye, Maikol	182.6	1.01	1.3	0.9	4.0
632	Maikol, Maikol	189.1	1.01	1.3	1.1	4.0
633	Kostomar, Maikol	189.6	1.01	1.5	0.9	4.2
634	Sheptikol	210.8	1.01	1.4	0.8	3.9
635	Lake without name, Elizavetinka village	207.4	1.01	1.3	1.0	4.9
636	Malyi Zhanat, Alekseevka village	169.4	1.03	1.6	0.6	4.4
637	Maimak, Sergeevskoye village	196.0	1.01	1.3	1.1	4.9
638	Koszhaltyr, Sergeevskoye village	202.8	1.04	1.3	1.1	4.6

(continued)

Table 3.10 (continued)

No.	Lake and its location	Altitude (m a.s.l.)	Lake area (km^2)	Length (km)	Max width (km)	Length of coastline (km)
639	Malyi Korzhynkol, Semenovka village	200.9	1.04	2.3	0.7	5.9
640	Kolkovatoye, Dobrynovka village	199.0	1.03	1.2	1.1	3.8
641	Maikol, Dobrynovka village	202.8	1.03	1.7	0.8	4.4
642	Lake without name, Dobrynovka village	201.5	1.02	1.2	1.1	3.9
643	Zhaltyrkol, Dobrynovka village	196.2	1.03	1.4	0.9	4.0
644	Obalykol, Klinovka village	272.4	1.02	1.3	1.0	3.9
645	Letpa, Kushmuryn	104.7	1.02	2.2	0.7	6.2
646	Shagalakkol, Smailovskoye village	209.9	1.01	1.4	0.9	6.9
647	Lake without name, Moskalevskii	188.0	1.01	2.0	0.7	5.0
648	Lake without name, Moskalevskii	209.2	1.02	1.3	1.0	4.3
649	Kokpekty, Moskalevskii	209.9	1.02	1.7	0.8	4.5
650	Mondybaysor, Timofeevka village	199.1	1.04	2.0	0.6	7.4
651	Ushkol, Philipovka village	262.1	1.02	1.4	1.0	4.2
652	Sazkol, Philipovka village	271.2	1.05	1.8	0.7	5.6
653	Shukyrkol, Bestau	260.8	1.03	1.8	1.0	4.1
654	Kungyrkol, Bestau	230.9	1.04	1.6	1.2	5.3
655	Kozhakol, Batpakty village	266.5	1.03	1.6	1.0	5.0
656	Lake without name, Kyzylkemer village	141.6	1.06	1.9	1.0	6.3
657	Lake without name, Sholkarasu village	98.6	1.03	1.5	0.9	5.3
658	Uzynsor, Kumkeshu village	125.8	1.03	1.3	0.6	13.3
659	Sulikti, Amangeldi village	167.7	1.02	1.5	1.2	4.1
660	Kan, Amangeldi village	159.6	1.01	2.1	0.9	5.5
661	Aminsor, Shubalan village	122.8	1.02	2.7	0.5	5.7
662	Lake without name, Shubalan village	95.8	1.02	1.7	0.9	4.3
663	Shoshkalykol, Amangeldi village	197.0	1.02	1.8	1.0	5.1
664	Lake without name, Amangeldi village	189.0	1.05	1.6	0.9	4.0

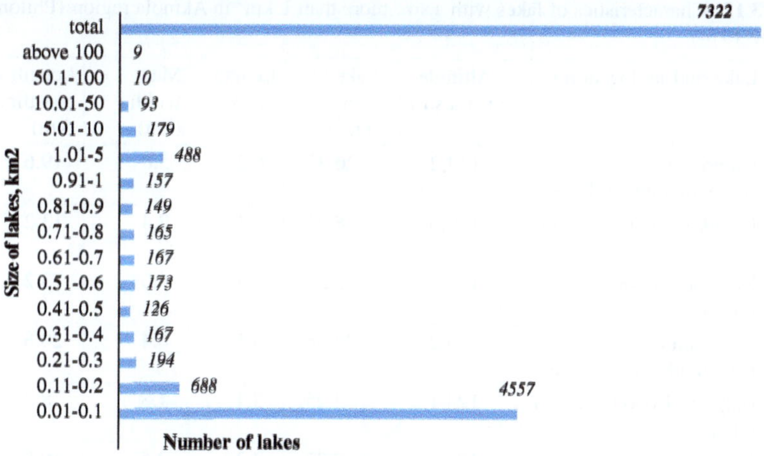

Fig. 3.10 Distribution of lakes in Akmola region by lake size

Table 3.11 Distribution of lakes in Akmola region by lake size and area (Philonets and Omarov 1974)

Area (km^2)	Number of lakes	Total area	
		km^2	%
0.01–0.1	4557	127.9	0.95
0.11–0.2	688	68.69	0.53
0.21–0.3	194	43.68	0.8
0.31–0.4	167	39.93	0.6
0.41–0.5	126	48.83	0.76
0.51–0.6	173	38.79	0.7
0.61–0.7	167	37.96	0.7
0.71–0.8	165	40.98	0.7
0.81–0.9	149	41.88	0.7
0.91–1	157	34.9	0.6
1.01–5	488	807.88	4.15
5.01–10	179	552.89	4.9
10.01–50	93	1478.9	14.1
50.1–100	10	331.89	27
Above 100	9	2842.89	43
Total	7322	6538	100

The lake water is suitable for household needs and is used for watering cattle. The tench, perch, crucian carp live in the lake. Fish are caught occasionally. Many ducks, the muskrat is found.

The lake water owns medicinal properties and is an excellent resting place.

Table 3.12 Characteristics of lakes with a size more than 1 km^2 in Akmola region (Philonets and Omarov 1974)

No.	Lake and its location	Altitude (m a.s.l.)	Lake area (km^2)	Length (km)	Max width (km)	Length of coastline (km)
1	Kumdykol, Bogodukhovka village	134.2	26.37	6.2	5.8	19.6
2	Uzynkol, Bogodukhovka village	129.6	8.21	4.1	2.4	13.2
3	Karasor, Bogodukhovka village	123.4	75.55	27.3	6.0	73.2
4	Zhamantuz, Bogodukhovka village	123.2	29.58	6.5	6.4	20.8
5	Balykty, Bogodukhovka village	124.1	3.25	3.1	1.8	8.2
6	Kishkenesor, Bogodukhovka village	124.7	7.77	4.2	2.5	10.8
7	Lake without name, Bogodukhovka village	128.5	1.26	1.85	1.0	5.0
8	Ulken-Toytansor, Bogodukhovka village	129.3	1.91	2.1	1.3	5.6
9	Kuduk, Kiali	138.0	1.24	2.5	0.8	7.0
10	Shagalalyteniz, Kialy village	135.6	267.36	42.9	12.5	95.8
11	Lake without name, Kiali	136.0	1.45	1.7	1.1	4.6
12	Lake without name, Kiali	136.0	1.1	1.3	1.1	4.0
13	Baskarakol, Kirov	132.0	3.59	5.2	1.2	11.8
14	Piketkol, Kirov	131.8	1.50	2.4	1.0	6.0
15	Sargyzkol, Kirov	130.6	3.48	3.9	1.4	11.2
16	Bazhenkol, Kirov	128.8	2.23	2.3	1.5	6.1
17	Konak-Koman, Kirov	129.2	1.98	2.1	1.4	6.2
18	Koskara, Kirov	129.2	1.38	1.6	1.5	4.6
19	Bol'shoy Burkit, Kirov	128.7	1.71	2.8	1.2	4.7
20	Shorkkol, Kirov	129.7	1.48	1.5	1.3	4.6
21	Taiynsha, Kirov	135.2	1.71	1.7	1.3	5.0
22	Uzynkol, Kirov	128.9	1.07	1.5	1.0	4.4
23	Akbas, Kirov	127.9	6.58/ (8.9)	4.9	1.8	14.8
24	Sarybalyk, Kirov	127.3	1.40	1.6	1.2	4.4
25	Akbas, Kirov	128.1	8.70	5.7	2.8	15.8
26	Kishi-Karoi, Vol'noye village	53.6	100.81	12.7	12.1	73.4
27	Ulken-Karoi, Kyzyltu village	56.8	305.53	67.7	11.9	126.2
28	Zharkent, Yasnovki village	174.3	2.21	2.3	1.2	6.4

(continued)

Table 3.12 (continued)

No.	Lake and its location	Altitude (m a.s.l.)	Lake area (km^2)	Length (km)	Max width (km)	Length of coastline (km)
29	Sasykkol, Letovochnoye village	156.7	12.47/ (13.6)	5.6	2.8	16.1
30	Zhaiylma, Kellerovka village	176.1	2.54	2.9	1.3	6.4
31	Sholakkaiyn, Yasnaya Poliana village	126.0	1.18	1.7	1.0	4.4
32	Kalibek, Sevastopol'	85.6	95.61	17.0	7.3	70.4
33	Ushsai, Sevastopol'	85.6	12.23	12.3	2.2	38.8
34	Kopasor, Sevastopol'	86.0	1.37	2.3	1.0	7.2
35	Alabota, Sevastopol'	100.2	48.25/ (22.4)	9.6	7.3	28.3
36	Borlykol, Sevastopol'	102.7	3.29	2.4	1.9	8.7
37	Zhirenat, Kyzyltu village	114.1	1.38	2.2	0.8	5.0
38	Ashykol, Kyzyltu village	123.8	2.05	2.2	1.4	5.6
39	Shalsyronkol, Kyzyltu village	112.0	2.19	1.85	1.75	5.4
40	Lake without name, Kyzyltu village	111.7	1.67	1.7	1.2	4.6
41	Kundybay, Kyzyltu village	123.3	2.89/ (1.3)	2.3	1.8	6.4
42	Kulkay, Gromoglasovo village	105.5	1.21	1.4	1.1	4.0
43	Teke Kyzyltu village	28.0	256.62/ (265.0)	33.3	20.1	155.2
44	Kamyshnoye, Stepanovka village	124.0	1.54	1.7	1.6	4.6
45	Kumdykol, Russkaya Poliana village	113.1	5.38	2.7	2.4	8.3
46	Uak-Kamyshnoye, Kachilovka village	252.3	3.90	2.75	1.9	8.2
47	Zhetykol, Kachilovka village	234.0	10.6	4.1	3.6	12.2
48	Akbas, Kachilovka village	220.4	11.08	5.95	3.15	19.0
49	Kumdykol, Kachilovka village	223.6	6.34	3.1	2.85	9.1
50	Karakamys, Lavrovka village	205.9	1.91	1.8	1.5	5.1
51	Bol'shoy Koskol, Lavrovka village	257.1	14.31	6.1	3.35	15.2
52	Zhaltyr, Karasevka village	224.5	1.21/1.1	2.0	1.3	5.8
53	Terenkol, Krasnaya Zaria village	258.7	5.82	3.5	2.55	9.5

(continued)

Table 3.12 (continued)

No.	Lake and its location	Altitude (m a.s.l.)	Lake area (km^2)	Length (km)	Max width (km)	Length of coastline (km)
54	Zharken, Krasnaya Zaria village	243.4	1.47	1.85	1.4	5.8
55	Baratay, Krasnaya Zaria village	289.0	2.61	2.2	1.5	6.0
56	Alabota, Krasnaya Zaria village	232.1	7.26	4.5	2.65	11.2
57	Zhaltyrkol, Krasnaya Zaria village	286.5	2.8	2.1	1.8	6.0
58	Zholdybay, Gor'kii	233.0	25.17	6.3	6.1	21.0
59	Zhanaozek, Gor'kii	239.0	1.25	1.6	4.0	4.1
60	Myrzakol, Alekseevka village	228.7	3.39	2.15	1.9	6.8
61	Esdaulet, Alekseevka village	189.5	2.89	2.3	1.7	6.2
62	Kopasor, Alekseevka village	207.4	1.17/0.4	1.6	1.1	4.8
63	Kenzhekarasor, Alekseevka village	208.9	2.54	2.6	1.7	6.9
64	Akhmedzhansor, Alekseevka village	229.2	7.75	3.78	2.95	10.8
65	Zhaltyr, Alekseevka village	184.8	2.89/2.6	2.3	1.6	6.2
66	Zhaltyr, Alekseevka village	221.2	1.09	1.6	1.6	4.2
67	Mezgilsor, Borovoye village	219.8	6.13/2.5	3.3	2.8	9.1
68	Baltaysor, Chkalovo village	187.0	2.71	2.7	1.5	7.2
69	Seksembaysor, Chkalovo village	177.8	8.42	4.3	3.0	15.1
70	Matsor, Chkalovo village	195.8	4.49	2.7	2.2	7.8
71	Alakol, Alabota	156.1	3.14	2.3	2.0	6.4
72	Zhangildisor, Alabota	158.9	5.16	2.8	2.6	9.1
73	Taldykol, Alabota	162.9	1.19	1.4	1.3	4.0
74	Zhaksybaysor, Chistokovka village	201.3	1.95	2.0	1.4	5.2
75	Kirtoge, Chistokovka village	211.0	1.49/0.7	1.6	1.2	4.5
76	Kombaysor, Talshik village	103.3	7.27	3.7	2.7	10.2
77	Zhantaysor, Talshik village	84.9	3.57/1.0	2.45	2.05	6.8
78	Kulykol, Talshik village	71.2	2.16/1.3	1.9	1.55	5.4

(continued)

Table 3.12 (continued)

No.	Lake and its location	Altitude (m a.s.l.)	Lake area (km^2)	Length (km)	Max width (km)	Length of coastline (km)
79	Kopa, Kishkenekol village	93.2	1.34	1.9	0.9	4.6
80	Zharykkol, Kishkenekol village	119.0	1.21	1.75	1.1	4.5
81	Karasor, Kishkenekol village	118.2	21.37	15.1	2.3	41.8
82	Sarankol, Kishkenekol village	124.7	1.98	1.6	1.6	5.0
83	Shumektykol, Kishkenekol village	125.7	1.08	1.35	1.0	3.8
84	Sarman, Kishkenekol village	124.0	1.00	1.2	1.0	3.6
85	Takyrkol, Kishkenekol village	126.0	1.18	1.9	1.0	3.9
86	Bozshasor, Kishkenekol village	116.0	2.62	2.10	2.0	6.4
87	Ashylykol, Kyzylty village	119.6	1.45	1.4	1.2	4.4
88	Zharykkol, Murtuk village	120.0	1.70	2.4	1.3	7.0
89	Koszharyk, Murtuk village	117.0	2.92	2.4	1.9	6.8
90	Aupildek, Murtuk village	118.8	1.39	1.6	1.1	4.3
91	Tastykol, Murtuk village	88.8	1.49	1.75	1.1	4.3
92	Uzynkol, Murtuk village	120.0	1.21	1.4	1.0	4.1
93	Zhamankol, Andagul village	226.8	1.38	1.6	1.2	4.4
94	Alakol, Karimovka village	281.8	4.36	3.15	2.1	8.1
95	Bayan, Kirillovka village	297.7	9.56	4.1	3.0	11.4
96	Bayantay, Voskresenka village	298.1	2.67	2.3	1.6	6.3
97	Saumalkol, Volodarskoye village	277.5	25.24/24.3	8.2	4.5	22.3
98	Imantau, Volodarskoye village	324.4	49.35/48.9	13.3	4.4	35.5
99	Shalkar, Shalkar village	309.7	35.54/35.4	15.0	3.7	39.6
100	Malyi Koskol, Antonovka village	274.8	9.04	4.1	2.7	11.1
101	Lobanovo, Volodarskoye village	369.5	3.9/4.0	2.6	2.1	7.6
102	Chebachki, Lobanovo village	365.5	1.62/1.8	1.6	1.3	4.7
103	Baysary, Lobanovo village	368.7	3.44/3.6	2.5	1.7	7.1

(continued)

Table 3.12 (continued)

No.	Lake and its location	Altitude (m a.s.l.)	Lake area (km^2)	Length (km)	Max width (km)	Length of coastline (km)
104	Gor'koye, Shalkar village	304.1	1.98	1.9	1.4	5.4
105	Beloye, Shalkar village	375.3	18.3	5.6	4.4	17.4
106	Barchiv, Shalkar village	339.0	4.28	2.75	2.0	7.5
107	Ashykol, Shalkar village	334.5	2.19	1.8	1.5	5.3
108	Karkykol, Shalkar village	335.8	3.31/3.2	2.5	1.7	7.3
109	Ashykol, Elenovka village	291.2	2.78	2.05	1.85	6.0
110	Kenetkol, Shalkar village	348.9	1.32	1.9	0.9	5.0
111	Basurman, Elenovka village	304.5	2.70/2.9	2.9	1.4	6.9
112	Kumdykol, Shalkar village	287.9	6.76	3.2	2.8	9.4
113	Aupildek, Kokshetau	321.7	1.13	1.45	1.1	4.0
114	Kurkol, Kokshetau	285.6	1.29	1.4	1.2	4.1
115	Kopa, Kokshetau	223.5	14.2/ 13.1	5.35	3.6	15.2
116	Koskol, Kokshetau	295.2	1.45/1.4	1.6	1.25	4.5
117	Zheltau, Kokshetau	345.4	5.48	3.0	2.4	9.4
118	Soldatkol, Kokshetau	337.4	1.30/1.3	1.8	1.0	4.5
119	Ashykol, Kokshetau	337.2	2.26	1.9	1.65	5.6
120	Kyzylsor, Aleksandrovka village	338.0	2.07	2.1	1.2	5.4
121	Akshasor, Kokshetau	257.4	2.55	2.25	1.5	5.9
122	Zhamantuz, Aleksandrovka village	284.7	7.28	3.6	2.7	10.2
123	Balyktykol, Aleksandrovka village	281.3	5.94/5.8	3.85	1.8	11.1
124	Maloye Chebach'e, Borovoye village	307.4	21.03	13.6	2.7	29.2
125	Kulak, Borovoye village	236.6	1.04	1.8	1.2	5.8
126	Ulkensor, Borovoye village	239.1	3.39	3.0	1.5	9.4
127	Balkash, Borovoye village	239.2	1.51/1.2	1.55	1.50	6.0
128	Bol'shoye Chebach'e, Borovoye village	305.0	25.18	3.3	5.1	31.0
129	Borovoye, Borovoye village	320.9	7.90/ 10.5	4.5	3.9	13.6
130	Maldybay, Borovoye village	211.5	3.06	2.3	2.0	6.5
131	Tekekol, Borovoye village	2206.2	1.16	1.95	1.0	4.2
132	Bazarbay, Borovoye village	214.7	2.81	2.1	1.8	6.3
133	Akkol, Borovoye village	304.9	1.95	2.8	1.2	5.9

(continued)

Table 3.12 (continued)

No.	Lake and its location	Altitude (m a.s.l.)	Lake area (km^2)	Length (km)	Max width (km)	Length of coastline (km)
134	Zhaynak, Borovoye village	296.4	1.72	2.85	0.95	6.2
135	Zhanasu-Kopa, Borovoye village	254.9	1.34	2.7	1.2	7.9
136	Bolatshalkar, Borovoye village	250.5	3.73	2.9	1.75	7.4
137	Karasor, Borovoye village	246.7	1.75	2.2	1.55	6.7
138	Koshubayshalkar, Borovoye village	232.3	5.96	3.8	2.8	10.2
139	Tasshalkar, Madeniat village	293.9	10.80	4.35	3.7	14.5
140	Karasor, Madeniat village	267.5	7.37	3.6	2.8	10.6
141	Koksor, Zhanalyk village	222.4	6.28	3.4	2.55	12.2
142	Maylysor, Zhanalyk village	230.7	8.27	4.3	2.6	13.1
143	Kindikti, Zhanalyk village	227.6	2.52	2.7	1.7	7.5
144	Sasykkopa, Aysary village	187.2	5.99	2.9	2.7	9.4
145	Sasyk, Aysary village	187.2	4.58	3.7	1.6	10.2
146	Shotsor, Aysary village	176.5	8.22	4.0	3.1	13.4
147	Koksengirsor, Aysary village	158.8	46.34/ 45.0	12.4	7.2	45.1
148	Sholaksor, Aysary village	170.7	5.11	5.8	2.5	21.7
149	Shorman, Aysary village	166.9	1.49	2.2	1.3	7.6
150	Lake without name, Aysary village	177.7	3.42	2.65	1.9	11.6
151	Lake without name, Karakoga village	65.2	1.70	1.8	1.7	6.9
152	Lake without name, Karakoga village	65.2	1.90	4.2	1.0	11.4
153	Lake without name, Karakoga village	65.2	2.43	3.0	1.4	7.8
154	Lake without name, Karakoga village	66.0	2.30	3.0	1.45	7.2
155	Lake without name, Karakoga village	66.0	1.03	1.25	1.0	3.8
156	Lake without name, Karakoga village	65.2	1.71	2.2	1.25	5.4
157	Sarykol, Alimsere village	66.6	1.23	2.0	1.1	5.1
158	Zharkol, Karakoga village	65.5	1.95	1.6	1.4	6.6
159	Siletyteniz, Karakoga village	64.7	750.28/ 777.0	64.7	22.15	273.6
160	Ulykol, Ruzaevka village	235.5	19.09/ 16.4	5.65	4.3	16.8

(continued)

Table 3.12 (continued)

No.	Lake and its location	Altitude (m a.s.l.)	Lake area (km^2)	Length (km)	Max width (km)	Length of coastline (km)
161	Kusbek, Akan	354.0	7.74/7.8	3.3	2.95	10.2
162	Arykbalyk, Arykbalyk village	373.0	2.06/2.2	1.7	1.65	5.2
163	Zerendi, Zerendi village	371.2	11.91/ 10.7	6.05	3.85	21.3
164	Aydabol, Viktorovka village	426.8	15.33/ 35.0	5.7	3.9	15.2
165	Karagaychik, Zerendi village	436.9	1.95/2.0	1.85	1.25	5.0
166	Shemuldi, Zerendi village	432.6	2.77	2.4	1.5	6.4
167	Maloye Tuktynskoye, Dorogovka village	441.0	2.08/1.9	2.2	1.6	6.3
168	Daulet, Dorogovka village	365.5	3.85/3.0	2.5	2.3	7.2
169	Karaungir, Dorogovka village	368.1	11.6	4.5	4.1	15.7
170	Ashykol, Dorogovka village	362.6	2.14	2.0	1.45	5.7
171	Ashysor, Dorogovka village	366.6	2.16	1.9	1.5	5.4
172	Alakol, Zlatopol'e village	387.3	1.41	1.8	1.1	4.7
173	Karagaykol, Zlatopol'e village	365.9	1.04	1.3	1.05	3.6
174	Kumdykol, Zlatopol'e village	379.6	10.19	4.05	3.9	12.7
175	Urymkay, Dmitrievka village	423.2	6.95	3.0	2.4	10.8
176	Shuch'e, Shuchinsk	397.5	18.32/ 18.6	7.2	3.7	20.3
177	Balykty, Shuchinsk	395.8	9.01	4.05	3.15	11.2
178	Kotyrkol, Kotyrkol village	439.9	4.69/4.5	3.4	1.65	9.8
179	Zhuken, Stepniak village	382.7	19.26/ 19.1	5.45	4.6	17.3
180	Kotyrkol, Stepniak village	356.2	18.68	7.2	3.9	20.4
181	Ulykol, Stepniak village	318.0	1.63	1.8	1.3	5.3
182	Alyshimbay, Terek village	246.0	1.46	1.9	1.1	5.6
183	Atansor, Terek village	221.2	26.66/ 26.7	7.7	6.1	27.2
184	Alabaskol, Karasu village	239.8	1.33	1.3	1.25	4.3
185	Lake without name, Karasu village	213.0	2.72	2.7	1.4	8.1
186	Shamgan, Karasu village	201.5	2.33	2.0	1.5	6.0

(continued)

Table 3.12 (continued)

No.	Lake and its location	Altitude (m a.s.l.)	Lake area (km^2)	Length (km)	Max width (km)	Length of coastline (km)
187	Zhambaysor, Terek village	225.0	10.74	6.8	2.8	20.8
188	Lake without name, Bogembay village	203.0	2.31	2.1	1.5	6.0
189	Matay, Bogembay village	206.0	1.56	2.2	1.7	7.0
190	Zhamantuz, Zhuantobe village	65.9	29.77	9.9	4.8	26.8
191	Zhaksytuz, Bestobe village	69.5	15.9	5.1	3.8	15.0
192	Lake without name, Karasu village	77.0	1.15	1.6	1.2	4.4
193	Sorkopa, Sladkovodsk village	79.0	1.55	2.1	1.7	6.3
194	Zharkol, Druzhba	218.5	1.77/1.8	2.0	1.2	5.4
195	Kalmakkol, Chistopol'e village	309.0	53.14	10.7	8.2	32.5
196	Zhaksy-Zhalgyztou, Yakshy-Yangiztau village	357.8	41.28/ 43.2	8.9	7.9	31.7
197	Borkol, Isakovka village	407.3	1.70	1.65	1.4	5.0
198	Sarykol, Boyevoye village	405.6	2.53	1.95	1.6	6.1
199	Solenoye, Boyevoye village	388.3	1.00	1.25	1.0	3.6
200	Shybyndykol, Raigorodok village	395.1	4.57	2.6	2.3	8.0
201	Koiandykol, Arsen'evka village	389.9	5.09	3.05	2.2	8.6
202	Shoshkaly, Danilovka village	311.9	22.08/ 22.1	6.9	4.1	19.6
203	Atalyk, Yablonovka village	315.3	2.70	1.9	1.75	7.6
204	Sor, Lidovka village	295.6	1.44	1.8	1.1	4.8
205	Borlikol, Lidovka village	303.5	1.0	1.4	0.95/0.8	3.6
206	Mamay, Lidovka village	280.6	43.48	8.5	7.8	35.9
207	Tastykol, Karasu village	258.8	1.49	2.0	1.2	5.0
208	Kokbaysor, Bogembay village	207.3	7.32	4.6	2.6	12.4
209	Mau, Bogembay village	218.9	3.2/3.3	2.25	2.0	6.6
210	Altaysor, Bogembay village	205.4	17.82/ 5.9	7.95	3.0	27.2
211	Kyzylsor, Bogembay village	194.8	3.19	2.6	1.75	7.0
212	Karbaysor, Bogembay village	194.1	1.17	1.6	1.0	4.6

(continued)

Table 3.12 (continued)

No.	Lake and its location	Altitude (m a.s.l.)	Lake area (km^2)	Length (km)	Max width (km)	Length of coastline (km)
213	Lake without name, Bogembay village	205.0	2.10	2.75	1.65	8.7
214	Tastykol, Bogembay village	199.4	1.77	2.1	1.5	5.6
215	Kazmakorak, Liubimovka village	257.4	2.26	2.0	1.6	5.7
216	Salkynkol, Kirov	289.9	20.60	6.05	4.2	16.3
217	Kishkenekol, Kirov	293.3	3.34	2.4	2.2	7.2
218	Biesoigan, Koibagar	202.8	3.93	3.05	1.4	8.6
219	Seraly, Kirov	128.0	1.92	1.6	0.9	4.0
220	Lake without name, Vinogradovka village	145.0	1.04	7.6	0.3	16.2
221	Kondykol, Novo-Nikol'skoye village	429.2	8.28/8.6	4.4	3.0	13.4
222	Baytubet, Boyarka village	369.8	1.08	1.5	0.95	4.0
223	Shuakkol, Lidovka village	259.6	2.13	2.2	1.45	6.1
224	Kyzylsor, Bestobe village	87.9	2.22	1.85	1.5	5.4
225	Zhamanshal, Bestobe village	158.5	2.56	2.2	1.5	6.2
226	Tamsor, Bestobe village	156.1	7.00	3.7	2.7	12.0
227	Ushkamys, Bestobe village	177.0	1.19	1.8	0.7	5.6
228	Kishkenesor, Bestobe village	173.5	4.45	3.2	1.7	15.5
229	Alkasor, Bestobe village	132.1	3.38	4.0	1.5	11.1
230	Kamys, Bestobe village	148.0	1.07	1.7	0.9	4.3
231	Lake without name, Naumovka village	352.3	1.42	1.7	1.2	4.6
232	Kamyshnoye, Naumovka village	371.7	1.18	1.5	1.1	4.6
233	Ortakol, Nikol'skoye village	389.5	1.39	1.8	1.0	4.7
234	Kemerkol, Vinogradovka village	384.4	1.9	2.15	1.1	5.9
235	Zharlykol, Nikol'skoye village	374.2	5.43	2.9	2.4	8.6
236	Kishkenekol, Nikol'skoye village	373.8	1.15	1.55	0.95	4.0
237	Shortandy, Nikol'skoye village	374.8	1.75	1.8	1.45	5.1
238	Esenbek, Nikol'skoye village	374.4	2.09	1.9	1.4	5.6

(continued)

Table 3.12 (continued)

No.	Lake and its location	Altitude (m a.s.l.)	Lake area (km^2)	Length (km)	Max width (km)	Length of coastline (km)
239	Baran, Alekseevka village	371.2	7.49	3.75	2.8	10.6
240	Zholdybay, Priozernoye village	301.8	1.26	1.55	1.15	4.3
241	Zhartykol, Zhartykol village	331.4	1.15	2.1	0.95	5.2
242	Zharsor, Zhaltyrkol village	317.2	6.21	3.55	2.4	10.0
243	Dualbay, Kna village	321.2	1.23	1.6	1.1	4.3
244	Kna, Kna village	328.4	2.32	2.45	1.2	6.9
245	Borlykol, Bogembay village	276.2	5.24	3.5	2.5	8.9
246	Alakol, Kna village	259.7	7.58	4.95	2.15	14.1
247	Sor, Bestobe village	203.7	2.04	1.9	1.5	6.4
248	Ushsor, Bestobe village	200.6	2.05	2.0	1.95	5.9
249	Songasy, Bestobe village	192.0	3.87	2.55	2.2	7.1
250	Maysor, Bestobe village	180.0	3.74	3.9	2.25	9.7
251	Kumdykol, Bestobe village	162.5	1.82	2.55	1.15	7.0
252	Zhylandykol, Esil city	285.0	1.24	1.4	1.0	4.2
253	Obalykol, Komsomol	275.0	2.62/2.9	2.2	1.6	7.8
254	Borlykol, Atbasar city	294.5	1.49	1.6	1.3	4.7
255	Uzynkol, Atbasar city	281.5	1.61	1.5	1.4	4.8
256	Ashykol, Atbasar city	268.0	1.03	1.6	0.9	4.0
257	Taberkol, Novoaleksandrovka village	308.0	1.14	1.3	1.1	3.9
258	Lake without name, Marinovka village	280.9	1.05	1.5	1.2	4.6
259	Kazkoskol, Marinovka village	281.5	1.18	2.2	0.9	5.6
260	Lake without name, Marinovka village	286.5	1.72	2.0	1.2	5.2
261	Zhamankul, Zhuravlevka village	290.8	1.48	1.6	1.3	4.6
262	Zharlykol, Zhuravlevka village	306.7	2.07	2.1	1.2	5.9
263	Sornoye, Zhuravlevka village	298.9	1.02	1.4	1.0	3.9
264	Taldykol, Zhuravlevka village	198.2	5.58/9.0	3.0	2.1	10.6
265	Bisembay, Zhuravlevka village	294.8	1.04	9.0	0.3	20.8

(continued)

Table 3.12 (continued)

No.	Lake and its location	Altitude (m a.s.l.)	Lake area (km^2)	Length (km)	Max width (km)	Length of coastline (km)
266	Ashykol, Zhuravlevka village	293.0	1.13	1.5	1.0	4.1
267	Ortakol, Zhuravlevka village	293.0	1.03	1.7	0.9	4.2
268	Ivan Vachilevich, Zhuravlevka village	292.0	1.00	1.6	0.9	3.9
269	Borlykol, Zhuravlevka village	238.8	1.03	1.4	1.0	3.8
270	Kamyshatoye, Kamyshenka village	302.1	1.05	1.5	1.0	4.2
271	Shurmana, Oksanovka village	289.6	2.07	2.2	1.3	6.6
272	Egindy, Oksanovka village	295.4	1.95	2.5	1.2	6.8
273	Shoshkaly, Novoromenka village	299.0	1.11	1.4	1.0	4.3
274	Lake without name, Oksanovka village	324.0	1.13	1.5	1.1	4.2
275	Gniloye, Guliay-Pole village	366.7	1.15	1.5	1.0	4.2
276	Batantay, Guliay-Pole village	364.8	1.55/1.1	2.0	1.1	5.4
277	Balykty, Guliay-Pole village	364.6	2.75/2.5	2.2	1.6	6.4
278	Konyrkol, Guliay-Pole village	364.6	1.38	1.95	0.8	5.4
279	Kamyshnoye, Alekseevka village	379.6	2.81	2.2	1.7	6.6
280	Zharlykol, Alekseevka village	366.4	1.24	1.5	1.2	4.4
281	Zharsor, Alekseevka village	347.2	2.07	2.6	1.5	6.9
282	Kanzhygaly, Alekseevka village	370.2	9.98	4.3	3.1	11.5
283	Kungenkol, Alekseevka village	374.9	1.18	1.3	1.1	4.8
284	Shubarkol, Alekseevka village	376.5	1.40	1.7	1.2	4.5
285	Laykol, Alekseevka village	358.2	1.80	1.9	1.4	5.4
286	Ashykol, Alekseevka village	373.4	6.72	4.7	2.3	11.4
287	Kotyrkol, Konkrynka village	372.0	1.49	1.85	1.2	4.6

(continued)

Table 3.12 (continued)

No.	Lake and its location	Altitude (m a.s.l.)	Lake area (km^2)	Length (km)	Max width (km)	Length of coastline (km)
288	Shandakkol, Netsvetayevka village	301.2	5.18	2.9	2.2	8.7
289	Karshin, Novomarkovka village	252.1	1.32	1.5	1.3	4.2
290	Solenoye, Turgay village	242.5	1.50	2.1	1.2	5.4
291	Malyi Sharykty, Turgay village	284.0	1.85	1.8	1.4	5.2
292	Bol'shoy Sharykty, Turgay village	288.0	6.6	3.5	2.55	9.2
293	Kaskaat, Ermentau village	334.9	1.70	2.0	1.3	5.1
294	Tamdykol, Ermentau village	334.4	1.48	1.8	1.4	5.3
295	Taygakkol, Karabulak village	331.4	1.57	1.8	1.25	5.1
296	Kurbetkol, Karabulak village	261.2	1.35	1.4	1.2	4.4
297	Teniz, Karabulak village	250.5	29.45	7.9	4.3	34.3
298	Bozshasor, Karabulak village	264.2	3.01	2.3	2.3	6.8
299	Kobeytuz, Azhe village	253.8	6.92	3.5	3.1	11.0
300	Zharsor, Azhe village	236.7	7.88	4.1	2.7	14.6
301	Zhamantuz, Azhe village	231.8	2.62	2.1	2.0	6.2
301	Karagaydykol, Shas-Utkul village	282.9	2.28	2.6	1.35	6.2
302	Karakoga, Karakuga village	283.6	1.26	2.5	0.8	6.0
303	Elton, Karakuga village	262.7	1.22	1.5	1.0	4.0
303	Zharkol, Sergeevka village	259.4	6.62	3.0	2.9	10.6
304	Magdalinovskoye, Sergeevka village	275.0	1.24	1.7	1.0	4.2
305	Uzynkol, Tolkynkul village	284.5	6.87	4.2	3.7	12.4
306	Sabakty, Tolkynkul village	285.9	2.81	2.7	1.7	11.2
307	Tolkynkol, Tolkynkul village	287.4	1.42	1.3	0.9	10.1
308	Kudaybergen, Tavolzhanka village	288.7	2.13	2.0	1.7	5.6
309	Zhaltyrkol, Khmelevka village	367.0	1.16	1.5	1.2	4.2
310	Bozaygyr, Elizavetinka village	396.6	2.30/2.4	1.9	1.8	5.5

(continued)

Table 3.12 (continued)

No.	Lake and its location	Altitude (m a.s.l.)	Lake area (km²)	Length (km)	Max width (km)	Length of coastline (km)
311	Sasykkol, Elizavetinka village	352.4	3.78/4.7	3.2	1.8	8.4
312	Kusak, Zhalbashe village	388.6	1.25	1.7	1.0	4.4
313	Kablankol, Ermentau village	347.6	1.04	2.2	0.6	6.0
314	Korzhynkol, Ermentau village	267.9	5.94	3.7	2.6	10.0
315	Bortekol, Ermentau village	265.7	1.39	1.7	1.1	4.6
316	Maykol, Ermentau village	297.6	3.00	2.9	1.4	8.0
317	Karasor, Boztal village	291.5	1.66	3.2	1.0	8.0
318	Ushtagantuz, Boztal village	370.3	2.64	2.5	1.6	6.6
319	Tomen, Boztal village	288.3	1.12	1.5	1.1	4.2
320	Kyzylsor, Boztal village	308.4	1.01	2.4	1.0	5.8
321	Shunkyrkol, Stepnoye village	324.9	2.31	2.4	1.4	6.3
322	Samarbay, Sarymsakty village	326.0	3.89	2.5	2.2	7.9
323	Lake without name, Pervomay village	324.8	1.15	1.4	1.1	3.9
324	Lake without name, Pervomay village	310.6	1.50	1.9	1.3	5.0
324	Kamyshnoye, Semenovka village	317.2	1.81	2.0	1.9	6.0
325	Zhalanash,	340.6	8.83	4.7	2.5	11.8
326	Bol'shoy Taldykol, Astana city	342.2	1.50	4.0	0.7	9.9
327	Talkol, Sorokovaya	353.3	2.29	2.1	1.4	5.5
328	Tyshkankol, Sorokovaya	412.0	1.37	1.9	1.2	4.6
329	Alakol, Aleksandrovka village	392.6	3.29	2.4	1.8	6.8
330	Zhakankol, Saryoba village	382.3	1.18	1.5	1.2	4.0
331	Ulken-Saryoba, Saryoba village	388.9	15.17	6.2	3.6	15.6
332	Kishi-Saryoba, Saryoba village	399.8	5.25	3.4	2.1	8.6
333	Shankan, Saryoba village	414.9	1.50	1.7	1.2	4.5
334	Taksary, Nikolaevka village	414.8	2.12	2.4	1.5	6.0
335	Lake without name, Nikolaevka village	432.6	1.12	2.2	0.7	5.2

(continued)

Table 3.12 (continued)

No.	Lake and its location	Altitude (m a.s.l.)	Lake area (km^2)	Length (km)	Max width (km)	Length of coastline (km)
336	Solenoye, Nikolaevka village	424.8	1.77	1.8	1.2	4.9
337	Manten, Blagodatnoye village	419.4	1.06	1.3	1.1	4.0
338	Koskol, Lazhydenka village	356.1	1.76	2.5	0.8	6.5
339	Aydarly, Lazhydenka village	368.0	1.07	1.6	1.1	4.4
340	Kobykol, Lazhydenka village	348.5	8.02/ 10.1	4.0	3.9	13.0
341	Kataukol, Lazhydenka village	369.0	1.60	2.0	1.1	5.2
342	Syntas, Symtas village	316.4	1.38	1.8	1.1	4.9
343	Zhalmankulak (Zhamankol), Kalinin	336.4	15.43	5.2	4.2	16.2
344	Zhosaly, Armavir	338.4	4.84	4.0	1.5	10.6
345	Shurba, Armavir	342.1	1.33	1.5	1.3	4.3
346	Egindikol, Armavir	351.6	1.42	1.9	1.1	5.0
347	Baybuta, Zhanaarka village	316.5	5.01	3.3	2.2	8.6
348	Temirastau, Zhanaarka village	322.0	4.28	2.6	2.2	7.9
349	Ashykol, Zhanaarka village	324.9	3.60	3.8	1.4	12.7
350	Kumkol, Zhanteke village	330.7	6.24	4.2	1.7	10.4
351	Lake without name, Zhanteke village	338.0	1.03	1.6	1.0	4.0
352	Uzynkol, Zhanteke village	332.8	1.64	2.0	1.2	5.9
353	Shiishalkar, Zhanteke village	328.2	15.53	5.8	3.8	18.6
354	Nauetek, Karakuga village	348.3	1.12	1.7	0.7	4.8
355	Zormankol, Karakuga village	348.3	4.13	3.7	1.8	9.1
356	Egindy, Egindy village	318.0	1.19	1.4	1.2	4.2
357	Burtash, Sabyndy village	339.4	5.56	3.8	2.8	10.2
358	Alakol, Egindy village	339.4	1.37	2.8	0.9	7.2
359	Karatomar, Rozhdestvenka village	349.2	1.15	1.7	0.8	4.6
360	Shenet, Rozhdestvenka village	344.8	5.64	4.7	1.5	14.4
361	Uzynsor, Rozhdestvenka village	355.1	5.57	6.7	1.0	21.0

(continued)

Table 3.12 (continued)

No.	Lake and its location	Altitude (m a.s.l.)	Lake area (km^2)	Length (km)	Max width (km)	Length of coastline (km)
362	Aganas, Rozhdestvenka village	346.0	1.88	2.3	1.1	6.8
363	Tuz, Rozhdestvenka village	357.0	3.85	3.9	1.6	11.2
364	Bol'shoy Sarykol, Rozhdestvenka village	359.8	1.42	1.9	1.1	5.0
365	Dombay, Rozhdestvenka village	358.6	2.73	2.8	1.9	7.3
366	Koskol, Rozhdestvenka village	360.1	1.68	2.1	1.0	5.4
367	Borlykol, Borlykol village	416.2	3.60	2.4	2.1	7.4
368	Zhaltyrkol, Martynovka village	374.2	1.38	1.5	1.3	4.4
369	Shoptykol, Vishnevka village	429.3	1.43	2.4	1.0	5.6
370	Batpakkol, Vishnevka village	419.7	2.05	2.0	1.5	5.6
371	Ashykol, Vishnevka village	416.2	2.13	1.8	1.5	5.6
372	Teniz, Urkendeu village	304.4	1161.54	74.4	32.2	488.0
373	Kokay, Abay	307.5	24.04	8.1	4.5	22.6
374	Soltankeldi, Ur	307.5	33.66	13.1	5.4	35.4
375	Lake without name, Urkendeu village	307.5	3.43	2.1	1.8	7.1
376	Lake without name, Urkendeu village	307.8	1.22	2.4	0.7	6.8
377	Sholak, Urkendisu village	318.0	31.91	19.7	3.0	68.2
378	Shalkar, Urkendeu village	318.0	24.52	6.5	5.5	22.0
379	Birtaban, Urkendeu village	317.8	13.54	4.8	4.0	15.6
380	Obaly, Korgalzhyn village	328.5	2.60	2.3	1.7	6.8
381	Lake without name, Korgalzhyn village	332.6	1.20	1.8	1.0	4.7
382	Koralay, Korgalzhyn village	337	2.25	5.0	0.7	16.0
383	Lake without name, Korgalzhyn village	342.7	1.03	1.9	0.95	4.7
384	Zhandyshalkar, Korgalzhyn village	328.2	15.79	5.1	4.8	16.6
385	Bestobe, Korgalzhyn village	330.3	1.62	2.1	1.5	6.9
386	Lake without name, Korgalzhyn village	354.9	1.17	1.8	0.8	4.8

(continued)

Table 3.12 (continued)

No.	Lake and its location	Altitude (m a.s.l.)	Lake area (km^2)	Length (km)	Max width (km)	Length of coastline (km)
387	Sulukurkol, Kumkol village	354.1	7.10	6.9	1.9	17.0
388	Bolgan, Kumkol village	357.0	1.46	2.0	1.0	5.4
389	Otebaysor, Arykty	350.4	1.27	1.9	1.0	5.0
390	Lake without name, Arykty	347	1.94	2.2	1.2	5.6
391	Alakol, Kumkol village	350.8	8.31	4.4	3.7	13.4
392	Ashykumkol, Arykty	356.4	5.39	4.6	2.7	15.6
393	Uzynkol, Arykty	357.8	4.25	5.9	1.3	15.9
394	Zharlykol, Arykty	334.5	12.10	7.8	2.1	19.6
395	Terektysor, Arykty	348.9	8.64	5.3	2.0	14.0
396	Kochkovatoye, Anar village	453.8	2.05	2.8	1.3	8.0
397	Baydaly, Anar village	439.2	2.63/2.7	2.1	1.7	6.2
398	Rybnoye, Anar village	423.6	1.74	1.7	1.4	5.2
399	Balyktykol, Anar village	335	1.42	2.5	1.0	6.7
400	Anarkol, Anar village	434.1	2.17	2.6	1.5	6.6
401	Konakay, Shenber village	329.9	2.3	4.0	0.9	13.4
402	Kyzylkol, Intemak village	309.9	3.97	3.9	1.3	10.6
403	Zhekekol, Zharapsay village	373.4	2.98	2.5	1.9	7.2
404	Usuiun', Zhaynakkaramola village	344.9	4.38	5.4	1.4	20.9

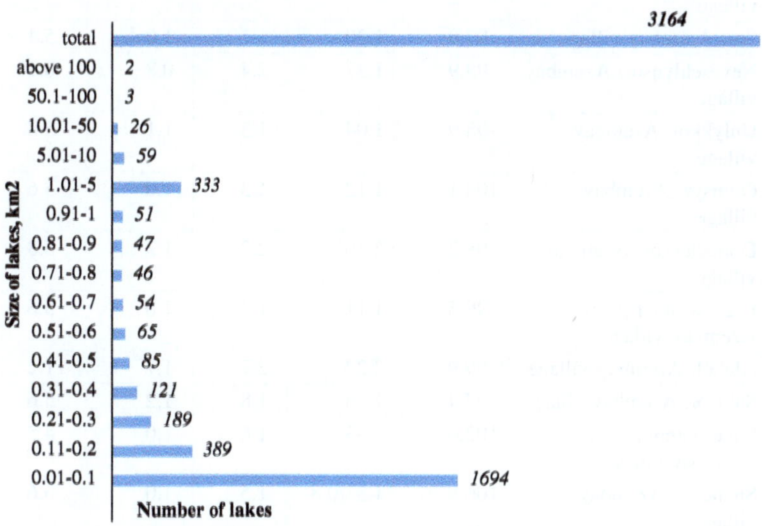

Fig. 3.11 Distribution of lakes in Pavlodar region by lake size

Table 3.13 Distribution of lakes in Pavlodar region by lake size and area (Philonets and Omarov 1974)

Area (km^2)	Number of lakes	Total area	
		km^2	%
0.01–0.1	1694	71.87	2.8
0.11–0.2	389	57.76	2.3
0.21–0.3	189	47.38	1.8
0.31–0.4	121	43.92	1.7
0.41–0.5	85	38.85	1.5
0.51–0.6	65	35.96	1.4
0.61–0.7	54	35.91	1.4
0.71–0.8	46	35.05	1.4
0.81–0.9	47	40.03	1.5
0.91–1	51	48.63	1.9
1.01–5	333	723.93	28.3
5.01–10	59	394.5	15.4
10.01–50	26	423.76	16.6
50.1–100	3	216.22	8.5
Above 100	2	345.67	13.5
Total	3164	2559	100

Table 3.14 Characteristics of lakes with a size more than 1 km^2 in Pavlodar region (Philonets and Omarov 1974)

No.	Lake and its location	Altitude (m a.s.l.)	Lake area (km^2)	Length (km)	Max width (km)	Length of coastline (km)
1	Zhalpinkol, Akhmechi village	100.4	2.32	2.3	1.6	1.10
2	Sor, Azembay village	97.9	1.20	1.7	1.0	5.4
3	Neveseldynsor, Azembay village	98.9	1.37	2.4	0.8	6.1
4	Unlykkol, Azembay village	98.9	1.04	1.3	1.1	5.8
5	Otynsyz, Azembay village	104.1	1.12	1.3	1.0	5.6
6	Dongeleksor, Azembay village	98.2	2.20	2.2	1.3	5.8
7	Lake without name, Azembay village	99.5	1.11	1.4	1.0	5.6
8	Alakol, Azembay village	99.9	3.22	2.7	1.7	11.2
9	Sakisor, Azembay village	97.4	1.53	1.8	1.2	5.6
10	Lake without name, Azembay village	102.0	1.33	1.6	1.0	4.5
11	Shandyk, Azembay village	100.8	1.35/0.8	1.5	1.0	6.6

(continued)

Table 3.14 (continued)

No.	Lake and its location	Altitude (m a.s.l.)	Lake area (km^2)	Length (km)	Max width (km)	Length of coastline (km)
12	Mukybaysor, Azembay village	98.0	2.01	1.7	1.6	5.1
13	Lake without name, Azembay village	99.1	1.40	1.5	1.2	4.4
14	Lake without name, Azembay village	100.1	2.60/1.4	2.5	1.3	11.0
15	Zhalpakkol, Novokuz'minka village	101.4	4.13	3.1	2.2	13.8
16	Zhaltyr, Novokuz'minka village	101.2	2.65	3.8	0.8	13.0
17	Shabakty, Novokuz'minka village	100.7	1.94	2.5	1.0	9.0
18	Vasil'evo, Novokuz'minka village	100.6	4.52	2.7	1.8	12.0
19	Andas, Novokuz'minka village	100.4	2.30	2.3	1.2	5.4
20	Alaaguly, Novokuz'minka village	99.9	1.35	1.7	1.0	4.2
21	Bol'shoy Shagalaly, Novokuz'minka village	100.0	2.19	2.1	1.3	6.2
22	Aktoshinek, Novokuz'minka village	99.8	2.90	2.8	2.2	8.2
23	Baytan, Novokuz'minka village	98.2	1.64	2.8	0.9	6.8
24	Talassor, Novokuz'minka village	99.5	1.87	1.9	1.2	5.8
25	Temirbaysor, Novokuz'minka village	99.9	4.82	2.7	2.3	8.8
26	Utinoye, Novokuz'minka village	99.8	1.70	2.6	1.2	7.2
27	Romankol, Novokuz'minka village	99.5	1.65	1.8	1.3	5.6
28	Lake without name, Novokuz'minka village	95.1	1.57	1.7	1.1	5.2
29	Pen'ki, Novokuz'minka village	96.9	1.16	1.6	0.8	5.0
30	Kostykkol, Bayte village	99.6	1.14	2.0	0.9	5.6
31	Uzynsor, Bayte village	97.8	1.47	3.4	0.6	8.2
32	Lake without name, Bayte village	99.6	1.16	2.0	0.7	5.6
33	Moyla, Bayte village	98.6	1.51	2.6	0.8	7.6
34	Zhalmandy, Bayte village	98.5	2.18	2.4	1.1	5.2

(continued)

Table 3.14 (continued)

No.	Lake and its location	Altitude (m a.s.l.)	Lake area (km^2)	Length (km)	Max width (km)	Length of coastline (km)
35	Karas', Karaterek village	102.1	1.09	3.0	0.7	8.0
36	Uzynkol, Karaterek village	101.0	6.12	7.2	1.0	15.8
37	Zharagash, Karaterek village	99.9	11.91	6.7	2.9	15.6
38	Lake without name, Zhonazhuldyz village	99.6	2.05	1.8	1.4	5.1
39	Kuria tret'ia, Kopylukha village	97.8	2.00	4.00	0.6	7.2
40	Kuria, Kopylukha village	97.2	1.30	2.8	0.7	5.2
41	Tobeles, Zhonazhuldyz village	102.1	2.30	2.3	1.1	5.5
42	Uzynkol, Kopylukha village	101.5	1.87	2.2	1.0	5.6
43	Aupildek, Enbekshi village	101.3	1.09	1.6	1.2	3.8
44	Basagashsor, Enbekshi village	100.0	2.91	2.4	1.5	6.8
45	Shoshkaly, Enbekshi village	104.8	1.88	2.1	1.2	5.0
46	Saygynsor, Enbekshi village	101.2	2.86	2.5	1.5	6.6
47	Saubaysor, Enbekshi village	100.4	2.24	1.8	1.6	5.4
48	Balkashsor, Dzhabul' village	101.6	2.04	2.2	1.4	6.2
49	Bulaktysor, Dzhabul' village	101.6	2.06	1.8	1.2	5.6
50	Malyi Zhalpak, Gruzdevka village	99.4	1.48	1.9	1.0	4.8
51	Kossor, Dzhabul' village	98.0	2.00	1.9	1.0	4.8
52	Eskara, Slavianka village	101.8	1.88	1.9	1.2	5.6
53	Muzdykol, Slavianka village	99.3	2.19	2.4	1.2	5.6
54	Zhamankol, Troitsk village	101.1	1.23	1.4	1.0	4.0
55	Muzdykol, Troitsk village	102.4	1.70/1.4	2.0	1.3	7.6
56	Kaskaatsor, Talapker village	104.8	1.92	2.7	1.3	7.6
57	Sorgus, Mikhaylovka village	97.9	3.32	3.3	2.0	10.2
58	Sor, Blagodatnyi village	98.0	2.10	3.3	1.3	7.8

(continued)

Table 3.14 (continued)

No.	Lake and its location	Altitude (m a.s.l.)	Lake area (km^2)	Length (km)	Max width (km)	Length of coastline (km)
59	Kyzyltuz, Blagodatnyi village	93.5	23.85	10.8	5.8	41.6
60	Korchikantuz, Mikhaylovka village	97.7	1.45	1.2	1.2	5.2
61	Taudysor, Mikhaylovka village	99.8	1.25	1.7	0.9	4.6
62	Zhaylau, Zholtaptyk village	98.3	1.11	1.6	1.0	4.2
63	Moynak, Golubovka village	113.9	1.68	1.9	1.2	5.0
64	Kyzylkak, Zapadnyi village	42.6	174.65	18.8	14.0	106.5
65	Ortakol, Golubovka village	109.6	2.92/2.6	2.3	1.8	7.2
66	Lake without name, Kaimanachikha village	84.0	1.30	1.6	1.1	4.4
67	Bol'shoy Koskol, Selety	107.0	1.89/1.5	1.7	1.5	5.0
68	Ashykol, Synty	123.4	1.23	1.4	1.1	4.6
69	Zharsor, Synty	121.5	3.05	2.1	1.9	6.6
70	Aralsor, Golubovka village	122.0	1.20	1.7	1.0	4.2
71	Ayteke, Golubovka village	122.3	1.40/ 0.74	1.6	1.1	4.5
72	Sor, Golubovka village	123.0	2.64	2.1	1.6	7.2
73	Takty, Grabovo village	99.1	3.33/ 2.45	2.4	1.7	7.2
74	Ashakol, Grabovo village	99.4	1.98	1.9	1.5	5.4
75	Kolbaysor, Kornilovka village	99.0	3.31	2.4	1.8	7.0
76	Solenoye, Kornilovka village	94.0	3.35	2.5	1.6	7.2
77	Uzynsor, Kornilovka village	91.4	4.96	4.6	1.6	11.2
78	Kangeldy, Zhanadauir village	95.4	1.05	1.4	1.0	3.8
79	Kanaysor, Zhanadauir village	91.7	2.91	2.5	1.5	9.6
80	Orlova, Zhaskayrat village	89.4	1.99	1.8	1.2	5.1
81	Uzynkol, Ivanovka village	98.7	2.02	2.5	1.1	6.2
82	Lake without name, Ivanovka village	99.0	1.46	2.7	1.0	4.4

(continued)

Table 3.14 (continued)

No.	Lake and its location	Altitude (m a.s.l.)	Lake area (km^2)	Length (km)	Max width (km)	Length of coastline (km)
83	Strelka, Kachiry village	94.8	1.12	1.6	0.9	4.0
84	Karasu, Kachiry village	96.7	5.12	7.2	1.8	14.6
85	Karasu, Kachiry village	93.6	2.63	3.1	1.3	8.6
86	Karasu, Kachiry village	94.2	5.05	7.6	1.0	18.6
87	Ormanik, Frument'evka village	103.4	1.33	1.8	1.1	4.6
88	Zhamantuz, Frument'evka village	94.6	7.24	4.5	2.6	13.8
89	Kyzyltuz, Frument'evka village	98.4	1.69	3.4	0.7	7.2
90	Shubarterek, Frument'evka village	99.2	1.09	3.6	0.5	7.2
91	Balkazy, Belotserovka village	91.7	3.76	3.4	2.4	12.2
92	Lake without name, Belotserovka village	90.4	1.06	1.4	0.9	3.8
93	Kossor, Belotserovka village	97.3	2.00	2.3	1.0	8.2
94	Lake without name, Lozovaya village	96.3	1.51	2.1	1.0	4.8
95	Kulaksor, Lozovaya village	94.0	4.80	3.2	2.0	8.1
96	Malyi Azhbulat, Lozovaya village	93.1	7.74	3.9	2.8	10.9
97	Stekliannoye, Lozovaya village	94.5	1.23	2.0	1.0	5.0
98	Zhambaysor, Lozovaya village	97.1	1.39	2.1	1.2	6.3
99	Aynak, Ekaterinoslavskii village	98.9	8.32	6.6	2.9	21.0
100	Lake without name, Birlik village	74.1	2.40	3.6	0.8	11.4
101	Ashybay, Chapaev	89.7	2.20	3.2	1.0	8.6
102	Zhalauly, Leninshil village	71.9	171.02	21.0	15.0	81.0
103	Lake without name, Leninshil village	74.0	1.23	1.4	1.2	5.2
104	Lake without name, Leninshil village	73.0	1.75	2.6	1.2	6.6
105	Lake without name, Leninshil village	82.7	1.84	2.0	1.5	5.0
106	Lake without name, Leninshil village	83.1	2.25	3.2	0.9	9.2
107	Ornek, Leninshil village	86.0	4.55	5.1	0.7	17.0

(continued)

Table 3.14 (continued)

No.	Lake and its location	Altitude (m a.s.l.)	Lake area (km^2)	Length (km)	Max width (km)	Length of coastline (km)
108	Lake without name, Leninshil village	84.0	1.92	3.2	1.0	10.8
109	Tuz, Leninshil village	84.2	2.81	2.0	1.5	6.1
110	Uzynsor, Leninshil village	82.6	1.39	2.2	1.0	4.8
111	Lake without name, Shilikty village	98.8	1.93	1.8	1.2	7.0
112	Liubomirskoye, Shilikty village	90.0	1.44	1.7	1.2	7.0
113	Lake without name, Shilikty village	88.5	1.29	1.7	1.0	5.0
114	Kodarsor, Shilikty village	89.0	1.21	1.4	1.1	4.4
115	Kishkenesor, Shilikty village	93.3	2.55	2.5	1.2	10.8
116	Mynshunkyr, Shilikty village	88.0	6.75	4.4	1.9	12.2
117	Lake without name, Shilikty village	93.7	1.02	1.2	1.1	4.2
118	Zhanysbay, Shilikty village	93.2	1.17	1.4	1.2	4.8
119	Kishibaysor, Novoalekseevka village	95.4	1.70	1.8	1.1	7.4
120	Lake without name, Novoalekseevka village	96.8	1.50	1.6	1.3	5.6
121	Dongeleksor, Novotroitskii village	88.0	1.21	1.5	0.9	4.6
122	Sasyksor, Novotroitskii village	86.0	7.08	4.0	3.0	14.0
123	Lake without name, Novoalekseevka village	98.4	3.44	2.1	2.0	7.4
124	Zhamantuz, Andrianovka village	93.1	9.35	5.3	3.2	15.4
125	Koserip, Razumovka village	96.4	3.22	2.6	1.7	8.2
126	Karkaraly, Andrianovka village	93.2	5.37	3.7	2.2	9.2
127	Lake without name, Andrianovka village	96.2	2.64	1.9	1.5	6.1
128	Sor Kurto, Khar'kovka village	97.6	2.90	2.4	1.5	8.2
129	Karabulak, Khar'kovka village	106.2	1.86	2.0	1.3	5.6
130	Taykonyr, Khar'kovka village	88.7	14.93	6.5	3.7	20.2

(continued)

Table 3.14 (continued)

No.	Lake and its location	Altitude (m a.s.l.)	Lake area (km^2)	Length (km)	Max width (km)	Length of coastline (km)
131	Baklanskoye, Zhana Tal village	93.3	1.47	6.0	0.3	14.6
132	Mashinsor, Natashino village	95.4	3.11	2.7	1.6	6.6
133	Lake without name, Ravnopol' village	95.1	2.48	2.8	1.2	6.6
134	Malyi Tavolzhan, Ravnopol' village	91.8	10.88	7.2	2.8	18.2
135	Kadyrbay, Ravnopol' village	94.0	3.69	2.8	2.0	8.0
136	Bol'shoy Tavolzhan, Tavolzhan village	92.0	12.50	7.7	2.1	17.4
137	Klevkino, Obraztsovka village	98.2	4.05	3.3	1.4	10.2
138	Kozhybayevo, Obraztsovka village	98.8	3.52	4.9	0.8	15.8
139	Kupal'noye, Obraztsovka village	100.7	1.28	2.0	0.7	4.4
140	Bozhmanovo, Krupskoye village	101.1	2.03	1.9	1.2	6.8
141	Uzynsor, Sladkovodsk village	90.8	4.55	4.3	1.8	11.2
141	Sholaksor, Novotroitskii village	65.6	28.26	7.9	4.6	21.2
142	Shyganak, Novotroitskii village	84.8	1430	8.7	2.3	34.4
143	Lake without name, Novotroitskii village	82.9	1.62	1.9	1.0	5.6
144	Bastuz, Shuga village	87.8	7.62	4.4	2.7	14.6
145	Koybagar, Shuga village	90.2	6.88	3.5	2.5	10.6
146	Aktobesor, Shuga village	103.2	1.37	1.8	1.0	4.2
147	Lake without name, Shuga village	111.1	1.45	1.5	1.3	4.8
148	Zhamansor, Baskamys village	99.7	2.15	1.6	1.6	6.4
149	Lake without name, Baskamys village	108.6	3.37	3.0	1.5	9.8
150	Bozshasor, Baskamys village	105.7	3.07	2.4	1.2	14.6
151	Zhamansor, Baskamys village	103.5	5.50	3.5	2.2	15.4
152	Zhamantuz, Baskamys village	103.0	5.01	3.4	2.2	16.6

(continued)

Table 3.14 (continued)

No.	Lake and its location	Altitude (m a.s.l.)	Lake area (km^2)	Length (km)	Max width (km)	Length of coastline (km)
153	Zharsor, Baskamys village	97.5	12.81	9.5	2.4	1.2
154	Lake without name, Baskamys village	102.5	1.16	1.3	1.0	7.8
155	Sholakbulaksor, Baskamys village	104.0	10.60	9.5	2.5	36.4
156	Borlysor, Baskamys village	98.4	11.49	5.7	3.5	22.6
157	Sulusor, Baskamys village	103.4	5.03	5.0	2.2	15.0
158	Begaly, Baskamys village	103.7	1.98	1.9	1.2	6.8
159	Sarykol, Airmola village	106.1	2.76	2.7	1.0	10.4
160	Bylkyldak, Sarychevka village	98.1	1.28/0.7	2.0	0.7	7.6
161	Koriakovka, Koriakovka village	92.2	7.70	7.4	1.3	20.6
162	Kabantakyr, Orlovka village	123.6	2.59/3.3	2.5	1.5	6.8
163	Ashytakyr, Kakhovka village	125.6	4.39	3.3	1.7	8.0
164	Kostakyr, Nazarovka village	127.3	2.40	2.1	1.4	6.2
165	Bura, Severnyi village	120.9	4.72	2.6	2.3	8.6
166	Mambetkol, Nikolaevka village	131.9	2.10	1.6	1.6	5.8
167	Bozaygyr, Karazhar village	124.2	2.97	2.3	1.8	9.4
168	Basentin, Karazhar village	124.4	1.13	1.7	1.1	6.4
169	Auliskol, Karazhar village	126.0	7.48	3.7	2.9	11.4
170	Sarykamys, Sarykamys village	114.5	3.55	3.4	1.5	10.0
171	Lake without name, Sarykamys village	114.5	1.30	2.0	0.8	5.8
172	Sasykkol, Sarykamys village	119.2	5.01	3.2	2.1	12.2
173	Shajdar, Sarykamys village	119.8	1.81	4.6	0.6	14.4
174	Sarykol, Markovka village	102.4	3.58	4.1	1.1	20.0
175	Lake without name, Markovka village	102.9	2.32	2.4	1.2	11.0

(continued)

Table 3.14 (continued)

No.	Lake and its location	Altitude (m a.s.l.)	Lake area (km^2)	Length (km)	Max width (km)	Length of coastline (km)
176	Lake without name, Markovka village	102.5	2.26	2.8	0.9	8.4
177	Shureksor, Markovka village	101.5	80.94	25.0	10.2	157.1
178	Lake without name, Markovka village	101.4	2.73	2.0	1.5	10.0
179	Zhamantuz, Markovka village	103.5	19.85	4.8	3.1	20.0
180	Senkese, Kalininskii village	104.2	1.50	1.9	1.1	5.8
181	Esentykol, Kalininskii village	104.3	2.63	3.8	1.0	12.0
182	Uzynbulak, Markovka village	107.0	1.10	2.5	0.7	6.2
183	Kalkamantuz, Ermak village	109.6	3.93	3.7	1.4	9.0
184	Kalkamantuz, Ermak village	108.6	12.63	5.0	3.6	16.6
185	Karamyrza, Ermak village	111.0	1.00	1.8	0.8	4.6
186	Yakbolshek, Zhertumsyk village	119.7	1.09	2.1	0.7	7.4
187	Zhambyoba, Zhertumsyk village	134.8	3.29	2.4	1.9	7.0
188	Maraldy, Romanovka village	60.3	54.75	9.5	8.3	27.0
189	Sor, Galkino village	139.5	1.38	1.9	1.0	4.4
190	Niaz, Chigirinovka village	142.6	1.26	1.9	0.9	4.2
191	Maysor, Bozshakol village	194.2	4.85	2.7	2.1	8.6
192	Karasor, Bozshakol village	223.3	6.76	4.0	2.2	11.6
193	Bozshasor, Bozshakol village	223.5	2.70	2.5	1.7	7.4
194	Kindikty, Ekibastuz-Ugol' village	204.7	4.82	3.1	2.2	14.8
195	Amanbaysor, Ekibastuz-Ugol' village	210.9	2.51	2.1	1.	5.8
196	Karabidaiyk, Ekibastuz-Ugol' village	191.8	6.77	3.4	3.0	10.2
197	Ekibastuz, Ekibastuz-Ugol' village	178.0	3.72	2.6	1.8	8.2
198	Karasor	68.1	44.30	17.2	5.0	78.7

(continued)

Table 3.14 (continued)

No.	Lake and its location	Altitude (m a.s.l.)	Lake area (km^2)	Length (km)	Max width (km)	Length of coastline (km)
199	Zhamantuz	123.9	7.20	3.5	2.8	11.4
200	Kyzylsor	126.6	4.21	2.9	2.1	12.0
201	Kalatuz, Yamyshevo village	95.6	7.38/8.2	3.9	2.4	12.2
202	Bestuz, Ol'ginka village	135.4	1.55	1.9	1.0	6.0
203	Aydorsha, Dongelek village	121.2	4.01/2.4	2.8	1.8	9.6
204	Borly, Karagaly village	115.7	15.99/ 15.9	5.6	4.0	18.0
205	Seyten, Baysak village	136.8	15.85	5.4	4.4	19.2
206	Kazy, Baysak village	116.3	6.03/5.4	3.0	2.7	8.9
207	Chaldayskoye, Chalday village	154.1	1.70	2.4	0.9	5.9
208	Sasyksor, Kirov	289.1	4.32	2.6	2.2	8.0
209	Balakeskensor, Kirov	316.9	2.98	2.5	1.5	7.0
210	Sarykol, Kirov	264.1	2.08	2.0	1.4	7.6
211	Shykyldak, Kirov	248.2	1.08	1.3	1.0	4.4
212	Kutayaksor, Kirov	233.6	2.14	1.9	1.8	7.0
213	Kantay, Kirov	250.3	1.10	1.4	1.0	4.3
214	Sarykol, Kirov	227.7	2.02/ 0.02	2.0	1.6	8.0
215	Maysor, Kirov	262.2	7.25	5.0	2.3	11.3
216	Zharsor, Kirov	271.4	1.64	3.2	0.7	7.6
217	Kulboldy, Kirov	282.7	3.25	2.8	1.5	6.6
218	Shansor, Kirov	299.8	8.04	4.0	3.0	10.4
219	Maysor, Maykaiyn village	216.1	3.09	2.2	1.6	6.9
220	Karasor, Maykaiyn village	259.6	4.68	3.2	1.8	7.8
221	Akbidaiyk, Maykaiyn village	192.4	1.97	2.2	1.5	5.2
222	Angren, Maykaiyn village	204.9	7.65	4.0	2.9	18.0
223	Zhaltyrkol, Maykaiyn village	248.3	1.24	1.6	1.0	5.2
224	Zhangozha, Maykaiyn village	169.0	1.19	1.5	1.2	4.2
225	Altybaysor, Koytas village	108.6	13.56	6.2	3.2	16.0
226	Bozshasor, Koytas village	129.1	4.82	2.3	1.8	11.4
227	Lake without name, Koytas village	161.9	2.45	2.0	1.3	5.8

(continued)

Table 3.14 (continued)

No.	Lake and its location	Altitude (m a.s.l.)	Lake area (km^2)	Length (km)	Max width (km)	Length of coastline (km)
228	Kaban, Saskuduk village	133.8	1.19	2.0	0.8	4.8
229	Kyzyltuz, Saskuduk village	136.0	5.75/6.0	3.2	2.4	8.8
230	Zhaltyr, Malybay village	139.0	9.58	8.7	1.8	22.2
231	Malybay, Malybay village	142.5	12.22	9.5	2.2	23.4
232	Ashykol, Kerey village	127.2	1.64	4.2	0.7	6.6
233	Kal'cha, Sharbakty village	122.0	2.53	3.8	1.1	8.0
234	Sharbakty, Sharbakty village	131.3	6.78/6.5	3.8	2.6	9.6
235	Tileuberdi, Sharbakty village	142.6	2.30	2.4	1.1	6.4
236	Akkol, Koktauymsyk village	163.8	1.34	2.0	0.8	4.6
237	Boranshoky, Tamdy village	362.0	3.71	2.8	1.8	7.6
238	Lake without name, Tamdy village	353.4	1.11	2.2	0.8	5.0
239	Esenkeldikol, Tamdy village	333.6	1.48	1.4	1.2	4.6
240	Toresor, Tamdy village	279.6	1.44	1.6	1.3	7.4
241	Zhosaly, Tamdy village	322.2	2.68	2.3	1.3	10.2
242	Ashykol, Tamdy village	261.5	2.93	2.4	1.6	13.2
243	Taldykol, Shoptykol village	233.3	1.20	1.5	0.8	6.5
244	Sarykol, Shoptykol village	236.0	1.27	2.9	0.6	8.0
245	Maukoben, Shoptykol village	238.1	5.85	3.0	2.5	11.4
246	Tuzkol, Shoptykol village	228.7	4.18	3.8	1.4	10.2
247	Zharkol, Shoptykol village	228.2	6.35	3.5	2.9	15.8
248	Sulusor, Shoptykol village	193.1	12.60	4.9	3.1	19.0
249	Karaadyrsor, Shoptykol village	191.6	2.98	3.5	1.0	11.2
250	Kokuirum, Zhamantuz village	151.8	5.72	5.9	2.2	14.6
251	Kaiyndy, Zhamantuz village	177.4	2.76	3.0	0.9	17.4
252	Lake without name, Zhamantuz village	187.2	2.12	1.9	1.0	6.8

(continued)

Table 3.14 (continued)

No.	Lake and its location	Altitude (m a.s.l.)	Lake area (km^2)	Length (km)	Max width (km)	Length of coastline (km)
253	Lake without name, Zhamantuz village	179.5	7.50	5.7	2.1	16.6
254	Lake without name, Zhamantuz village	179.0	2.10	1.9	1.4	6.2
255	Lake without name, Zhamantuz village	188.1	5.02	4.0	1.6	12.4
256	Alkamergen, Zhamantuz village	179.6	20.03	7.4	5.1	26.6
257	Lake without name, Zhamantuz village	187.9	3.47	3.5	1.0	11.8
258	Lake without name, Zhamantuz village	196.7	1.05	1.3	1.0	5.4
259	Aiykmalaysor, Zhamantuz village	199.0	5.08	6.5	1.4	26.6
260	Lake without name, Zhamantuz village	207.2	1.26	1.7	1.0	5.4
261	Lake without name, Zhamantuz village	209.4	1.95	2.7	0.6	11.2
262	Kemertuz, Tarakty village	136.9	5.55/4.9	3.2	2.3	10.0
263	Aktomar, Tarakty village	145.5	5.21	4.4	2.0	22.0
264	Karasor	144.2	28.35	9.3	6.0	46.0
265	Uialy, Krivinskoye village	164.5	3.76	3.0	1.7	7.0
266	Pishendy, Krivinskoye village	161.8	1.13	2.3	0.6	5.0
267	Tuz, Tuzagash village	134.5	10.63	5.8	2.6	13.4
268	Sor, Tuzagash village	155.0	3.08	2.4	1.6	7.6
269	Lake without name, Tuzagash village	155.2	1.75	2.3	1.0	6.4
270	Chushkalinskoye, Tuzagash village	155.0	4.83	2.9	2.0	8.4
271	Saumalkol, Kyzyltas village	426.1	7.15	5.3	2.1	13.4
272	Zavodskoye, Alekseevka village	429.1	1.63	2.1	1.4	6.6
273	Zhasybay, Bayanauyl village	397.1	3.12/4.0	3.2	2.4	10.8
274	Sabyndykol, Bayanauyl village	452.7	7.08/6.7	4.5	2.6	15.0
275	Espentuz, Bayanauyl village	213.6	10.52	5.3	3.2	17.0
276	Kambobasor, Bayanauyl village	319.0	2.44	2.2	1.2	9.6

(continued)

Table 3.14 (continued)

No.	Lake and its location	Altitude (m a.s.l.)	Lake area (km^2)	Length (km)	Max width (km)	Length of coastline (km)
277	Lake without name, Bayanauyl village	308.9	1.28	2.1	0.8	5.6
278	Babystytuz, Bayanauyl village	251.3	1.53	2.3	1.0	7.0
279	Lake without name, Dzhusaly village	262.0	2.32	3.4	1.2	11.0
280	Koshkarbay, Semiarsk village	172.5	2.89	2.0	1.7	7.2
281	Sulusor, Mayskoye village	149.3	7.50	7.5	2.4	26.6
282	Dogobasor, Mayskoye village	167.8	1.32	1.8	1.0	7.9
283	Sor, Tendyk village	405.3	4.35	2.8	2.0	8.8
284	Sor, Zhanatlek	444.1	1.08	1.8	0.9	5.8
285	Tastykol, Zhanatlek	446.2	1.26	1.7	1.1	5.0
286	Tuz, Zhanatlek	423.8	10.84	4.2	3.5	14.8
287	Omakokol, Zhapauium village	468.9	2.14	1.7	1.5	5.8
288	Temissor, Dzhusaly village	313.9	3.47	2.5	2.0	8.4
289	Lake without name, Dzhusaly village	314.7	2.05	1.8	1.0	9.0
290	Teresor, Mayskoye village	238.0	5.19	3.9	2.1	15.2
291	Kotansor, Mayskoye village	237.0	2.40	3.2	1.2	10.8
292	Zhamantuz, Mayskoye village	249.1	2.52	2.3	1.4	6.6
293	Zhaksytuz, Mayskoye village	262.1	6.81	3.8	2.6	11.2
294	Lake without name, Mayskoye village	218.3	1.40	3.8	0.5	9.8
295	Shandakkol, Mayskoye village	221.4	1.18	2.2	0.7	6.0
296	Kumkol, Izvestkovyi village	203.8	1.14	1.2	1.0	6.2
297	Oynaksor, Izvestkovyi village	202.4	2.14	2.3	1.5	8.0
298	Shalkarkol, Zhapauium village	629.0	5.68	3.9	2.0	10.2
299	Lake without name, Mayskoye village	345.8	1.83	1.4	1.1	5.8
300	Akmolsor, Mayskoye village	330.7	1.42	2.3	0.9	5.4

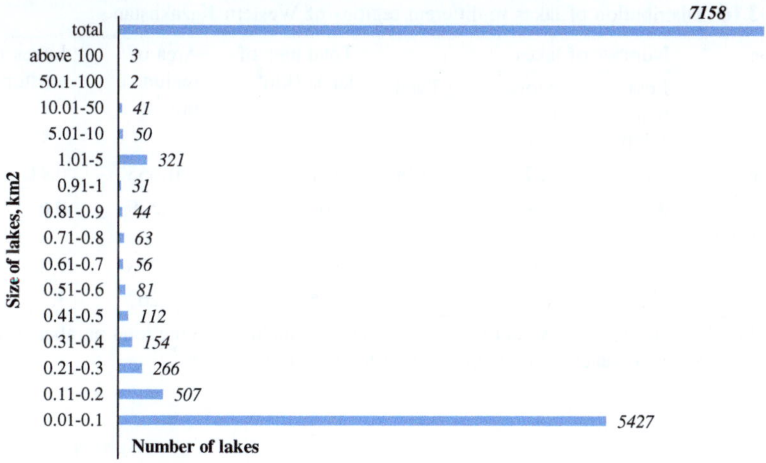

Fig. 3.12 Distribution of lakes in the Western Kazakhstan by lake size

Table 3.15 Distribution of lakes in West Kazakhstan by lake size and area (Philonets and Omarov 1974)

Area (km^2)	Number of lakes	Total area	
		km^2	%
0.01–0.1	5427	147.83	4.97
0.11–0.2	507	75.6	2.54
0.21–0.3	266	66.89	2.25
0.31–0.4	154	54.3	1.82
0.41–0.5	112	50.78	1.71
0.51–0.6	81	74.68	2.51
0.61–0.7	56	36.56	1.23
0.71–0.8	63	48.08	1.62
0.81–0.9	44	37.29	1.25
0.91–1	31	29.27	0.98
1.01–5	321	695.6	23.38
5.01–10	50	336.2	11.30
10.01–50	41	730.79	24.57
50.1–100	2	150.66	5.07
Above 100	3	440.41	14.80
Total	7158	2975	100

Notice Ponds, water reservoirs, and plexus lakes are not included to the total number of lakes

There are 579 plexus lakes in the region with a total area of 9.77 km^2

Lake Alka (Korzhynkol) is located east of the Algabas village, at an altitude of 828 m a.s.l. Its area is 1.17 km^2, the length is 2.9 km, the width is 0.7 km, and the length of the coastline is 6.8 km (Fig. 3.19).

Table 3.16 Distribution of lakes in different regions of Western Kazakhstan

Region name	Number of lakes			Total area of lakes (km^2)	Area of region (km^2)	Lakes in the territory (%)
	Less than 1 km^2	More than 1 km^2	Total			
Aktobe	1521	227	1748	1211	301,655	0.41
West Kazakhstan	3176	84	3260	908	151,339	0.6
Atyrau	2041	109	2150	865	284,273	0.3
Total	6738	420	7158	2975	737,267	1.31

Notice Ponds, water reservoirs, and plexus lakes are not included to the total number of lakes There are 579 plexus lakes in the region with a total area of 9.77 km^2

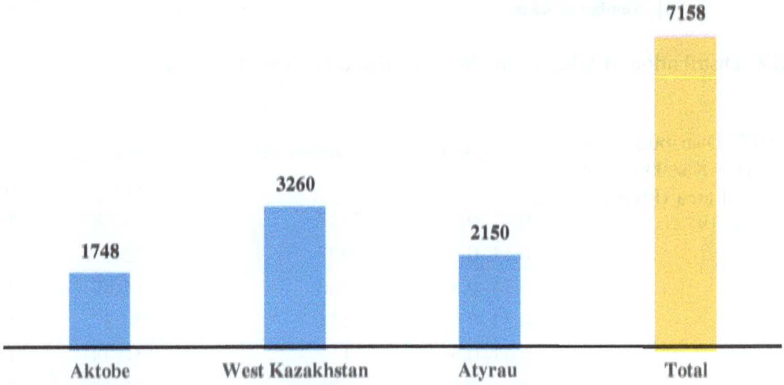

Fig. 3.13 Lakes in West Kazakhstan by different regions

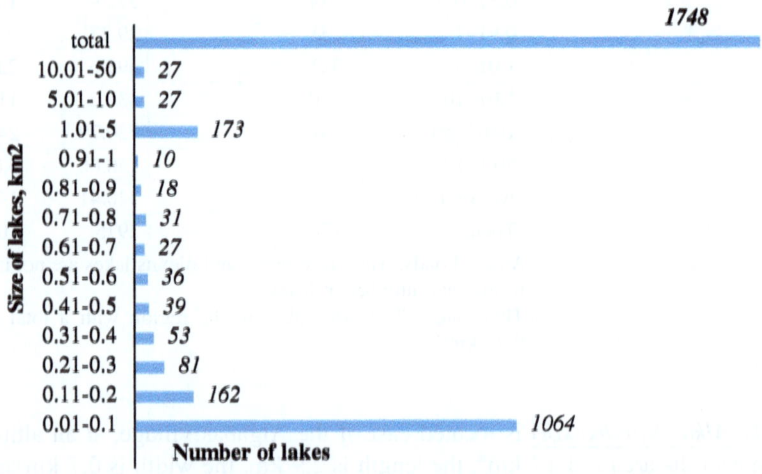

Fig. 3.14 Distribution of lakes in the Aktobe region by lake size

Table 3.17 Distribution of lakes in Aktobe region by lake size and area (Philonets and Omarov 1974)

Area (km^2)	Number of lakes	Total area	
		km^2	%
0.01–0.1	1064	33.55	2.77
0.11–0.2	162	24.11	1.99
0.21–0.3	81	20.22	1.67
0.31–0.4	53	18.82	1.55
0.41–0.5	39	17.59	1.45
0.51–0.6	36	19.81	1.63
0.61–0.7	27	17.72	1.46
0.71–0.8	31	23.61	1.95
0.81–0.9	18	15.40	1.27
0.91–1	10	9.40	0.77
1.01–5	173	371.13	30.64
5.01–10	27	187.67	15.50
10.01–50	27	452.29	37.35
–	–	–	–
Total	1748	1211	100

Notice Ponds, water reservoirs, and plexus lakes are not included to the total number of lakes

Table 3.18 Characteristics of lakes with a size more than 1 km^2 in Aktobe region (Philonets and Omarov 1974)

No.	Lake and its location	Altitude (m a.s.l.)	Lake area (km^2)	Length (km)	Max width (km)	Length of coastline (km)
1	Kul'	437.5	1.01	1.1	1.0	3.8
2	Taboba	222.6	4.79	5.8	1.8	16.4
3	Lake without name	224.1	1.48	3.3	1.5	11.0
4	Zhylandy	257.6	1.80/1.4	1.7	1.5	5.4
5	Koskol	342.2	1.14	1.6	0.9	4.4
6	Aitastykol	339.7	3.12/3.0	2.2	1.9	7.0
7	Karakul	337.8	1.44	1.5	1.2	4.4
8	Sorkol	259.7	2.59	3.1	1.4	11.6
9	Sulukol	204.9	1.25/1.1	1.7	1.1	6.6
10	Kopa	204.2	2.24	1.8	1.6	8.0
11	Lake without name	143.0	1.97	2.4	1.0	12.6
12	Lake without name	139.4	2.98	4.6	1.2	10.0
13	Lake without name	145.7	1.15	2.0	0.8	5.4
14	Tenizsor	104.8	18.26	12.1	2.6	29.6
15	Lake without name	106.7	1.56	1.7	1.0	9.6
16	Tuzdykol	366.3	1.22	1.8	1.1	4.8
17	Saumalkol	199.7	1.64	1.6	1.2	6.8

(continued)

Table 3.18 (continued)

No.	Lake and its location	Altitude (m a.s.l.)	Lake area (km²)	Length (km)	Max width (km)	Length of coastline (km)
18	Sorkol	89.9	1.42	1.8	1.0	5.0
19	Solenoye	356.7	2.07	1.9	1.4	6.0
20	Sorkol	353.7	2.50	2.2	1.4	6.4
21	Sorkol	284.5	1.64	1.7	1.0	5.2
22	Mukhtarkol	76.2	13.20	8.0	2.3	23.0
23	Ushkol	77.3	3.05	3.0	1.5	8.5
24	Akkol	78.0	4.0	2.3	2.2	9.4
25	Airkol	76.2	43.70	33.0	2.7	82.5
26	Lake without name	91.6	1.23	2.7	0.5	6.5
27	Buraiashykol	76.3	5.01	3.7	1.8	9.6
28	Aksarykol	71.0	1.06	1.6	0.9	4.6
29	Lake without name	76.6	1.82	3.2	0.9	8.8
30	Shuiakkol	76.3	5.02	4.6	2.1	13.6
31	Tokerekkol	76.3	10.01	7.4	2.2	22.1
32	Iirkol	76.3	3.78	4.2	1.3	9.2
33	Zharmakol	76.3	12.21	9.8	1.6	25.4
34	Mamyrkol	76.4	1.23	1.8	0.7	5.6
35	Shozhikol	76.3	9.18	9.7	1.4	26.5
36	Lake without name	76.2	9.25	11.5	1.3	22.0
37	Zhangildikol	76.1	12.32	7.8	2.6	20.3
38	Lake without name	76.1	7.20	5.8	1.8	16.2
39	Lake without name	75.8	1.55	2.1	1.1	5.3
40	Sarykol	75.8	1.42	2.0	0.8	5.2
41	Lake without name	82.5	1.47	1.7	1.3	5.3
42	Bukynkol	76.2	11.28	5.2	2.6	18.0
43	Lake without name	76.2	1.55	4.1	1.0	9.1
44	Karpykkol	76.1	7.72	6.2	2.1	14.3
45	Zharkol	76.2	15.00	9.5	2.8	30.7
46	Baitakkol	75.8	45.21	28.0	3.3	66.0
47	Zharkamys	75.2	11.62	11.8	1.6	33.1
48	Akkol	162.4	2.29	2.0	1.5	6.2
49	Sarykol	158.0	9.72	4.2	3.0	12.8
50	Kandykol	219.0	3.44	2.	1.8	7.8
51	Kishkenezhalanash	93.2	1.65	2.4	1.1	4.8
52	Ulken Zhalanash	88.0	6.64	4.2	2.3	11.2
53	Sorkol	85.5	4.03	3.2	1.3	8.4
54	Kozhakol	78.1	8.12	7.2	1.6	16.0
55	Lake without name	74.2	1.07	2.0	0.8	5.0
56	Kogakol	74.0	10.46	11.3	1.5	24.6

(continued)

Table 3.18 (continued)

No.	Lake and its location	Altitude (m a.s.l.)	Lake area (km^2)	Length (km)	Max width (km)	Length of coastline (km)
57	Lake without name	73.3	1.72	2.7	0.7	6.7
58	Lake without name	73.3	2.01	2.1	1.5	5.2
59	Lake without name	73.4	2.65	2.7	1.2	6.0
60	Lake without name	73.2	3.25	4.8	1.4	11.0
61	Lake without name	73.0	1.62	1.7	1.1	6.8
62	Lake without name	72.4	1.64	2.6	0.8	6.9
63	Lake without name	73.4	1.09	3.0	0.5	10.1
64	Lake without name	71.8	1.18	1.8	0.8	4.7
65	Lake without name	71.5	1.16	1.8	0.9	5.0
66	Keltekol	71.6	5.35	8.1	1.2	18.7
67	Maikol	71.6	1.95	3.8	0.9	9.6
68	Rynkol	71.5	2.15	2.7	0.9	7.0
69	Sorkol	73.7	3.43	3.5	1.1	10.0
70	Karakol	51.2	4.68	3.9	2.2	15.6
71	Zhomart	168.7	1.74	2.1	1.2	9.0
72	Shalkar	167.0	5.65/7.9	7.0	1.8	18.0
73	Atkaikol	54.6	2.05	2.4	0.9	5.2
74	Bakhshakol	52.9	1.05	2.2	0.5	5.8
75	Kurdym	51.1	22.58	10.5	3.0	31.5
76	Koskol	315.0	1.01	1.5	0.9	4.0
77	Shonai	233.3	1.22/1.6	1.6	0.8	8.4
78	Lake without name	86.0	1.04	2.0	0.6	4.6
79	Konyrkol	169.3	1.03	1.6	0.9	4.0

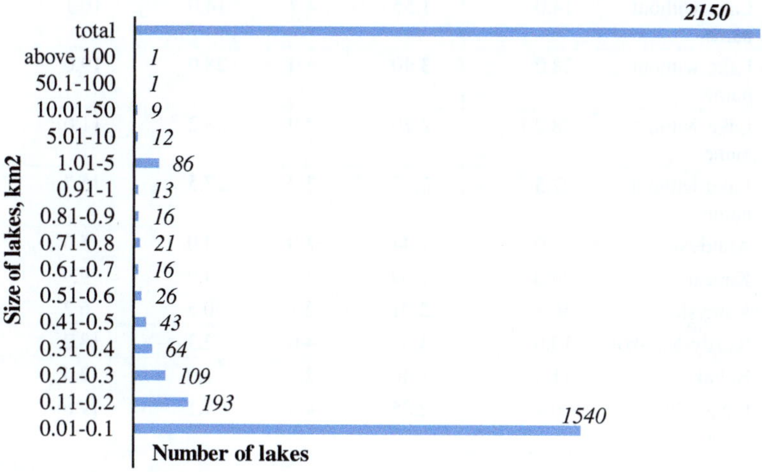

Fig. 3.15 Distribution of lakes in the Atyrau region by lake size

Table 3.19 Distribution of lakes in Atyrau region by lake size and area (Philonets and Omarov 1974)

Area (km^2)	Number of lakes	Total area	
		km^2	%
0.01–0.1	1540	47.11	5.8
0.11–0.2	193	28.64	3.5
0.21–0.3	109	27.30	3.3
0.31–0.4	64	22.16	2.7
0.41–0.5	43	19.54	2.4
0.51–0.6	26	14.44	1.7
0.61–0.7	16	10.49	1.3
0.71–0.8	21	16.17	1.9
0.81–0.9	16	13.40	1.6
0.91–1	13	12.26	1.5
1.01–5	86	180.07	21.8
5.01–10	12	72.82	8.8
10.01–50	9	158.18	19.1
50.1–100	1	92.80	11.2
Above 100	1	110.48	13.4
Total	2150	856	100

Notice Ponds, water reservoirs, and plexus lakes are not included to the total number of lakes

Table 3.20 Characteristics of lakes with a size more than 1 km^2 in Atyrau region (Philonets and Omarov 1974)

No.	Lake and its location	Altitude (m a.s.l.)	Lake area (km^2)	Length (km)	Max width (km)	Length of coastline (km)
1	Lake without name	34.7	1.94	2.1	34.7	5.9
2	Lake without name	14.0	1.55	4.7	14.0	10.1
3	Lake without name	28.0	3.40	5.9	28.0	19.0
4	Lake without name	28.2	2.20	2.9	28.2	11.0
5	Lake without name	27.5	2.00	3.7	27.5	15.7
6	Araldykol	13.0	1.34	2.0	13.0	5.0
7	Karasu	24.0	1.09	3.4	24.0	9.9
8	Kamyskol	30.5	2.01	2.1	30.5	6.2
9	Kozdy-Karakol	13.0	4.35	4.0	2.3	18.2
10	Baitak	11.0	1.40	2.7	1.0	12.4
11	Lake without name	−0.4	2.55	4.4	1.1	19.6

(continued)

Table 3.20 (continued)

No.	Lake and its location	Altitude (m a.s.l.)	Lake area (km²)	Length (km)	Max width (km)	Length of coastline (km)
12	Mangyt	47.9	1.08	1.8	0.8	4.2
13	Aktobe	−5.0	5.56	5.1	2.8	17.4
14	Inder	23.8	110.48	13.5	11.0	40.0
15	Saralzhin	34.8	2.37	2.1	1.4	5.8
16	Lake without name	−4.0	3.50	3.6	0.9	22.2
17	Lake without name	−14.0	1.91	2.9	0.9	7.6
18	Ermek	4.3	1.92	2.3	1.2	12.7
19	Zhan-Terek	46.3	2.12	2.2	1.5	6.9
20	Tomak	−20.9	1.61	3.3	09	8.7
21	Karasu	−23.0	1.07	2.9	0.6	10.0
22	Lake without name	−26.5	1.77	5.8	0.7	15.0
23	Zhaltyr	−25.3	92.80	13.4	8.3	51.7
24	Bartyldakty	−22.7	6.46	5.0	1.9	13.5
25	Mashe	−20.7	11.65	6.5	2.8	20.6
26	Lake without name	−24.4	2.73	2.6	2	8.8
27	Lake without name	−25.5	1.13	2.6	1.1	10.6
28	Kamyskol	−13.8	2.81	3.2	1.3	9.6
29	Lake without name	−24.4	5.90	7.2	1.3	19.2
30	Lake without name	−25.3	14.77	11.9	2.3	32.8
31	Il'men Zharsuat	−24.5	1.11	3.4	0.7	8.2
32	Lake without name	−27.0	1.55	2.8	0.8	8.6
33	Il'men Gordeevskii	−26.6	1.97	2.3	1.1	7.8
34	Bol'shoye	−27.8	9.84	17.9	1.3	46.0

The lake has an elongated shape and occupies two dissimilar in size circus basins from the group of the Sebi lakes: one of them that is the northern one is extensive and oval in the meridional direction, and the other is southwestern, which is located in a small flat circus. The basin is surrounded by granite rocks on three sides and is open to the southeast.

The southern coast of the lake is shallow, sandy loamy and overgrown with fescue, feather grass, and white wormwood (in some places the shore is stony,

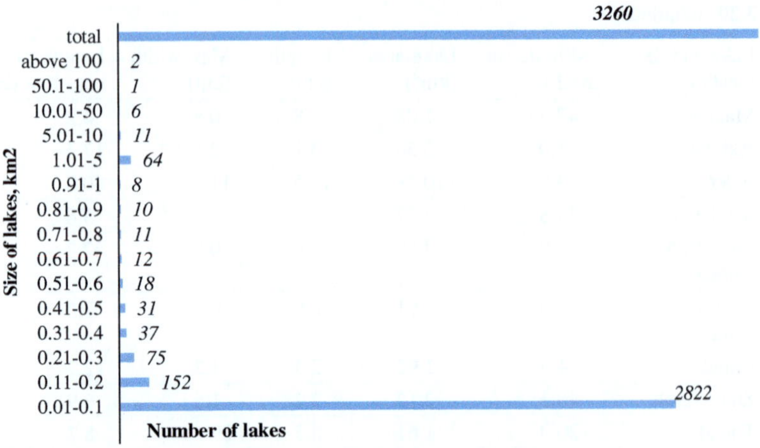

Fig. 3.16 Distribution of lakes in the West Kazakhstan region by lake size

Table 3.21 Distribution of lakes in West Kazakhstan region by lake size and area (Philonets and Omarov 1974)

Area (km^2)	Number of lakes	Total area	
		km^2	%
0.01–0.1	2822	67.17	7.3
0.11–0.2	152	22.85	2.5
0.21–0.3	75	19.37	2.1
0.31–0.4	37	13.32	1.5
0.41–0.5	31	13.65	1.5
0.51–0.6	18	10.43	1.1
0.61–0.7	12	8.35	0.9
0.71–0.8	11	8.30	0.9
0.81–0.9	10	8.49	0.9
0.91–1	8	7.61	0.8
1.01–5	64	144.35	15.7
5.01–10	11	75.71	8.2
10.01–50	6	120.32	13.1
50.1–100	1	57.86	6.3
Above 100	2	329.93	36.0
Total	3260	908	100

Notice Ponds, water reservoirs, and plexus lakes are not included to the total number of lakes

overgrown with reed ordinary). A narrow strip of sandy deposits descends gently to the edge of the water at the northern shore of the large basin. The eastern and western shores are steep and composed of granite blocks. On the slopes, there are small patches of grassy and shrubby (predominantly dog-rose) vegetation,

Table 3.22 Characteristics of lakes with a size more than 1 km^2 in West Kazakhstan region (Philonets and Omarov 1974)

No.	Lake and its location	Altitude (m a.s.l.)	Lake area (km^2)	Length (km)	Max width (km)	Length of coastline (km)
1	Akkol	78.0	1.19	1.6	1.0	4.4
2	Sorkol	83.9	3.48	4.4	1.9	15.5
3	Koskol	104.0	1.52	1.6	1.2	4.7
4	Lake without name	30.8	1.14	1.9	1.3	8.3
5	Birkazan	22.6	1.31	2.3	0.9	8.3
6	Birkazan	19.6	1.75	1.8	1.5	6.6
7	Birkazan	19.6	1.21	1.4	1.1	4.4
8	Kopakostino	21.4	1.92	3.5	1.1	9.7
9	Malyi Satei	19.8	1.25	1.7	1.2	8.1
10	Shalkar	16.7	205.80	18.4	14.7	57.1
11	Sulukol	159.0	1.10	1.7	1.0	4.6
12	Al'zhan	24.1	1.70	1.5	1.4	4.9
13	Sorkol	138.0	2.99	2.2	1.6	7.4
14	Egindykol	99.0	1.16	2.4	0.7	5.7
15	Balykty	5.5	7.64	5.5	2.7	13.9
16	Almalyk	6.7	4.03	2.5	2.1	8.6
17	Sorkol	88.4	2.75	1.9	1.8	6.4
18	Lake without name	8.1	1.49	3.9	0.8	10.4
19	Bilshoy Kyzyloba	0.5	2.21	3.1	1.4	12.0
20	Rybnyi Sakryl	4.8	16.67	6.5	4.8	23.0
21	Besoba	0.0	3.80	2.5	2.4	8.3
22	Sorshyganak	0.5	18.67	5.9	4.1	20.2
23	Etbatyr	0.4	4.91	4.2	2.3	19.6
24	Lake without name	0.5	2.02	2.9	1.3	9.6
25	Oryskopa	0.1	2.94	2.4	1.5	7.5
26	Lake without name	0.1	2.25	3.2	1.4	11.8
27	Birkazankol	0.3	4.51	5.7	1.4	13.9
28	Kinzhekara	2.0	1.49	2.6	0.7	6.4
29	Kartashova	−6.0	1.10	2.7	0.7	6.9
30	Staritsa	−14.0	1.11	5.5	0.3	12.1
31	Zhaltyrkol	−7.0	19.09	8.5	3.7	29.4
32	Shoptykol	148.0	1.02	1.7	0.8	4.0

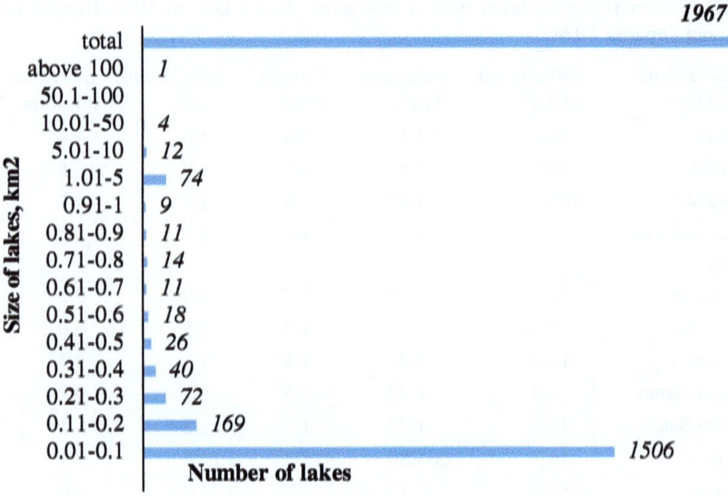

Fig. 3.17 Distribution of lakes in the Eastern Kazakhstan by lake size

Table 3.23 Distribution of lakes in Eastern Kazakhstan by lake size and area (Philonets and Omarov 1974)

Area (km²)	Number of lakes	Total area	
		km²	%
0.01–0.1	1506	51.43	5.7
0.11–0.2	169	24.39	2.7
0.21–0.3	72	17.74	2.0
0.31–0.4	40	13.98	1.6
0.41–0.5	26	11.87	1.3
0.51–0.6	18	9.49	1.1
0.61–0.7	11	7.27	0.8
0.71–0.8	14	10.6	1.2
0.81–0.9	11	9.37	1.0
0.91–1	9	8.62	1.0
1.01–5	74	140.18	15.7
5.01–10	12	79.38	8.9
10.01–50	4	61.72	6.9
50.1–100	–	–	–
Above 100	1	449.83	50.20
Total	1967	896	100

Notice Ponds, water reservoirs, and plexus lakes, as well as Alakol and Sasykkol lakes are not included to the total number of lakes

Table 3.24 Characteristics of lakes with a size more than 1 km^2 in East Kazakhstan region (Philonets and Omarov 1974)

No.	Lake and its location	Altitude (m a.s.l.)	Lake area (km^2)	Length (km)	Max width (km)	Length of coastline (km)
1	Dubygaly	343.7	1.84	2.4	1.2	6.8
2	Sebi (Ablay: Kashkerbay	835	0.22	0.7	0.5	–
3	Sebi: Alka (Korzhynkol)	828	1.17	2.9	0.7	6.8
4	Sebi: Ulmeis (Shalkar)	825.4	2.51	3.0	1.3	7.8
5	Sebi: Duisen (Tortkarakol)	770.0	1.01	1.4	1.1	4.2
6	Sebi: Istykpa (Kunakkol)	708.0	0.65	1.3	0.6	–
7	Maral'e	1763.8	1.92	3.5	1.0	7.8
8	Chernoye	1694.0	1.47	4.0	0.6	8.2
9	Yazovoye	1656.2	1.59	3.0	0.7	6.6
10	Rakhmanovskoye	1760.0	1.14	2.6	0.6	5.6
11	Bukhtyrminskoye	2056.1	4.23	5.3	1.1	12.0
12	Markakol	1449.3	455.0	38.0	19.0	95.5
13	Kemerkol	839.7	1.72	2.3	1.1	6.8
14	Kozenkol	369.3	1.39	2.0	1.0	5.2
15	Takyrkol	665.0	1.01	2.5	0.6	6.0
16	Aiyr	505.7	0.9	1.5	0.7	4.2
17	Lake without name	355.0	0.21	0.7	0.4	–
18	Ulkensor	163.8	1.21	1.9	0.9	4.0
19	Shunkyrsor	205.1	1.78	1.8	1.1	7.8
20	Lake without name	158.9	1.92	6.2	0.4	12.5
21	Bol'shoye	211.3	1.02	1.6	1.1	4.8
22	Kamyshki	257.0	1.02	1.8	0.7	4.0
23	Presnoye	257.1	1.36	2.0	1.1	4.4
24	Bol'shoye	308.8	12.3	11.8	2.4	27.0
25	Kol'	173.2	7.22	4.5	2.3	15.2
26	Sozyksor	185.5	5.12	4.2	2.3	17.3
27	Balyktykol	217.5	6.08	5.8	1.5	15.8
28	Lake without name	295.7	1.18	1.4	1.1	4.6
29	Sasykkol	259.2	2.48	2.4	1.3	7.8
30	Alzhan	254.9	2.90	2.3	1.2	13.6
31	Zhyngyldy	442.3	1.20	1.9	1.0	5.4
32	Kamyshnoye	390.0	1.39	3.0	0.7	7.6
33	Chistoye	280.0	4.12	3.5	1.5	12.2

(continued)

Table 3.24 (continued)

No.	Lake and its location	Altitude (m a.s.l.)	Lake area (km²)	Length (km)	Max width (km)	Length of coastline (km)
34	Ashykol	336.2	2.21	1.9	1.5	7.8
35	Iirsaikol	449.0	1.85	2.1	1.3	6.0
36	Koskabulakkol	409.4	6.3	5.1	1.8	15.4
37	Ospankol	488.9	2.17	2.3	1.4	7.6
38	Bakshoky	522.1	1.60	1.3	1.5	4.6
39	Nizh.Balyktykol	501.5	2.96	2.3	1.3	9.4
40	Verkh. Balyktykol	505.0	2.41	2.1	1.7	6.4
41	Burligan	812.2	21.4	8.5	4.1	24.0
42	Karakol	754.8	2.11	2.7	1.1	6.4
43	Koldar	369.0	6.51	6.0	1.6	19.2
44	Lake without name	346.8	2.85	5.6	0.9	14.6
45	Lake without name	347.7	1.95	2.6	1.5	13.8
46	Aksu	348.5	1.50	2.0	1.3	7.6
47	Kenenbay	347.4	1.82	2.1	1.0	7.1
48	Balyktykol	277.3	1.01	1.8	1.0	6.0

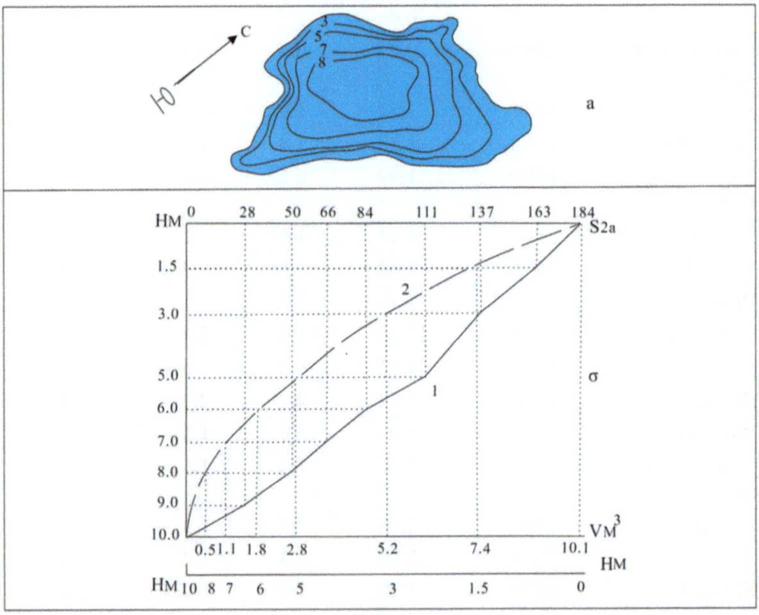

Fig. 3.18 Lake Dubygaly: a) plan and coastal areas; σ) curved lines: (1) bathymetric and (2) three-dimensional

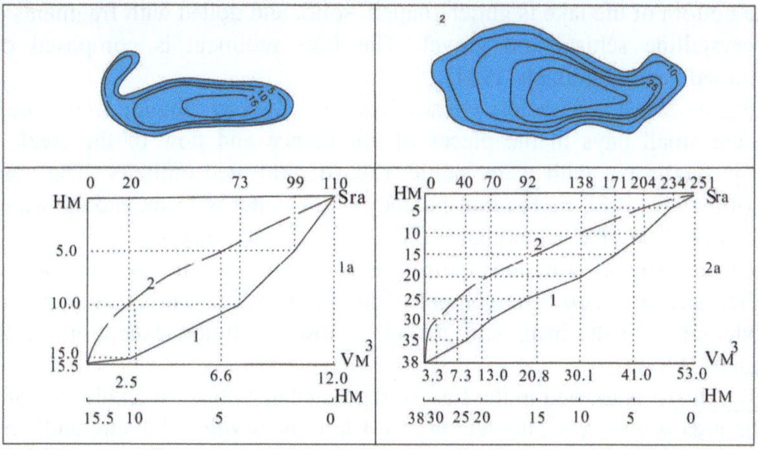

Fig. 3.19 Lake Alka: 1-plan: σ) 1 a) curved lines: bathymetric (1), three-dimensional and coastal areas (2); 2, 2a-the same for Lake Ulmeis

individual trees of birch, poplar, aspen, bird cherry, and others. The lake bottom is sandy and hard.

In the southeast, an unnamed stream flows into the lake and flows out of it along the canyon (more than 10 m deep) in the southwest. The maximum depth of the lake is 15.5 m, the water color is brownish-yellow, and the transparency is 2 m. Lake Alka accumulates 12 million m^3 of water (Philonets 1981).

The water is fresh, soft, and slightly alkaline. It refers to the hydrocarbonate class with the calcium group. The content of iodine, bromine, boron, zinc, copper, manganese, molybdenum, cobalt, iron, and nickel is low, with the exception of fluorine.

Lake Ulmeis (Shalkar) is located in the East of the Algabas village, at an altitude of 825.4 m m a.s.l.. Its area is 2.51 km^2, the length is 3 km, the width is 1.3 km, and the length of the coastline is 7.8 km (Fig. 3.19).

The Lake Ulmeis is the largest lake of the group of the Sebi lakes. It has a basin elongated in the meridional direction, extending to the south. Like other lakes, the hollow in the south is open. The southern coast of the lake is flat and sandy loamy. The lake is separated from a small swampy coastal strip by a coastal shaft of 0.7-m high. The terrain gradually rises further south. The coast has become overgrown with fescue, feather grass, and white wormwood, and often there are bushes of caragana, yellow acacia, vines, and dogrose. The northern coast rises above the water level by more than 600 m. Sand–gravel deposits stretch along the water by a narrow strip. The western and eastern shores are steep and composed of granites. There are bushes of yellow acacia, bird cherry, caragan, and dog rose, individual trees of poplar, aspen, pine, and birch on these shores.

The bottom of the lake is gutter-shaped, solid, and dotted with fragments of dark gray crystalline schists and gravel. The lake sediment is composed of gray fine-grained sand (Philonets 1981).

In the southeast, a nameless stream flows into the bay, flowing in the southwest. There are small bays in the places of confluence and flow of the creek, which partially overgrown with water nettle, rhdasts, and reed ordinary. The maximum depth of the lake is 38 m, the transparency of the water is 25 m, and the water color is yellowish. The lake accumulates 53 million m^3 of water.

The water is fresh, soft, and slightly alkaline. It refers to the hydrocarbonate class with the calcium group. The content of iodine, bromine, boron, zinc, manganese, molybdenum, cobalt, iron, and nickel is low, with the exception of fluorine (Philonets 1981).

Lake Duisen is located in the East of the Algabas village, at an altitude of 770 m a.s.l. Its area is 1.01 km^2, the length is 1.4 km, the width is 1.1 km, and the length of the coastline is 4.2 km (Fig. 3.20).

The lake is located in a circus basin that stretched from north to south, and is surrounded by granite rocks on three sides. The bottom of it is cup-shaped, solid, and dotted with fragments of granite and gravel. The lake sediment is composed of gray fine-grained sand (Fig. 3.20).

The southern coast of the lake is open, flat, sandy loamy, and turfed. The coast is a meadow. Reed ordinary, pondy, lily grows along the coast in the water.

The western and eastern banks are granite that steeply descend to the water; in some places, the cliffs reach 20 m. The coastal slopes are overgrown with stellate, and there are individual aspen trees, birches, currant bushes, bird cherries, and ishpony.

The northern coast is gently sloping but closed. Along the edge of the water stretches a coastal shaft of gray sand up to 2 m high. The lake coast is overgrown with aspen, vine, bird cherry, currant and rosehip, sedge and reed ordinary grow near the water.

In the southeast, a nameless stream flows into the bay, which flows in the southwest. The maximum depth of the lake is 19 m, water transparency is 6 m, and water color is yellowish. More than 12 million m^3 of water accumulate in the lake (Philonets 1981).

The lake water is fresh and soft alkaline (pH = 7.82–8.60). The content of dissolved oxygen is high (from 5.49 to 12.48 mg/l) during the year. Permanganate oxidability is 13.6–17.8 mg O_2/l. Dissolved carbon dioxide is 0.97–2.2 mg/l and at the bottom is 3.52–16.28 mg/l. The lake water belongs to the hydrocarbonate class with the calcium group. The content of iodine, bromine, boron, zinc, copper, manganese, molybdenum, cobalt, iron, and nickel is low, except for fluorine (Philonets 1981).

Lake Istykpa is located in the East of the Algabas village, at an altitude of 708 m a.s.l. It is the westernmost lake of the Sebi lake group. Its area is 0.65 km^2, the length is 1.3 km, the width is 0.6 km, and the length of the coastline is 2.6 km (Fig. 3.20).

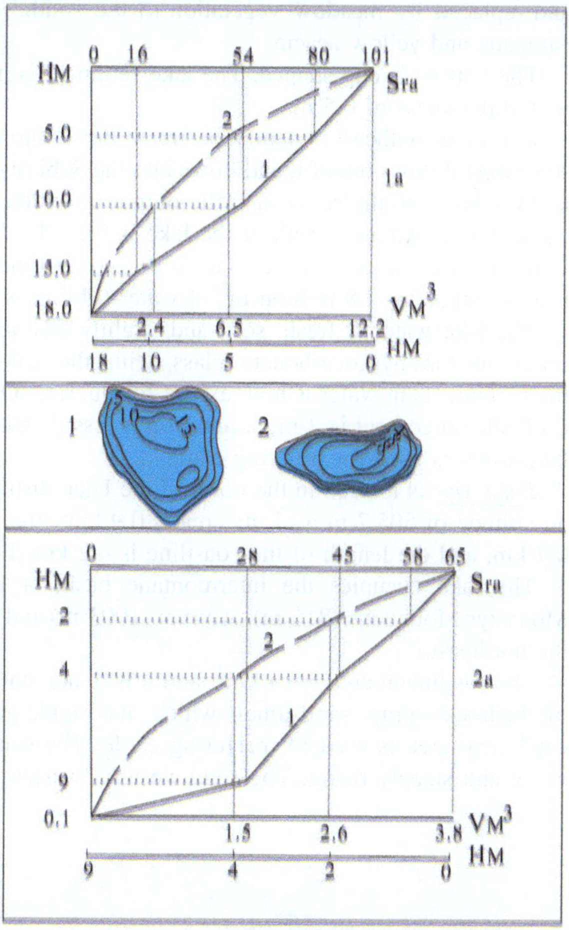

Fig. 3.20 Lake Duisen: 1-plan: σ) 1 a) curved lines: bathymetric (1), three-dimensional and coastal areas (2); 2, 2a-the same for Lake Istykpa

The lake is located in a circus basin, which stretched from the southwest to the northeast. From the west, north, and east it is surrounded by mountains; from the south it is open. The coasts are less steep than other lakes and rich in vegetation (Philonets 1981).

The northern shore is flat. At the edge of the water, there is a coastal shaft in gray sand 2-m high and 8–25-m wide washed. The sand is partially fixed by vegetation. Behind the shaft, close to the mountains is a lake 50 m in diameter, around a common reed covered with reeds, sedge, a white water lily, the vine, birch, and aspen. In 0.5 km from the water, the northern shore folded by gray granite and steeply rises. There are some small trees of birch, aspen on the slopes. The eastern and western shores are steep and also composed of gray granite. There are individual trees of birch, aspen, yellow acacia bushes, and cherry trees along the coast.

The southern coast is gently sloping, sandy loamy, and rugged. A significant part of it is swamped and covered with reed grass, sedge, reds, and white water lily and replaced by meadow vegetation to the south. There are separate bushes of caragana and yellow acacia.

The bottom is cup-shaped. The lake sediment is in gray color and fine-grained, with a thickness of 0.5 m.

From the southeast, a nameless stream flows into the bay of the lake, and in the southwest it flows out of it and flows into the Sebi river. The lake is replenished due to river flow, groundwater, and the output of which is observed near the northern coast. The maximum depth of the lake is 9 m. The water level fluctuations reach 1 m. The transparency of the water is 4–6 m, and the water color is yellowish. The lake accumulates 3.8 million m^3 of water (Philonets 1981).

The lake water is fresh, soft, and slightly alkaline. The chemical composition refers to the hydrocarbonate class with the calcium group. The content of microelements in water is low, except for fluorine. Oxidability of water is 2.56 mg O_2/l, silicon content is 2 mg/l, and iron is less than 0.1 mg/l (Philonets 1981). The lake water is used for watering cattle.

Lake Aiyr is located in the north of the Ulan district, East Kazakhstan region, at an altitude of 505.7 m a.s.l. Its area is 0.9 km^2, the length is 1.5 km, the width is 0.7 km, and the length of the coastline is 4.2 km (Fig. 3.21).

The lake occupies the intermontane basin at the northwestern foot of the Monastyr Mountains (816 m), Aiyrtau (1003 m) and extends from the southwest to the northeast.

The catchment area is 13 km^2, and it is a mountainous, with numerous outcrops of bedrock—gray sandstone—which has undergone strong weathering. The catchment area is used for pasturing cattle. The southeastern shore of the lake is stony and steeply rising. The southwestern, western, and northeasterly shores are

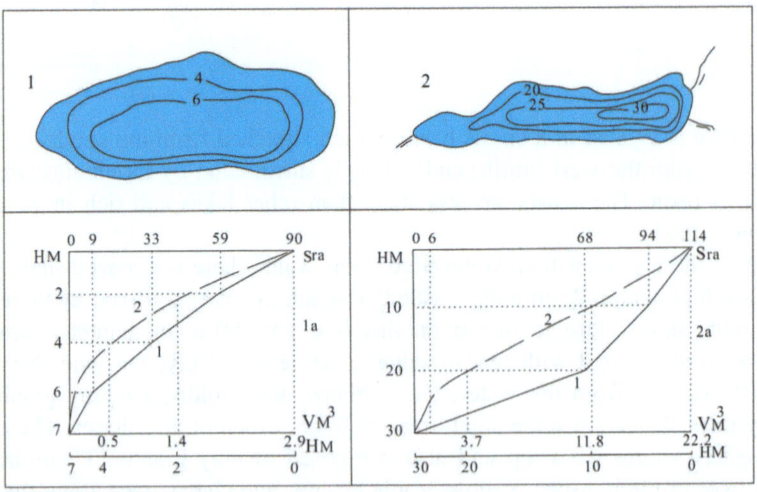

Fig. 3.21 Lake Aiyr: 1-plan: σ) 1 a) curved lines: bathymetric (1), three-dimensional and coastal areas (2); 2, 2a-the same for Lake Rakhmanovskoye

gently sloping, open, and seared. Along the shores of the lake, an ordinary cane grows in an intermittent stripe, and in the southeast there is a water horsetail. The coasts are overgrown with fescue and white wormwood. There are bushes, wheat grass, and licorice found in the lake shores.

The lake is replenished due to spring snowmelt, ground runoff, and precipitation. Its maximum depth is 7 m, the water transparency is 1.5 m, and the water color is greenish. It accumulates about 3 million m^3 of water (Philonets 1981).

The lake water is slightly saline, slightly alkaline, and moderately hard. The water mineralization varies little by year. By chemical composition, it refers to the hydrocarbonate class with the sodium group. The content of iodine, bromine, boron, copper, zinc, manganese, molybdenum, cobalt, nickel, and iron in water is low, except for fluorine. The bottom of the lake is solid. The lake sediment is in yellowish color, and it is composed of coarse-grained sand (Philonets 1981). There are roach, perch, carp, and pike in the lake, and ducks nest in the reeds.

The lake water is used for watering cattle. The scenery of the lake and the healing properties of its water (suitable for the treatment of eczema and other skin diseases) attract many tourists.

Lake Rakhmanovskoye is located in the northeast of the Katonkaragay region, at an altitude of 1760 m a.s.l. Its area is 1.14 km^2, the length is 2.6 km, the width is 0.6 km, and the coastline length is 5.6 km (Fig. 3.21). The lake occupies a small gutter-like pond basin in a narrow valley, located between the slopes of the mountains. It is composed of the oldest Paleozoic rocks and extends from the northwest to the southeast (Philonets 1981).

The shores of the lake are closed, and they are composed of biotite granites, sometimes covered with clay and sodden. The high mountain ranges surrounding the lake on three sides (northern, eastern, and southern) from the water's edge steeply rise and 2 km from the shore reach a height of 2558 m. There are abundant exits to the surface of bedrock rocks—granites throughout the catchment area. The mountains are covered with slender larch, cedar, spruce, and in low parts—birch, willow. The catchment area is 69 km^2, almost 80% of it is covered with forest, and only mountain peaks with a height of more than 2200 m are devoid of vegetation. The vegetation of the shoreline of the lake is especially majestic: Century-old cedars and fir trees encircle the mirror surface of the lake with a dark green border. Most of the trees are covered with gray beards of a bearded lichen.

Lake Rakhmanovskoye is flowing one, from the east the Arasan river and the nameless stream and the nameless stream from the north flow into it (Philonets 1981).

The Arasan river follows from the lake in the northwestern part of the lake. The lake replenishes due to river flow, snow, and groundwater. Its maximum depth is 30 m. The deepest part is the southwestern part of the lake. The transparency of the water is 6 m, and the water color is blue. The average annual fluctuation of the water level in the lake is 1.5 m. A high level is observed in early May and low level in September. More than 20 million m^3 of water accumulate in the Lake.

The lake water is ultra-fresh, very soft, and weakly acidic. It refers to the hydrocarbonate class with the sodium group. The content of iodine, fluorine, bromine, boron, copper, manganese, cobalt, molybdenum, nickel, and iron in water is very low, with the exception of zinc. The bottom of the lake is solid, sandy and abundantly covered with needles. The lake sediment is composed of fine-grained sand in gray color (Philonets 1981).

Ducks nest on the lake. There is the balneological sanatorium on the northwest coast of the lake. The sanatorium is open all year round. High healing properties of water (radon springs) and the presence of a good mountain road and picture sequences of the surrounding area in the summer attract many tourists.

Lake Markakol is located in the extreme east of the Kazakhstan, in the mountains of the Southern Altay, at an altitude of 1449.3 m. Its area is 455 km^2, the length is 38 km, the width is 19 km, and the length of the coastline is 106 km (Fig. 3.22). The catchment area of the lake is mountainous, and it is 1180 km^2. The height of the catchment ranges from 1450 to 3304 m (Aksubas Mountain), and an average height is 2500 m (Mitrofanov 1961). The terrain is crossed, difficult to traverse, and is riddled with numerous streams. The entire territory is overgrown with forest, except for mountain ridges and peaks (with a height of more than 2000 m). There are distributed mountain-tundra, peaty podzolized, mountain-meadow, mountain-forest, and mountain-meadow-steppe soils.

The lake is located in the Markakol Basin, between the Azutau Ranges (from the south) and the Kurshim Ranges (from the north). The basin has an oval form that tapers from west to east. It was formed during the interglacial epoch of the Quaternary period under the influence of endogenous processes (Nekhoroshev 1936). The depth of the lake is uneven: In the west, the lake bottom decreases gradually from shore to the middle part of the lake, except for the southwestern part, where the maximum depth is −27 m. The depth distribution near the southern coast is abrupt; 30–40 m from the coastal edge, the depth reaches 15–16 m. The northern and eastern parts of the lake are shallow. The bottom is a wavy plain (Philonets 1981).

The coastline is indented slightly, and only insignificant cuts are noticeable at the confluence of the rivers. There is a shallow and silted bay in the northeast of the lake. The southern shores are steep and composed of solid bedrock and descend into the water with one or two ledges. In some places, the ledges are flat and swampy.

The steep banks are composed of solid bedrock. The northern shores of the lake are poorly divided. Their height reaches 7 m, and they are composed of bedrock. There are wetlands that can be traced in the estuaries of the rivers such as Tikhushka, Topolevka, Multykbai, Yelovka, and Tastybulak. Almost all the banks of the lake are overgrown with larch, spruce, fir, birch, aspen, mountain ash, bird cherry. Currants, honeysuckle, tavolga, and willow grow on the fringes.

The lake basin is located in the region of high humidity and characterized by a large amount of precipitation, insignificant evaporation, and preservation of snow on mountain peaks and in river valleys for a long period.

There are 27 small rivers and streams in the lake. The largest ones are the Topolevka, Karabulak, Matabay, and Zhiren-Baytal rivers. The flow of water from the lake is through the Kalzhir river in the southwest.

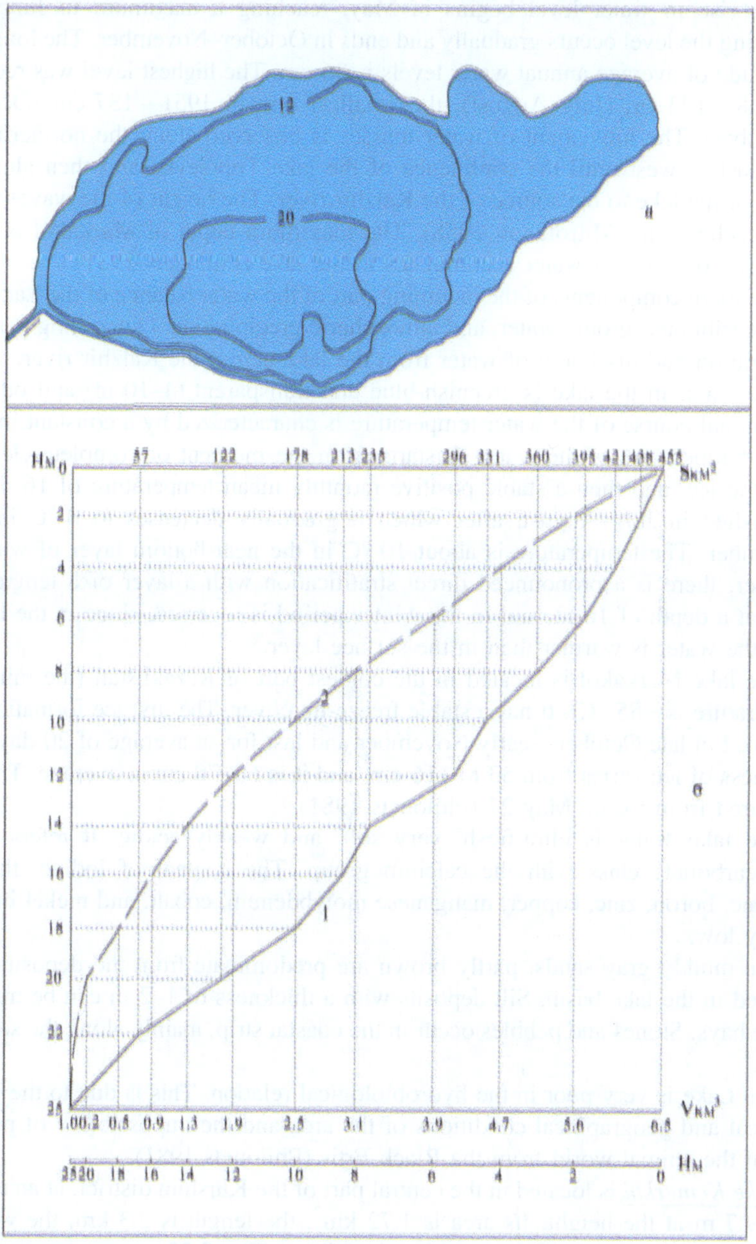

Fig. 3.22 Lake Markakol: a) plan and coastal areas; σ) curved lines: (1) bathymetric and (2) three-dimensional

The rise in water level begins in May, reaching a maximum in June–July. Lowering the level occurs gradually and ends in October–November. The long-term amplitude of average annual water levels is 36 cm. The highest level was recorded in 1958—173 cm (July–August), the smallest one in 1951—137 cm (October–December). The movement of water masses is observed along the northern coast from east to west until the confluence of the lake Topolevka and then along the center of the lake to the source of the Kalzhir river. The height of the waves on the lake reaches 2 m (Mitrofanov 1958). The maximum depth of Markakol is 27 m. There are 6.5 km^3 of water accumulates in the lake (Philonets 1981).

The main components of the incoming part of the water balance of the Lake are a surface tributary, groundwater, and atmospheric precipitation. Outcoming part is an evaporation and discharge of water from the lake across the Kalzhir river.

The water in the lake is greenish-blue and transparent (4–10 m) and odorless. The annual course of the water temperature is characterized by a constant increase in the temperature of the water. It starts from the moment of complete cleansing from the ice, and then a stable positive monthly mean temperature of 16–7 °C is established in July–August, after which it gradually decreases to 0 °C in early November. The temperature is about 10 °C in the near-bottom layer of water. In summer, there is a pronounced direct stratification with a layer of a temperature jump at a depth of 10 m, and in the winter period is reversed, since in the bottom layer the water is warmer than in the surface layer.

The lake Markakol is located in the coldest part of Kazakhstan (the minimum temperature is −55 °C). It has a stable freeze-up cover. The first ice formations are observed in late October—early November and last for an average of 20 days. The thickness of ice varies from 53 to 116 cm, and it is 60–70 cm at average. The lake is cleared from ice on May 27 (Philonets 1981).

The lake water is ultra-fresh, very soft, and weakly acidic. It refers to the hydrocarbonate class with the calcium group. The content of iodine, fluorine, bromine, boron, zinc, copper, manganese molybdenum, cobalt, and nickel in water is very low.

The muddy gray sands, partly brown are predominate from the deposits accumulated in the lake basin. Silt deposits with a thickness of 1–2 m can be traced in all the bays. Stones and pebbles occur in the coastal strip, mainly along the southern shore.

The Lake is very poor in the hydrobiological relation. This is due to the unique physical and geographical conditions of the area and the impossibility of penetration of the animal world from the Black Ertis (Philonets 1981).

Lake Kemerkol is located in the central part of the Kurshim district, at an altitude of 839.7 m at the height. Its area is 1.72 km^2, the length is 2.3 km, the width is 1.1 km, and the length of the coastline is 6.8 km (Fig. 3.23).

The lake occupies a deep-seated hollow of complex shape in the western spurs of the Kurchim Range, which extends from the northwest to the southeast. The catchment area of the lake is 19 km^2, and it is a mountainous with numerous outcrops of bedrock such as gray granite and sandstone (Philonets 1981). The catchment area is used for grazing livestock.

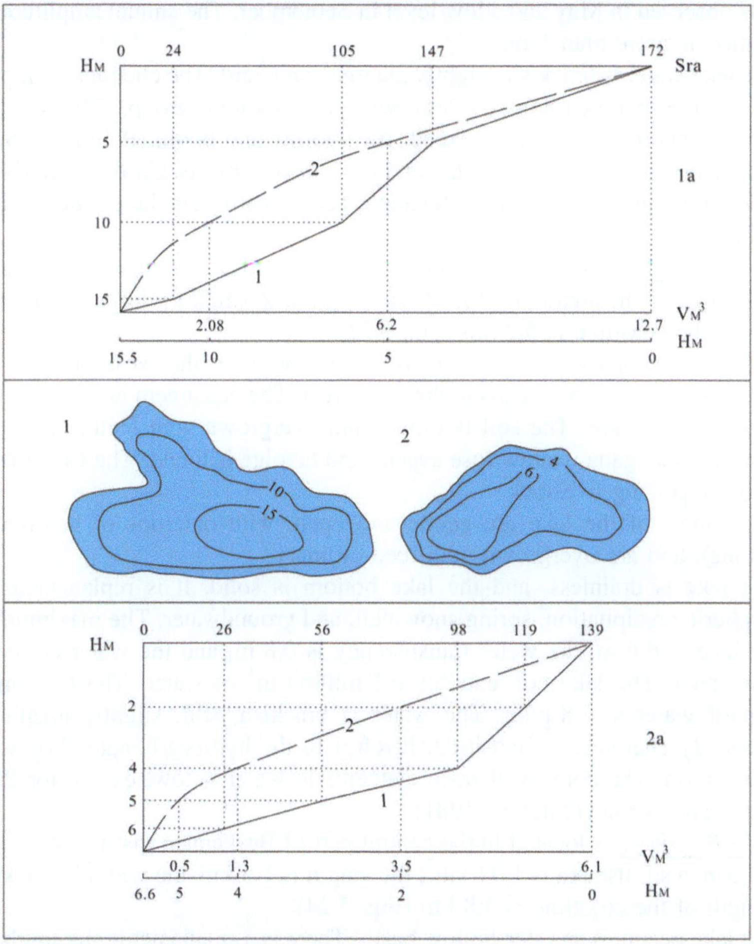

Fig. 3.23 Lake Kemerkol: 1-plan: σ) 1 a) curved lines: bathymetric (1), three-dimensional and coastal areas (2); 2, 2a-the same for Lake Kozenkol

There are brown soils where feather grass, fescue, white wormwood, licorice, caragana bushes and yellow desert acacia, and vines and birch grow.

The northern and southern shores of the lake are high and steep; sometimes the cliffs reach 50 m. The eastern and western shores are gently sloping. The eastern shore is enlivened by several springs with freshwater. The coast of the lake was overgrown with caragana and yellow desert acacia. The lake bottom is sandy, solid, and in the eastern part is partially silted. The lake sediment is in gray color with a thickness of 0.3 m. The coastal zone is covered with reed and pondy.

The Lake is drainless. It replenishes due to precipitation and groundwater. Its maximum depth is 15.5 m, the transparency of water is 3 m, and the color is greenish-yellow. The lake accumulates 12.7 million m^3 of water. A high water level is observed in May and a low level in September. The annual amplitude of the oscillation is more than 1 m.

The lake water is brackish, slightly alkaline, and hard. The chemical composition refers to the hydrocarbonate class with the sodium group. The content of microelements in water is low, except for fluorine and boron (Philonets 1981).

Ducks and gulls nest on the lake, and the muskrat dwells. Carp, roach, dace, and gudgeon are found in the lake. Fish caught occasionally. The lake water is used for watering cattle.

Lake Kozenkol is located in the central part of the Kurchim district, at an altitude of 369.3 m a.s.l. Its area is 1.39 km^2, the length is 2 km, the width is 1 km, and the length of the coastline is 5.2 km (Fig. 3.23).

The lake occupies a shallow intermontane basin in the western spurs of the Kurchim Range, which extends to the northwest. The catchment area is 45 km^2 and is a stony hilly plain. The soil is brown and overgrown with feather grass, black wormwood, caragana, yellow rose acacia, and tavolga is found. The catchment area is used for grazing livestock.

The shores of the lake are gentle and open, with outcrops of bedrock (gray sandstone), and are overgrown with reed ordinary.

The lake is drainless, and the lake bottom is solid. It is replenished due to atmospheric precipitation, spring snowmelt, and groundwater. The maximum depth of the lake is 6.6 m, the water transparency is 0.8 m, and the water color is yellowish green. The lake concentrates 6.1 million m^3 of water. The total mineralization of water is 3.8 g/kg. The water is brackish, stiff, slightly alkaline, and odorless. By chemical composition, it refers to the hydrocarbonate class with the sodium group. The content of trace elements in water is low, except for fluorine, bromine, and boron (Philonets 1981).

Lake Bol'shoye is located in the central part of Beskaragai district, at an altitude of 211.3 m a.s.l. Its area is 1.02 km^2, the length is 1.6 km, the width is 1.1 km, and the length of the coastline is 4.8 km (Fig. 3.24).

The lake occupies an inter-hollow basin. There is a small islet in the southeastern part of it. The bottom of the lake near the Kanonerka village is sandy and solid, and the rest of the banks are hot, overgrown with water nettles and silted. The lake sediment is in gray color, and its thickness is 0.2–0.5 m (Philonets 1981).

The banks are sandy loamy, and the southwest part is marshy. There are abundant outcrops of groundwater (more than 10 springs) along the coast.

Astrip of 20–30 m stretches of vines, aspens, and viburnum are distributed along the rest of the banks and a pine grows along the sandy beaches. The coast of the lake is overgrown with fescue, white wormwood, and chiem.

The area of the catchment is hilly, and its larger part is plowed and used for sowing cereals and vegetable gardens. The lake is replenished due to groundwater and spring snowmelt. A high water level is observed in April and a low level in October. The amplitude of the oscillation reaches 0.7 m. The water level is almost constant. Transparency of water is 1 m, and water color is yellowish. The greatest depth is 3.1 m. There are 1.9 million m^3 of water in the lake (Philonets 1981).

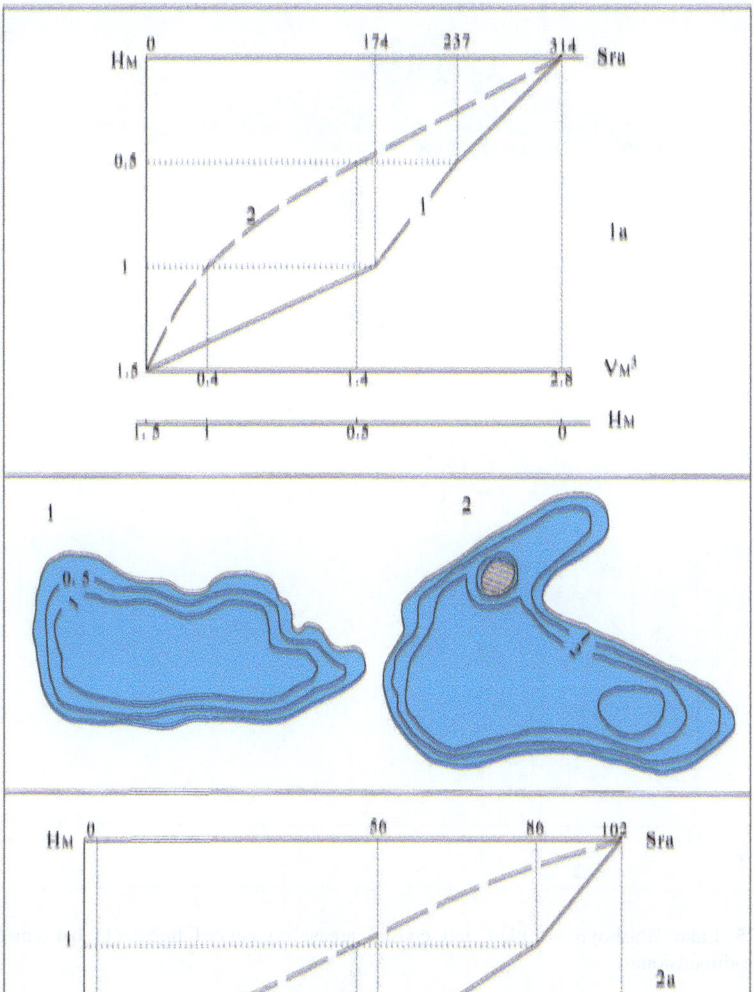

Fig. 3.24 Lake Ushkol II: 1-plan: σ) 1 a) curved lines: bathymetric (1), three-dimensional and coastal areas (2); 2, 2a-the same for Lake Bolshoye

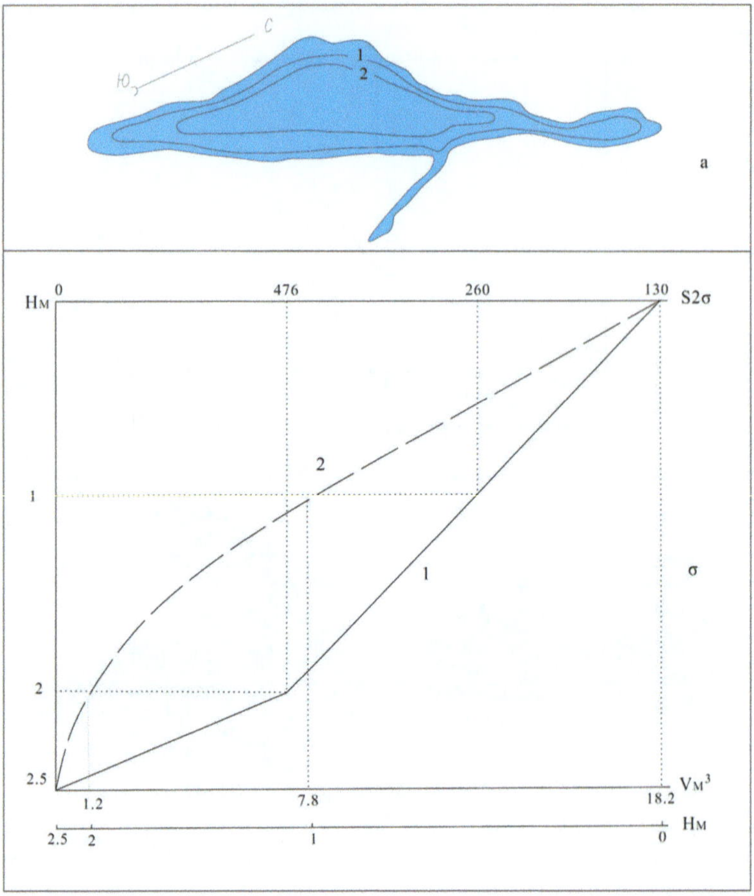

Fig. 3.25 Lake Bolshoye: a) plan and coastal areas; σ) curved lines: (1) bathymetric and (2) three-dimensional

The lake water is ultra-fresh, soft, and neutral. The chemical composition refers to the hydrocarbonate class with the calcium group. The content of microelements is low. The lake is inhabited by carp, tench, pike, perch, chebak, and ide.

Lake Bol'shoye is located in the southeastern part of Borodulikha district, at an altitude of 308.8 m a.s.l. Its area is 12.3 km^2, the length is11.8 km, the width is 2.4 km, and the length of the coastline is 27 km (Fig. 3.25).

The lake occupies a shallow, but long basin and stretches to the northeast. The lake is slightly silty, and the lake sediment is in gray color. The ordinary reed in individual spots grows along the banks of the lake.

The banks are shallow, sandy and sometimes are marshy. A pine forest adjoins the lake in the west, north, and northeast.

The catchment area is 310 km^2, and it is hilly and mostly forested. The coast is covered with fescue, chi, white bush, halophytes, and meadow vegetation. The area is used for grazing livestock.

The lake is replenished with thawed and groundwater, as well as atmospheric precipitation. From the south into the lake flows nameless stream, usually dry in the dry season. Transparency of water is 0.1–0.2 m, and the water color is yellowish-brown. The greatest depth is 2.5 m. It contains about 18 million m^3 of water. A high water level is observed in April and a low level in October. The average annual amplitude of the oscillation reaches 0.6 m, the long-term 1–1.5 m. In 1974, the water level fell sharply, and the area of the lake was reduced by half (Philonets 1981).

The lake water is brackish, slightly alkaline, and hard. According to the chemical composition, it refers to the chloride class with the sodium group.

Lake Balyktykol is located in the northwest of Zhanasemey district, at an altitude of 217.5 m a.s.l. Its area is 08 km^2, the length is 5.8 km, the width is 1.5 km, and the length of the coastline is 15.8 km (Fig. 3.26).

The lake is located in the inter-hollow basin in the floodplain of the Shagan river. Depth from the shore is growing rapidly. The bottom is wavy and sometimes muddy (silt of 0.3 m thick).

The banks are open, gently sloping, and loamy. The southern shore is steep (1.5 m), and the cliffs reach 6 m in the northwest. The coast is overgrown with fescue, white wormwood, stamping grass, and chi.

The catchment area is 94 km^2, and it is hilly and mostly plowed; the rest part is used for pasturing cattle.

The lake is replenished due to meltwater and Shagan river runoff. Transparency of water is 0.5 m; water color is greenish-yellow. The maximum depth of the lake is 6.8 m; it contains more than 24 million m^3 of water. Low water level is observed in October and high one in April (Philonets 1981).

The lake water is brackish, neutral, and very hard. The chemical composition refers to the chlorine class with the sodium group. The content of microelements in water is low. In the spring, water from the lake is widely used for watering cattle. There are sazan, crucian carp, perch, chebak, pike, minnow, and tench found in the lake.

Lake Alzhan (Kyzyl-Chilik) is located in the northeastern part of Zhanasemey district, at an altitude of 254.9 m a.s.l. Its area is 2.9 km^2, the length is 2.3 km, the width is 1.2 km, and the length of the coastline is 13.6 km (Fig. 3.27).

The lake occupies an equalized basin near the northwestern slopes of the Delbegatei Mountain. The basin is in a bizarre shape, and the lake bottom is flat, partly silted. The lake sediment is in gray color with inter-layers of black one. Along the banks, an ordinary cane and rhodia grow in a daisy stripe at 30–50 m.

The banks are shallow and loamy. There is a small and plump solonchak to the north and outcrops of bedrock—gray sandstone near the southeast. The northern, southern, and western coasts are covered with tamarisk bushes, caragans, in some places hips, abundantly overgrown with fescue–feather grass and wormwood vegetation.

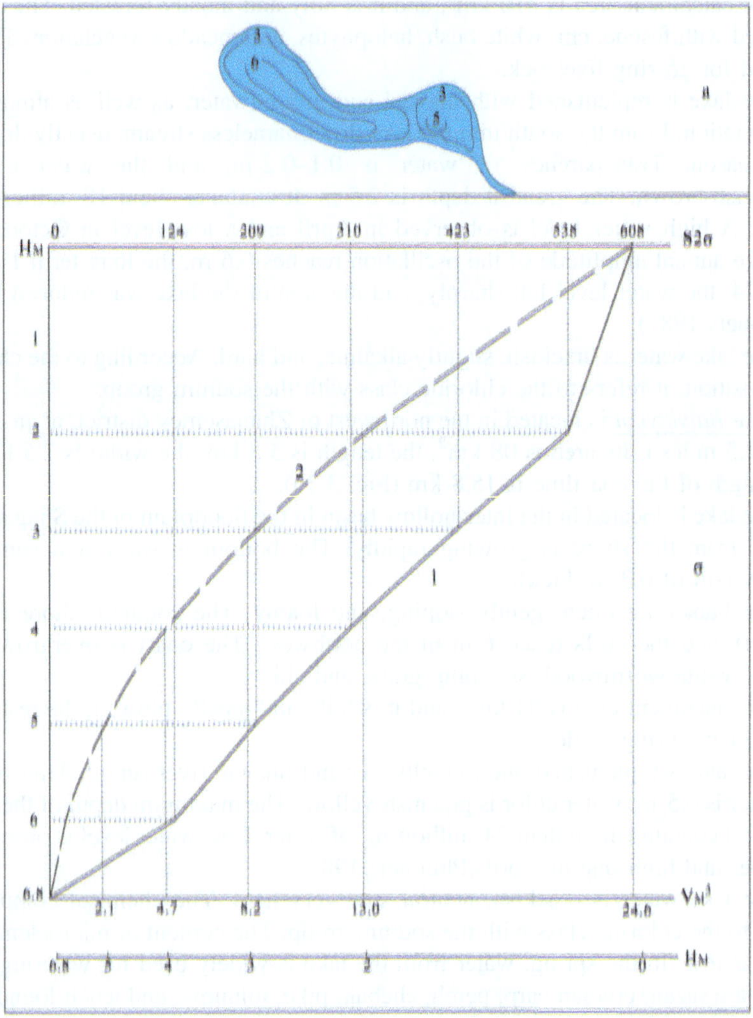

Fig. 3.26 Lake Balyktykol: a) plan and coastal areas; σ) curved lines: (1) bathymetric and (2) three-dimensional

The catchment area is 301 km^2, crossed, mostly plowed, the rest part is overgrown with fescue–feather grass, wormwood, and saltwort vegetation. There are bushes of caragan, dogrose, and tamarisk. It is used for grazing livestock.

The lake replenishes with meltwater. A high level is observed in April and a low level in October. The fluctuation amplitude of the average annual level is 0.6 m. The water transparency is 0.8 m; the water color is greenish-yellow. The greatest depth is 3.7 m, and more than 7 million m^3 of water is concentrated in the lake.

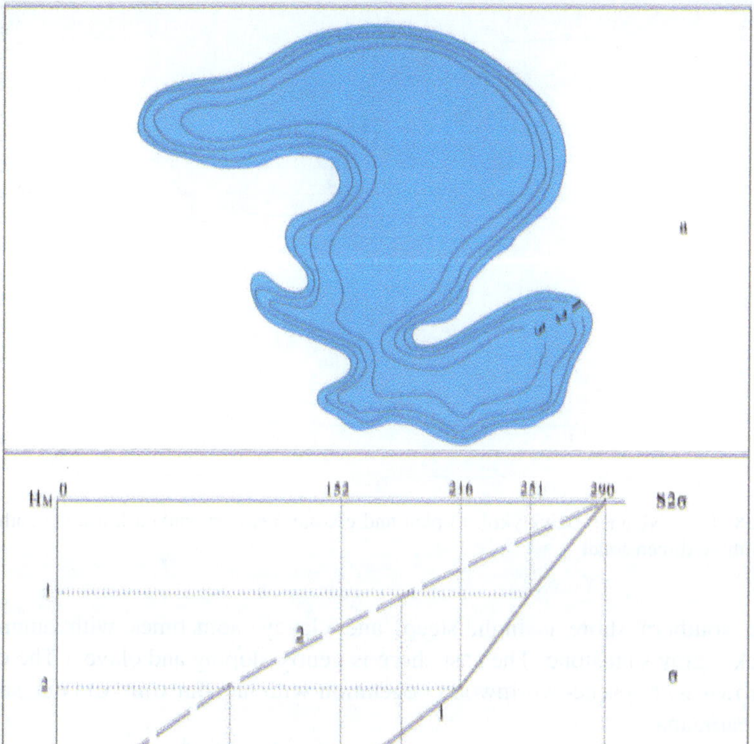

Fig. 3.27 Lake Alzhan: a) plan and coastal areas; σ) curved lines: (1) bathymetric and (2) three-dimensional

The lake water is brackish, neutral, and very hard. According to the chemical composition, it belongs to the chloride class with the sodium group.

Lake Nizhnee Balyktykol is located in the south of the Char district, at an altitude of 501.5 m a.s.l. Its area is 2.96 km^2, the length is 3 km, the width is 1.3 km, and the length of the coastline is 9.4 km (Fig. 3.28).

The lake occupies the inter-mountain hollow in the flood plain of the middle reaches of Zharma, the river. The bottom of the basin is wavy with an insignificant inclination to the west. The basin is silty along the coast. The lake sediment is in gray color with a thickness of up to 0.4 m. On the banks, an ordinary reed grows along the banks with an intermittent stripe of 30–100 m (Philonets 1981).

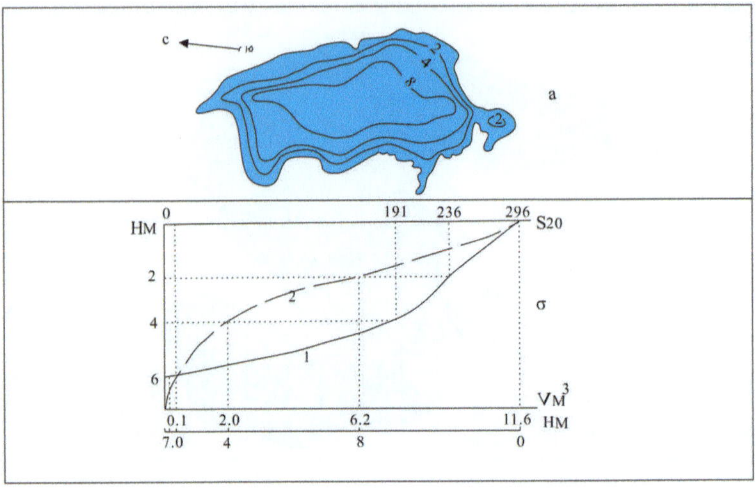

Fig. 3.28 Lake Nizhnee Balyktykol: a) plan and coastal areas; σ) curved lines: (1) bathymetric and (2) three-dimensional

The southern shore is high, steep, and clayey, sometimes with outcrops of bedrock—gray sandstone. The east shore is gently sloping and clayey. The coast is overgrown with fescue–wormwood vegetation with an admixture of chia, saltwort, shrub caragana.

The catchment area is crossed and partially plowed. The rest part is used for pasturing cattle.

The lake is replenished due to the flow of the Zharma river and spring mèlting snow. A high water level is observed in April and a low level in October. The annual amplitude of the oscillation reaches 0.8 m. The transparency of the water is 1.7 m, and the color is brownish-yellow. The maximum depth of the lake is 7.5 m. The lake accumulates more than 11 million m^3 of water (Philonets 1981).

The lake water is slightly salty, neutral, and hard. The chemical composition refers to the sulfate class with the sodium group. The water from the lake is used for watering cattle. There are pike, carp, tench, seroglazka, crucian carp in the lake. There are several species of ducks, seagulls, waders, and other birds, and there is a muskrat in the lake.

Lake Verkhneye Balyktykol is located in the southern part of the Char district, at an altitude of 505 m a.s.l. Its area is 2.41 km^2, the length is 2.2 km, the width is 1.8 km, and a coastline length is 6.4 km (Fig. 3.29).

The lake occupies the inter-mountain hollow in the flood plain of the middle reaches of the Zharma river. The basin is flat and shallow, with a slight slope to the west, and partially silted. The lake sediment is in gray color, and the thickness of it in places reaches 0.3 m.

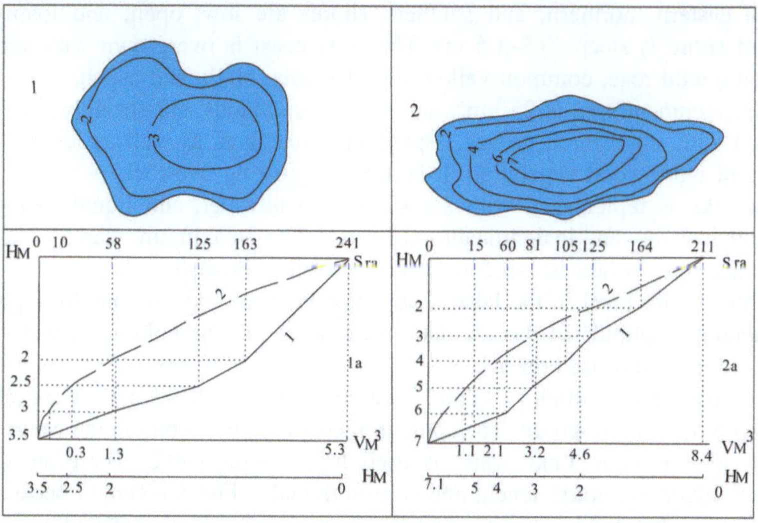

Fig. 3.29 Lake Verkhneye Balyktykol: 1-plan: σ) 1 a) curved lines: bathymetric (1), three-dimensional and coastal areas (2); 2, 2a-the same for Lake Karakol

The banks are shallow, open, and loamy. The coast is overgrown with fescue, feather grass, white wormwood; in places there are chi, saltworts, and meadow vegetation (Philonets 1981).

The catchment area is crossed and partially plowed up. It is overgrown with fescue–feather grass and wormwood vegetation with an admixture of saltworts, chia, shrub caragana. It is used mainly for pasturing cattle.

The lake is replenished due to the flow of the Zharma river and spring snowmelt. The annual amplitude of the oscillation reaches 1 m; the long-term amplitude is about 2 m. The transparency of the water is 1.5 m; the color is brownish yellow. The maximum depth of the lake is 3.5 m. More than 5 million m^3 of water is concentrated in the lake (Philonets 1981).

The lake water is slightly salty, alkaline, and very hard. The chemical composition refers to the sulfate class with the sodium group. The lake water is widely used for watering cattle.

Crucian carp, tench, carp, pike, and quail are found in the lake. Fish are caught occasionally. There are several species of ducks, sandpipers, seagulls, and muskrat in the lake.

<u>*Lake Karakol*</u> is located in the southwest of Zharma region, at an altitude of 754.8 m a.s.l. Its area is 2.11 km^2, the length is 2.7 km, the width is 1.1 km, and the length of the coastline is 6.4 km (Fig. 3.29).

The lake occupies a narrow intermontane basin, which stretched along the northeast foot of the rocky mountains of Karakoltas (1161.7 m). The hollow is of a grooved form, the lake bottom with an insignificant inclination to the northeast. The lake sediment is in gray color, and its thickness in some places reaches 0.3 m (Philonets 1981).

The eastern, northern, and southern shores are low, open, and loamy. The western shore is steep (0.5–1.5 m). The west coast is overgrown with shrubs of caragana, wild rose, common yellow desert acacia, birch, and aspen.

The catchment area is 25 km^2, and it is mountainous and strongly intersected, covered with small birch groves, aspen, shrub caragan, as well as fescue–feather grass and wormwood vegetation. It is used for grazing livestock.

The lake is replenished with snow and groundwater, and liquid atmospheric precipitation in an insignificant quantity. The maximum depth is 7.1 m. Transparency of water is 1–1.5 m; water color is yellowish.

A high water level in the lake is observed in April and low one in September. The annual amplitude of the oscillation reaches 1 m. The volume of water in the lake is more than 8 million m^3.

The lake water is slightly saline, hard, and alkaline. It belongs to the chloride class with the sodium group. The content of copper, zinc, cobalt in the water is low, and fluorine is high. Lake water is used for watering cattle. There are chebak, crucian (silver and gold), tench, and carp in the lake. Fish are caught occasionally. The lake is inhabited by ducks, seagulls, waders, and other waterfowl and wading birds, as well as a muskrat.

Lake Koldar is located in the southwestern part of the Ayagoz district, at an altitude of 369 m a.s.l. Its area is 6.51 km^2, the length is 6 km, the width is 1.6 km, and the length of the coastline is 19.2 km (Fig. 3.30).

The lake occupies the lowest part of the vast intermontane basin, which extends from the southwest to the northeast and has a bizarre shape. Its bottom is flat and

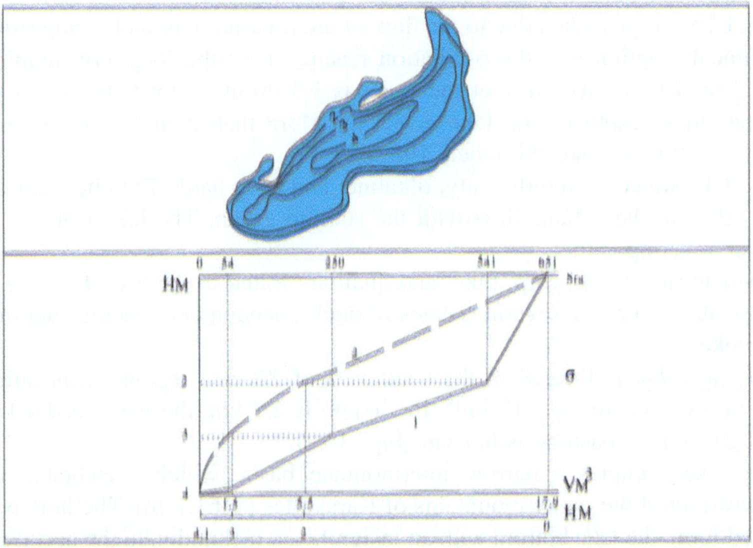

Fig. 3.30 Lake Koldar: a) plan and coastal areas; σ) curved lines: (1) bathymetric and (2) three-dimensional

sandy. The lake sediment is in gray color, and its thickness in the coastal part in some places reaches 0.3 m. The entire water area of the lake is abundantly overgrown with rdest.

The banks are low, open, and loamy. A plump solonchak adjoins them in the north, northeast, and west. The southern shore is precipitous, and in some places the cliffs reach 1–5 m. The shore of the lake is overgrown with hodgepounds, tamarisk bushes, white wormwood, and thinned cane. There is desert poplar in the south-eastern part of the lake.

The catchment area is 1787 km^2, and it is hilly with the presence of plump solonchaks. It is overgrown with feather grass and wormwood and used for grazing livestock.

The lake feeds on snowmelt waters and liquid precipitation to the water area. Transparency of water is 0.4 m, and the water color is yellowish. The water high level is observed in April and the low level in October. The amplitude of the oscillation reaches 0.8 m; the long-term is more than 2 m. The maximum depth of the lake is 4.1 m. It contains about 18 million m^3 of water.

The lake water is brackish, neutral, and very hard. It belongs to the sulfate class with the sodium group (Philonets 1981).

Lakes of the Southern Kazakhstan. There are 10,842 lakes in Almaty region with a total area of 1598.1 km^2 (Fig. 3.31), including 10,775 lakes with size less than 1 km^2 and take about 5% of the total area of the region. There are 67 lakes with a size of more than 1 km^2 and take 2% of the total water surface (Fig. 3.31). In addition, there are 571 plexus lakes and water reservoirs in the region.

Lake Koshkarkol (Uialy) is located northeast of Usharal city (Alakol district) in the Alakol depression, between the Sasykkol and Alakol lakes, which are separated by isthmuses 4.5 and 5.5 km wide. The lake has an elliptical shape, and it stretched from north to south. It is a flowing lake (Fig. 3.32).

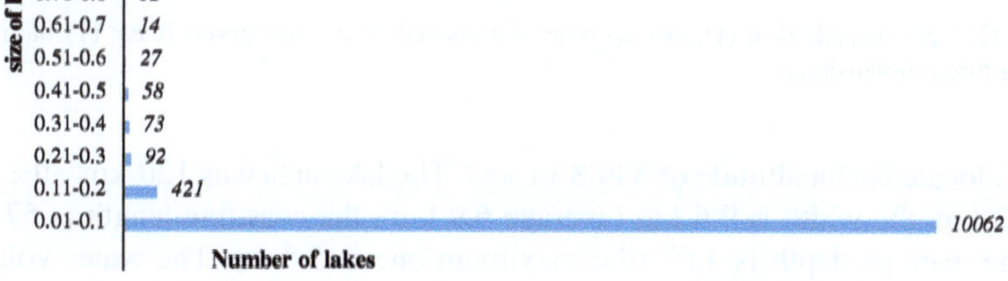

Fig. 3.31 Distribution of lakes in Almaty region by lake size. Notice: Balkash and Alakol lakes, plexus lakes, ponds, and water reservoirs are not included to the list. Plexus lakes are 571 with a total area of 128 km^2

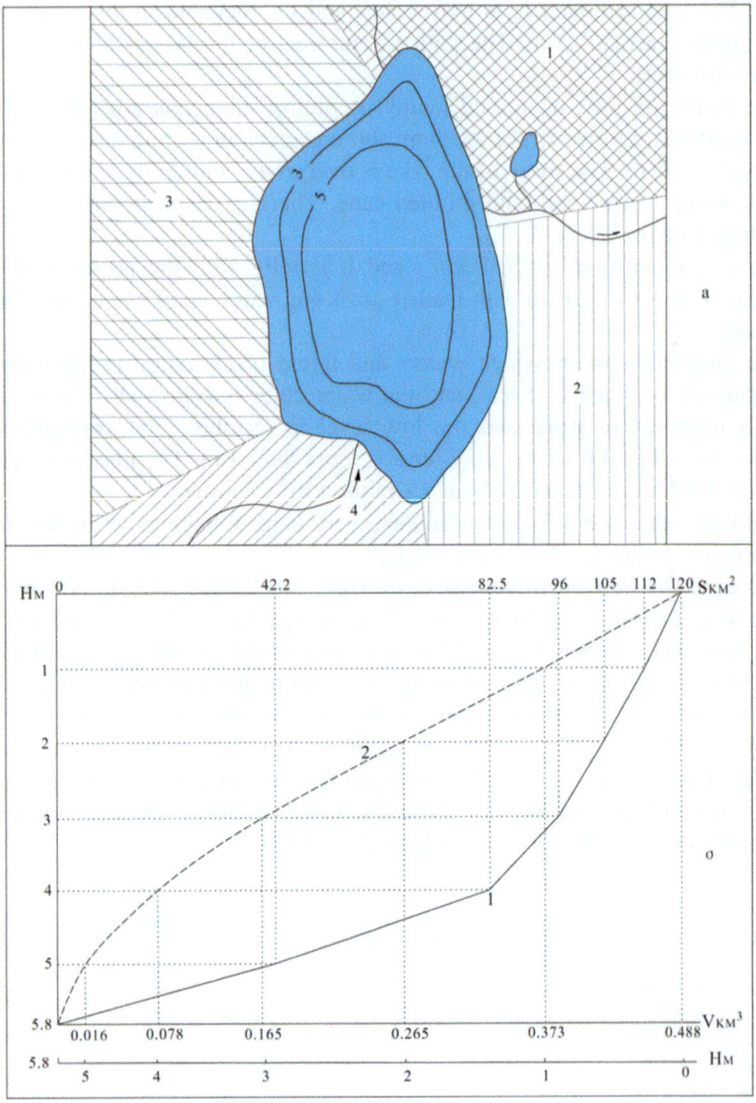

Fig. 3.32 Lake Koshkarkol (Uialy): a) plan and coastal areas; σ) curved lines: (1) bathymetric and (2) three-dimensional

It is located at an altitude of 349.8 m a.s.l. The lake area was 120 km^2, the length is 18.3 km, the width is 9.6 km (average 6.9 km), the coastline length is 57.3 km, and the average depth is 4.07 (the maximum one is 5.8 m). The water volume is 488.3 million m^3 (Table 3.25). Its catchment area with the Sasykkol lake is 20,800 km^2 (Philonets 1981).

Table 3.25 Water resources of the large lakes in the Southern Kazakhstan (Philonets 1981)

Lake name	Lake area (km^2)	Max depth (m)	Water volume (mln m^3)	Length of coastline (km)
Almaty region				
Balkash lake	18,200	27	106	2383
Alakol lake	2650	54	58.56 billion	384
Sasykkol lake	736	4.7	2.43 billion	182
Koshkarkol (Uialy)	120	5.8	488.3	57.3
Zhalanashkol lake	37.5	3.25	104	23.8
Bol'shoi Semizkol	10.91	3.9	31.7	
Bugut (bol'shoye)	4	3.9	10	
Bol'shoye Almaty lake	0.44	40	13.5	
Beloye	11.8	3.5	17.7	
Sinee	6.86	3.5	14.1	
Nauryzbayevskoye (bol'shoye)	5.06	6	20.1	
Ushkol	9.49	6	37	11.2
Ushkol	3.14	1.5	28	7.8
Altay	4.13	3.5	5	17.2
Zhangirlikol	1.6	2.5	1.8	10
Zhambyl region				
Akkol lake	52	5	166	40
Biilekol lake	86.5	4.5	182	53
Ashykol lake	88.5	6	314	56
Kyzylauyt lake	2.6	–	3.6	11
South Kazakhstan region				
Kyzylkol	16.2	8.5	89	18.6
Kaldykol lake	13.3	2.5	6	–
Shevykbeldi lake	10.49	4.5	5.2	–
Kyzylorda region				
Kamyslybas lake	178	9.5	955	115.6

The depth of the lake increases gradually, and it is in the southwestern part and reaches a maximum of 5.8 m. The lake bottom is flat. Bottom deposits are very monotonous. There are three main biotopes: Sand, sandy, and gray silt are distinguished depending on the nature of the bottom sediments. The gray silt biotope occupies the most lowered parts of the lake basin that is 62% of the water area. The sandy biotope begins from the water's edge and extends to a depth of 1.5–2 m. Its area is 20–30% of the water area. The sandy silty biotope is transitional.

The shoreline of the lake is little rugged. The shores with the exception of the eastern (from the village of Uialy to the village of Alakol) are low and marshy with separate small lakes overgrown with reeds. The submerged reed vegetation, drowning out the waves, protects the coast from erosion and contributes to sedimentation in the coastal zone of sediments, as a result of which the banks are gradually increasing. There are four coastal regions (Kazanskaya 1966):

1. The Northern and northeastern coastal region are accumulative and drained. It is formed under the influence of the Zhenishkesu river flowing into the lake and the Uialy river that flows from it. The basin of the lake is inclined to the southeast. The whole area is overgrown with reeds.
2. The eastern coastal region is accumulative with the coastal shaft, which is a watershed between the lakes of Uialy and Alakol. The relief is undulating that intersected by old riverbeds of the Uialy river in the direction of the lake. It is composed of sand and pebble material.
3. The Western coastal region is also accumulative and phytogenic, which is an isthmus between the lakes of Uialy and Sasykkol. In the northern part, eastern slope of the Araltobe Mountain descends to the lake that representing a deluvial proluvial train, gradually turning into a low accumulative lakeside plain. The rest part of the shore is a marshy lowland. Its surface is intersected by a weakly expressed hollow, which in the high water years serves as a drainage place from the Sasykkol lake.
4. The Southwestern coastal region is an accumulative and deltaic. Influence on its structure is provided by the Sukhaya river (the right tributary of the Tentek river). It formed a wedge-shaped delta from alluvial outcrops, with an area of more than 13 km^2. The surface of the delta is partly swamped (Philonets 1981).

Hydrography and water regime. The lake does not have own tributaries and is replenished mainly due to the Sassykkol lake, by filtration through the coastal shaft and surface overflow, as well as along the Zhenishkesu river. In the period of low standing of the Sassykkol lake level, the surface tributary ceases. A small seasonal inflow (less than 0.5 m^3/s) in the spring–summer period is provided by the Sukhaya river by draining groundwater in its lower part. The flow from the lake is carried through the Uialy into the Urzhar river and further into the Alakol lake.

There are 488.3 million m^3 of water accumulated in the Uialy lake. Spring–summer heating of water in the lake begins in March and lasts until the second half of July. In the period of the exposed lake, the water temperature varies from 0.2 to 3.0 °C, and during the cleansing period from ice it ranges from 1–3 to 7–10 °C. The highest water temperature is recorded at the end of July, and at the shore is 28–29 °C (Philonets 1981).

The eastern, southeastern, and northwestern winds are dominated over the water area of the lake. The average annual wind speed is 2–4 m/s, and the maximum is 40–45 m/s in the autumn–winter period. The wave height reaches 1.3 m, and the maximum is 2.1 m (Philonets 1981).

The first ice formations appear at the end of November, and the freezing process lasts 7–11 days. The largest ice thickness (more than 60–62 cm) reaches in the second half of February. In winter, the height of snow on ice reaches 5–60 cm. At the end of February and beginning of March, the thickness of the ice cover begins to decrease. The beginning of April is the period of complete cleansing from ice.

Water balance and levels. For the lake, the arrival is equal to the flow, approximately. The average long-term surface inflow in the Uialy lake is about 165 million m^3, underground inflow was 20 million m^3, and atmospheric precipitation on the mirror was 33 million m^3. The average long-term evaporation from the lake was 119 million m^3, and the surface and underground runoff was 99 million m^3 of water (Philonets 1981).

Rise in water level at the Uialy lake begins in March. The annual maximum is observed in April–June. The summer–spring decline of the levels continues until October–November. The annual amplitude of the levels varies from 41 to 90 cm, and average was 68 cm (Philonets 1981). Running–surging fluctuations in the water level reach 10 cm. The secular amplitude is 6–8 m.

Lake hydrochemistry. Mineralization of the Uialy lake water ranges from 0.85 to 1.16 g/kg. The annual mineralization over a long period was 0.94–1.28 g/kg, and the average was 1.1 g/kg. According to the chemical composition, the water of the Uialy lake belongs to the hydrocarbon class. Sodium prevails among the cations (Philonets 1981).

The lake water transparency is 0.2–1 m. The pH value is 7.3–8.6. The content of dissolved phosphorus during the year varies from 0 to 0.061 mg/l, nitrate nitrogen from 0 to 0.5 mg/l, iron from 0.04 to 0.96, and silicon from 0 to 10.2 mg/l. Dissolved oxygen was 66–69% saturation, and it is less in August. Permanganate oxidizability is 9–29 mg O_2/l. The lake water is hard (5–7 meq/l), and the aggressiveness was 1.36–10.12 mg CO_2/l (Philonets 1981).

There is a muskrat on the west and southwest coasts of the lake. The lake is inhabited by various species of ducks, geese, cormorants, pelicans, swans, seagulls, and waders.

Lake Zhalanashkol ("Naked Lake") is located in the southeastern part of Kazakhstan, at the beginning of the Dzungar Gate, at an altitude of 372.5 m a.s.l. The lake is oval in shape and extends to the northwest. The area of the lake's water surface was 37.5 km^2, the length of the lake is 9 km, the width is 5.8 km, the length of the coastline is 23.8 km, the average depth is 2.6 m (the maximum is 3.25 m), and the water volume was 104 million m^3 (Fig. 3.33) (Philonets 1981).

The basin of the lake has a slight deviation to the north, where it is supported by the shore shaft "Taskuly". The maximum depth is recorded in the northwestern part of the lake. The relief of the lake bottom is weakly toed. The bottom is composed of gravel–pebbles; along the east coast, it is sand–gravel in places. The lake sediment (silt) is in gray and black color in some places and oily with remnants of vegetation.

The coastline is lined. The beaches are mostly low and flooded, sometimes swampy, and overgrown with reeds. There are four coastal regions of the lake (Kazanskaya 1970).

(1) Western region is a flat and abrasion-accumulative and composed of loose pebble–gravel sediments. Due to lake erosion of loose deposits, a coastal step of 1–3 m was formed.

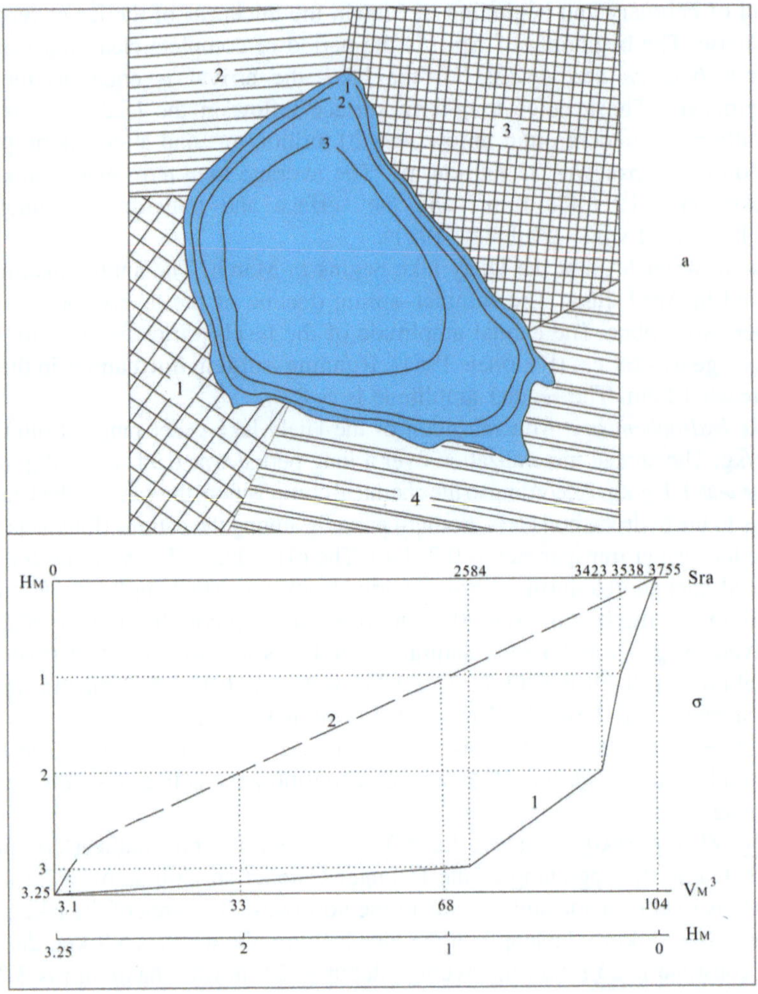

Fig. 3.33 Lake Zhalanashkol: a) plan and coastal areas; σ) curved lines: (1) bathymetric and (2) three-dimensional

(2) Northwestern region is also accumulative with a coastal shaft of 1–3 m high and up to 100 m wide. The underwater coastal slope of the shaft slopes gently toward the lake basin.
(3) Eastern region is an abrasion-accumulative, which represents a hollow-inclined rocky foothill plain. Rubbly-pebbly deposits in the coastal zone are covered with a layer of loamy yellowish rocks, in which a ledge with a height of up to 5 m was formed. Stunted trees: Karagach, saxaul, and shrubs grow on the platform ledge. The underwater coastal slope is shallow and covered with gravel–pebble deposits. There are abundant outcrops of groundwater in the region.

(4) Southern region is also accumulative, low, and swampy with reed vegetation. There are abundant outcrops of groundwater in the region. Here is the narrowest part of the mountain pass of the Dzungar Gate (12–14 km). From the west, the coastal region is cut off by a ledge to the side of the lake, up to 3 m high, which forms a floodplain terrace.

Hydrography and water regime. Lake Zhalanashkol has no river tributaries, replenished by underground flow, and a small slope drain of thawed and rainwater. The average precipitation falling on the water mirror is 288 mm. Evaporation from the water surface (March–November) ranges from 1225 to 1334 mm, an average of 1313 mm. The lake concentrates 104 million m^3 of water is in (Philonets 1981).

Spring–summer heating of water in the lake begins in late February—early March. The water temperature during the cleansing period varies from 1–3 to 7–10 °C. The highest water temperature is recorded in late June—early July—and reaches 24–26 °C in the lake coast.

The east, southeast, and northwest winds predominate over the water area. The average annual wind speed is 7 m/s, and the maximum (60–70 m/s) is observed in the autumn–winter period. The height of the wave in the lake is 1.5–2 m (Philonets 1981).

The lake freezes for 1–2 days in early November. The ice thickness reaches 90–100 cm in February. The opening and purification of the lake from ice take place at the end of March.

After a low autumn water-level period, the water level in the lake rises until February to March. When the lake is filled, the water flows over the bank of the shore to the northwest coast into the Zhamanotkel canal, with the gradual development of the temporary watercourse and an increase in the size of the runoff, which results in the generation of part of the water reserves accumulated over the winter. The annual amplitude is 42–89 cm (Philonets 1981).

Hydrochemistry. The mineralization of water in the lake varies in different years from 1.2 to 5 g/kg. The chemical composition of water is predominantly sulfate-hydrocarbonate-sodium; the pH varies from 7.6 to 8.6 (the highest in January–February). The transparency of the water increases with depth (from 0.2 to 1.2 m). The water contains (mg/l): nitrate nitrogen 0–0.75, silicon 8.4–12.0, iron 0–0.96, phosphorus 0.009–0.05. Dissolved oxygen is 79–80% saturation; permanganate oxidability is 19–20 mg O_2/l.

The lake water is stiff and very hard (5–9 meq/l). There are fish such as perch and sazan in the lake.

There is a common reed along the eastern and southern shores. It occupies a strip with a width of several to 100 m. The total area is 30 ha. There are many ducks, gulls, geese, waders, and other waterfowl and wading birds on the lake.

The Lake Ushkol I (Urtakol) is located in the eastern part of the Karatal region, at an altitude of 359.9 m a.s.l. Its area is 9.49 km^2, the length 4.2 km, the width is 3 km, and the shoreline length is 11.2 km (Fig. 3.34).

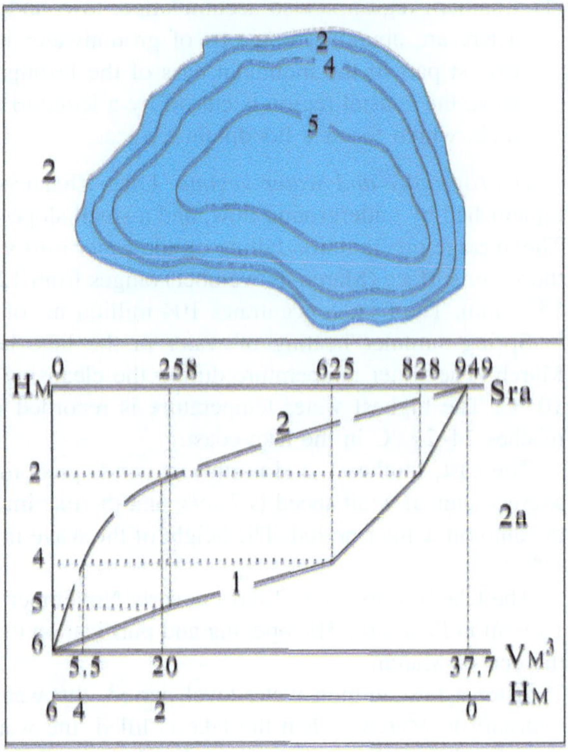

Fig. 3.34 Lake Ushkol I (Urtakol): a) plan and coastal areas; σ) curved lines: (1) bathymetric and (2) three-dimensional

The lake occupies an inter-grown sandy depression near the northeast slopes of the Ushkara mountains. The catchment area is 2500 km^2, and it is hilly and used for grazing cattle.

The lake banks are shallow, open, and sandy. The coast was overgrown with depressed thinned reeds and wormwood, as well as there are saxaul (Philonets 1981).

The lake has a rounded shape. The depth increases gradually and in the center reaches 6 m. Lake silt is in black color with plant residues and the smell of hydrogen sulfide. The thickness of it is 0.2 m. Along the banks, a strip of 20–50 m stretches thickets of reed ordinary with an admixture of lake cane. Its total area is 15 ha (Philonets 1981).

The lake is replenished by meltwater, and also water along the canal from the Aksu river. Transparency of water is 0.7 m, the water color is greenish-yellow, and the maximum depth is 6 m. The lake accumulates 37 million m^3 of water. The highest water level is observed in late March—early April—and the lowest level in October. The long-term amplitude of water level fluctuations in the lake reaches 1.5–2 m. The lake water is slightly saline, hard, and alkaline. It belongs to the sulfate class with the sodium group (Philonets 1981). The content of trace elements in water is low; iodine and molybdenum are not defined in the lake water (Philonets 1981).

There are fish such as carp, bream, marinka, crucian, and minnow in the lake. In addition, there are many ducks, gulls, waders in the lake.

Lake Ushkol II is located in the eastern part of the Karatal region, at an altitude of 365 m a.s.l. The lake area is 3.14 km^2, the length is 3 km, the width is 1.5 km, and the shoreline length is 7.8 km.

The lake occupies an inter-mountainous depression near the northeastern slopes of the Ushkara mountains. Its catchment area is 3860 km^2, and it is used for pasturing cattle.

The lake basin is in saucer-shaped. The lake bottom is even and partly silted. The silt thickness in some places reaches 0.5 m and is in black color with plant residues and the smell of hydrogen sulfide. Thickets of ordinary reed with an admixture of lake cane stretches along the coasts a strip of 10–50 m.

The banks are shallow, open, and sandy. The coasts were overgrown with bushes of saxaul, loch, windmills, and desert poplar (Philonets 1981).

The lake is replenished by melting water. Transparency of water is 0.4 m; the water color is yellow. The greatest depth is 1.5 m. The lake concentrates 28 million m^3 of water. The long-term amplitude of water fluctuation reaches 1.5 m. The highest level is observed in April, and the lowest level is in October.

The water is fresh, slightly alkaline, and moderately hard. It refers to the hydrocarbonate class with the sodium group. The content of microelements in water is low. There are fish such as carp, bream, marinka, crucian carp, chibak and gudgeon in the lake. There are many ducks, gulls, waders, and other waterfowl and swamp birds in the lake (Philonets 1981).

Lake Altay is located in the Burlitobe district, northwest of the Zhanalyk village, at an altitude of 465 m a.s.l. Its area is 4.13 km^2, the length is 4.2 km, the width is 2.5 km, and the length of the coastline is 17.2 km (Philonets 1981).

The lake occupies a shallow basin in the inter-barhan depression of the middle reaches of the Baskan river and has the form of a triangle. Its banks are shallow, open, rugged, sandy, and loamy sandy. There are many small islets on the lake. The depth from the coasts is increasing rapidly, reaching to the middle of 3.5 m. The bottom of the coast is sandy silty. Lake silt is in gray color with vegetable residues and the smell of hydrogen sulfide. The basin of the lake is abundantly overgrown with ordinary reed with an admixture of lake cane, rdestom, a water lily. The area of reed thickets is about 150 ha (Philonets 1981).

The catchment area is a hilly terrain with loamy, clayey, and sandy soils and partially plowed under crops, the rest part being used as pastures.

The lake feeds on the waters of the Baskan river, which flows into the lake in the southeast, and in the southwest flows out of it. The lake is partially replenished due to thawed waters. The greatest depth of the lake is 3.5 m; it accumulates 5 million m^3 of water. Transparency of water is 1.5 m; the color is yellow and brownish-yellow. A high water level is observed in April and a low level in October. The annual amplitude of the oscillation is about 1 m (Philonets 1981).

The lake water is ultra-fresh, soft, and neutral. The chemical composition refers to the hydrocarbonate class with the calcium group. The content of microelements in water is low; iodine, boron, and molybdenum are not detected. There are carp, perch, bream, and marinka in the lake. As well as the muskrat was found in the lake. There are many waterfowl and swamp birds: ducks, geese, seagulls, sandpipers, and herons in the lake (Philonets 1981).

Lake Zhangirlykol is located in the southeastern part of the Karatal region, at an altitude of 410 m a.s.l. Its area is 1.6 km^2, its length is 2.5 km, its width is 1.3 km, and the length of the coastline is 10 km (Philonets 1981).

The lake occupies an inter-low depression. The water catchment area is hilly. Rice fields adjoin to it from south.

The basin of the lake is a bizarre form and almost one-third overgrown with ordinary reed. The lake bottom is silted. Lake silt is in black color with the smell of hydrogen sulfide. The thickness of it is 0.2–0.5 m.

The shores of the lake are gentle, open, sandy, and sandy loamy. The coast is covered with fescue–wormwood and feather grass vegetation.

The lake is replenished by spring meltwater and sewage water from rice fields. Transparency of water is 1 m; the water color is brownish-yellow. The average depth is 1.1 m (the largest is 2.5 m). The lake has 1.8 million m^3 of water. The amplitude of the water level fluctuation reaches 2 m (Philonets 1981).

The lake water is slightly saline, alkaline, and moderately hard. It refers to the sulfate class with the sodium group. The content of microelements in water is low.

There are many waterfowl and swamp birds: ducks, seagulls, waders on the lake.

There are 1171 lakes in **Zhambyl region** with a total area of 468 km^2 (Fig. 3.35), including 1133 lakes with size less than 1 km^2 and take about 18.8% of the total area of the region. There are 38 lakes with a size of more than 1 km^2 and takes 81.2% of the total water surface.

Lake Ashykol is located in the south part of the Sarysu district, and northwest of the Akkol village, in the lowest part of the Talas-Chui depression, in an

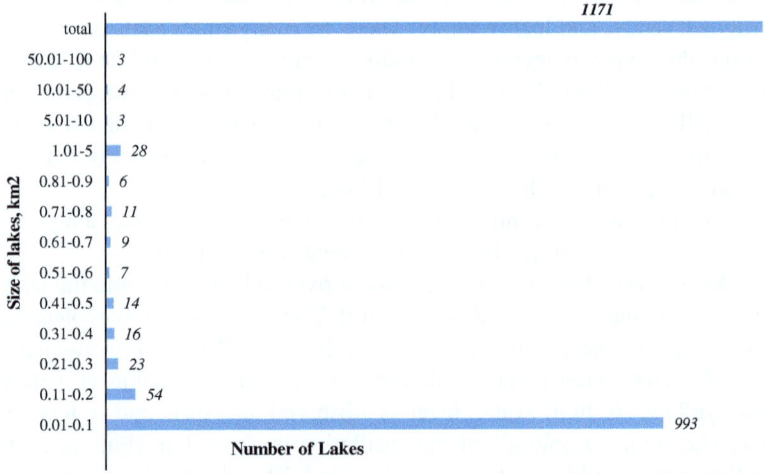

Fig. 3.35 Distribution of lakes in Zhambyl region by lake size. Notice: Balkash lake, ponds, and water reservoirs are not included to the list

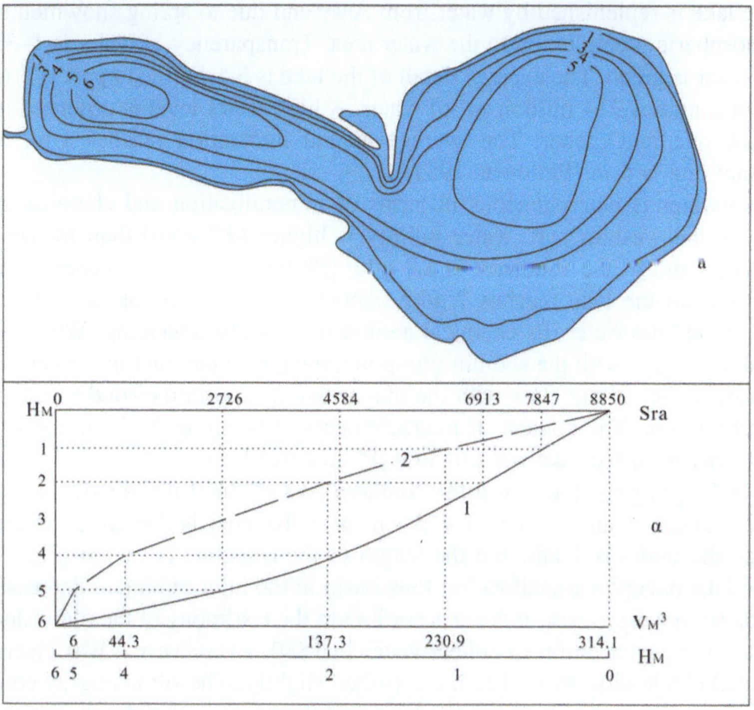

Fig. 3.36 Lake Ashykol: a) plan and coastal areas; σ) curved lines: (1) bathymetric and (2) three-dimensional

inter-lowland depression at an altitude of 379 m a.s.l. It has the shape of a fungus and extends to the west. Two independent reaches with a low water level are formed. The eastern reach is called Ashykol, and the western one is called Tuzkol. The area of the lake is 88.5 km^2, the length is 18.5 km, the width is 10 km, and the length of the coastline is 56 km (Philonets 1981) (Fig. 3.36).

Along the coast the lake is silted. The silt is in gray color, and its thickness in places up to 0.4 m. Along the eastern and southern coasts, a strip of 50–100 m grows reed ordinary in height which is more than 3 m.

The northern shore of the eastern part of the lake is precipitous in 3–4 m high and the western shore is sloping, and composed of red clays, pebbles in some places. The whole coastline is high (4–6 m), with a gradient to the west (Philonets 1981).

The southern coast is low and composed of loam, clay, sandy loam, and pebbles. The western and eastern shores are shallow and loamy; in places the plump solonchaks adjoin them. The eastern coast is cut by the Assy river.

The catchment area is hilly with a general slope from the southeast to the northwest, and overgrown with fescue–feather grass and wormwood vegetation, there are bushy shrubs. It is used as grazing for cattle.

The lake is replenished by water from Assy and due to spring snowmelt, as well as atmospheric precipitation to the water area. Transparency of water is 1–3 m; the water color is green. The average depth of the lake is 3.5 m (the largest is 6 m). The lake concentrates 314 million m^3 of water. A high water level is observed in April and low one in October. The average annual fluctuation reaches 1 m, and the perennial one is 6 m (Philonets 1981).

Lake water is heterogeneous in terms of mineralization and chemical composition. In the western part, water salinity is higher (9.3 g/kg) than in the eastern (6.2 g/kg), and in the southeast is 1.3 g/kg (Philonets 1981). In general, the mineralization on the lake reaches 7 g/kg. About 2 million tons of salts are concentrated in the lake water. Its chemical composition is also changing. Water refers to the chloride class with the sodium group in the western part and in the center of the east part, to the sulfate class with the magnesium group in the southeastern part of the eastern area. The content of microelements of bromine, barium, copper, manganese, cobalt in the lake water is low (Philonets 1981).

Lake Kyzylauyt is located in the southern part of the Talas district, west of the Akkol village, at an altitude of 453.4 m a.s.l. Its area is 2.6 km^2, the length is 4.4 km, the width is 1 km, and the length of the coastline is 11 km (Fig. 3.37).

The lake occupies a shallow but long basin in the inter-mountain depression and extends from west to east. It arose according to the testimony of the old residents, at the site of the excavation to collect water in 1880. Every year it was cleaned and expanded (Philonets 1981). The lake is silted slightly. The silt is in gray color. The sedge grows along the shores of the lake, and the basin is abundantly covered with algae.

The banks are shallow, open, and sometimes marshy. The banks are made up of red granite, loam, and sandy loam. The coasts are overgrown with fescue–feather grass and wormwood vegetation (Philonets 1981).

The catchment area is hilly and crushed with outcrops on the surface of red granite. It is overgrown with fescue–feather grass vegetation. It is used as grazing for cattle.

The lake replenishes with water from the Koktal river and also due to spring snowmelts. The water the color is bluish-green. The average depth of the lake is 1.4 m, and it contains about 3.6 million m^3 of water. The level is unstable: high level in April and low one in October. The average annual fluctuation reaches 0.7 m, perennial 1–2 m (Philonets 1981).

The lake water is weakly mineralized, fresh, and soft, but alkaline. By chemical composition, it refers to the sulfate class with the magnesium group. The content of trace elements in the water is low; fluorine, bromine, barium, copper, manganese, and cobalt are not detected (Philonets 1981).

Lake Akkol is located in the south of the Talas district of the Zhambul region, southwest of the Akkol village, at an altitude of 404 m a.s.l. Its area is 52 km^2, the length is 14 km, the width is 7.6 km, and the length of the coastline is 40 km (Fig. 3.38).

Earlier, the lake had an area of no more than 2 km^2, and then a dam was built in the northeastern part of the lake (length up to 200 m, width 40 m, height 30 m),

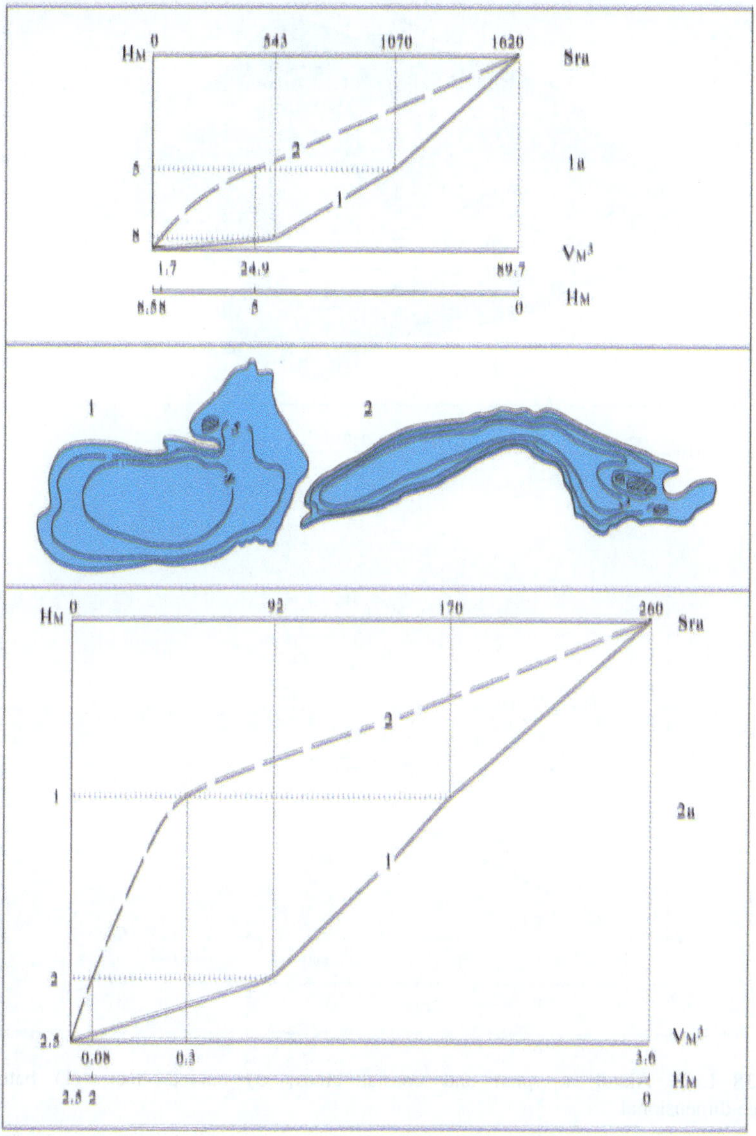

Fig. 3.37 Lake Kyzylkol: 1-plan: σ) 1 a) curved lines: bathymetric (1), three-dimensional and coastal areas (2); 2, 2a-the same for Lake Kyzylauyt

which blocked the Assa river, and in the west—dam (length more than 1.5 km, width 15 m). The resulting hollow gradually filled with water. The bottom of the lake near the coast is muddy; in the central part it is solid. The silt is sandy, in gray color, (thickness up to 0.5 m), with plant residues and the smell of hydrogen sulfide. Along the eastern, northern, and western coasts, a strip of 10–50 m

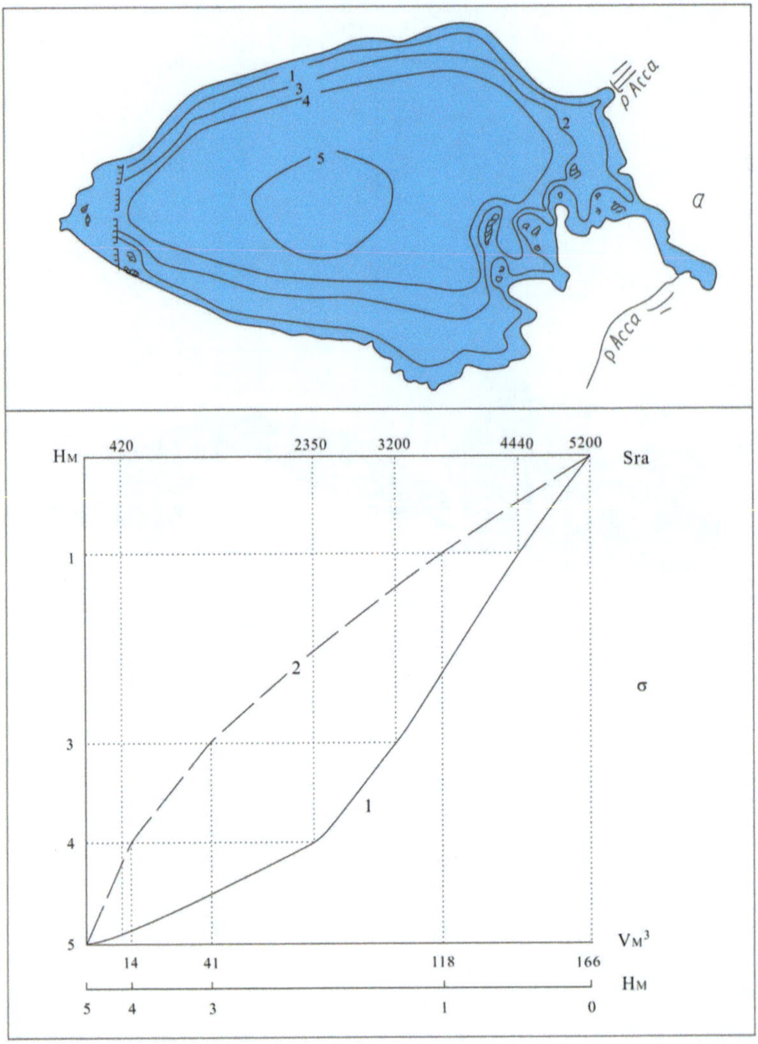

Fig. 3.38 Lake Akkol: a) plan and coastal areas; σ) curved lines: (1) bathymetric (2) three-dimensional

stretches thickets of ordinary reed with the total area of more than 50 ha. More than half the area of the bottom of the basin has been overgrown with charred algae mixed with other water-loving vegetation (Philonets 1981).

The shores are open, gently sloping, and only the northwest and southwestern precipitous (height 1–5 m) is composed of sandy loam, loam, and clay. Plump solonchaks adjoin to the lake from the south-east and southwest.

The catchment area in the south is mountainous, and in the region of the lake there is an inclined crushed stone plain. The soils are clayey, loamy, and sandy loamy. Part of the catchment area is plowed, and the rest part serves as a grazing land. Feather grass, fescue, and white wormwood grow; sometimes there are bushes of tamarisk, saxaul, and Jing (Philonets 1981).

The lake Akkol is a flowing lake and is replenished by river flow, the spring thaw, and liquid precipitation on the waters. In the southeast, the Assa river flows into the lake, and in the northeast it flows from. The average depth of the lake is 3 m (the maximum is 5 m). It accumulates 166 million m^3 of water. Its transparency is 0.1–1 m, and the color is green. The high water level in the lake is observed in April and the low level in September. The annual amplitude of the level fluctuation reaches 2 m; the long-term amplitude exceeds 5 m. In 1918, 1928, and 1942, the lake dried up, and in 1921, the water level in it was the highest (Philonets 1981).

The lake water is fresh, hard, and alkaline. The maximum mineralization in May–June is half than that in September–October. The chemical composition refers to the sulfate class with the sodium group. The content of microelements in water is low, and bromine, barium, copper, zinc, manganese, and cobalt are not found (Philonets 1981).

Lake Biilekol ("Dancing lake") is located in the far south of the Talas district of the Zhambul region, south of the Zhanautkol village, at an altitude of 438.6 m a.s.l. Its area is 86.5 km^2, the length is 18.5 km, the width is 7.3 km, and the length of the coastline is 53 km (Fig. 3.39).

The lake occupies the intermontane basin at the northern foot of the Karatau Ridge. In form, it resembles the figure height and extends to the north. The narrowing up to 2 km wide divides it into two unequal parts (the southern one is twice large). The basin of the lake is shallow, and the lake bottom is even, clean, and near the shore is silted. The thickness of the silt is not more than 0.5 m; the silt is in gray color. Along the southern coast, a strip of 1.5–3 km of the lake was overgrown with reed ordinary. Soft water vegetation (hari and rdest) is abundantly developed in bays, ducts, and places protected by thickets of reeds (Philonets 1981).

Shores of the lake are open, steep, and gently sloping; only in the northern and western parts are steep. The southern shore is pebbly, and the rest parts are composed of loam, sandy loamy, and sandy.

The catchment area in the south and west is mountainous; the rest is hilly. South of the lake, Karatau Ridge (1610.3 m) extends in the northwestern direction, in the southeast—the Ulken-Burultau (843.4 m), in the west of the northern lake—mountains such as Bultun, Zhetimshoky, Kebekashkan with a height of 550–678 m. The soils are clayey, loamy, and sandy loamy and covered with feather grass, fescue, white wormwood. Part of the catchment area is plowed, and the rest is used for pasturing cattle (Philonets 1981).

In the east, the Assa river flows into the lake, and in the northeast it flows from. Basically, the lake replenishes due to river flow, as well as the spring melting of snow and liquid precipitation to the water area. Since the end of April, the flow of water into the lake has stopped, because the water of the Assa river is used for irrigation (Philonets 1981). The average depth of the lake is 2.1 m (the maximum is 4.5 m), and it contains 182 million m^3 of water. Transparency of water varies from 0.2 to 1.5 m, and the water color is brownish-yellow. A high level is observed in

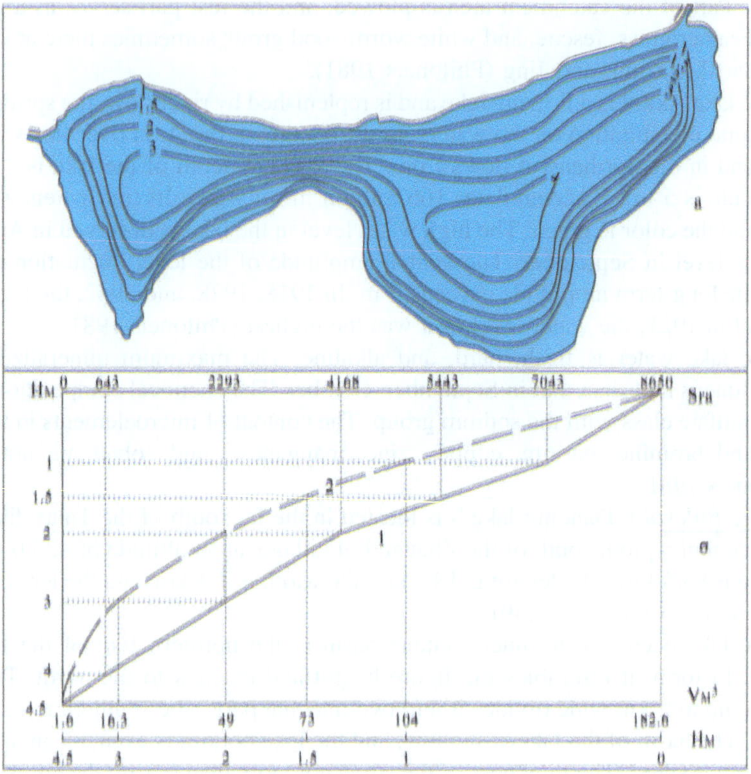

Fig. 3.39 Lake Baikal: a) plan and coastal areas; σ) curved lines: (1) bathymetric and (2) three-dimensional

March–April and low one in September. Annual amplitude of the oscillation reaches 3 m, and long-term one up to 5 m (Philonets 1981).

The lake water is fresh, moderately hard, and alkaline. The maximum mineralization in May–June is half than that in September–October. The lake water refers to the sulfate class with the sodium group. The content of microelements in water is low; iodine, bromine, barium, copper, manganese, and cobalt are not detected (Philonets 1981).

There are 1049 lakes in **South Kazakhstan region** with a total area of 502 km^2 (Fig. 3.40), including 975 lakes with size less than 1 km^2 and take about 23.6% of the total area of the region. There are 74 lakes with a size of more than 1 km^2 and take 76% of the total water surface.

Lake Kyzylkol is located in the extreme southeast of the Suzak region of the South Kazakhstan region, west of the Kumkent village, at an altitude of 335 m a.s.l. Its area is 16.2 km^2, the length is 6.1 km, the width is 2.8 km, and the length of the coastline is 18.6 km.

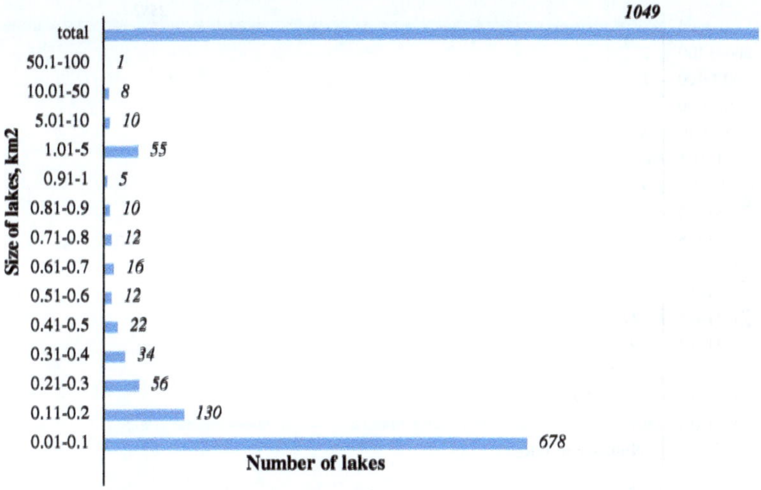

Fig. 3.40 Distribution of lakes in South Kazakhstan region by lake size. Notice: Ponds and water reservoirs are not included to the list

Lake Kyzylkol occupies a drainless inter-mountain hollow with a depth of more than 10 m and extends from west to east. There is a small island in the northeastern part of it, which is flooded at a high water level. The lake bottom relief is hilly. The bottom is solid in the center of the lake; the banks are silty (silt thickness up to 0.3 m). The silt is in black color, sometimes gray in color with a strong smell of hydrogen sulfide (Philonets 1981).

The north and northwestern coasts are steep for 2 km, sometimes up to 20 m, and they are composed of red clays. The depth of the lake grows rapidly; at 10 m from the shore it reaches 5 m. The western and southern shores are flat and composed of red loam. The southeastern coast is steep (up to 50 m high) and stretches a 1.5 km ridge and is composed of red clay. The eastern and northeastern coasts are gently sloping and composed of gray loam. The coast of the lake is overgrown with feather grass, white wormwood, in places licorice. As well as there are separate bushes of saxaul and caragana.

The catchment area is mostly mountainous and overgrown with fescue–feather grass and wormwood vegetation, in places shrubs of caragana and saxaul. It is used for grazing livestock.

The lake is replenished due to river flow, melted waters, and a small part of liquid precipitation to the water area. In the southeast, the Ushbas river (width at the mouth of 50 m, depth up to 2 m) flows into the lake. There is a 2-m waterfall in the deepwater canyon on the river in 1 km from the lake. The water discharge at the waterfall is 1 m^3/s. The river dries up in dry years. The average depth of the lake is 3.7 m (the maximum is 8.5 m). It accumulates 89 million m^3 of water. Its transparency is from 0.5 to 2 m, and the water color is greenish. Frequent strong winds blowing along the Karatau Ridge mix well the water in the lake. A high water level

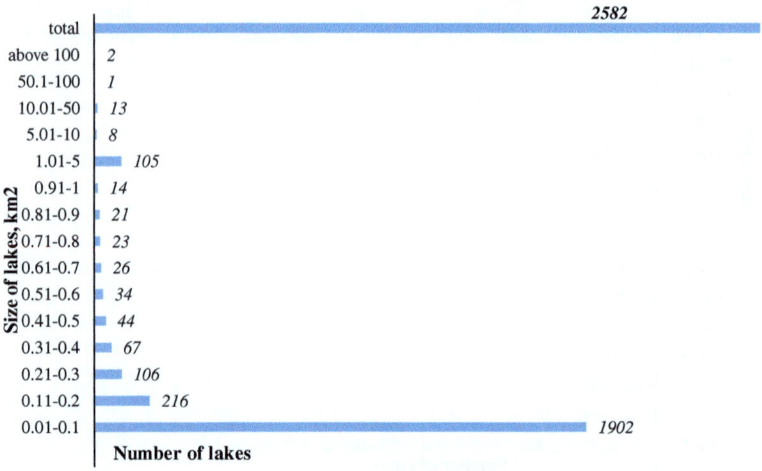

Fig. 3.41 Distribution of lakes in Kyzylorda region by lake size. Notice: Aral Sea, ponds, and water reservoirs are not included to the list

is observed in April and a low level in October. The annual amplitude of the water level reaches 1 m and long-term one more than 3 m (Philonets 1981).

The mineralization and chemical composition of the water are not uniform. In the eastern part, the mineralization is higher (4.8 g/kg) than in the western (4.5 g/kg). In general, the lake water is brackish, very hard, and alkaline. The chemical composition refers to the sulfate class with the sodium group (Philonets 1981).

The content of microelements in water is low, and bromine, copper, manganese, plumbum, and cobalt are not detected. In spring and summer, water from the lake is used for watering cattle.

There are 2582 lakes in **Kyzylorda region** with a total area of 1164 km^2 (Fig. 3.41), including 2453 lakes with size less than 1 km^2 or 95% of lake. In general, the lake area is low in the region. There are 0.53 km^2 of water surface for 100 km^2 of land.

There are 129 lakes with a size of more than 1 km^2 and take 79.5% of the total water surface in the region. They represent the greatest economic interest. There are 16 large lakes (size is more than 10 km^2) such as Kamyslybas, Arys, Zhaksykylysh, Domalakkol, Akshatau (Bolshoe), Birkazan, Tibes, Tuschibas, Kapshinkol, Kilemzhaigan.

Lakes are mainly located in the Syrdarya river valley. A considerable part of lakes is concentrated on the northern and eastern coast of the Aral Sea.

The basins of most lakes are oblong in shape. The slopes are shallow, strongly dissected and composed of clayey and loamy soils.

Lake Kamyslybas is located in the northeast of the Aral Sea region (Kyzylorda region). The lake Kamyslybas is the largest in the Kamyslibas lake system, which is represented by five water bodies connected with each other: Raimkol, Zhalanash,

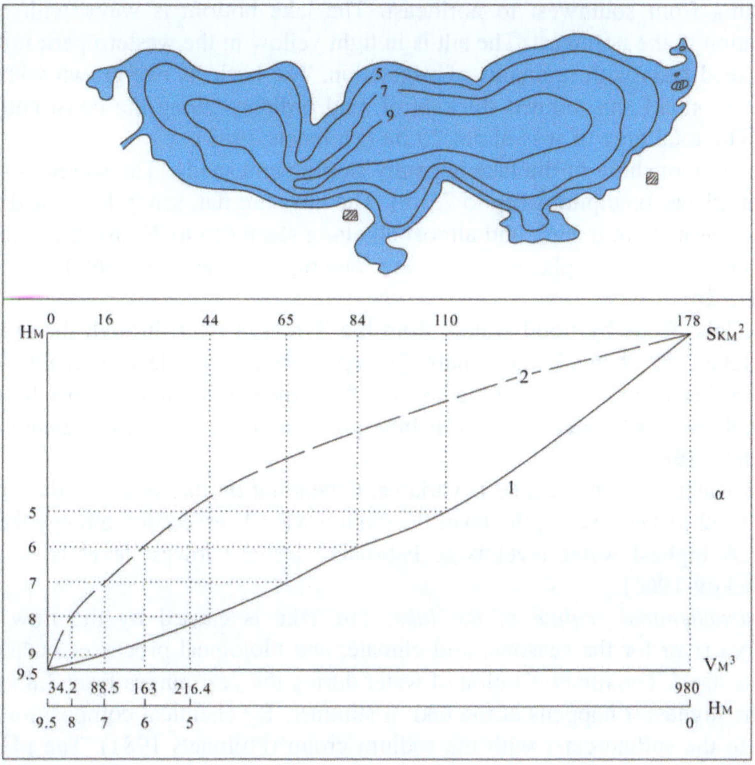

Fig. 3.42 Lake Kamyslybas: a) plan and coastal areas; σ) curved lines: (1) bathymetric and (2) three-dimensional

Koyazdy, Laikol, and Kamyslybas. These lakes have relict origin and are the remnants of the sea bay (Berg 1908). They feed on the flood waters of the Syrdarya river.

The terrain surrounding the lake is characterized by a wide spread of clay and loamy dining outlets up to 70 m high and folded horizontally by the underlying tertiary deposits. The slopes of the remnants are dug by deep ravines. There is a rare solonchak vegetation, in some places shrubs (gingil, comber) (Philonets 1981).

The lake is located at an altitude of 58.1 m a.s.l. The lake area is 178 km^2, the length is 27.6 km, the width is 9.5 km, and the length of the coastline is 115.6 km (Philonets 1981) (Fig. 3.42). It is located in the inter-lowland depression on the right bank of the Syrdarya river and in its floodplain. The lake is gutter-like that extending from southwest to northeast. The lake bottom is wavy with a slight inclination to the northeast. The silt is in light yellow in the western part; it is black in the reed beds with remnants of vegetation. The basin is overgrown with algae. There are small and thinned thickets of reed ordinary along the coast and in the bays. The total area of it is about 20 ha (Philonets 1981).

The eastern shore of the lake is gently sloping and sandy. The southern shore is high, in places precipitous (up to 23 m). The bays are flat, sandy loam, and loamy. The northern shore is high and almost all along steep (up to 30 m). It is composed of loamy; in some places, there are outcrops of gypsum (Philonets 1981) (Table 3.26).

The lake feeds by flood waters from the Syrdarya river through the system of small lakes, the channel, and canals. The appropriate water level is maintained by the dam. The average depth of the lake is 5.5 m (the maximum is 9.5 m). It contains 980 million m^3 of water. Its color is blue-green, and the water transparency is 6 m (Philonets 1981).

The water level in the lake is variable, depending on the Syrdarya river regime. From 1852 to 1962 years, the level fluctuation was observed in the range from 2 to 11 m. A highest water level is in February, and the lowest level is in August (Tlenbekov 1966).

Hydrochemical regime of the lake. The lake is caused by the flow of the Syrdarya river for the seasons, arid climate, and biological processes occurring in the lake itself. The mineralization of water during the year ranges from 2.8 to 4.3 g/kg. The highest it happens at the end of summer. By chemical composition, water refers to the sulfate class with the sodium group (Philonets 1981). The pH of the water fluctuates between 7.3 and 8.2. The oxygen content is 4.3–12.3 mg/l, and the saturation is 35.1–93.9. Carbon dioxide is 8.8–4.4 mg/l, average phosphorus content is 0.01 mg/l, nitrogen is 0.2–10.7 mg/l, and silicon is 0.46 mg/l. The content of microelements: copper, zinc, and manganese is insufficient (Table 3.27).

There are fish such as carp, bream, catfish, asp, pike perch, perch, and barbel in the lake. About 700 tons of commercial fish were harvested annually (Philonets 1981). Many waterfowl and marsh birds: ducks, seagulls, sandpipers, geese, herons, and muskrat are found in the lake.

Table 3.26 Chemical composition of lake water and water reservoirs in the South Kazakhstan region, mg/l (Philonets 1981)

No.	Lake name	pH	Iodine	Fluorine	Bromine	Copper	Zinc	Manganese	Plumbum	Molybdenum	Aluminum	Cobalt	Iron
1	Shevykbeldi	7.2	–	0.4	–	–	0.04	–	0.01	0.005	0.2	–	<0.1
2	Akzhar	7.2	0.03	0.5	–	–	–	–	–	0.0025	0.05	–	<0.1
3	Kyzylkol	8.2	0.01	0.9	–	–	0.06	–	0.01	0.0075	0.03	–	<0.1
4	Tuzkol	8.0	0.015	0.6	–	–	–	–	–	0.005	–	–	<0.1
5	Iirkol	7.4	0.01	0.6	–	–	–	–	–	0.005	0.15	–	<0.1
6	Kaldykol (Keldikol)	7.4	0.01	0.6	–	–	0.01	–	–	0.005	0.02	–	<0.1
7	Sinakol	7.4	0.01	0.6	–	–	0.01	–	–	0.005	0.02	–	<0.1
8	Sarykamys	7.1	–	0.4	–	–	0.07	–	–	0.0025	–	–	<0.1
9	Lake without name	7.3	–	0.3	–	–	0.02	–	0.01	–	–	–	<0.1
10	Lake without name	7.3	–	0.3	–	–	0.02	–	0.01	–	–	–	<0.1
11	Shardara water reservoir	–	–	0.3	–	–	–	–	0.01	–	0.05	–	<0.1
12	Bugun water reservoir, northern	–	–	0.2	–	–	–	–	–	–	–	–	<0.1
13	Bugun water reservoir, southern	–	–	0.5	–	–	–	–	–	–	0.03	–	<0.1
14	Burzhar water reservoir	–	–	0.2	–	–	0.02	–	0.02	–	0.05	–	<0.1

Table 3.27 Content of trace elements in lake waters of Kyzylorda region, mg/l (Philonets and Omarov 1974)

Lake number	Lake name	Iodine	Fluorine	Bromine	Boron	Copper	Iron	Zinc	Manganese	Molybdenum
1	Kamyslybas	0.01	1.5	–	1.0	0.002	<0.10	0.04	0.007	0.06
2	Laikol	0.01	1.5	0.4	1.0	<0.002	<0.10	0.2	0.004	0.03
3	Makpal	0.01	3.5	1.7	4.0	<0.001	<0.10	0.06	0.03	0.05
4	Saryozek	0.02	0.5	0.3	0.5	<0.001	<0.10	0.15	0.036	0.025
5	Kenzhebai	–	1.0	0.3	0.5	<0.001	<0.10	0.1	0.24	0.025
6	Tuzkol	<0.01	0.5	0.3	0.25	<0.001	<0.10	0.02	0.03	0.02
7	Kelalikol	0.01	0.5	0.3	0.1	0.02	<0.10	0.03	0.007	0.025
8	Zhashinkol	0.01	0.5	0.1	0.25	<0.001	<0.10	0.015	0.01	0.02
9	Shukiroi	<0.01	0.7	0.5	0.25	<0.001	<0.10	0.01	0.084	0.025
10	Borkol	0.01	0.6	0.1	0.25	<0.001	<0.10	0.01	0.018	0.025
11	Arys	0.05	1.0	149.0	10.0	<0.001	<0.10	0.075	0.24	<0.0025
12	Keshkensu	0.02	1.5	0.9	2.0	<0.002	<0.10	0.1	0.011	0.03
13	Koltugan reservoir	<0.01	0.5	<0.1	0.1	0.002	<0.10	0.005	0.048	0.015
14	Aral Sea—Saryshyganak Gulf	0.02	2.5	2.0	2.5	0.001	<0.10	0.06	0.007	0.06

References

Kazanskaya EA (1966) Structure of the shores of the Sasykkol Lake. In: Arid zones of Kazakhstan. Alma-Ata, Nauka, KazSSR (in Russian)

Mitrofanov VP (1961) Fishes of the Lake Markakol. In: Proceedings of the conference on fisheries of the middle Asian countries and Kazakhstan. Frunze (in Russian)

Nekhoroshev VP (1936) Explanatory note to the geological map of the Upper Ertis river basin at scale 1:500 000. Moscow-Leningrad (in Russian)

Philonets PP (1981) Essays on the geography of inland waters of the Central, Southern and Eastern Kazakhstan (lakes, water reservoirs and glaciers). Nauka, KazUSSR, Alma-Ata, 283 pp (in Russian)

Philonets PP, Omarov TR (1973) Lakes in the Central and Southern Kazakhstan (reference book). Nauka, KazUSSR, Alma-Ata, 192 pp (in Russian)

Philonets PP, Omarov TR (1974) Lakes in the Northern, West, and Eastern Kazakhstan (reference book). Hydrometeoizdat, Leningrad, 121 pp (in Russian)

Tlenbekov OK (1966) About hydrology of Kamyslybas lake system. In: Fish resources of water bodies in Kazakhstan and their use. Alma-Ata, Nauka, KazSSR (in Russian)

Chapter 4
Lakes in the Central Kazakhstan

4.1 Morphology and Genetic Types of the Lakes in the Central Kazakhstan

The Central Kazakhstan includes the Karagandy region. There are 1910 lakes in Karagandy region with a total area of 926 km^2 (Fig. 4.1), including 1779 lakes with size less than 1 km^2 and takes about 14% of the total area. The hilly relief with a rare low water river network and climate aridity contribute a large number of low-capacity lakes and most of the drainless lakes.

There are 131 lakes with a size of more than 1 km^2 and takes 86% of the total water surface, and these lakes are of the greatest economic interest (Fig. 4.1). In addition, there are 460 ponds and water reservoirs in the region.

Small lakes are concentrated along rivers, streams, and channels; medium and large ones are mainly confined in the ancient river valleys. About 90% of the lakes are concentrated in the northern part of the region. So, there are 108 lakes: 85 lakes with a size of 10 ha, 17 lakes with a size of 10–100 ha, and over 1 km^2 is 6. They are located to the south of 48 parallels dividing the region in half. Balkash Lake lies in the southeast part of the region. Distribution of lakes in Karagandy region is insignificant. There is 0.23 km^2 of water surface for 100 km^2. The Nura district is the richest in lakes and takes 1.3% of the lakes.

The catchment area of lakes is small: from 4 to 200 km^2. Per 1 km^2 of water surface is 29 km^2 of the basin area on average. Most of the watersheds of the lakes in the northern half of the region are largely plowed (from 30 to 80%).

The basins of many lakes are oval in shape. Their slopes are shallow and weakly dissected and are composed of loamy soils. Lakes such as Karasor, Balyktykol, Saumalkol, Katynkol have a small part of the coast which is steep and composed of gray sandstone. They are very rarely found in the region.

The maximum depth of the lakes varies from 1.5 to 8.5 m (average 1–5 m). 576 million m^3 of water are accumulated in lakes over 1 km^2, including 270 million m^3 with mineralization up to 6 g/l. A sharp rise in the level from the inflow of water

J. Abuduwaili et al., *Hydrology and Limnology of Central Asia*, Water Resources Development and Management, https://doi.org/10.1007/978-981-13-0929-8_4

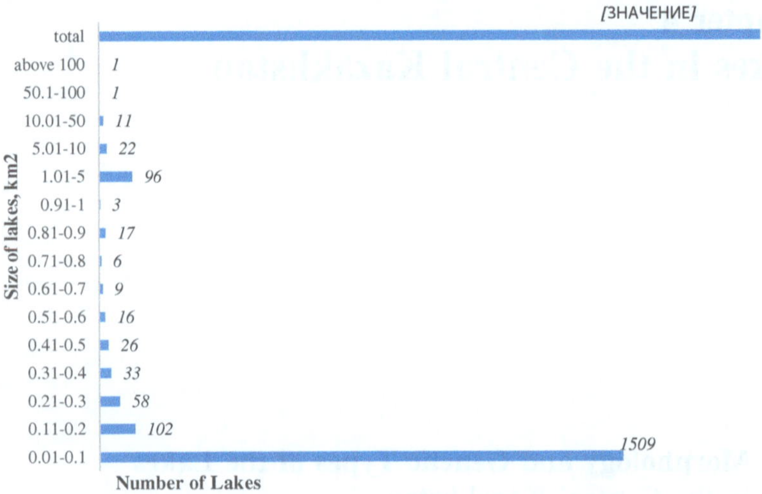

Fig. 4.1 Distribution of lakes in Karagandy region by lake size *Notice* Balkash lake, plexus lakes, ponds and water reservoirs are not included to the list. Plexus lakes are 3290 with a total area of 47 km^2. The total number of ponds and water reservoirs is 460 in the region

during the spring snowmelt is a characteristic feature of the regime of lakes. The water level increase on almost all lakes begins at the end of March and the first ten days of April and ends mainly in 10–15 days. The height of the spring rise of the level strongly fluctuates in the years and can exceed 2 m. The lowest level in the lakes is established in October. The annual amplitude of the level fluctuations averages 0.6–0.8 m. Besides seasonal fluctuations, long-term ones are observed. The volume of water inflow into the lake increases sharply during large water years. The high water periods last two to three years; the low water periods are longer. Even the medium lakes often completely dry up by the end of the low water period.

4.2 Hydrological Characteristics of the Lakes in the Central Kazakhstan

Lake Karasor is located in the southwest of the Yegendybulak region and north of Karkaraly town (east of Karagandy city). Its area is 155 km^2, the length is 43.5 km, the maximum width is 7.3 km (the smallest is 1.4 km), and the maximum depth is 2.5 m (average is 1.1 m). The length of the coastline is 103 km, and it is located on 621.5 m above sea level (Fig. 4.2).

The Karasor lake is the largest one in the Karagandy region. It has S-shaped form and occupies the lowest part of the Karasor depression of subaqueous origin, without drainage.

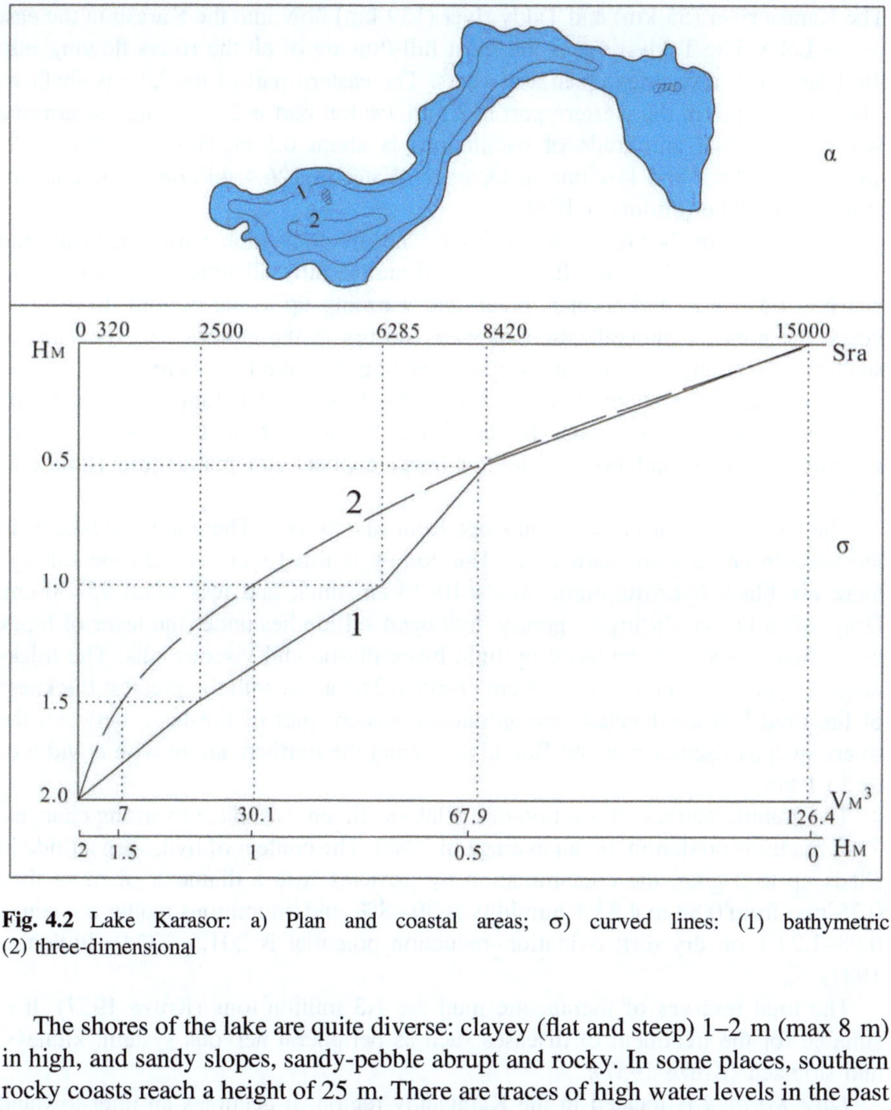

Fig. 4.2 Lake Karasor: a) Plan and coastal areas; σ) curved lines: (1) bathymetric (2) three-dimensional

The shores of the lake are quite diverse: clayey (flat and steep) 1–2 m (max 8 m) in high, and sandy slopes, sandy-pebble abrupt and rocky. In some places, southern rocky coasts reach a height of 25 m. There are traces of high water levels in the past (about 640 m abs). At the same height, traces of high water levels on the banks of the lakes such as Saumalkol and Katynkol in the Karasor Depression are noted. This gives grounds to assert that earlier area of the lake Karasor was approximately 740 km^2, the depth reached 22–25 m, and its waters were fed by the Nura river (Philonets 1981).

The catchment area of the lake is 8750 km^2 and partially plowed. The relief is rugged; the soils are clayey and loamy. Most of the catchment area is overgrown with feather grass, fescue, and wormwood vegetation, caragan shrubs.

The lake feeds on melted and river waters. There are 14 rivers and streams in it, which are temporary watercourses operating mainly during the spring flood period.

The Karasu river (55 km) and Taldy river (159 km) flow into the Karasu in the east of the Lake. The Taldy river is the most full-flowing of all the rivers flowing into the lake, but it also dries up in arid years. The eastern part of the lake is shallow. Maximum depth in the western part is 2.5 m, central part is 2.3 m, and eastern one is 0.9 m. Annual amplitude of oscillations is about 0.5 m. High water level is observed in April and low one in October. Water in 126.4 million m^3 is concentrated in the lake (Philonets 1981).

The lake water is highly *mineralized.* The mineralization varies in years and seasons from 40 to 75 g/kg. It is very hard and slightly alkaline. Transparency of water is 1.5–1.8 m and color is green and warming up to the bottom. In drought years, the greatest mineralization of water occurs in the eastern part (because of shallow water), although the main rivers feeding the lake flow here.

According to the chemical composition, the lake water belongs to the chloride class with the sodium group. It has a high content of microelements: iodine, fluorine, bromine, and boron, but low zinc, copper, and manganese (Philonets 1981).

The lake bottom has a common slope from east to west. There are two islands in the western and eastern parts of it. The hollow is filled with silt. On the surface, there are black hydrosulphuric muds 10–15 cm thick and less often 25–30 cm. Gray, often brownish-tinged, greasy, hydrogen sulfide lies under the layer of black ooze. Below, they are replaced by light black plastic and viscous silts. The thickness of light gray mud reaches 70 cm or more. The areas with the greatest thickness of the mud bed are located throughout the eastern part of the lake, between the rivers such as Esen-Aman and Barak, and along the northern shore with a width of up to 1 km.

The granulometric composition of the lake sediment is subject to strong changes. Thin fractions predominate (an average of 77%). The content of hydrogen sulfide in silt is up to 3 g/kg, the contamination by particles with a diameter of more than 0.25 mm from 0.88 to 4.35% humidity is 39.48%, and bituminous matter content is 0.98–1.24% on dry dirt; oxidation–reduction potential is 201.5–235.9 (Philonets 1981).

The total reserves of therapeutic mud are 1.3 million tons (Kruze 1957). It is suitable for the treatment of diseases such as peripheral nervous system, kidneys, and stomach (Kruze 1957).

Lake Karakol is located in the Karagandy region. It occupies an inter-lowland depression near the northeastern slopes of the Niaz Mountains. The lake extends from the northwest to the southeast. Its area is 7.15 km^2, the length is 4 km, the width is 2.3 km, and the length of the coastline is 12.5 km. The lake is drainless, located at an altitude of 512.3 m above sea level (Fig. 4.3).

The northeastern coast of the lake is steep and composed of limestone; the rest coast is shallow and sandy loamy. The western and southern coasts at the water's edge are covered with red pebbles. The coasts (except for the eastern, where the vegetable gardens are located) are overgrown with fescue–feather grass and wormwood vegetation (Philonets 1981).

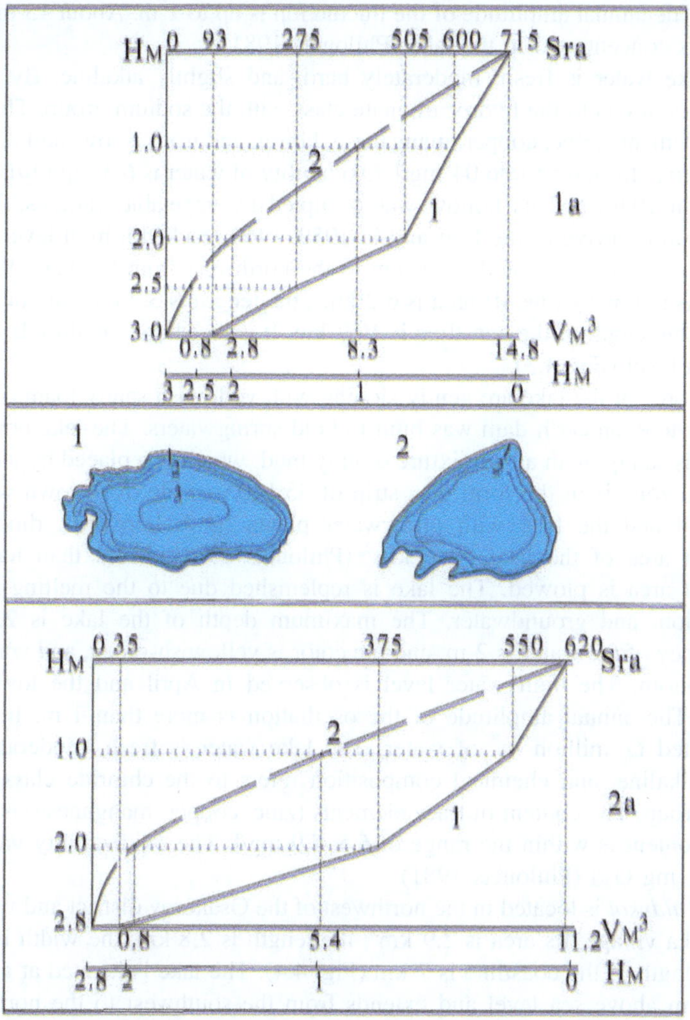

Fig. 4.3 Lake Karakol: 1-plan: σ) 1 a) Curved lines: bathymetric (1), three-dimensional and coastal areas (2); 2, 2a the same for Lake Toksumak

In the coastal part, there are thickets of reeds and cattail with an admixture of Cane lake. The lake bottom is silt, partially overgrown with reddest, and black ooze.

The catchment area of the lake is 160 km^2 and it is more than 20% plowed, and the rest part is overgrown with fescue–feather grass and wormwood vegetation. The relief is crossed.

The lake feeds on the spring melting of snow and groundwater. Its maximum depth is 3 m, the transparency of water is 1.5 m, the color is brownish-yellow, and the warming up to the bottom. The high water level is in April and the low level in

October. The annual amplitude of the fluctuation is up to 1 m. About 15 million m^3 of water is concentrated in the lake (Philonets 1981).

The lake water is fresh, moderately hard, and slightly alkaline. By chemical composition refers to the hydrocarbonate class with the sodium group. The content of microelements (zinc, copper, manganese, boron, and iron) is low, and the oxygen content varies from 6.55 to 6.04 mg/l. Oxidability of water is 6.5 mg O_2/l. The lake dried up in 1946, and its hollow was occupied by vegetable gardens. Then, the water began to arrive to the lake and in 1959–1960 reached a high level.

Lake Toksumak is located in 10 km to the northwest from Osakarovka village. The lake is a flowing one, its area is 6.2 km^2, the length is 3.2 km, and the width is 2.8 km. The length of the coastline is 10.7 km. It is located at an altitude of 440 m above sea level (Fig. 4.3).

The shores of the lake are gently sloping with yellowish sandy loam. Along the northern shore, an earth dam was built to hold springwaters. The lake bottom near the coast is sandy, with an admixture of gray mud, which is replaced by a dark one.

Coastal zone is in the form of a strip of 35,150 m wide overgrown with reeds and cattail and the lake with underwater plants (urot, hornwort, rhodes). The catchment area of the lake is 68 km^2 (Philonets 1981). More than half of the catchment area is plowed. The lake is replenished due to the melting of snow, precipitation, and groundwater. The maximum depth of the lake is 2.8 m, the transparency of the water is 2 m, and the color is yellowish-green, and warming up to the bottom. The high water level is observed in April and the low level in October. The annual amplitude of the oscillation is more than 1 m. In the lake accumulated 12 million m^3 of water. The lake water is fresh, moderately hard, slightly alkaline, and chemical composition refers to the chloride class with the sodium group. The content of trace elements (zinc, copper, manganese) is low, and oxygen content is within the range of 6.5–7.9 mg/l. The oxidizability varies from 4.9 to 6.1 mg O_2/l (Philonets 1981).

Lake Taldykol is located in the northwest of the Osakarov district and west of the Skobelevka village. Its area is 2.9 km^2, the length is 2.8 km, the width is 1.5 km, and the length of the coastline is 7 km (Fig. 4.4). The lake is located at an altitude of 466.8 m above sea level and extends from the southwest to the northeast. Its banks are shallow, loamy, and overgrown with fescue, feather grass, and wormwood. The basin of the lake is wavy and strongly silted. The basin is overgrown mainly with reed ordinary with an admixture of reeds, kugi, and sedges. The area of the thickets is more than 100 ha (Philonets 1981).

The lake is replenished due to the spring melting of snow and groundwater. Its catchment area is 37 km^2. It accumulates more than 2 million m^3 of water. High water level is observed in April and low one in October. The annual amplitude of the oscillation reaches 0.7 m. The greatest depth of the lake is 1.5 m, the transparency of the water is to the bottom, and the color is brownish-yellow. The water in the lake is fresh, moderately hard, and neutral. Oxidizability is 12.6 mg O_2/l. The silicon content is 2.6 mg/l; no iron was found. According to the chemical composition, the lake water belongs to the chloride class with the sodium group. A low content of zinc

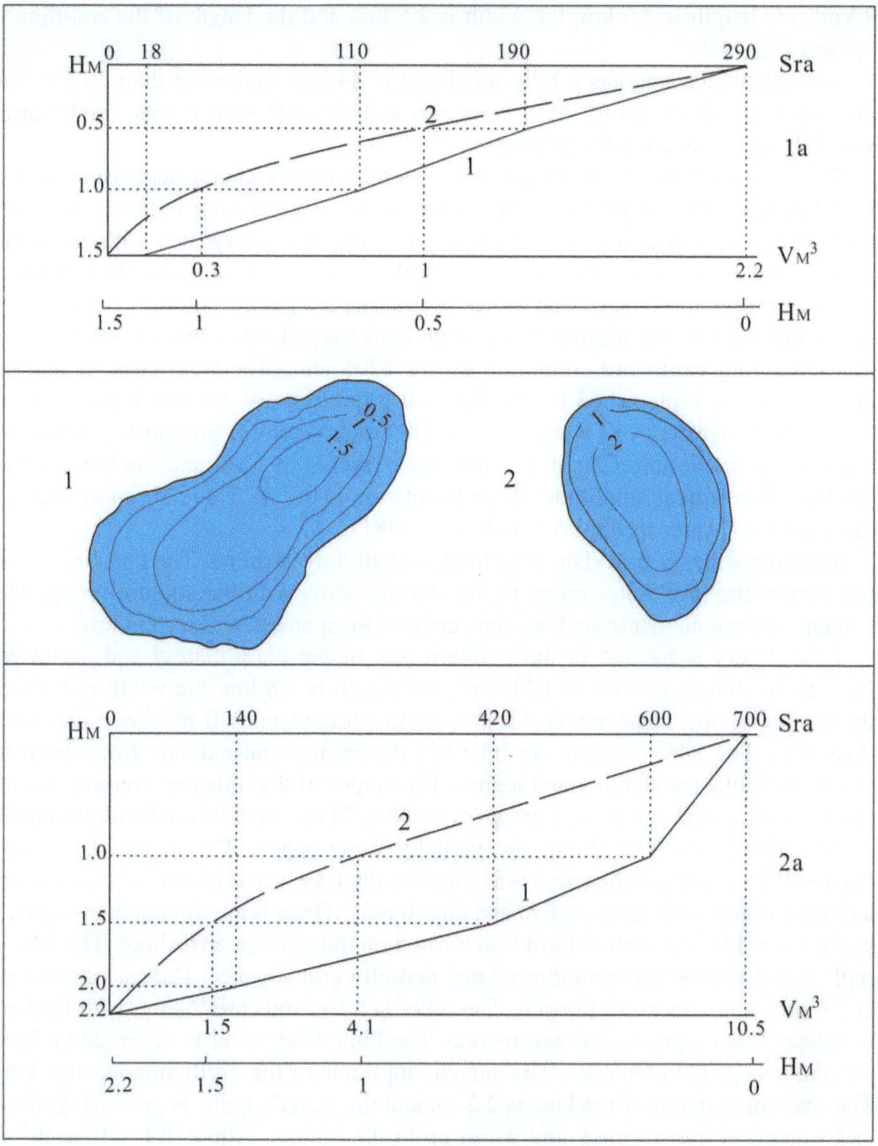

Fig. 4.4 Lake Taldykol: 1-plan: σ) 1 a) Curved lines: bathymetric (1), three-dimensional and coastal areas (2); 2, 2a the same for Lake Batpakkol

and copper was observed in water. In the lake is found the silver and goldfish. Ducks and gulls, waders and geese live on the lake (Philonets 1981).

Lake Batpakkol is located in the northwest of the Ulyanov district, on the border with the Pavlodar region, at an altitude of 467.2 m above sea level. Its area is

7 km^2, the length is 3.6 km, the width is 2.5 km, and the length of the coastline is 9.4 km (Fig. 4.4).

The catchment area has a hilly relief and is 84 km^2 and more than 20% of the plowed fields. Soils are loamy, overgrown with fescue, feather grass, and wormwood. It is widely used for grazing.

The lake is ellipsoidal in shape, occupies the inter-mountain depression, and is stretched from the northwest to the southeast. The lake bottom is level, along the banks silted in gray color. Near the coast, the lake was overgrown with reeds and rdest. In the center and in the southern part, there are two small islets (the first is 1.5 ha, and the second is 3 ha). The shores of the lake are gentle, and only the northern one is high (5 m); the southeast and south banks are made of grayish sandy loam. The remaining banks are composed of grayish loams. The west coast is plowed up. The lake is replenished by melting and groundwaters. Its maximum depth is 2.2 m, the transparency of water is 1.5 m, the water's color is brownish-yellow, and warm up to the bottom. The maximum water level is in April and the low level in October. The annual amplitude of the fluctuation is 0.7 m. There are more than 10 million m^3 of water in the lake (Philonets 1981).

The lake water is brackish, very hard, and slightly alkaline. The chemical composition of the lake water refers to the chloride class with the sodium group. The content of trace elements such as zinc, copper, manganese, and iron is low.

Lake Arykty is located in the northern part of the Nura district and northwest from Kiev village. Its area is 1.51 km^2, the length is 1.8 km, the width is 1.3 km, the length of the coastline is 4.5 km, at an altitude of 390 m above sea level (Fig. 4.5). The lake occupies an inter-hill depression and extends from north to south. Its banks are shallow and loamy. The slopes of the hills are overgrown with fescue, wormwood and sometimes grow stubble. There are thicketed willows on the southeastern coast. The shores are partially waterlogged. The depth of the mud reaches 0.7 m. Almost the entire water area of the lake is overgrown with reeds, and along the coasts with reeds, sedge and duckweeds. There is amphibian buckwheat, a white water lily. An animal farm was located on the northeastern shore. The lake is replenished with water from thawed and partially groundwater. The catchment area is 58 km^2, and almost all plowed. The relief is hilly; soils are loamy, with thickets of fescue, feather grass, and wormwood. The high water level is observed in April and the low level in October. The annual amplitude of the oscillation reaches 1 m. The maximum depth of the lake is 2.2 m, and the water's color is greenish-yellow, and water and transparency and warm up to the bottom. About 1.4 million m^3 of water is accumulated in the lake. The lake water is fresh, soft, and neutral. Chemical composition refers to the hydrocarbonate class with sodium group. The content of microelements of zinc, copper, and manganese is low in the lake water. The lake water is used year-round for watering cattle. The water level in the lake until 1921 was significantly higher than now. Water was used for irrigation. The lake was dried up in 1932 and 1943 (Philonets 1981).

Lake Kurgankol is located in the northwest of the Ulyanov district, in the inter-substitute depression at an altitude of 570.6 m above sea level. Its area is 2.9 km^2, the length is 2.3 km, the width is 1.8 km, and the length of the coastline is

Fig. 4.5 Lake Arykty: 1-plan: σ) 1 a) Curved lines: bathymetric (1), three-dimensional and coastal areas (2); 2, 2a the same for Lake Kurgankol

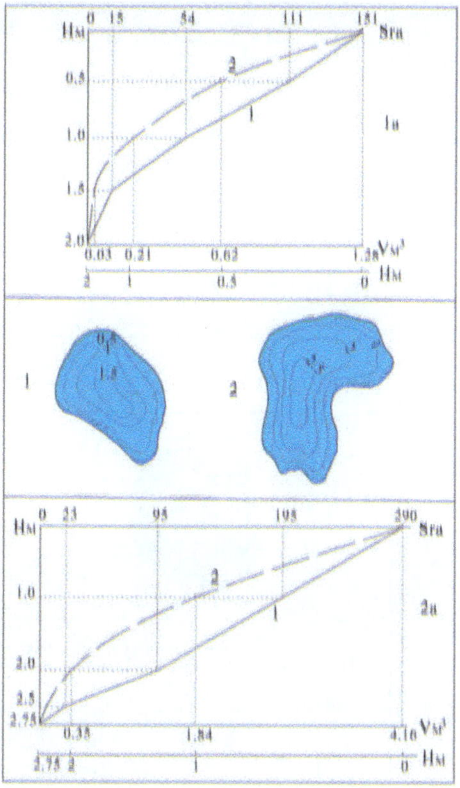

6.8 km (Fig. 4.5). The catchment area is 69 km^2, and fully plowed. The relief is hilly (Philonets 1981). The shores of the lake are open, gently sloping, loamy, and overgrown with fescue, chi, quinoa, wormwood. There are spots of saltwort in the lowlands. The lake bottom is firm and level. Along the banks grows reed, cattail, and in the lake the rdest. The area of reed thickets is more than 5 ha, and the height is up to 2 m.

The lake is replenished by thawing and groundwater. Its maximum depth is 3 m. Transparency of water is 2 m, water's color is brownish-yellow and warming up to the bottom. The high water level is observed in April and the low level in October. The annual amplitude of the fluctuation is about 0.7 m. The volume of water in the lake is about 5 million m^3. The lake is without drainage.

The water in the lake is salty, hard, and slightly alkaline. The chemical composition of the lake water refers to the chloride class with the sodium group. The content of trace elements such as zinc, copper, manganese, and iron is low. Oxygen varied from 3.05mg/l to 4.18 mg/l, oxidizability 32.8 mg O_2/l. The lake water was used for watering cattle. In 1937, the lake dried up (Philonets 1981).

Lake Kaiyndy is located in the northeast of the Ulyanov district. The area of the lake is 2 km^2, the length is 1.8 km, the width is 1.4 km, and the length of the

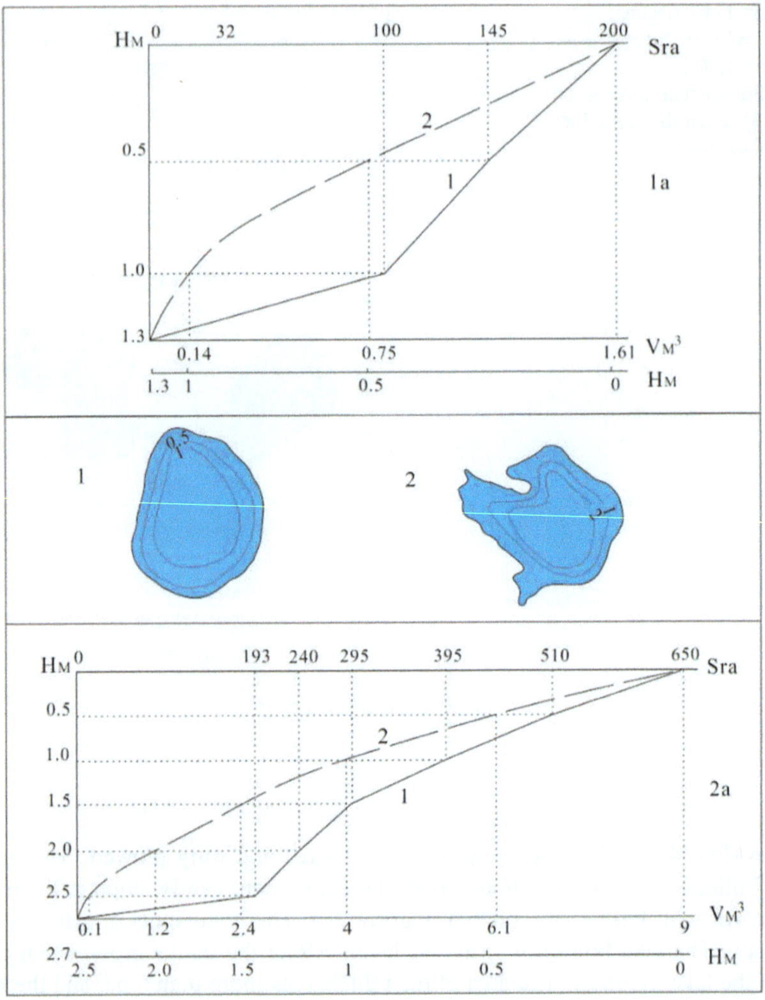

Fig. 4.6 Lake Kaiyndy: 1-plan: σ) 1 a) Curved lines: bathymetric (1), three-dimensional and coastal areas (2); 2, 2a the same for Lake Saumalkol

coastline is 4.9 km, at an altitude of 663.3 m above sea level (Fig. 4.6). The lake occupies a hilly depression on the southeastern slopes of the Saryktau Mountain. Its catchment area is 40 km^2, crossed and half plowed. The northern and southern shores are steep with a height of 3–8 m above the water's edge, and the slopes are not crooked. The western coast is flat. The coast is overgrown with fescue–wormwood vegetation (Philonets 1981).

The basin of the lake is saucer-shaped, and the lake bottom is even and partly silted, and overgrown with pond. Along the banks grows reed, cattails, and reeds, whose height sometimes reaches 1.5–2 m, the area is 9 ha.

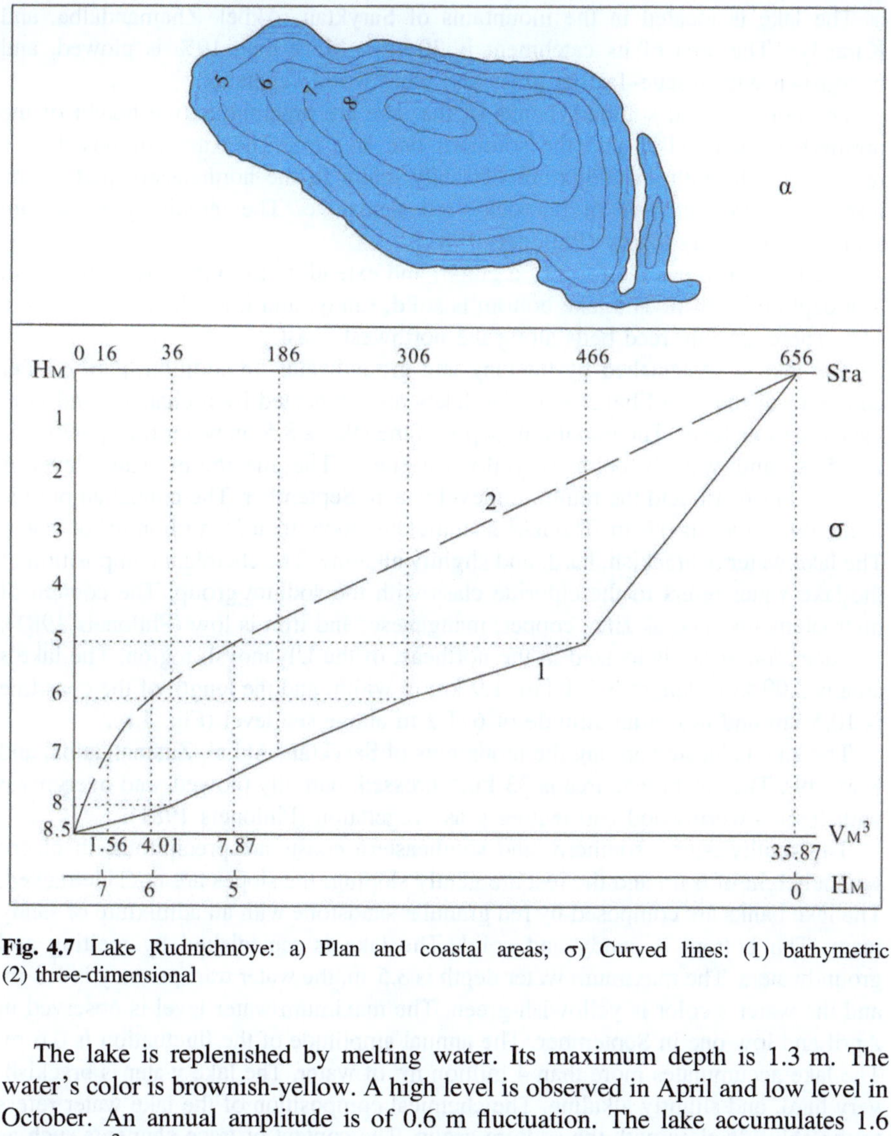

Fig. 4.7 Lake Rudnichnoye: a) Plan and coastal areas; σ) Curved lines: (1) bathymetric (2) three-dimensional

The lake is replenished by melting water. Its maximum depth is 1.3 m. The water's color is brownish-yellow. A high level is observed in April and low level in October. An annual amplitude is of 0.6 m fluctuation. The lake accumulates 1.6 million m^3 of water. The lake water is brackish, moderately stiff, and slightly alkaline. The chemical composition of the water refers to the chloride class with the sodium group. The content of trace elements such as zinc, copper, manganese, and iron in water is low (Philonets 1981). There is gold and silver crucian carp in the lake. The lake water is used for watering cattle.

Lake Rudnichnoye is located in the northeast of the Ulyanovsk region, northeast of the Semiz-Bugu village. Its area is 6.56 km^2, length is 4.2 km, width is 2 km, and length of the coastline is 15 km, at an altitude of 671.2 m above sea level (Fig. 4.7) (Philonets 1981).

The lake is located in the mountains of Saryktau, Akbel, Zhamandalba, and Karaadyr. The area of its catchment is 30 km^2. More than 10% is plowed, and overgrown with fescue–feather grass and wormwood vegetation.

The northern and southern banks of the lake are precipitous (the height of the northern bank is 11 m and the southern one is 3 m). They are composed of a reddish pebble with an admixture of sandy loam. In the northeastern part of the lake, there are outcrops of bedrock—red sandstone. The remaining banks are shallow, and sandy loamy (Philonets 1981).

The lake basin has the form of a gutter, and extending from the east to the west, to a depth of 7–8 m. The lake bottom is solid, sandy, and it is silted in the central part. There are rare reed beds along the northwest coast.

The lake is replenished by thawing and groundwater. In addition, partly water comes from the lake Shalkarkol. The lakes are connected by a channel, and both have the same level. The maximum depth of the lake is 8.5 m, water transparency is 2–2.5 m, and water's color is yellowish-green. The maximum water level is observed in April, and the minimum level was in September. The annual amplitude of the fluctuation is 0.6 m. The lake accumulated more than 35 million m^3 of water. The lake water is brackish, hard, and slightly alkaline. The chemical composition of the lake water refers to the chloride class with the sodium group. The content of microelements such as zinc, copper, manganese, and iron is low (Philonets 1981).

Lake Shalkarkol is located in the northeast of the Ulyanovsk region. The lake's area is 3.09 km^2, length is 2.1 km, 1.9 km in width, and the length of the coastline is 10.5 km and lies at an altitude of 671.2 m above sea level (Fig. 4.8).

The lake is located among the mountains of Saryktau, Akbel, Zhamandalba, and Karaadyr. The catchment area is 33 km^2, crossed, partially plowed, and overgrown with fescue–wormwood and feather grass vegetation (Philonets 1981).

The southwestern, southern, and southeastern coasts are precipitous, in places with a height of 6 m, and the rest are gently sloping; the slopes are not sidetracked. The lake banks are composed by red granular sandstone with an admixture of sandy loam. The bottom is sandy and solid. The lake is replenished by melting and groundwaters. The maximum water depth is 3.5 m, the water transparency is 1.2 m, and the water's color is yellowish-green. The maximum water level is observed in April and low one in September. The annual amplitude of the fluctuation is 0.6 m. The lake accumulates more than 4 million m^3 of water. The lake water is brackish, very hard, and slightly alkaline. The chemical composition of the lake water refers to the chloride class with the sodium group. The content of trace elements such as zinc, copper, manganese, and iron is low in water (Philonets 1981).

Lake Agashty is located in the southwestern part of Osakarovsky district, at an altitude of 376.6 m above sea level. The area of the lake is 2.75 km^2, the length is 2.8 km, the width is 1.4 km, and the shoreline length is 7.3 km (Fig. 4.9).

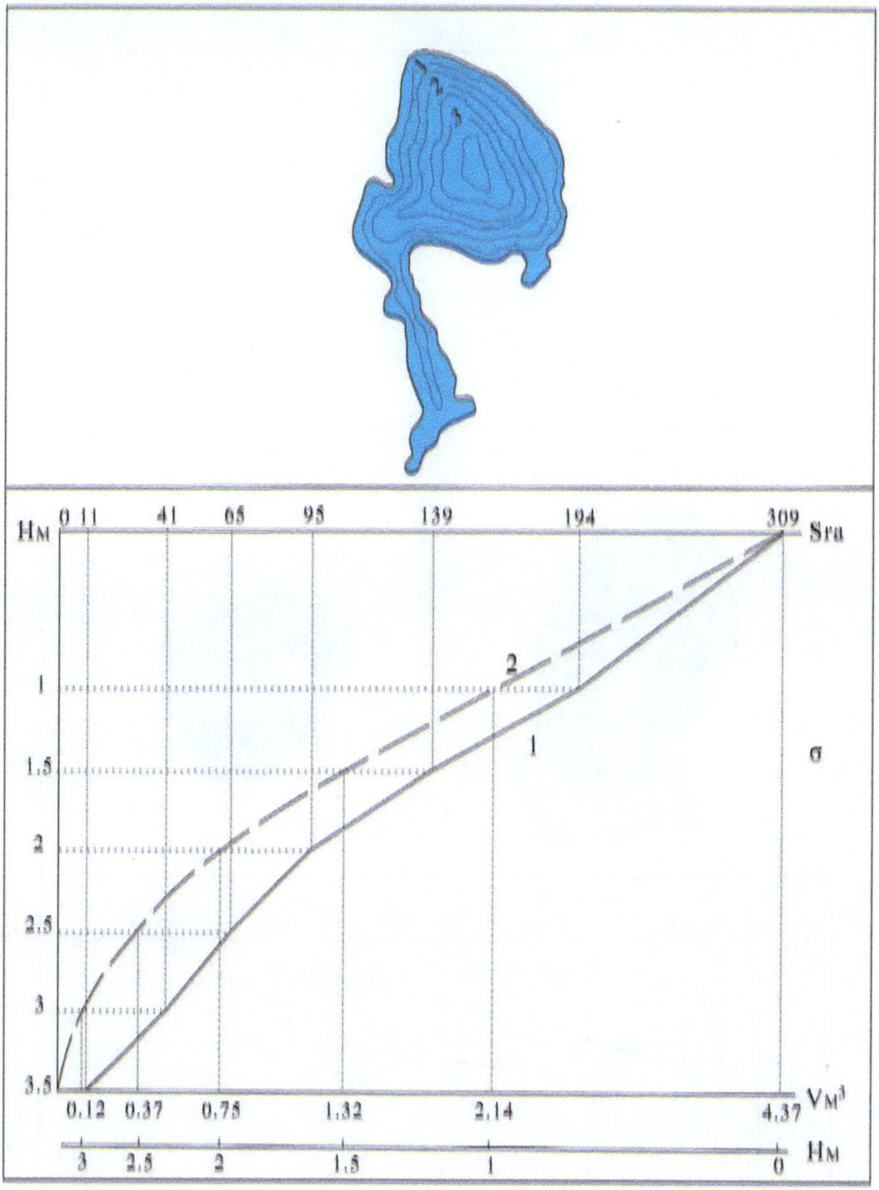

Fig. 4.8 Lake Shalkar: a) plan and coastal areas; σ) Curved lines: (1) bathymetric (2) three-dimensional

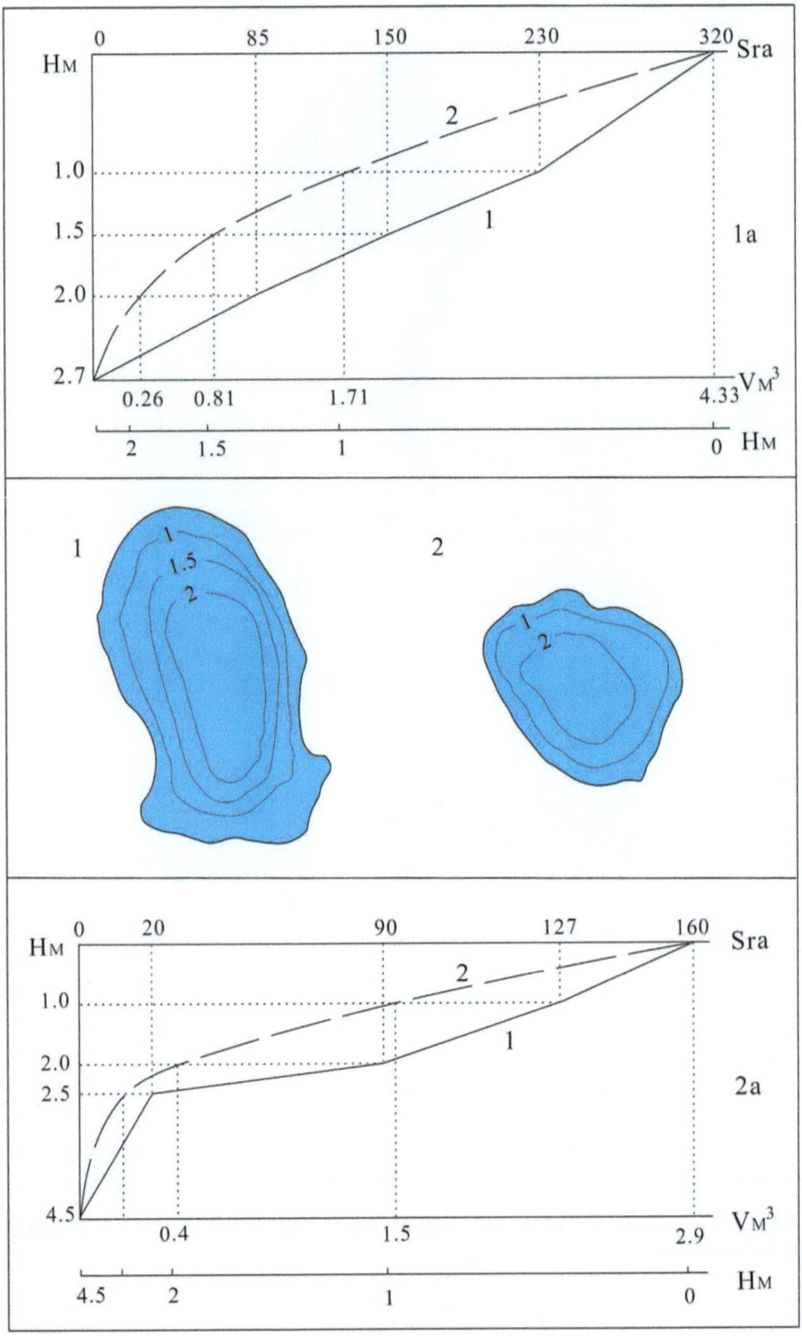

Fig. 4.9 Lake Agashty: 1-plan: σ) 1 a) Curved lines: bathymetric (1), three-dimensional and coastal areas (2); 2, 2a the same for Lake Aryktykol

The lake occupies an inter-mountain depression and extends from north to south. Its catchment area is 96 km^2, and more than half part of it is plowed. The relief is hilly, soils are loamy, and overgrown with feather grass–fescue and white wormwood vegetation.

The shores of the lake are gentle and composed of yellowish loam. The lake bottom is wavy, marshy, and two-thirds part overgrown with reeds, rdest.

The lake feeds on meltwater. Its maximum depth is 2.7 m, the water transparency is up to 1.5–2 m, and the color is brownish-yellow. The high water level is observed in April and the low level in October. The annual amplitude of the fluctuation is up to 1 m. The lake accumulates more than 4 million m^3 of water.

The lake water is fresh, hard, and slightly alkaline. The chemical composition refers to the chloride class with the sodium group. The content of trace elements such as fluorine, boron, zinc, copper, manganese, and iron is low in the water.

Lake Arykty is located in the northwest of the Telman district, at an altitude of 535.3 m above sea level. Its area is 1.6 km^2, the length is 1.7 km, the width is 1.4 km, and the length of the coastline is 4.8 km (Fig. 4.9). The lake occupies an inter-mountain depression. The catchment area is 80 km^2, and about one-third of it is plowed. Soils are loamy, and the topography is hilly.

The lake shores are gentle, open, and loamy. The coast is plowed up. The basin of the lake is saucer-shaped and strongly silted. In some places, the lake bottom is covered with rare thickets of reeds.

The lake is replenished due to snow melting waters, as well as due to the canal and the retaining dam on the lake. Its maximum depth is 4.5 m, the annual amplitude of the level fluctuation is more than 2 m, the water transparency is 1 m, and the water's color is brownish-yellow. The lake accumulates 3 million m^3 of water.

The lake water is fresh, moderately stiff, and neutral. The chemical composition refers to the hydrocarbonate class with the calcium group. The content of trace elements such as zinc, copper, and manganese is low in the water.

Lake Saumalkol is located in the eastern part of the Ulyanovsk region. The area of the lake is 4.5 km^2, the length is 2.8 km, the width is 2.2 km, and the length of the coastline is 10 km, at an altitude of 625.7 m above sea level. The lake is irregularly shaped, and it occupies an intermontane depression and is stretched from the northwest to the southeast. The catchment area is 98 km^2. More than 20% of the territory is plowed. Soils are loamy and vegetation is fescue–feather grass, wormwood, caragana (Philonets 1981).

The shores of the lake are gentle and open. The northern shore is high (4 m) and steep. The coast has become overgrown with fescue, wormwood, feather grass, spurge, chiy, shrub kargan. The lake bottom is even; in the central part, it is solid and sometimes overgrown with pond. Along the banks, stretch reed beds are 1.5 m high and their area is of 20 ha.

The lake is replenished by melting and groundwater. Its maximum depth is 3.5 m, the transparency of water is 1 m, and the color is brownish-yellow. The high water level is in April, and the low level is in October. The annual amplitude of the fluctuation is more than 0.5 m. The lake accumulates more than 7 million m^3 of water (Philonets 1981). The lake water is brackish, very hard, and slightly alkaline. The chemical composition refers to the chloride class with the sodium group. The content of trace elements such as zinc, copper, manganese, and iron is low in water. Water is used for household purposes.

Lake Karaukamys is located in the Nura district, at an altitude of 385 m above sea level. Its area is 6.2 km^2, the length is 4.5 km, the width is 2.5 km, and the length of the coastline is 12.8 km (Fig. 4.10). The lake is drainless, occupying a shallow basin of the northern strike. The catchment area is 94 km^2, about half of it is plowed, and the rest is overgrown with feather grass and wormwood. The terrain is hilly and soils are loamy.

The shores of the lake are gentle, open, and composed of yellowish-brown loam. The lake is completely overgrown with reeds. Reed thickets occupy an area of more than 5 km^3.

The lake is replenished with melting snow waters. The maximum depth is 2 m. It accumulates more than 6 million m^3 of water. The water is fresh, soft, and slightly acidic. The chemical composition refers to the chloride class with the sodium group. The content of trace elements such as zinc, copper, iron, and manganese is low in the water.

4.3 Hydrological and Hydrochemical Regimes of the Lakes in the Central Kazakhstan

An analysis of the water balance in lakes shows that the influx from the surface of the watersheds usually gives more than half of the incoming part, amounting to 50–60% in the more humid northeast part of the region and reaching 80–85% in arid regions. The influx is highly variable. It can exceed the norm by several times in high water years, in shallow water years is 5–10 times lower than normal. Summer precipitation gives from 12 to 39% of the entire influx, and their role increases in areas with an increased rainfall rate of the warm period. Solid precipitation accumulated on ice in the form of snow makes up 5–15% of the incoming part of the balance. Their portion is higher on lakes which are overgrown with semi-submerged plants and where the norm of solid precipitation is higher. The flow rate of the water balance in the drainless lakes is almost entirely determined by evaporation from the water surface.

The water temperature in the lakes increases to 7–15 °C by May, reaching 25–30 °C by July. The water in lakes begins to cool gradually in August, and its temperature on the surface is usually reduced to 0 °C by the end of October—beginning of November. The water mass is heated from the bottom after the formation of the ice cover in winter. A winter reverse stratification of the temperature with a zero value at the bottom surface of the ice and 2–4 °C at the bottom is established in the reservoirs (Lezin 1982).

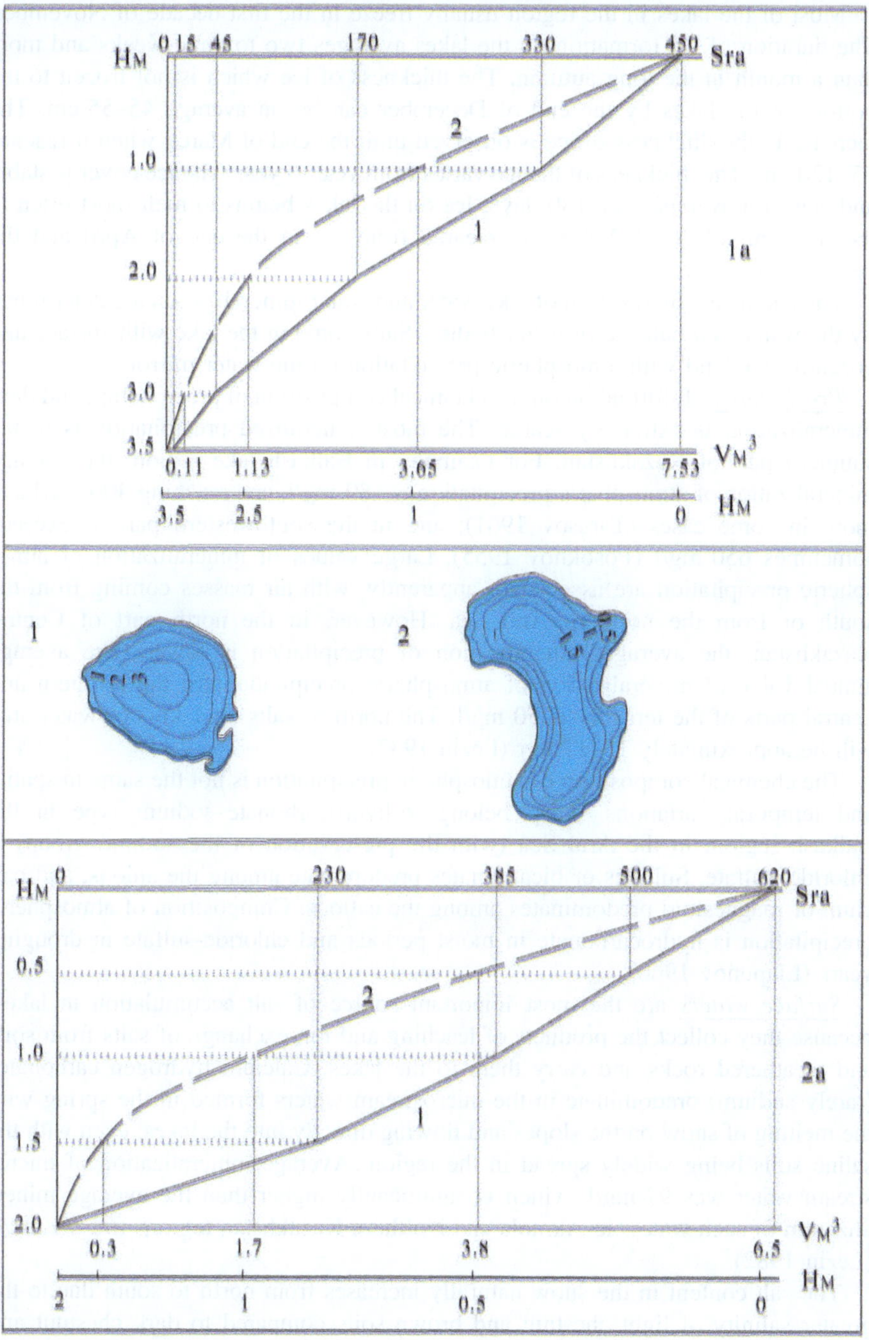

Fig. 4.10 Lake Saumalkol: 1-plan: σ) 1 a) Curved lines: bathymetric (1), three-dimensional and coastal areas (2); 2, 2a the same for Lake Karaukamys

Most of the lakes in the region usually freeze in the first decade of November. The duration of ice formation on the lakes averages two to three weeks and more than a month in the long autumn. The thickness of ice which is not frozen to the bottom of the lakes by the end of December can be on average 45–55 cm. The increase in the thickness of ice is observed until the end of March when it reaches 95–120 cm. The thickness of the ice varies from year to year. The ice cover is stable and stays on average 165–180 days. Ice on the lakes begins to melt most often in the first half of April. Lakes are cleared from ice in the end of April and the beginning of May.

The chemical composition of lake water and water mineralization are determined by the water–salt balance of water bodies. Salts come in the lake with surface and groundwaters and with atmospheric precipitation on the water mirror.

Precipitation. Information on the chemical composition of precipitation and their mineralization is extremely scarce. The most mineralized precipitation is in the southern part of Kazakhstan. For example, in Balkash lake region, the average mineralization of atmospheric precipitation is 180 mg/l, and reaching 400 mg/l and more in some cases (Tarasov 1961); and in the southwestern part, it exceeds sometimes 650 mg/l (Posokhov 1955). Large values of mineralization of atmospheric precipitation are associated, apparently, with air masses coming from the south or from the northern Aral Sea. However, in the north part of Central Kazakhstan, the average mineralization of precipitation is lower. The average annual value of mineralization of atmospheric precipitation for the northern and central parts of the territory is 50 mg/l. The norm of salts on 1 km^2 of water area will be approximately 15–17 t/yr (Lezin 1982).

The chemical composition of atmospheric precipitation is not the same in spatial and temporal variations. They belong to hydrocarbonate–sodium type in the Balkash region, in the Aral Sea (with the preservation of the sodium group) is chloride–sulfate. Sulfates or bicarbonates predominate among the anions, and calcium or magnesium predominates among the cations. Composition of atmospheric precipitation is hydrocarbonate in moist periods and chloride-sulfate in droughty years (Liapunov 1960).

Surface waters are the most important source of salt accumulation in lakes, because they collect the products of leaching and ion exchange of salts from soils and weathered rocks and carry them to the lakes. Calcium hydrogen carbonates (rarely sodium) predominate in the microstream waters formed in the spring with the melting of snow on the slopes and flowing directly into the lakes, even with the saline soils being widely spread in the region. Average mineralization of microstream water was 97 mg/l, which is significantly higher than the average mineralization of such waters in Akmola and Northern Kazakhstan regions (68–75 mg/l) (Lezin 1982).

The salt content in the snow naturally increases from north to south due to the greater salinity of light chestnut and brown soils compared to dark chestnut and southern chernozems. In addition, there are more solonchaks and dry salt lakes containing easily weathering salt in the south of the Central Kazakhstan. The thickness of the snow cover decreases to the south. Under conditions of strong winds and sparse vegetation cover, it facilitates snow mixing with loose soil and enrichment with salts.

The most important role in the water–salt feeding of lakes is played by rivers and temporary watercourses, in the valleys of which the vast majority of small lakes are located, which periodically (during the spring flood) are replenished with river waters.

According to the hydrochemical map of the USSR rivers, for the period of low levels of warm seasons, in the territory of Central Kazakhstan, there are waters of all three classes with high and high mineralization (>0.5 g/l). During periods of high water availability of those classes, mineralization decreases to 0.1–0.3 g/l in the Ulytau region and in the basins of the Nura, Sherubai-Nura, Kulanotpes, and Esil rivers (SWR 1966).

All rivers are characterized by a large range of fluctuations in the degree of mineralization and inconsistent chemical composition of water over time.

The lowest values of mineralization are on the period of spring flood, when the river water discharge is the greatest. With a decrease in water flow, mineralization continuously increases until the end of August; in September, the curve drops somewhat due to a slight increase in runoff due to liquid precipitation and a decrease in evaporation. In October, it again increases and reaches its maximum values at the end of winter. In low water years, when springwater discharge is low, the salinity of rivers is much higher than in years with high spring floods (Lezin 1982). The chemical composition of river waters is very different and varies with time. Mineralization of the Nura river varies in the flood from 0.17 to 1.5 g/l. In the upper reaches of the Nura river, depending on the water content of the year and the degree of mineralization, water during the spring flood is more often chloride, less often hydrocarbonate or mixed anionic composition. From cations dominate calcium (with mineralization up to 0.3 g/l) and sodium (with a mineralization above 0.3 g/l). In the rest of the time, chlorides and sodium sulfates predominate. It is noteworthy that the second type of water is constant here at any time of the year. However, Downstream, after the tributaries of the Sherubai-Nura and Ulken-Kundyszdy rivers (with water of hydrocarbonate, calcium, or sodium, with sulfates following bicarbonates), the composition slightly changes in the direction of increase Bicarbonates (Lezin 1982).

The *mineralization* of the lake waters in Karagandy region which ranges from 0.159 to 335.8 g/l. 85 lakes with area of more than 100 ha was surveyed, of which 24 lakes are fresh and 42 lakes are slightly saline up to 5 g/l and brackish, salty is 6 and 13 is saline. Mineralization does not exceed 0.5 g/l in the lakes of Arykty, Ulykol, Kurpesh, Izendy, Zhamankol, Karakol, Tapalkol, Karaukamys, Aryktykol, Sasykkol, and Burshiktykol.

Water mineralization fluctuates from 103 to 335.8 g/l in the lakes of Karasor, Sarykol, Dogalan-Karasor, Tuzkol, Sarykaska, Karatuz, Karakol, Kokdombak, and Biesoigan.

According to the chemical composition of water, 66 lakes belong to chloride (77.7%), 16 is bicarbonate (18.8%), and 3 is sulfate (3.5%) class (Alekin 1970)

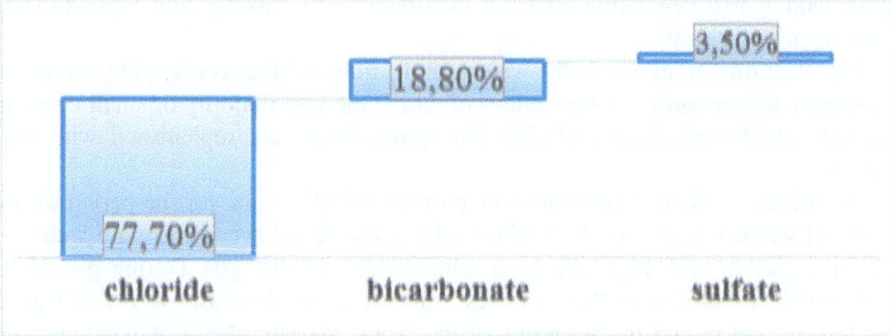

Fig. 4.11 Chemical composition of water in lakes of central Kazakhstan

(Fig. 4.11). Mostly soft or moderately hard neutral water is accumulated in fresh lakes, and it is very hard and slightly alkaline in brackish and salt lakes.

The content of trace elements in lake water (mg/l) is as follows: Zn-0.005-0.15; Mn less than 0.01; Cu is less than 0.002; Fe is less than 0.1; J is 0.003-0.7; F-0.5-10.0; Br-0.13-272.4; B-0.01-60.0.

The concentration of metals decreases when salinity (10 g/kg and more) of water in lakes increases. Then, the content of fluorine, iodine, bromine, and boron increases.

The content of zinc, manganese, and copper is not sufficient in fresh and slightly brackish waters; but fluorine content is high, i.e., it exceeds the maximum permissible concentration for drinking water. A low percentage of some microelements in the lake water is due to climate aridity and a small catchment area of lakes, very weak feeding by groundwaters, and also a marginal reserve of microelements in the soils.

Thus, the mineralization and chemical composition of atmospheric precipitation, surface and groundwaters are different in territory and in time. When all these waters come into contact with lake water, they do not just mingle, and chemical processes occur in them, which lead to the formation of a certain chemical composition of the lake waters. Consequently, the chemical composition of the water in the lakes is determined by the complex of physical–geographical and hydrogeological conditions of the basin and by those processes (physical, physical–chemical, chemical, and biochemical) that occur in the lake itself.

References

Alekin OA (1970) The basics of hydrochemistry. Hydrometeoizdat (in Russian), Leningrad

Kruze VV (1957) Physical and chemical properties of the water and mud of the Karasor lake, Vol 1, No 5. Karaganda State medical University (in Russian)

Liapunov MF (1960) Formation of the chemical composition of brine of the Lake Bol'shoy Azhbulat and its hydrochemical regime. Abstarct of the dissertation for candidate in Geography. Novocherkassk, p 23 (in Russian)

Lezin VA (1982) Lakes of central Kazakhstan: comprehensive typological characteristics of the regime and resources. Alma-Ata: Nauka, pp 188 (in Russian)

Philonets PP (1981) Essays on the geography of inland waters of the Central, Southern and Eastern Kazakhstan (lakes, water reservoirs and glaciers). Nauka, KazUSSR, Alma-Ata, pp 283 (in Russian)

Posokhov EV (1955) Salty lakes in Kazakhstan. Academy of Sciences of the USSR (in Russian), Moscow

SWR (Surface water resources of the USSR) (1966) Leningrad, Hydrometeoizdat. Part 13. Central and Southern Kazakhstan, Volume 1. Karagandy region, p 482 (in Russian)

Tarasov MN (1961) Hydrochemistry of the lake Balkash. Academy of Sciences of the USSR (in Russian), Moscow

Chapter 5
Morphometry and Genesis of Lakes in Kazakhstan

5.1 Morphometry and Genesis of Lake Basins in Kazakhstan

There are 48,262 lakes in Kazakhstan, except the Caspian and Aral Seas, ponds, water reservoirs, and pliable lakes. The size of these lakes are more than 1 ha with a total surface area of 45,002 km^2. Small lakes with size of less than 1 km^2 make up 94%, by the area 10%, and large lakes (more than 1 km^2) counts 3014 with an area of 40,769 km^2 (90%), including 21 lakes larger than 100 km^2 in size with a total area of 26,886 km^2, i.e., 59% of the water surface of all lakes in Kazakhstan (Table 5.1) (Philonets and Omarov 1974).

Lakes are distributed unevenly on the territory of Kazakhstan. They are located from each other for hundreds of kilometers, or they are distributed so dense that form lake regions.

Thus, there are 21,680 lakes (45%) in Northern Kazakhstan, with a total area of 15,623 km^2 (35%), while in Central and Southern Kazakhstan there are 17,554 lakes (36%) with a total area of 4658 km^2 (10%), excluding Balkash Lake. In Northern Kazakhstan, there are 2.61 km^2 of water area of lakes for 100 km^2 of territory and in Central and Southern Kazakhstan is 0.23 and 0.53 km^2, respectively. The largest lakes are located in the southwest and southeast parts of Kazakhstan. These are the Caspian and Aral Seas and the Balkash and Tengiz Lakes in Central Kazakhstan; Alakol and Sasykkol are near the Dzhungar Gate, Markakol is in the Altai Mountains.

Most of all lakes are distributed in the forest–steppe and the northern part of the steppe zones, and also there are many in the floodplains of large and delta areas of drainless rivers which loose in the sands. Small lakes are located mainly in river floodplains; medium and large ones are mainly confined to ancient river valleys and tectonic depressions. The main mass of water reservoirs is concentrated at absolute altitudes from 100 to 350 m, mainly in loose (friable) Cenozoic deposits. The area of watersheds (catchment area) of lakes varies from 10 to 320 km^2; however, they

J. Abuduwaili et al., *Hydrology and Limnology of Central Asia*, Water Resources Development and Management, https://doi.org/10.1007/978-981-13-0929-8_5

Table 5.1 Distribution of lakes in Kazakhstan by area

Area (km^2)	Number of lakes	Total area	
		km^2	%
0.01–0.1	35,763	1025	2.3
0.11–0.2	3435	467	1
0.21–0.3	1764	438	1
0,31–0,4	1196	420	1
0.41–0.5	879	396	0.9
0.51–0.6	656	395	0.9
0.61–0.7	477	310	0.7
0.71–0.8	448	338	0.7
0.81–0.9	362	308	0.7
0.91–1	268	254	0.6
1.01–5	2357	4900	10.9
5.01–10	361	2424	5.4
10.01–50	252	4869	10.8
50.1–100	23	1572	3.5
Above 100	21	26,886	59.6
Total	48,262	45,002	100

Notice 1. Caspian and Aral Seas, lakes, ponds, and reservoirs are not included to the list. 2. Plexus lakes are 6941 with a total area of 235 km^2

are much larger in semideserts and deserts zones. The coefficient of lacustrine lakeiness varies from 0.2 to 4.4% in administrative regions of Kazakhstan. In general, for Kazakhstan it is 1.65%. The total volume of water in the lakes exceeds 190 km^3.

It should be noted that the information on the morphometry of many lakes of Kazakhstan, especially for small ones, can significantly have changed and will change in the future. This is evidenced by data on the hydration of lake basins and also the level regime of lakes in various natural zones of Kazakhstan.

The lake basins of Kazakhstan belong to different genetic groups, which creates a wide variety of lakes in size, shape, depth, regime, hydrochemical features, etc. Differences in the genesis of the basins depend both on geological factors such as latest tectonic movements, seismicity, structure, and lithology of rocks and from exogenous factors as actions of river erosion, wind, ice, karst, suffusion, gravity processes. In addition, the changes caused by human activities and the inconstancy of climatic conditions cannot be ignored. Aridity of the climate leads to the fact that many depressions are not occupied by water or are not fully occupied. Small water reservoirs disappear, and large ones grow shallow. The increase in humidity contributes to the overflow of the lake basins and their descent.

Rhythmic climate fluctuations explain many peculiarities of lakes during Holocene period. Currently, the direction of the natural rhythm is generally unfavorable for water supply of lakes, although the age-old changes smooth this trend in their development. Steppe lakes owe their existence to a dry climate, which determines the insufficient development of erosion, although this sounds paradoxical (Bykov 1933). Thus, the extraordinary growth of lakes in Kazakhstan finds an explanation.

Several genetic types of lake depressions or hollows can be distinguished on the territory of Kazakhstan:

I. Tectonic depressions can be divided into mountain regions and plain areas.

A. Mountain regions includes: (a) intermontane depressions; (b) intra-montane depressions; (c) foothill depressions; (d) dams–lakes; (e) gravel-fissured depressions of granite massifs.

B. Plain areas include: (a) platform bends (flexure); (b) dams–lakes; (c) compensatory troughs of salt domes.

II. Exogenous depressions can be divided into erosive, soro (coarse)-deflationary, and deflationary depressions, karst depressions, suffosive depressions, gravitational lacustrine depressions, glacial lakes.

A. Erosive depressions: (a) residual plasy or plaques (baldheads)-плёсы; (b) floodplain lakes; (c) delta lakes; (d) intra-continental delta (limestone) lakes.

B. Soro-deflationary and deflationary depressions; C. suffasion depressions; D. karst depressions; E. gravitational lacustrine depressions; F. glacial lakes (Fig. 5.1).

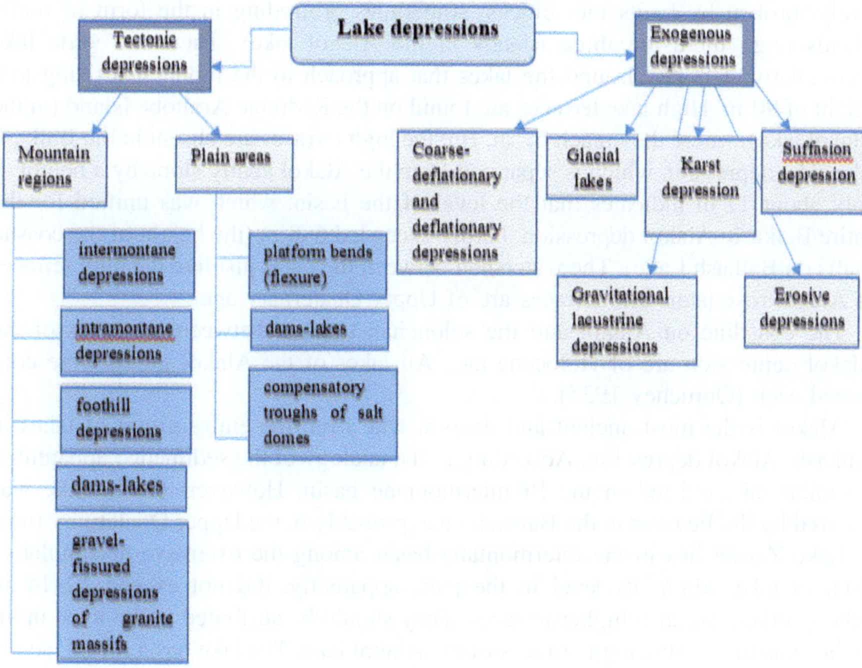

Fig. 5.1 Different genetic groups of lake basins in Kazakhstan

5.2 Tectonic Depressions

A. Mountain regions. Tectonic depressions are widespread in mountain areas, but the lakes in them are mostly run down by rivers and therefore are currently scarce.

(A) Intermontane depressions. The South Caspian uncompensated deflection or bends (flexure) are the largest depression which is concluded between the mountain systems of the Greater and Lesser Caucasus in the west and Kopet Dag and Elbrus in the east. It forms the southern part of the Caspian Sea and separated from the Middle Caspian Basin by the Absheron threshold. The depth of the depression reaches 975 m, there is no granite layer in its deepest part, and the thickness of loose deposits lies directly on the basaltic layer. According to the data, the Black Sea and South Caspian basins are residual from the former Tethys Ocean in the Mesozoic, which blocked as a result of the drift to the north of the southern continental blocks. The Caspian reservoir became isolated after the Middle Pliocene, since then its level has experienced repeated oscillations of large amplitude. As a result, numerous flooded and elevated coastal relief forms have occurred that fix the transgressions and regressions of the sea.

The Alakol intermontane basin divides the mountain systems of Tarbagatai and Zhetysu (Dzhungar) Alatau and represents a vast plain formed by the sediments of ancient Alakol basin, surrounded by powerful cones of river discharges flowing into the group of lakes such as Alakol, Sasykkol, Koshkarkol (Uialy), and Zhalanash located in it. The foundations of the Alakol depression are formed comprehensively, broken by faults into blocks, sometimes protruding in the form of rocky islands (e.g., there are three islands in the Alakol lake). There are wide lake accumulative terraces around the lakes that approach to the mountains rising to a height of 90 m. High lake terraces are found on the Kishkine Araltobe Island (in the Alakol lake), where they reach 67 m. But the high terraces are absent in the Balkash lake. The depression which is separated from the Alakol gently slope by a height of only about 17 m indicates that the level of the basin, which was unified for the entire Balkash–Alakol depression, hardly exceeded 6–8 m (the height of the coastal shafts on Balkash Lake). The watershed between the lakes just like the high terraces of Alakol rose later. The terraces are of Upper Quaternary age.

The coastline on Alakol and the solonchak terraces between the lakes of the Alakol depression are of Holocene age. All lakes of the Alakol group were connected each (Obruchev 1935).

Alakol is the most ancient and deepest lake from the entire group of lakes in Balkash–Alakol depression. According to the analogy of the sediments, apparently, a similar lake existed in the Ili intermontane basin. However, the Ili lake was lowered by the Ile river in the Balkash lake, probably in the Upper Quaternary time.

Lake Zaisan lies in the intermontane basin among the extensive accumulative plain of lake origin. Its level in the past, apparently, did not exceed 12–14 m, although there are also higher terraces. They should be attributed to the local uplift of the coastline, as the high terraces of the Alakol lake. The lake has a runoff, so the lake did not overflow with water even in the era of the maximum glaciation of the depression. Coastal shafts and swampy areas are widespread that fix the higher levels of the lake (they were at an altitude of 7–8 m before it was flooded) (Philonets and Omarov 1974).

(B) Intra-mountain depressions. Only Markakol lake in the East Kazakhstan region (South Altai) belongs to the lakes of the intra-mountain depressions. However, the traces of now deflated lakes have been preserved in a number of places in the mountains. Usually, these depressions are less bent than the deeply lowered inter-mountain deflection or bends (flexure). Therefore, they are easily opened by rivers. They arose later intermontane depressions only from the beginning of the manifestation of new movements.

Markakol Lake is fresh and flowing, and its depression is asymmetric. The lake is close to the mountainous steep slopes in the south, and in the north the shore is accumulative. It is located in a depression, apparently connected with the basin of the Kara-Kaba river that lying to the east. At first, the Kalzhir river flowed in the depression, later it was piled up, and Markakol Lake was appeared due to the tectonic uplift of the block which played the role of a threshold.

(C) Foothill depressions. As a rule, they are filled with powerful sediments of the cones of removal and rarely there is the possibility of the formation of lakes. But in Kazakhstan, the piedmont deflection located between the Kazakhstan shield and the Zhetysu (Dzhungar) Alatau was not filled with sediments in northern part and as a result Balkash lake was formed. Foothill Balkash depression merges/connected with the Alakol intermontane one in the eastern direction.

In the west, it is closed by uplifts of the Shu-Ili Mountains and Eastern Betpakdala.

Balkash Lake is young and arose after the breakthrough of the Ile river through the Kapchagai gorge only in the Upper Quaternary period, when apparently the Ili lake was discharged. His youth is evidenced by the intensive nature of the coast and its insignificant processing by wave processes, low salinity especially in the western part which desalinated by Ile river. The structure of the foundation has been little disturbed/broken by the latest movements, and the surface of the bottom is comparatively flat (Philonets and Omarov 1974).

Accumulation of undermountain plumes of the Zhetysu (Dzhungar) Alatau gradually pushed the Balkash Lake to the north (Kostenko 1946), and now its southern shores are asymmetric, mostly shallow, low lying, easily changing the configuration at the slightest fluctuations in the lake level, while the northern ones are more stable, mostly shallow, and are limited by coastal valleys (modern and Upper Quaternary) 3–5 m in height. Sometimes, there are parts of a steep bank, and their steepness is not due to abrasion. It is due to the latest discharges that continue on land which limit the elevated areas of the hills. The newest uplifts are also shown by lake sands on fault banks which occur at altitudes from 9–12 to 18 m.

In the Middle Quaternary period, the foothill depression was filled with a layer of sands carried out by rivers from the slopes of ice-covered mountains. The origin of the lake belongs to the Lower Quaternary period (Kostenko 1946). The broad development of aeolian processes in the xerothermic (interglacial) period is indicated by the sediments of the Alakol depression (Svarichevskaya 1952, 1961) and the findings of flooded peatlands in Balkash (Sapozhnikov 1954). Apparently, the Balkash lake significantly dried out at that time and then it was restored in the Upper Quaternary era. Thus, desert conditions were preceded to the appearance of the lake in the Upper Quaternary time. The powerful strata of alluvial sands and perhaps alluvial–lacustrine origin ones accumulated in the era of the Maximum and Middle Quaternary glaciation of the Zhetysu (Dzhungar) Alatau that were undergoing a waving process. The remnants of this desert have survived to this day in the Southern Balkash region. Young Ile river paves valley, crossing ancient ridge sands. Balkash Lake ingressed into inter-ridge depressions in the stage of maximum flooding of the Upper Quaternary glaciation period.

Balkash was connected to the united Alakol lake at the maximum transgression. This is evidenced by a smooth, slightly convex lake plain between the lakes. It reaches a height of 17 m above Balkash. Its rise is confined to the intersection of the Balkash–Alakol depression with the main Dzungarian fault and appeared in the Holocene.

(D) Dams–lakes. in the mountains usually arose as a result of the springs of mountain rivers by caving or moraine formations. Kostenko (1971) studied the section of the dam broken by a flattened Issyk lake in the Ili Alatau, and he showed that it was only covered with detrital material on top. The main body was created by blocks protruding along faults of sublatitudinal strike (according to the general folded structure of the ridge). The Big Almaty Lake dam has a similar structure.

(E) Gravel-fissured depressions of granite massifs. Their formation is associated with the emergence or extrusion of intrusive, mainly granite bodies from focal volcanic chambers. Typical depressions are occupied by fresh lakes of the granite island mountains of Kokshetau, Bayanaul, Zerendi, and others.

The lakes of the Kokshetau mountains are most studied. They are fresh and deep (Bolshaya Chebache, Shchuchye, etc.), begin immediately from the rocky slopes of mountains devoid of foothill trails (plumes). The banks are shallow; there are about five terraces along the mountain slope (but not always clearly); the coastal shafts are more often along the lower coastline. There are up to five more or less large lakes in all around the Kokshetau Mountains. Almost all of them were connected together in the period of the Upper Quaternary flooding.

B. Plain areas. Lakes of plain areas, despite the low intensity of tectonic movements, also bear traces of dependence on them.

(a) Platform deflections/flexure. These depressions include the basin of the Aral Sea, the Middle Caspian, and the North Caspian basins.

The Aral Sea is young Sea. It arose in the Upper Quaternary period, after the final turn of the Amudarya river to the north (Tetyukhin 1970). Prior to this, at the end of the Neogene, the Aral Sea was part of a sea basin that stretched to east from the Caspian Sea as evidenced by marine sediments of Apsheron age.

The sea regressed to the west in the Lower Quaternary time, and sediments of the lake lagoon type began to be deposited in its place in a limited area. There was no lake basin in the Middle Quaternary period.

Intensive aeolian development takes place at the end of the Middle Quaternary and the beginning of the Upper Quaternary time. An aeolian sandy relief is formed that surrounds modern Aral Sea which traced especially in the south, on its bottom. For the first time, a large amount of water in the Aral Sea appeared in the Upper Quaternary time, when Amudarya high water river broke through Kyzylkum, using probably a small river valley or a system of synclinal narrow depressions (Svarichevskaya 1965). It arose not earlier than 22 thousand years ago, and its runoff fluctuated for a long time between the Aral and Sarykamysh basins (Tolstov and Kes' 1960). The Amudarya eventually turned into the Aral basin only in the Holocene as evidenced by bottom sediments with *Cardium edule L.* of Late Holocene time (Tetyukhin 1970).

The occurrence of a basin in the central part of the Turan epipaleozoic platform is explained by the tectonic deflection of this part of the earth's crust. The Paleozoic base is lowered to a depth of more than 3 km, and there is a whole system of meridional short folds with which the basement is crushed, and they are reflected not only in the structure of the Mesozoic–Cenozoic cover, but also in the modern relief. They are confined to a narrow deflection along the western shore of the Aral Sea, islands, bays/gulfs, and peninsulas of the northern coast. But it is likely that during the period of the existence of solonchaks and in the desert conditions of xerothermic time, the tectonic depression experienced also a sable/coarse deflation that deepened the depression.

The Middle Caspian and North Caspian basins, whose water areas are located on the territory of Kazakhstan, present flat-bottomed deflections. The Middle Caspian Depression is delineated from the deep and ancient South Caspian deflection by a transverse shaft—the Apsheron spur (Svarichevskaya 1965); it is separated from the North Caspian basin by a less distinct Karpinsky shaft which extended along the continuation of Mangyshlak and Donbas. The deepest part of the depression reaches 190 m and lies in its southern half.

The Middle Caspian depression is geologically structurally heterogeneous. Its large northeastern part is confined to the epihercine platform with a shallow (1–3 km) bedding of the basement that is covered by the Mesozoic–Cenozoic cover. The smaller part which is southwestern wing is located on the northeastern wing of the meganticlinoria of the Greater Caucasus that composed of Paleogene–Neogene and Quaternary sediments.

Flooded coastal lines that present terraces and bank shafts such as Krasnovodskaya—68–77 m, Bektashskaya—58–69 m, Mangyshlak—48–59 m, then 40–42 m and 32–34 m have been found at the bottom of the Middle Caspian basin. Coastlines are also observed at great depths, up to 300 m, but they are less pronounced. The bottom of the basin continues to bend. The differentiated local movements predetermine the formation of positive and negative forms of underwater relief and deformation of the shelf edge.

The North Caspian depression is confined to the southern outskirts of the Precambrian in the northern part and in the southern part to the epihercine platform. The relief of the bottom differs in leveling with depths from 5–7 to 20 m and by the predominant development of accumulative forms. A chain of beams such as

Chasovaya, M. Zhemchuzhnaya, Rakushechnaya stretches along the southern part of the depression on the Karpinsky threshold and relatively large islands (Kulaly, Tyuleny, etc.) representing growing brachyanticlines. Sandy forms in the form of underwater barkhans are observed on the bottom of the Kulaly island and to the north of the Tubkaragan Peninsula. The Urals trough/hollow/gully is clearly expressed in the relief of the bottom and clearly formed in the epoch of post-Khvalyn regression.

The terraces stretching along the coasts of the Caspian Sea are transferred to the vast plains in the north that were formed during the period of marine transgressions (elevations above the Caspian Sea level): ***(1)*** Newcaspian, 1929 with 2–2.5 m in height; Newcaspian, the beginning of nineteenth century in 4–4.5 m; Newcaspian of maximum phase in 6–6.5 m (4000–5000 years ago); ***(2)*** Late Khvalynian in 10–12, 18, 26 m; ***(3)*** Early Khvalyn in 36–38, 42–43, 47–51, 61–62, 75 m; ***(4)*** Bakin in 90 m; and ***(5)*** Akchagyl in 114–124 m.

The Tengiz–Korgalzhin depression also arose as a result of the flat deflection of the Kazakhstan shield. It is located in the central part of a large superimposed Paleozoic basin, the pre-Paleozoic basement of which is lowered to a depth of 2000–2500 m. Its lowering continued in the Cenozoic when there were conditions for the existence of lakes.

Filling of modern lakes was preceded by erosion dismemberment of depression surface by valleys of small rivers that flowed into the Esil river. The depression has been filled with water and lakes such as drainless salt Tengiz lake, and flowing fresh Korgalzhyn began to form in the Quaternary period. Deflection of the depression intensified in the Middle Quaternary period. As a result, one of the eastern narrow deflections of the northeastern strike was intercepted by the Nura river, and drain into the basin began only in the Upper Quaternary period.

The origin of large but shallow and flat-bottomed lakes such as Kalmakkol, Ulukol, Mamai lying on the watershed plains of Northern Kazakhstan is not entirely clearly studied. It is very likely that they also arose as a result of smooth tectonic flexure. Small synclinal and monoclinal depressions of solonchaks described a right bank of the lower reaches of Shu river (Aleksandrova 1952).

(b) The dams–lakes. There are lakes that tucked up/stiffened by growing tectonic folds on the plains of Kazakhstan. Usually, small rivers in low-power and often with a temporary flow are sprung. They are unable to overcome the rising, albeit poorly expressed. Thus, usually the ancient erosion valleys collapse with the appearance of basins occupied by lakes.

For instance, Karasor lake in Central Kazakhstan which has a twisty shape belongs to the dams–lakes of the tectonic type (Svarichevskaya 1965). It falls into two rivers: Karasu and Taldy which were the upper reaches of the Motak river that flows into the Nura river. A dam 23 m high in the form of a very gentle swelling/flatus is located on the extension of the Karkaraly new fold broken by a fault (northeastern direction). The lake is young and does not yet have terraces. There are two terraces with river origin in 1.5 and 6 m on the shore of the lake, near the Koyandy village (Philonets and Omarov 1974).

There is an Analogue group of lakes formed on the watershed of the tributaries of the Terisakkan and Tasty-Turgai rivers. The group of lakes include Koskakol, Palvankol, Shoindykol which arose as a result of the separation of the upper reaches of the rivers by growing anticlines and forming subterranean strata (Mikhailov 1957). These anticlines have been inherited from Paleozoic times. The revival of their uplift occurred after the establishment of the valleys in the Quaternary period. The subterranean strata are expressed by gentle shafts in 5–6 m of height. The lakes are shallow, and their depth rarely exceeds 2 m.

(c) Compensatory troughs are located at the salt domes which rise above the low plain of the Caspian Sea. These include the troughs occupied by the lakes such as Baskunchak, Aralsor, Shalkar, Inder. The rise of the salt domes corresponds to the lowering of the lake shores (Meshcheryakov 1960). So, Big Bogda lake rises at a speed of 0.5–0.6 mm/year, and the coast of the Baskunchak lake experiences lowering at a speed of 1 mm/year.

5.3 Depressions of Exogenous Origin

Depressions of exogenous origin include erosion, soro-deflation (aeolian), suffosive and karst depressions which are specific for the plains and gravitational and glacial depressions for the mountains.

A. Erosive depressions

(a) Residual plasy or plaques (baldheads)-плёсы are depressions in ancient valleys. They have a linear, elongated gully shape, and they represent relicts of individual valleys of the ancient river network that found on the watershed surface of the northern Kazakh shield (Shantser et al. 1967). The vast cup-shaped basins of the upper reaches of the Zhabaya and Archaly rivers are referred to as more complex formations and considered as whole ancient watersheds consisting of several converging valleys covered with loose sediments. The basins of the Kaindykol, Karagayshyk-Aidabul, Shebundy-Kamysh lakes can be an example of the erosive depressions (Shantser et al. 1967).

The flowing Biylikol lake located in the valley of the Assy river at the south-eastern foothills of the Karatau ridge referred to depression with denudation–erosional origin. The Biylikol lake is quite large; there are up to two or three lake terraces around the lake and the lowland plain north of the lake (Krieger and Semenov 1953). However, perhaps, the latest tectonic movements are also involved in the origin of the lake basin. The valley from the Kainazar lake stretches on the northwestern extension of the Biylikol Lake Plain. Further, the Akkol, Kyzylautkol, Aschykol, Tuzkol, and other lakes are distributed to the northeast on the same longitudinal valleys.

Some lacustrine depressions of Central Kazakhstan and the northeastern part of the Kazakhstani shield should also be classified as complex formations. These are the Alkasor, Sangasy, Maisor, Tengiz, and Korzhynkol located in the ancient erosive–tectonic meridional valley north of the Erementau Mountains. Tengiz and

Korzhunkol lakes have four lacustrine sculptured terraces, and only the third one is characterized by powerful coastal shafts. The upper terraces are common for both lakes. The two lower ones are formed in the conditions of a reservoir located. The process of depression deepening continues and is now under the influence of (soro) coarse deflation (Svarichevskaya 1961).

A similar complex genesis is formed by hollow lakes located in the valleys of the now dry right tributaries of the Ertis river. Age of the valleys is Upper Quaternary and Holocene. Most of the lake basins experienced a soro-deflation depression and is now clearly incised into the bottoms of the gully. They are drainless and surrounded by lake terraces. These include the lakes such as Koryakovskoe, Tavolzhanskoe, Bolshoi Azhbulat, Studenoe, Kyzyltuz, Krasnovishnevoe. The depressions of Arys lake are complex but similar with the Quaternary age in genesis located to the west of the lower reaches of the Sarysu river and the Dabusyntuz lake to the east of it are confined to the ancient Sarysu river. The system of lakes in the Torgai Valley also includes: Ubagan (Kushmurun), Aksuat, Sarykopa, Yarkamysh, and the extensive Shalkar–Tengiz solonchak which closes the chain of depressions in the south and is the Turgai river basin. The Shalkar–Tengiz lake periodically dries up turning into a solonchak.

(b) Floodplain lakes are located in the modern floodplains of rivers. The extensive Buralkenyntuz solonchak in the low reaches of the Shu river belongs to the floodplain lakes, although its vast size allows us to think that other processes (tectonic, soro-deflationary ones) could have taken part in the formation of the depression occupied by it.

(c) Delta lakes are widely developed in the present delta of the Ile river. There are also old lakes that appeared on the site of the abandoned meanders of the riverbed. The lakes usually have a curved shape, and end/terminal lakes where the peripheral riverbeds of the Ile river, and its streams such as Topar and Zhideli, as well as sickle-shaped lakes and chains of lakes which were the remnants of ancient river beds even before the formation of ancient barkhans (Shtegman 1952). There are such chains of lakes that have arisen and formed in the Southern Balkash before the penetration of the Ile river because of the flooding of inter-ridge depressions dividing the ancient aeolian ridges. They are located next to the modern delta.

(d) Intra-continental delta lakes include the already noted Shalkar–Tengiz lake solonchak which accepts the system of the Turgai river. This vast depression has traces of higher levels in the form of terraces on the slopes.

There is an extensive intra-continental delta composed of sediments of the Talas, Shu, and Sarysu rivers in the lower reaches of the Shu and Sarysu rivers. In general, this is a lowered area with various forms of relief, deflated depressions, dry beds, remnants, delta lakes. The delta is located in the tectonic deflection. In the middle Quaternary period, Shu and Sarysu rivers merged into the Syrdarya river; and Shu and Sarysu rivers have a common mid-quarter terrace with Syrdarya river. But tectonic uplifts of the Karatau shafts already were occurred in the Upper Quaternary time. The shafts were stretched in the northwestern direction on the extension of the Karatau ridge. The ridge is the southern part of the deflection and separated by the lower reaches of the Shu and Sarysu rivers from the Syrdarya river, and the rivers end in lake plains of Aschykol, Saumalkol. Now, the lakes are small and solonchakous; some of them is a self-shrinking and widely extended. Aschykol lake reaches the dry riverbeds of Shu and Bakhtykaryn rivers that has a rugged coastline, a lot of islands which have arisen as a result of the flooding of the hilly relief. A large number of lakes are located on the continuation of the main riverbed of Sarysu river turning to the west (Sagis lake and others).

B. Soro-deflationary depressions are a characteristic group of depressions which surprise researchers with sizes and cause skepticism about their aeolian genesis. The vast depressions of lakes such as Selety, Teke, Zhalauly, Kyzylkak, Ulken-Karoi in the northeast of Kazakhstan belong to this group (Svarichevkaya 1961). A characteristic feature of the depressions is also their deep incision: For instance, the cut in the depression of the Teke lake is about 100 m. The incision of the Karakia basin on the Mangyshlak plateau is characterized by a slightly larger value (bottom mark is −132 m); i.e., the depression is deeper than the level of the nearest erosion bases (e.g., the Caspian Sea with a level of 28 m). They are closed with slightly elevated bank peaks which indicates the high water in the past. At present, they are mostly small salty lakes and sometimes even solonchaks. They are cut into rocks of any composition: in friable Paleogene–Neogene rocks (Teke, Selety, Maraldy lakes) and in solid Paleozoic sedimentary and intrusive formations (Betpakdala-Mankavra, Krasnyi takyr depressions, depressions of Northern Balkash region, Ustyurt, and Mangyshlak). Extensive puffy solonchaks were required for the appearance of depressions in this size. Deflation occurred from the solonchak surface. Therefore, they are either confined to primary depressions of tectonic origin (at least weak deflections, as in the Northern Ustyurt) or to erosive valleys (Teke, Kyzylkak, Seleti Lakes). But sometimes anticlinal uplifts—some Ustyurt depressions (Karakia)—are vulnerable areas for deflation. As a result of the process of coarse deflation, the deepest depressions turned out to be lake depressions in the ancient tributaries of the Ertis river.

In addition to the soro-deflation depressions, there are simply deflationary lake depressions at Kalba at the western foot of the Monastery Mountains. These basins blew in granites (Velikovskaya 1947). There are a large number of closed and rounded valleys of small size at the foot of the Malyi Karatau, and they filled with water in spring. Many of them dry up in summer, their bottom is occupied by takyrs or sors, and only a few, for example, Akchakur keeps the water.

C. Suffasion depressions are widely developed in Kazakhstan. They dot the aligned surfaces of interfluves and gentle slopes covered with loamy layers or folded with loess. Usually, they are randomly distributed over the surface, but aerial photographs of the northeast Kazakh shield make it possible to establish their elongation by strings along the strike of rocks. Birch twigs (curtains) are often confined to them in the northern and northeastern parts of the Kazakhstan shield. Sometimes, the bottom of the depression is moistened; there appears a small lake surrounded by reeds. Despite the small dimensions of the valleys (depth 1–3 m, diameter 20–200–150 m), they stand out clearly on the grayish-yellow background of steppe vegetation.

Some small depressions are connected, probably, not with the processes of suffasion, but with permafrost in northern Kazakhstan within the West Siberian Lowland. Tubercles (bumps) and hydrolaccolitis were formed during the permafrost aggradation (Boytsov 1961). Later the depressions developed undergoing to denudation in the summer seasons, and the frozen bottom subsided and the soil settled after the permafrost degradation when the punch stopped. A depression remained on the site of the former partially eroded bulge.

D. Karst depressions are not typical for Kazakhstan as the development of carbonate rocks depends on the solubility of the rock mass and on the amount of precipitation.

Carbonate deposits occur among Precambrian, Cambrian, Ordovician, Silurian, Upper Devonian, and Lower Carboniferous rocks.

Modern karst is poorly developed, but deep (up to 240 m) ancient karst funnels executed at the bottom by Paleocene sediments indicate favorable conditions for karst in the past (Sladkopevtsev 1963). Ancient karst craters are expressed by extensive flat depressions in the modern relief. They are often filled with water (e.g., in the Astana area, Bolshoe and Maloye Aksu Lake, etc).

Karst processes took an active part in the formation of the lake basins of the right bank of the Shu river. In the future, the karst cavities/holes were used by a valley network and then again disconnected.

E. Gravitational lacustrine depressions are mainly developed in the mountains, to a lesser extent on the plains. They arise because of the collapse of rock masses from slopes to valleys (landslides, screes, and rockfall) and the springs of rivers. Usually, they are located in the lower tier of the mountains where there are a deep erosion cut and steep slopes. These include Kaindy Lake, as well as the Saty and Kulsu Lakes in Kungei Alatau.

Balyktykol lake is the dam–lake tucked by the aeolian sands of the Kuluzhon and Laily rivers within the Kalba Range. The dam–lakes are deep and unstable and relatively easy to flush their dam.

F. Glacial lakes are located in the upper tier of the mountains. They are very numerous, especially karst and circus. They are usually stiffened by moraine accumulations and are fairly stable due to their small size. They are threatened with descent mainly when regressively moving upward, and the gorges open the depressions. Below are the lakes stiffened by terminal moraines. Usually, they are smaller than lakes and sprinkled with debris in the gorges of the lower tier of mountains. There are lakes of glacial genesis in the mountains of Tarbagatai (Kokkol), Altai (Yazevoye, Bukhtarminsky, and others).

It is possible to distinguish lake basins of different sizes in the listed genetic groups, but of the same genetic category. In addition, there are also lacustrine depressions of complex genesis.

Concluding the consideration of the genesis of the lake basins, it should be noted that in most cases the lakes are unstable, especially in the mountains. Therefore, it is necessary to carefully study natural dams to prevent catastrophes.

References

Aleksandrova MI (1952) About issue of basin genesis in Betpakdala. Geographical collection, Moscow-Leningrad, Izvestia Academy of Sciencse of the USSR 1:118–129

Bykov GE (1933) Geological structure and minerals in the western part of Atbasar region. In: Trudy of the All-Union geological prospecting association USSR, 283 (in Russian)

Boytsov MN (1961) About formation of relief in the condition of underground glaciation. Trudy VSEGEI 64:27–36 (in Russian)

Krieger NI, Semenov AI (1953) The role of young tectonics in geomorphogenesis of the Malyi Karatau and its Foothills. Izvestia VGO, 85:5 (in Russian)

Kostenko NN (1946) About history of the Balkash lake. Izvestia KazPHAN USSR, Geol Ser 8:96–104 (in Russian)

Mikhailov AE (1957) About issues on some lake genesis in the northern part of the Central Kazakhstan. 4:252–258 (in Russian)

Meshcheryakov YuA (1960) Morphostructure of the plain-flatform areas. Academy of Sciencse of the USSR (in Russian), Moscow

Obruchev VV (1935) The bays of the Northeastern Caspian Sea. In: Expeditions of Academy of Sciences of USSR. Moscow-Leningrad, pp 289–292 (in Russian)

Philonets PP, Omarov TR (1974) Lakes in the Northern, West, and Eastern Kazakhstan (reference book). Hydrometeoizdat, Leningrad, p 121 (in Russian)

Svarichevskaya ZA (1952) About history of the Balkash-Alakol basin. Vestn LGU, Geol Geogr Ser 7:107–112 (in Russian)

Svarichevskaya ZA (1961) The acient peneplen of Kazakhstan and the main period of its transformation. LGU, p 295 (in Russian)

Svarichevskaya ZA (1965) Geomorphology of Kazakhstan Middle Asia. Leningrad, LGU, p 296 (in Russian)

Sapozhnikov DG (1954) Formation of precipitation in lakes of the arid zones of the USSR. Lake Balkash. In: Formation of precipitation in modern water bodies. Moscow, pp 283–314 (in Russian)

Shantser EV, Mikulina TM, Malinovskii VYu (1967) Cenozoic of the Central part of Kazakh shield. Moscow, Moscow State University. pp 350 (in Russian)

Shtegman BK (1952) About history of formation of the Ile river delta. Geogr Collect USSR 1:133–150 (in Russian)

Sladkopevtsev SA (1963) About genesis and types of hollows in the Central Kazakhstan. Izv Acad Sci USSR Geogr Ser pp 84–90 (in Russian)

Tetyukhin GPH (1970) About paleogeography of the Aral Sea Basin. Izv Acad Sci USSR, Geol Ser 5:67–69 (in Russian)

Velikovskaya EM (1947) Pliocene and quaternary sediments and the development of the relief of Kalba. Moscow State University, Geological Series, Part 2, Issue 124 (in Russian)

Chapter 6
Water Balance and Physical and Chemical Properties of Water in Lakes of Kazakhstan

6.1 Water Balance and Fluctuation of Levels in Lakes of Kazakhstan

According to the conditions of water exchange, there are differentiated drainage, periodically sewage, sewage, and flowing lakes in the territory of Kazakhstan. Main of them are drainless.

The components of the water balance of the drainless lakes (surface runoff, precipitation falling on the surface of the lake, groundwater, filtration and evaporation) are subject to large intra-annual and long-term changes. They are in conditions of shallow water lead to a characteristic for Kazakhstan in inconstancy level, the drying out of a significant part of the water bodies in summer and autumn seasons, as well as in perennial low-water periods.

In the lakes of Central Kazakhstan, the sharp rise in the water level caused by the influx of large amounts of water during the spring snowmelt begins at the end of March: the first ten days of April and ends mainly in 10–15 days. Its height varies greatly over the years and can reach more than 2 m. The lowest level is set at the end of winter. The average annual amplitude of the level fluctuation is 0.6–0.8 m. In the water balance, the runoff from the surface of the catchment area is 50–60% of the total parish in the more humid northeast and 80–85% in arid areas. Summer precipitation is 12–39% of the incoming part of the balance; solid one accumulated on ice in the form of snow is 5–15% and more on lakes with thickets of higher aquatic vegetation. The flow rate of the water balance of the drainless lakes is almost entirely dependent on the evaporation of the water surface.

In the lakes of Northern Kazakhstan, the rise of water begins from the first half of April and continues on plain reservoirs from several days to one month. The height of the water level varies from 0.2 to 2 m in low-water years and from 3 to 6 m in the high water years. The level of lakes drops sharply, and some water reservoirs completely dry up by the middle of summer. The water balance of lakes is affected by surface inflow from the catchment area (60–80% of the incoming part), precipitation

J. Abuduwaili et al., *Hydrology and Limnology of Central Asia*, Water Resources Development and Management, https://doi.org/10.1007/978-981-13-0929-8_6

falls to the surface of the reservoir in an ice-free period (on average 30–40%); winter precipitation is 3–10%, and 5–20% in lakes with thickets. 40% of the incoming part of the balance gives a ground runoff in lakes located in granite massifs such as Kokshetau and Bayanaul mountains and some other areas. The expenditure part of the balance of the majority of lakes is 85–90% due to evaporation from the water surface and 10–15% losses to filtration. In the lakes of the Kokshetau Upland, the share of evaporation losses is reduced to 60–70% due to the increase in filtration.

On the lakes of Western Kazakhstan, the highest water level caused by the influx of snow meltwater is observed most often in the second half of April and the first half of May. Filling them with meltwater usually last for 20–25 days; then the water level begins to drop as a result of evaporation. The annual amplitude of water level fluctuations can reach 4.7 m. The portion of surface runoff in the incoming part of the water balance of lakes varies within fairly large limits. The surface influx is 10–40%, and for lakes without groundwater is 60–90% during abundant supply by groundwater. The portion of winter rainfall accounts for 10–15%. The expenditure part of the annual balance of drainless reservoirs is almost completely determined by losses due to evaporation from the water surface.

6.2 Physical and Chemical Properties of Water in Lakes of Kazakhstan

Physical properties of water. Transparency. Water has the property of absorbing sun rays. However, the light does not go to great depths in perfectly pure water. In fresh and muddy water, the sun's rays penetrate to a shallower depth than in the salty and pure water. In summer, a meter layer of water absorbs up to 4/5 of the light in freshwater bodies.

Transparency of water in the lakes of the Kazakhstan varies from 0.3 to 18 m. It is more in winter, in the spring is less when there are thawed (meltwater) waters in the lake.

As is known, the duration of a light day in water is much shorter than on land. For example, there is light only for 11 h in the summer at a depth of 20 m; this time rapidly decreases with increasing depth. Thus, "water day" lasts only 15 min at a depth of 40 m (Movchan 1966).

The depth of penetration of the sun's rays depends on the transparency of the water which plays an important role in the life of fish and plants. As a rule, the transparency of water is not high in highly productive lakes which is explained by the massive development of small algae and aquatic animals.

Water temperature and ice phenomena. The amplitude of temperature variation in freshwaters is from 0 to +93 °C, but if the water is highly mineralized it can be cooled to −20 °C or somewhat lower and not freeze.

Water slowly absorbs heat and slowly gives it away, so there is no such great difference in temperature at different times of the day in water bodies as on land.

The highest water temperature is recorded in the evening; the lowest one is in the morning.

In summer, the water in the upper layers is warmer than at the depth. Later in the autumn, it cools in the upper layers and becoming heavier and sinks to the bottom and mixes. When the ice cover is formed, the upper layer of water is the coldest due to contact with cold air and ice, but the water becomes warmer with increasing depth.

The solubility of various salts in water increases when the water is heated. On the contrary, the solubility of gases, in particular oxygen is decreases.

Fluctuation of water temperature strongly affects not only the life processes of living organisms in water, but also determines their habitat. The total temperature range for some marine and freshwater organisms is 100.75 °C (from minus 7.75 to 93 °C).

The body temperature in most fish is equal to the temperature of the surrounding water or differs from it by 0.5–1 °C. The optimal temperature that causes spawning is 13–22 °C for warm-water fishes such as beluga, carp, crucian carp, kutum, bream, sturgeon, carp, stellate sturgeon, and pike perch. Intensive growth of fish occurs at 18–25 °C.

Cold-water fish such as white salmon, salmon, vendace, whitefish, and trout spawn at a temperature below 10 °C, and intensively grow at 8–16 °C.

Increasing or decreasing the temperature (in relation to the optimal temperature conditions) disrupts the vital processes in the fish body. So, carp at 12 °C does not breed and consumes feed poorly; as a rule, they stop eating and growing at 2–4 °C, as well as breathing and circulation slow down.

Freezing over on most lakes in the north of Kazakhstan is established in the first half of November, i.e., a month earlier than in the south.

Ice melting begins most often in the first half of April and ends usually in late April or early May. It is 1–1.5 months earlier in the south. The average long-term thickness of ice can reach 140 cm at the northern lakes at the end of the snowless winter, at 120 cm in the southern winters, and in the average snow-accumulation conditions at 110 and 90 cm, respectively. The ice thickness is 160–170 cm during severe winters; it does not exceed 90 cm in warm winters.

Color, smell, and taste of water. The presence of all these signs indicates changes taking place in the water and often undesirable. Shades of water indicate its origin: River water is colorless or slightly yellow; bogs (marsh) water is yellow or brown; mountain water is slightly whitish or milky; water from polluted water bodies is gray or yellow.

Water in the lakes of Kazakhstan varies from bluish-green to yellowish-brown in color and it is bitter–salty in most lakes with the smell of hydrogen sulfide (Philonets and Omarov 1974).

Water chemistry. The chemical composition of water depends on the composition of the water tributaries and groundwater that feeding the lake. It is closely connected with the biological processes taking place in the lake and with a complex of physical–geographical conditions characterizing the catchment basin. The presence or absence of runoff its depth, volume, and area has an importance in the formation of the chemical composition of lake water.

The chemical composition of water is understood as the whole complex of gases, ions, colloids of mineral, and organic origin which dissolved in it under natural conditions. Most of the elements of the Mendeleev periodic table were found in the water.

Among the main components of water should be assigned ions such as chloride (Cl^-), sulfate (SO_4^{2-}), hydrocarbonate (HCO_3^{2-}), carbonate (CO_3^-), sodium (Na^+), potassium (K), calcium (Ca), magnesium (Mg), and gases: oxygen (O_2), carbon dioxide (CO_2), and hydrogen sulfide (H_2S).

Waters from the predominant anion are divided into three classes: hydrocarbonate or carbonate ($HCO_3 + CO_2$), sulfate (SO_4), and chloride (Cl) (Alekin 1970).

Each class is divided into three groups according to the predominant cation: calcium, magnesium, and sodium. Each group is divided into types determined by the relationship between the ions and measured in milligrams per equivalent. The first type of water is characterized by the ratio $HCO_3 > Ca + Mg$; the second type is $HCO_3 < Ca + Mg < HCO_3 + SO_4$; the third one is $HCO_3 + SO_4 < Ca + Mg$ or $Cl > Na$; fourth one is $HCO_3 = 0$.

The chemical composition of the lake waters of Kazakhstan is distinguished by a great variety and inconstancy of total mineralization and ionic composition. According to the chemical composition of the water, there are lakes of three classes: sulfate, carbonate, and chloride. More than 87% of the lakes (Balkash, Alakol, Kamyslybas, Tengiz, and a number of small lakes) belong to the sulfate class. The rest is almost equally distributed between the carbonate and chloride classes.

Mostly mild or moderately hard neutral water is accumulated in fresh lakes, and water in brackish and salt lakes is very hard and slightly alkaline.

Dissolved gases. Dissolved gases in water such as oxygen, carbon dioxide, and hydrogen sulfide have a great importance in the formation of the hydrochemical regime of lakes and the development of biological processes.

Dissolution of gases increases with decreasing in water temperature and its salinity, and with increasing pressure. There is a certain amount of gas in water in accordance with the temperature, salinity, and pressure. The gases can dissolve under the given conditions. This amount is taken as 100% saturation. But since gas exchange with the atmosphere, as well as the spread of gases within the reservoir require a certain time. The content and distribution of gases in the reservoirs do not correspond to 100% saturation, and they are uneven.

Oxygen is one of the most important elements that without life in the water is impossible. If its content does not reach a certain standard, then the fish feels bad and often dies. Oxygen is needed primarily for breathing. But digestion also slows down with a lack of oxygen. In addition, the fish more eats the feed if oxygen is less in the water.

Oxygen enters the reservoirs mainly from the atmosphere; some of it is released by aquatic plants that decompose carbon dioxide into carbon forming the living matter and oxygen released into the water under the influence of sunlight. The upper layers of lake water are most saturated with oxygen from where it penetrates to the depth under the influence of thermal circulation.

The oxygen content is not the same in different water bodies. More oxygen is contained in the water of uninhabited, flowing water bodies that open to winds, especially in large lakes, ponds, and rivers. However, the amount of oxygen in the water is considerable fluctuated in the same water body: It can increase or decrease.

The oxygen content especially noticeable decreasing in winter as its entry from the atmosphere is prevented by the ice cover and its isolation is weakened by plants. Sometimes the amount of oxygen in the reservoir is reduced by half during one winter month. Oxygen starvation usually comes to the end of winter when the process of putrefaction (decomposition) and decay of dead organisms is particularly intense. At this time, the oxygen supply can completely dry out and lead to death of fish.

Oxygen starvation is possible not only in winter, but also in summer with a tightly "planted" fish and high temperature; also in the spring, if a large amount of organic substances is introduced into the reservoir by meltwater, because of a lot of oxygen consumes for decomposition.

A bottom ground has a great importance in regulating the gas regime. If it is rich in organic substances, then there is an increased absorption of oxygen during their decomposition. Its content depends on the time of day: It is lower at night than in the daytime.

0.0069 g of oxygen in 100 g of water can dissolve at 0 °C; 0.0054 g at 10 °C; and at 20 °C is 0.0044 g. A different amount of oxygen is necessary for different fish. So, for salmon and sturgeon it is required 6–8 ml/l; for carp, bream, pike perch, and others is 4–5 ml/l. Goldfish and crucian can live in water with an oxygen content of 0.5 ml/l (1 ml of oxygen equal to 1.429 mg).

The need for oxygen depends on the age of the fish: The younger one needs more oxygen, and it takes per unit of body weight. The fish consumes more oxygen at a high temperature.

The content of soluble oxygen in natural waters ranges from 0 to 14 ml/l, rarely below. Its high content in the fastest flowing cold streams is during its flowering period.

Carbon dioxide is formed in water as a result of oxidation and decay of organic substances and respiration of living organisms. Pure water absorbs it from the air in an amount of 0.3–0.5 cm^3 per liter. The carbon dioxide content increases with increasing depth. Seawater contains up to 48 cm^3 of carbon dioxide per liter. The carbon dioxide content in the lakes rarely exceeds 20–30 mg/l. Its excess indicates the contamination of the reservoir by organic substances and leads to the death of aquatic animals. Carbon dioxide should not exceed 10–20 cm^3 per liter in water for normal life of organisms. However, if it is a poison for animals, then for most plants is a necessary condition for their existence.

Hydrogen sulfide is formed in heavily polluted water bodies when decaying of proteins. The amount of hydrogen sulfide reaches 0.5 cm^3 per liter of water in deep layers of lakes. For example, at depths of 300–500 m of the Black Sea, its content is 4 cm^3, and in the bottom layers is 6.5 cm^3.

Hydrogen sulfide dissolved in water even in small amounts is a poison for most animals and plants. The oxidation of hydrogen sulfide consumes oxygen which significantly worsens the gas regime of water.

Active reaction of water. Aqueous solutions always contain hydrogen ions. The concentration of hydrogen ions determines the active reaction of water. Most lakes with fresh and slightly mineralized water have a neutral or close to water reaction (pH 7.0).

The reaction of the water is alkaline at pH > 7.0, and it is acidic at pH < 7.0. Highly mineralized lakes are alkaline, and an acid reaction is characteristic for swamped lakes. Aquatic organisms are very sensitive to the active reaction of water. Almost all freshwater inhabitants are adapted to a neutral or slightly alkaline reaction. The acid reaction of water acts depressingly on animals (mollusks, crustaceans) and aquatic plants. Fish can die at pH below 6.0 and above 8.5 (Karpanin and Ivanov 1967).

Oxidizability of water. Organic substances of different chemical composition (humic) except dissolved mineral substances are also found in water. These are the products of decomposition of plants and animals that lived in the lake or got into it from the outside. A considerable amount of oxygen is consumed during decomposition, which is water is oxidized that leads to oxygen starvation of the organisms inhabiting in the reservoir.

Oxidability is usually expressed by the number of milligrams of oxygen which required to oxidize substances contained in a liter of water. It is especially high in heavily swamped lakes where it can reach more than 60 mg of oxygen per liter of water. The water is oxidizable by 10–30 mg of oxygen per liter of water when fish are diluted.

The high content of humic substances in the lake has a depressing effect on the development of aquatic plants and animals.

Mineralization of water is the most important indicator of its use. There are different waters by the degree of salinity:

(1) freshwater up to 1 g/kg (‰), the boundary between fresh and brackish waters is accepted by the average limit of the person's taste sensitivity.
(2) brackish is from 1 to 25 g/kg (25‰); 25 g/kg is chosen as the boundary between brackish and saline waters, because at this salinity (24.695‰) the freezing point and the highest density for seawater are the same. If the salinity is less than 24.7 g/kg, water first reaches the highest density with continuous cooling and then starts to freeze. If the salinity is greater than this limit, then the water will freeze under similar conditions before reaching the highest density;
(3) saline (with sea salinity) is from 25 to 50 g/kg (47‰—the greatest salinity of the World Ocean is the Red Sea);
(4) brines are above 50 g/kg (50‰).

Organic substances are most intensively formed when the water is mineralized from 0.1 to 1 g/l in the continental water bodies.

The productivity of the reservoir is reduced due to increasing of water mineralization. Thus, the loss of bream eggs is on average about 63% in mineralization of 10 g/l, and the mortality of larvae of four to five days old reaches 95%. The withdrawal of carp roe is higher (Konovalov 1950). The normal physiological functions of carp, perch, crucian carp are rapidly deteriorating in such mineralization. Evaporation of water decreases when the mineralization increases, while its transparency increases.

Water should differ not only in the absence of substances harmful to the body when it is consumed with food, and it should be in a small mineralization. The mineralization is not more than 1 g/l is considered as a good drinking water; satisfactory one is 2 g/l. Water for livestock is used with more mineralization.

The mineralization of water should not exceed 1 g/l for irrigation of fields. Because mineralized water can increase a soil salinization, as a result saline soils will have occurred.

The industry consumes almost distilled water.

Hardness of water. Ca and Mg are the ions that impart negative qualities to water. Such water is hard; vegetables, meat, and cereals are poorly cooked in such waters; but tea is brewed. It also causes an increased consumption of soap when washing laundry. In addition, the cloth quality deteriorates rapidly from such water. Using of hard water adversely affects industrial and technical purposes.

Natural waters are divided into the following groups depending on the amount of hardness: Very soft is up to 1.5 mg/eq, soft is up to 3 mg/eq, and moderately hard is up to 6 mg/eq, hard one to 9 mg/eq and very hard over 9 mg/eq (Alekin 1970). The soft water has the best quality.

Aggressiveness of water. Water ability that destroys various building materials by dissolved salts and gases is the aggressive action of water. For example, the following types of aggressiveness are distinguished for concrete materials: carbon dioxide, leaching, total acid, sulfate, and magnesia. The degree of aggression on concrete is determined by the chemical composition of the water.

Microelements are chemical substances. The content of microelements in the natural waters, soil, plant, and animal organisms is insignificant: from thousands to trillions of percentage proportion (10^{-3}–10^{-12}). About 75 microelements are found in the body of animals and humans, including 24 in the blood, in the brain is 15 and milk is 30.

In different natural landscapes, the chemical elements enter the human and animal organism with food, water, and air in an unequal number. This quantity satisfies the need of the organism in various ways (Table 6.1).

The lack of individual microelements causes painful reactions in humans, animals, and plants.

Iodine, copper, cobalt, zinc, fluorine, manganese, and iron belong to the group of irreplaceable trace elements for humans. They are a constant part of plant and animal organisms (Kolomiytseva 1966).

Table 6.1 Daily demand of microelements

Microelement	For human (mg)	For animals, in 1 kg of dry feed (mg)	Acceptable concentration for drinking water (mg/l)
Iodine	0.05–0.2	0.2–0.6	0.005
Fluorine	0.8–1.2	1 mg in 1 l of water	0.5–1.2
Manganese	5–10	30–70	0.1
Copper	2–5	6–10	0.5
Zinc	12–16	6–40	3
Cobalt	0.04–0.07	0.3–1	
Iron	15	100–200	0.3

References

Alekin OA (1970) The basics of hydrochemistry. Leningrad, Hydrometeoizdat. 444 p (in Russian)
Karpanin DP, Ivanov AP (1967) Fish farming. Moscow, 370 p (in Russian)
Kolomiytseva MG (1966) Results and perspectives in the development of the doctrine of microelements in medicine. In: Microelements in agriculture and medicine. Kiev, 145–151 pp (in Russian)
Konovalov PM (1950) Experiments to study the effect of salinity on the development of roach roach, bream and carp. Materials on Ichthyofauna and water regime of the Aral Sea Basin. Moscow, 70–82 pp (in Russian)
Movchan VA (1966) Life of fish and their breeding. Moscow, Kolos (in Russian)
Philonets PP, Omarov TR (1974) Lakes in the Northern, West, and Eastern Kazakhstan (reference book). Hydrometeoizdat, Leningrad. 121 pp (in Russian)

Chapter 7
Lake and Sea Basins in Kazakhstan

7.1 Caspian Sea

The Caspian Sea is a drainless inland water reservoir located at the junction of the Eurasian continent. The reservoir is extended in the meridional direction by 1200 km, and its width is from 240 to 440 km (average width is about 320 km). The length of the coastline is about 700 km, of which 230 km on the territory of Kazakhstan. The area of the water mirror is 376,000 km^2 at the level of 28 m abs. (390,000 km^2 at the level of 27 m abs; 405,000 m^2 at 26 m abs) and is almost equal to the area of the Baltic Sea and slightly less than the Black Sea (Medeu 2010).

There are 130 rivers and temporary watercourses fall to the Caspian Sea, and their runoff ranges from 205–215 to 450–460 km^3/year, or an average of 300 km^3/year (Baydina and Kosereva 1986), 80% of the river runoff from this the volume is Volga river, and 5% is Zhaiyk river. About 10% of the flow is provided by the rivers of the western coast: Terek, Sulak, Samur, Kura, and a number of other small rivers. The remaining 5% give the rivers of the Iranian coast. There is not a single permanent drainage on the east coast (Medeu 2010).

The climate of the Caspian Sea is moderately deserted (Caspian Sea 1992). The average amount of precipitation is 200 mm/year (ranging from 173 to 243 mm); the average annual air temperature of the water surface is 0 °C.

The basin of the Caspian Sea along the seabed relief (Mangystau and Absheron rapids) and by hydrological features is divided into three parts: northern, middle, and southern.

The Northern Caspian occupies the shallowest part with an average depth of 5–8 m maximum to 18–20 m in the southwestern part of the Zhaiyk (Ural) furrow, with an area of about 80,000 km^2 and a water volume of 500 km^3. Within the Middle Caspian, the maximum depth reaches 790 m and an average of 213 m with an area of about 140,000 km^2, and a water volume of 26,000 km^3 (Baydina and Kosereva 1986). The Shelf zone is bounded by an average depth of about 100 m in these parts, where the Caspian water is washing the coasts of Kazakhstan. Below this,

J. Abuduwaili et al., *Hydrology and Limnology of Central Asia*, Water Resources Development and Management, https://doi.org/10.1007/978-981-13-0929-8_7

crook begins the continental slope, which ends in the Middle Caspian approximately at a depth of 500–550 m.

The southern part of the Caspian Sea is already outside of Kazakhstan and is the deepest zone with a maximum depth of 1025 m at an average depth of about 330 m, the area of the water mirror is about 170,000 km^2, and the water volume is 50,000 km^3 (Medeu 2010).

The air temperature over the sea varies considerably. The average annual air temperature over a long period varies from 9 °C in the northeast to 17 °C in the southeast. The coldest month over the northern and eastern regions of the sea is January (continental climate type), in the southwestern and central regions is February (marine type). In these months, the position of the ice edge is delineated by 0 °C isotherm, and negative temperatures are observed above the fast ice and land at this latitude. The general warming up of the atmosphere begins in March, which is slowest in the north and east coast of the sea. The rise in temperature from March to July occurs in the north and east in a uniform manner; in the deepwater areas of the Middle and Southern Caspian, the maximum air temperature since July shifts to August. The maximum of summer temperatures over the sea is somewhat lagging relative to the temperature above the land. Autumn cooling is more intense than warming in the spring and the air over the sea cools faster than over land. The winter months of the northern and eastern coasts are colder than the western months, and in the north, under the influence of the ice cover, the air cools more intensively than in the southern and southwestern regions (Rodionov 1969).

The level regime of the Caspian Sea in the historical past was subjected to significant fluctuations, which is due mainly to tectonic and climatic, and in recent decades—anthropogenic factors. The amplitude of these fluctuations was about 7 m over the past 200 years. The sea level decreased by 3 m and amounted to 29.01 m from 1930 to 1977, and from 1978 to 1995, it increased by 2.5 m and amounted to 26.66 m. The intra-annual amplitude of water level variability is determined by the river flow and reaches 0.5–1.0 m (Medeu 2010).

As a result of a decline in sea level from 1930 to in 1977, the sea area was reduced by 48 thousand km^2, the length of the coastline was almost 200 km, and the volume of water decreased by more than 7 thousand km^3. The current rise in sea level after 1978 is different in that it turned out to be the most lasting for the last 175 years, ate 1820, despite a significant increase in ineffective water consumption in the Caspian Sea basin (Medeu 2010).

The northern and eastern coasts adjacent to the territory of Kazakhstan are low and uniform, open to free entry air masses. Hydrological and morphological sea processes are most affected by wind, temperature, and humidity.

The eastern and the western wind directions prevail over the water area of the eastern part of the North Caspian. The greatest number of storms and the highest wind speeds are recorded in these wind directions. Over the eastern part of the Middle Caspian (in the Kazakh part), winds from the southeast and northwest most often blow, there are southeastern storms lasting up to 100–140 hours. South winds have the least frequency, and windless weather for a year is about 15%.

Average monthly wind speeds during the year vary slightly: from 4 m/s to 7 m/s. The largest average monthly wind speeds are observed in winter (up to 7 m/s), and the smallest in the summer (4.5 m/s). Winds with the speed of 2–3 m/s have the greatest frequency. The wind speed of 10 m/s and more is noted in all seasons of the year, however, with the highest frequency in the winter. The maximum wind speeds reach 30–32 m/s during storms.

The currents of the water mass in the Northern and Middle Caspian are mainly determined by the direction of the wind over the sea mirror and only in the areas immediately adjacent to the mouths of the Volga and Zhaiyk rivers and their runoff.

The development of waves in the shallow part of the Northern Caspian is in good agreement with the wind. Zone of maximum excitement up to 3 m is the area of Zhaiyk furrow. At depths of 2 to 1.5 m, the wave height does not exceed 1.2 m. At depths of less than 1 m, the maximum wave height is 0.45 m. The north, northwestern, and southeastern waves prevail along the eastern Kazakhstan coast of the Middle Caspian. The highest wave height is observed in the southeast direction (Medeu 2010).

The average annual water temperature in the water area of the northern and middle parts of the Caspian Sea is generally positive. The temperature regime of seawater in the Northern Caspian stretches along the isobaths. The range of fluctuations in water temperature is 26–27 °C in Zhaiyk furrow and in shallow coastal regions is 30–35 °C. The minimum temperature of water decreases in January and February to -10 °C. Spring warming begins in the eastern shallow areas and then spreads west and south. The difference in the water temperature in the Zhaiyk furrow and in the coastal regions can reach 5–8 °C. The maximum warming of the whole water mass in the eastern part of the Northern Caspian occurs at the end of July, when the water temperature is equalized throughout the water area and reaches 25-–27 °C. Since August, the cooling of waters begins. First, the waters of the eastern shoal are cooled, and then the Zhaiyk furrow cools the western waters. In November, the temperature of the water in coastal areas becomes close to its freezing temperature (Rodionov 1969; Dobrovol'skii et al. 1969; Klige 1986).

Most part of the Northern Caspian Sea is annually covered with ice. The first appearance of ice is observed in mid-November. The appearance of ice shifts for several weeks forward and back in the anomalously cold and warm winters. In warm winters, fast ice lies along the northern and eastern coasts with a narrow strip of up to 30–50 km, and in severe winters, almost all of the North Caspian water area is covered with fast ice. The maximum thickness (up to 1 m) of fast ice reaches in severe winters in late February and early March. Strong winds in November–December break the fast ice transforming into floating ice. With large surges in the beginning of winter, the height of individual piles of blocks of ice on the ground is 10–12 m. Destruction of fast ice begins in late February and early March. The final cleansing from the ice takes place at the end of March and early April, and it in the second half of April (cold winter), and at the end of February (in the warm winter).

The water of the Northern Caspian Sea is considerably freshened in comparison with the water of the Middle and Southern Caspian because of the influence of the powerful inflow of fresh water from the Volga and Zhaiyk rivers. In this regard, in the mouth areas of these rivers, the smallest salinity of water (1–2‰) is observed almost throughout the year. The water salinity varies from 5-6 to 8‰ in the northeastern part of the Northern Caspian Sea in shallow water. Further to the south-west and west, as well as across the Kazakhstan coastal zone of the Middle Caspian, the water salinity reaches 12-12.5‰, and in the south is13‰ (Medeu 2010).

The gradual salinity of water in the northern shallow coastal area of the sea begins in July, and the desalinated areas are reduced. This process continues in August–September with an increase in the flow of the Middle Caspian waters, especially along the southern and southeastern part of the Northern Caspian Sea. In autumn and early winter period, the waters of the entire Northern Caspian Sea continue to become saline, and the western freshened "stream' of Volga water is shrinking. In winter, compared with the autumn, the salinity of the water increases by an average of 1–2.5‰ and more in some cases. During the formation of the ice cover, it increases by 1–1.3‰ in the eastern (Kazakhstan part) of the Northern Caspian under the influence of the brine that flows out of ice, and when it melts, it decreases by approximately the same amount (Rodionov 1969; Klige 1986; Dobrovol'skii et al. 1969; Leonov 1960).

In the long-term period, in extremely shallow years (1930–1941), with a drastic reduction in the runoff of the Volga and Zhaiyk, when the sea level dropped for 1.9 mm in 12 years, and the average water salinity of the North Caspian Sea increased up to 10–10.5 ‰, and in the 1940s, with increased water content of these rivers, its decrease (on average to 7 ‰) (Katunin 1975).

At present, in the conditions of increasing the level of the Caspian Sea, the general background of salinity in the sea is significantly reduced.

7.2 Aral Sea Basin

The Aral Sea Basin is located on the territory of Central Asia and in the center of Eurasian continent. Central Asia is a vast and landlocked region in Asia (Fig. 7.1). Despite some uncertainty of its borders, the general characteristics of this region are singled out; in particular, Central Asia has historically been associated with the nomadic peoples inhabiting there and the Great Silk Road. Central Asia has always acted as a place where people, goods, and ideas converged from different ends of the Eurasian continent: Europe, the Middle East, South and East Asia.

The Aral Sea basin is located between 34° 30′–45° 30′NL and 52° 20′–80° 20′EL. The territory is stretched from west to east by about 2400 km and from north to south by 1160 km and covers an area of 1280 thousand km^2. A significant part of Central Asian countries (Kazakhstan, Kyrgyzstan, Uzbekistan, Tajikistan, and Turkmenistan), as well as the northern provinces of Afghanistan and Iran, is located in the basin of the Aral Sea. The latitudinal position of the Aral Sea basin roughly corresponds to the position of Mediterranean states such as Spain, Greece, and Italy. However, the natural conditions differ from these subtropical countries. This is due to the fact that Aral Sea basin is one of the most remote basins from the seas. In addition, it is open from the north for the penetration of dry and cold arctic air masses; moist and warm air masses (monsoons) from the south are blocked by the highest

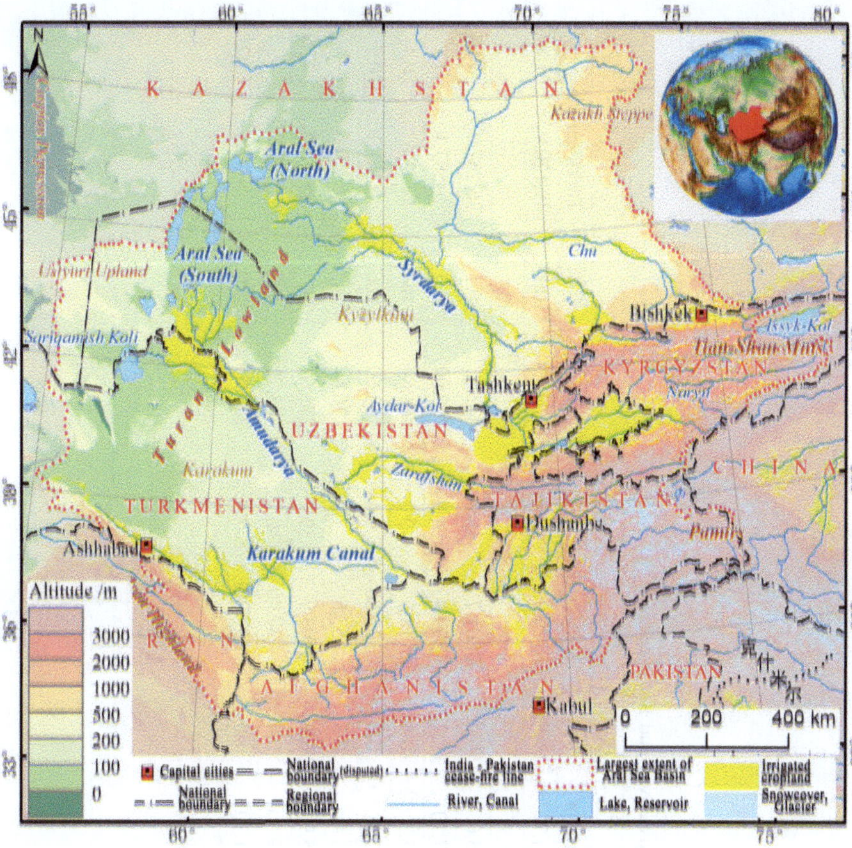

Fig. 7.1 Aral Sea basin

mountains. Therefore, the climate of the basin is subtropical and sharply continental. Most part of the Aral Sea basin stretching from the northwest to the southeast is flat and occupied by deserts and the southeast part by foothills and mountains.

The basin covers the entire territory of Tajikistan, Uzbekistan and most part of Turkmenistan, three regions (Osh, Zhalalabad, and Naryn) of Kyrgyzstan, southern part of Kazakhstan within two regions such as Kyzylorda and South Kazakhstan, and northern Afghanistan and Iran (Fig. 7.1).

However, in this chapter only the areas of the first five Central Asian countries located in the basin of the Aral Sea are considered. This territory extends between 56° and 78° east longitude (EL) and 33° and 52° north latitude (NL). It covers an area of 1.549 million km^2. About 0.59 million km^2 of these land is suitable for cultivation (Table 7.1) (Mal'tsev 1969).

The territory of the Aral Sea basin can be divided into two main zones: the Turanian plain and the mountainous zone. The western and northwestern parts of the Aral Sea basin within the Turanian Plain are covered by the Karakum and Kyzylkum deserts.

Table 7.1 Land fund of the Aral Sea basin (thousand ha) (Mustafayev and Kozykeeva 2009)

Total area (thousand ha)	Central Asian Countries									
	Kazakhstan		Uzbekistan		Kyrgyzstan		Tajikistan		Turkmenistan	
	24,969	100%	45,114.7	100%	8129.3	100%	14,254.9	100%	49,120.9	100%
Agricultural lands										
Total	16,487.8	66	26,815.0	59.3	3801.7	46.8	4342.5	30.5	36,767.0	74.8
Arable land	118.20	0.7	3467.10	7.7	399.90	4.9	779.90	5.5	587.1	1.2
Perennial plantings	1.6	–	195.3	0.4	18.5	0.2	61.8	0.4	30	0.1
Deposits	53.60	0.3	283.3	0.6	35.80	0.4	88.60	0.6	361.10	0.7
Hayfield	377.80	2.3	123.4	0.3	153.6	1.9	36.6	0.3	10.6	–
Pasture	15,936.6	86.66	22,745.9	50.3	3193.9	39.4	3375.6	23.7	35778.2	72.8
Homestead lands	3.80	–	169.50	0.4	37.90	0.5	43.70	0.3	23.20	–
Forests	1485.5	5.95	651.10	1.4	637.30	7.8	280.6	2.0	612.6	1.2

East and southeastern parts of the basin belong to the highland zone of the Tian Shan and the Pamir's ridges. The remaining part of the basin includes alluvial and inter-mountain–valleys, dry and semidry steppes. Different forms of relief in these countries have created certain conditions that are reflected in the interconnection between water, land, and the populated areas of the region.

About 90% of the territory of the Kyrgyzstan Tajikistan is mountains. This creates, on the one hand, a "monopoly" for these two countries to form water resources in the basin and, on the other hand, a deficit of suitable land for cultivation. The most important feature of the region is the oases (Fergana Valley, Khorezm).

The *water balance* of the Aral Sea was as follows: The incoming part is precipitation—8.7 km^3, and river flow is 5.5 km^3; consumable part: Evaporation from the water surface is 63.8 km^3.

The Aral is a closed natural drainage water pond. Two largest rivers of Central Asia such as Amudarya and Syrdarya fall into the Sea. The rivers played an important role in the trade and political relations of the ancient peoples and retained their importance to the present day.

7.2.1 Water Resources of the Aral Sea Basin

Water resources of the Aral Sea Basin are composed of large, medium, and small rivers, small number of lakes, temporary runoff due to atmospheric precipitation, and groundwater. To the plains of Central Asia, 153 km^3 of river water and 16 km^3 of groundwater come from the mountain regions. In addition, up to 40% of the surface runoff is filtered into the soil and evaporates (Shults 1965).

The proximity of high mountains with eternal snows and glaciers is favorable for the formation of rivers. Some of which reach only the peripheral parts of the deserts, and others can reach their inner regions.

In accordance with the formation region of the Aral Sea basin, there are the rivers of glacial-snow feeding (Amudarya, Syrdarya), snow–rain (Murgab, Tejen, Atrek), spring–rain (small rivers flowing from the Kopet Dag and other medium-altitude mountains) that are distinguished.

Water resources of the Aral Sea region consist of replenished surface and groundwater, as well as from return waters from anthropogenic uses (wastewater and drainage water). There are two large river basins in the basin of the Aral Sea: the Syrdarya in the north and the Amudarya in the south. Between these main rivers is the Zerafshan River, a former tributary of the Amudarya river.

The Syrdarya is the second river in terms of water content and the first longest river in Central Asia (Fig. 7.2). From the sources of Naryn, its length is 3019 km, and the basin area is 219 thousand km^2.

The sources of the Syrdarya river lie in the Central (Inner) Tian Shan. The Syrdarya river is formed by merger of Naryn with Karadarya rivers. Feeding the river is glacial and snow, with the predominance of snow.

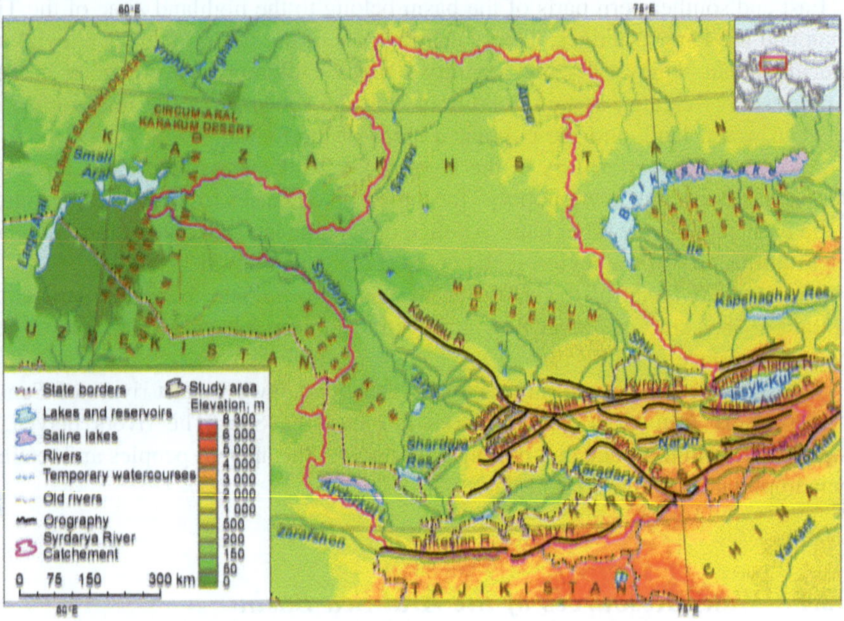

Fig. 7.2 Syrdarya river basin

The water regime is characterized by a spring–summer high water, which begins in April. The largest runoff is in June. About 75.2% of the Syrdarya flow is formed on the territory of Kyrgyzstan. Then, the Syrdarya crosses Uzbekistan and Tajikistan and flows into the Aral Sea on the territory of Kazakhstan. About 15.2% of the Syrdarya flow is formed on the territory of Uzbekistan, 6.9% in Kazakhstan, and 2.7% in Tajikistan.

The Amudarya river is the largest river in Central Asia. Its length from the sources of Panj is 2540 km, and the basin area is 309 thousand km^2.

The Amudarya river is formed by merger of Panj with Vakhsh rivers. In the middle reaches, three large right tributaries flow into the Amudarya (Kafirnigan, Surkhandarya, and Sherabad) and one left tributary (Kunduz).

About 13.9% of the Amudarya river flow is generated in Afghanistan and Iran, and 8.5% in the territory of Uzbekistan (Table 7.2).

The length of the Amudarya river is 2620 km (according to other sources 2540 km). It becomes the Amudarya after the confluence of the Vakhsh rivers of the collecting water from the Alai Valley and the Northern Pamirs, and the Panj with the main tributaries of the Gunt and Bartang draining the southeastern part of the Pamir plateau. From the point of confluence, its length is 1400 km.

The Amudarya river basin also includes the Kafirnigan and Surkhandarya rivers, flowing from the southern slopes of the Hissar Range and the Kunduz River, which forms a drain within Afghanistan.

The area of the basin is 465 thousand km^2, of which only the mountain part gives a runoff (about 217 thousand km). The average annual runoff varies from 48

Table 7.2 Surface water resources of the Aral Sea basin (average annual flow, km³/year)

Countries	River basin		For whole Aral Sea basin	
	Syrdarya	Amudarya	km^3	%
Kazakhstan	2.426	–	2.426	2.1
Kyrgyzstan	27.605	1.604	29.209	25.1
Tajikistan	1.005	49.578	50.583	43.4
Turkmenistan	–	1.549	1.549	1.2
Uzbekistan	6.167	5.056	11.223	9.6
Afghanistan and Iran	–	21.593	21.593	18.6
Total for basin	37.203	79.800	116.483	100

to 101 km^3 with an average of 63 km (Table 7.2). The waters of the river are characterized by a strong turbidity of 2500–4000 g/m^3. A strong erosion of the banks is common in the basin.

In the lower reaches, the river flows into the Aral Sea with several sleeves, forming a delta with an area of about 19 thousand km^2.

The average value of the total runoff over the entire observation period (1911–2000) is 112,609 km^3/year for the Aral Sea basin, including 77,093 km^3/year for the Amudarya river basin and 34.076 km^3/year for the Syrdarya river basin. The average annual flow for the Amudarya river basin is 79.280 km^3/year and for the Syrdarya river basin 37.203 km^3/year. Consequently, the total average annual resources of surface (river) water in the Aral Sea basin are 116,483 km/year (Mustafayev and Kozykeeva 2009).

Annual indicators of water resources, due to fluctuations in water content, change from low water years (95% of supply) to high water (5% of supply) in the following ranges: in the Amudarya river from 58.6 to 109.9 km^3 and in the Syrdarya river from 23.6 to 51.1 km^3; 25.1% of the total runoff in the Aral Sea basin is formed within Kyrgyzstan, 43.4% in Tajikistan, 9.6% in Uzbekistan, 2.1% in Kazakhstan, 1.2% in Turkmenistan, and 18.6% in Afghanistan and Iran (Mustafayev and Kozykeeva 2009).

The total annual flow of all rivers in the Aral Sea basin is 116.483 km. This volume includes 79.80 km of the Amudarya and 37.203 km^3 of the Syrdarya rivers. According to the probability distribution of the runoff, 5% (the water year) and 95% (the dry year), for the Amudarya river the annual runoff varies from 109.9 to 58.6 km and for the Syrdarya from 51.1 to 23.6 km.

7.2.2 *Environmental Impact of the Aral Sea Desiccation*

The Aral Sea belonged to the category of large water reservoirs on the Earth. The sea is located amid the great deserts of Central Asia. Its drainage basin covered 1.8 million km^2 within Kazakhstan, Uzbekistan, Turkmenistan, Tajikistan, Afghanistan, and Iran. Only Kazakhstan and Uzbekistan are riparian on the sea properly and possess an approximately equal length of shoreline. The entire Aral Sea coastline lies within Karakalpakstan in Uzbekistan (Micklin 2007). The terminal lake had a surface inflow but no surface outflow (Bortnik and Chistyaevaya 1990). The Aral Sea basin has the main rivers: Syrdarya and Amudarya, and water from those rivers is used mainly for irrigation purposes. As a result of long-lasting anthropogenic factors in the use of water resources of the basin, issues related to the reduction of water surface and volume of the sea have formed. Due to the uncontrolled and irrational extraction of water for irrigation in Central Asia and Kazakhstan, which began in the 1950s–1960s of the last century, the water flow has decreased in the Aral Sea since the 1980s has ceased. As a result, anthropogenic impacts on the basin ecosystem have led to several negative consequences such as soil salinization, desertification, and environmental changes. Since the early 1960s, the water balance, morphology, and ecology of the Aral Sea have changed dramatically. Consequently, the sea has steadily shrunk and been salinized (Micklin 2007). The river inflow decreased dramatically for the period after 1960 (Issanova and Abuduwaili 2017).

A desiccation of the seafloor affected the ecology as a whole including climate condition in the region. Due to shrinkage of the sea volume, microclimate condition has changed 100 km—wide along the former shoreline in Uzbekistan and Kazakhstan (Micklin 1991, 2007; Glazovskiy 1990; Abuduwaili et al. 2010). Maritime conditions of the sea have been replaced by more continental and desertic regimes. Summers have become warm and winters cool; the air humidity is lower; the growing (vegetation) season has become shorter; the autumn frosts come earlier and the spring frosts later. According to expert climate researchers from Uzbekistan, the surface radiation depends on atmospheric content; that is, the increasing level of salt and dust in the atmosphere reduces the surface radiation and thereby photosynthetic activity as well as increases the acidity of precipitation (Chub 1998). Environmental pollution associated with the use of toxic chemicals as pesticides and defoliants for cotton in irrigated agriculture causes serious health-related problems (Micklin 1992; Medicine sans Frontieres 2000).

The desiccation of the Aral Sea, caused mainly by human activities and water flow reduction, soil salinization, and pollution of influent rivers (Syrdarya and Amudarya rivers), has had severe negative effects for the environment (Micklin 2004). In addition, the consequences of these negative effects affected the area surrounding the sea including a population of several million (Khvorog 1992). The regions that suffered the most include some parts of the Kyzylorda region in Kazakhstan and Karakalpakstan in Uzbekistan. In Turkmenistan, the Dashauz region has been substantially affected although the region is not abutted on the sea (Micklin 2007). In addition, the extensive Amudarya River delta, which has a rich and diverse ecosystem primarily located in Karakalpakstan (Uzbekistan), has suffered considerable harm (Micklin 1991, 2004).

In Kazakhstan, the Syrdarya River delta has endured relatively less but still substantial damage. The river flows through the deltas have been greatly reduced by the virtual elimination of spring floods and construction of upstream storage reservoirs. As a result, the level of the Aral Sea has decreased and, consequently, groundwater levels have declined. All of these realities have led to a spreading and intensified desertification process in the Aral Sea basin (Issanova and Abuduwaili 2017). The vegetation community of the region has replaced by other types of vegetation. For instance, halophytes and xerophytes are rapidly replacing endemic vegetation communities (Novikova 1996, 1997). Moreover, salts have accumulated on the soil surface, which form solonchak (saline soils) in some places, and therefore these places have become worthless (non-fertile) land for growing. Vast areas with unique tugay vegetation communities, which are distributed along the main rivers and distributary channels, have been particularly damaged. For instance, in 1950 tugay communities covered 100,000 ha in the Amudarya River delta (Novikova 1996), but by 1999 they decreased to only 20,000–30,000 ha (Severskiy et al. 2005). Lakes and wetlands, as well as their associated reed communities, have significantly diminished due to the desiccation of the river deltas. The area of lakes in the Amudarya River delta decreased from 49,000 to 8000 km^2 between 1960 and 1980 (Chub 2002). Reed communities in the delta covered as much as 500,000 ha in 1965, but by the mid-1980s the area had declined dramatically (Palvaniyazov 1989). This destruction of natural resources (water, soil, vegetation, etc.) has resulted in serious ecological consequences and changes of the environment. The ecological consequences of these negative effects have affected even waterfowl (permanent and migratory) in these dessicated zones. A number of permanent and migratory waterfowl are currently endangered (Micklin 1991). In addition, the aggregate water surface area decreased, and the water bodies have been increasingly polluted primarily by irrigation runoff containing salts, pesticides, fertilizers, herbicides, and cotton defoliants, which adversely affect aquatic bird populations. However, since the late 1980s measures for improvement have been taken, and significant efforts have been made to restore wetlands, decrease environmental pollution, and improve habitat conditions (Chub 2002). Irresponsible water use of the Amudarya and Syrdarya rivers affected the irrigated agriculture in the deltas. Consequently, the irrigated agriculture has suffered from an inadequacy of water; that is, water inflow to the deltas has decreased because of heavy upstream consumptive use for irrigation. In addition, the leaching of salts, which is caused by repeated use in the middle and upper courses of the rivers, has elevated the salinity of the water reaching the river deltas (World Bank 1998; Micklin 2007). At times >2 g/l, these saline flows have decreased crop yields and, in association with poor drainage of irrigated fields, have promoted soil salinization. In addition to the reduction of pasture area and their declining productivity resulting from replacement of natural vegetation suitable for grazing by inedible species, decreasing groundwater levels and desertification process in general have led to damage to animal husbandry in the deltas and desert regions adjacent to the Aral Sea (Micklin 2007). As a result of the intensive development of irrigation, irresponsible use of water resources has led to serious soil and land degradation processes. Significant areas of secondary saline soils and anthropogenic solonchaks have appeared in the Aral Sea basin. These areas have become a source of salt, dust, and sand transfer. Dust, sand, and salt from the dried bottom of the Aral Sea are transported by strong and very strong winds. Salt and dust that blow from the dried bottom of the sea and from irrigated farmland in areas adjacent to the

Aral Sea are contaminated with heavy metals and pesticides, which would exacerbate the negative impacts on humans and other animals (O'Hara et al. 2000). Large portions of blown material are transported onto surrounding areas that are now a barren desert. Storms most seriously impact the Ustyurt Plateau in the west part of the sea and the Amudarya River delta in the south end of the sea (Bortnik and Chistyaevaya 1990). The Amudarya River delta is the most densely settled as well as economically and ecologically important region around the sea.

Shrinking and desiccation of the Aral Sea created an environmental problem with negative impact on the environment. The former seafloor was desiccated and formed the Aralkum, man-made desert. So, the driest part of the former seafloor is considered to be a source for the development of salt and dust storms. Salt and dust particles from the dried bottom of the sea started to be transported over a significant part of the basin. The salt and dust thus removed affect not only climate and landscapes but also the health, living condition, and economic activity of the local population (Micklin 2007; Prospero et al. 2002; Gills 1996). In addition, the intensive salt and dust transfers and deposits affect the quality of the environment as well as the local biodiversity and biological productivity. At present, the removal, transportation, and deposition of dust, sand, and salt—all of which arise due to human activity and natural factors—are some of the most negative phenomena experienced in the Aral Sea region and southern Pre-Balkash deserts. The problem of the Aral Sea is part of the general desertification process occurring in many areas of the world. The processes of desertification and soil degradation have occurred due to anthropogenic activity and level change of the Aral Sea and Balkash Lake in Kazakhstan (Medoev 1981; Kudekov 2002; Faizov and Tapalova 2003; Spivak et al. 2012; Issanova et al. 2014, 2015a, b). Regulation of the flow level of the Syrdarya and Ili rivers led to a reduction of the groundwater level, thus increasing their mineralization, soil salinization, and drying of pond, which lead to the promotion of deflation processes. All of these actions seriously destroyed the environmental processes and led to serious forms of rapid soil and land degradation.

7.2.3 Modern State of the Aral Sea

The Aral Sea was the second largest water reservoir after the Caspian Sea on the Earth. The Aral is not connected with the ocean and therefore is not a sea, but a lake. The sea is called because of its huge size and regime similar to the sea. As in the past, the reason for the modern retreat of the Aral Sea is a marked decrease in the inflow of surface water from the Syrdarya and the Amudarya rivers, the main sources of its nutrition, which significantly worsened the water balance in the sea.

Since 1961, the level of the Aral Sea began to fall rapidly, and the water reservoir began to dry up (Fig. 7.3; Tables 7.3 and 7.4). During the life of almost one generation, the largest ecological catastrophe occurred on the Earth.

The regime of the Aral Sea was closely monitored by specialists until the mid-1980s and was fairly well studied (Kes' et al. 1980; Zenkevich 1963; Micklin

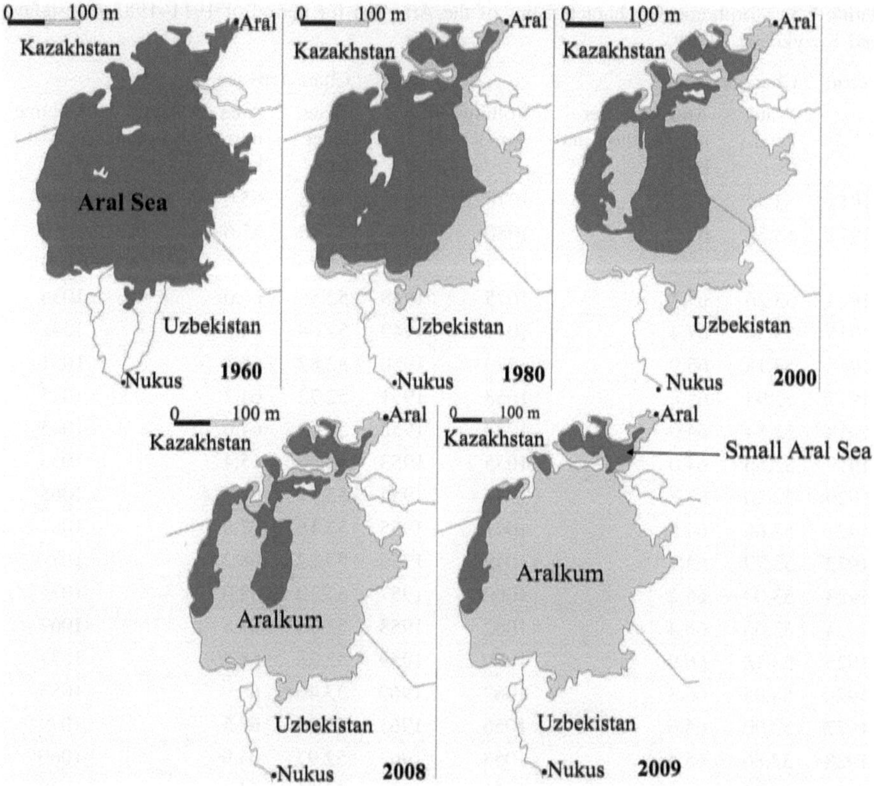

Fig. 7.3 Changing the Aral Sea profile and formation of Aralkum desert (1960–2009)—water covered area—dark gray, dry sea bottom—light gray (modified after Groll et al. 2012; Issanova et al. 2015a, b)

1990; Lunezheva et al. 1987; L'vovich 1971). Recently, data on changes in the natural conditions of the Aral Sea has been received much less.

Beginning in 1961, the level of the Aral Sea began to decrease with acceleration. For 1961–1970, the sea level dropped by 2.0 m; the average intensity of this decrease was 20 cm/year. In 1971–1980 and 1981–1990, the level fell by 5.7 and 7.2 m, respectively, with the average intensity of its fall increased to 57 and 72 cm per year. In some years, the level decreased by more than 1 m (Mustafayev and Kozykeeva 2009).

From 1961 to 1990, water parameters of the Sea were changed: The level of the Aral Sea decreased by 14.8 m. At the same time, the water volume in the Aral Sea decreased significantly (from 1093 to 330 km^3, that is by 763 km^3, or more than three times) and the area of the water reservoir decreased from 68,500 to 36,500 km^2, that is 32,000 km^2, or almost twice. The modern state of water parameters of the Aral Sea is shown in Fig. 7.4.

Table 7.3 Morphometric characteristics of the Aral Sea for period of 1911–1985 (Mustafayev and Kozykeeva 2009)

Years	Characteristics			Years	Characteristics		
	Water level (m)	Area of water mirror (thousand km^2)	Volume (km^3)		Water level (m)	Area of water mirror (thousand km^2)	Volume (km^3)
1911	53.32	67.5	1078	1945	52.78	65.0	1048
1912	53.35	67.7	1080	1946	52.90	65.6	1056
1913	53.24	67.2	1074	1947	52.79	65.1	1049
1914	53.26	67.3	1075	1948	52.56	64.0	1035
1915	53.30	67.4	1077	1949	52.68	64.6	1042
1916	53.18	66.9	1070	1950	52.82	65.2	1051
1917	52.94	65.8	1058	1951	52.72	64.7	1045
1918	52.54	64.0	1034	1952	52.69	64.6	1043
1919	52.56	64.0	1035	1953	52.86	65.4	1053
1920	52.50	63.8	1031	1954	53.12	67.7	1065
1921	52.66	64.5	1041	1955	53.16	67.8	1067
1922	52.79	65.1	1049	1956	53.22	68.2	1077
1923	53.03	66.2	1060	1957	53.19	68.0	1074
1924	53.06	66.4	1062	1958	53.16	67.8	1067
1925	53.18	66.9	1070	1959	53.28	68.4	1077
1926	53.05	66.3	1062	1960	53.40	68.9	1083
1927	52.90	65.6	1056	1961	53.29	68.5	1079
1928	52.86	65.4	1053	1962	52.97	65.9	1060
1929	52.89	65.5	1055	1963	52.61	64.3	1038
1930	52.76	64.9	1047	1964	52.49	64.8	1030
1931	52.76	64.9	1047	1965	52.30	62.38	972.47
1932	52.97	65.9	1062	1970	51.43	58.92	941.23
1933	53.07	66.4	1064	1971	51.06	57.73	902.43
1934	53.10	66.5	1065	1972	50.54	56.85	875.12
1935	53.25	67.2	1074	1973	50.22	56.17	845.47
1936	53.21	67.0	1071	1974	49.85	56.01	844.46
1937	53.10	66.5	1065	1975	49.01	54.67	802.74
1938	52.97	65.9	1060	1980	45.75	49.21	631.81
1939	52.87	65.4	1054	1981	45.18	48.63	625.78
1940	52.67	64.5	1042	1982	44.39	47.13	578.65
1941	52.67	64.5	1042	1983	43.55	46.07	532.58
1942	52.71	64.7	1044	1984	42.75	44.92	487.66
1943	52.79	65.1	1049	1985	41.94	43.08	444.58
1944	52.71	64.7	1044				

Table 7.4 Morphometric characteristics of the Aral Sea for period of 1985–2006 (Mustafayev and Kozykeeva 2009)

Years	Big Aral Sea			Small Aral Sea		
	Water level (m)	Area of water mirror (thousand km^2)	Volume (km^3)	Water level (m)	Area of water mirror (thousand km^2)	Volume (km^3)
1986	41.02	38.56	380.63	40.90	2.83	22.47
1987	40.19	37.13	343.17	40.80	2.81	22.39
1988	39.67	36.18	312.65	40.50	2.75	21.84
1989	39.10	35.30	306.92	40.20	2.71	20.28
1990	38.24	33.67	280.44	40.50	2.75	21.84
1991	37.66	32.02	257.16	40.40	2.73	20.92
1992	37.20	31.83	240.17	40.20	2.71	20.28
1993	36.95	31.42	231.70	39.37	2.57	18.43
1994	36.90	31.31	229.87	40.10	2.69	20.01
1995	36.50	30.04	217.25	40.50	2.75	21.84
1996	35.48	28.54	195.63	40.50	2.75	21.84
1997	34.80	26.91	173.44	41.20	2.91	22.67
1998	34.21	25.75	168.43	42.50	3.24	27.03
1999	33.98	24.12	147.62	36.80	2.09	12.03
2000	33.50	22.93	139.53	39.80	2.62	19.26
2001	32.40	21.00	131.16	39.20	2.55	17.97
2002	32.00	18.70	110.84	39.30	2.58	18.44
2003	31.50	17.30	97.23	40.00	2.65	19.77
2004	31.09	16.40	93.46	40.80	2.81	22.39
2005	30.70	15.77	89.79	41.00	2.86	22.52
2006	30.40	13.47	81.35	41.80	2.99	24.01

The average depth decreased from 16.0 to 9.0 m. By 1995, the level of the water reservoir fell by about 2 m. Thus, for 35 years the level of the Aral Sea has decreased by almost 17 m (Tables 7.3 and 7.4) (Mustafayev and Kozykeeva 2009).

In 1988–1989, the Aral Sea was divided into two parts: the smaller northern Sea—the Small Aral Sea, where a small flow of the Syrdarya flowed, and the large southern one—the Big Aral Sea that feeding on water of the Amudarya river. Since 1989, the area of the Small Aral Sea has changed insignificantly. This indicates that its level has stabilized. The area of the Big Aral Sea continued to decrease. The Berg Strait connecting the formerly Small and Big Aral Seas has turned into a small but fairly long channel through which surplus water from the Small Aral Sea was discharged into the Big Aral Sea.

Starting from 1960, the total river runoff, even in the zone of formation, has significantly decreased: from 117 km^3/year in 1961–1970 and up to 100 km^3/year in 1971–1975 (Gerasimov et al. 1983; Mal'kovsky 2003).

The total flow of the Amudarya and Syrdarya rivers in the zone of formation in the period 1911–1960 was 117 km^3/year, including 80 km^3/year in the Amudarya river and 37 km^3/year in the Syrdarya river, of which the actual inflow to the Aral Sea did not exceed 56 km^3/year, including 42 km^3/year in the Amudarya river and 14 km^3/year in the Syrdarya river(Table 7.4) (Mustafayev and Kozykeeva 2009).

A retrospective analysis shows that the share of natural river runoff in the Aral Sea water balance in the multi-year plan fluctuated rather widely, especially in the second half of the last century (Tables 7.5, 7.6 and 7.7).

As the level decreased, the Aral Sea rapidly changed its habitual shape. The coastline moved into the pond and leveled off, some islands were lapped to the shore, and many bays dried up. Prior to regulation of the river flow and direct sea level drop in the delta, 877.5 km^2 of land was flooded with flood waters and the area of lake systems was 517.73 km^2. From 1961 to 1970, the runoff along the Syrdarya River in the delta decreased to an average of 6.7 km^3/year with a minimum in 1965 to 3.2 km^3/year and a maximum 10.6 km^3/year in 1969. During the period 1971–1980, the average annual flow was 2.3 km^3/year and in 1981–1986 was 0.72 km^3/year. The total area of the water mirror of delta lakes in the mid-1970s decreased by 1.9 times and did not exceed 280 km^2, of which more than half was in the Kamyslybas lake system. By the beginning of the 1990s, the area of the water mirror of the lake systems in the delta of the Syrdarya River was 450 km^2 with a water volume of about 15 km^3 and flooded with water only 111 km^2 of land (Mustafayev and Kozykeeva 2009).

Since 1987, there has been a tendency to improve the water in the delta of the Syrdarya with an average water supply of 5.93 and 7.41 km^3/year, according to different sources. This allowed to preserve the area of the water mirror of the lake systems to 450 km^2. As a result, about 15% of their water mass was diluted annually with fresher river water, positively affecting the water–salt balance of water bodies. In subsequent years (1995–1996), as a result of a decrease in the total volume of river flow, flooding of lake systems in the lower and middle reaches of the Syrdarya River has become very problematic. A negative role was played by the unsatisfactory state of the channel network, through which the water–salt and level regimes of the lake systems are maintained.

The irrigation of the Kamyslybas lake system is currently carried out along four channels: Keragar in length 50 m, width 10 m, depth 2 m; Zhasulan (1.5 km, 10 m, 2.5 m); Kuya (0.6 km, 15 m, 5 m); and Taldyaral (60 m, 15 m, 2 m).

In the high water years (1993–1994), favorable conditions for natural water exchange of lake systems with channel flow were created in the middle river delta.

The Akshatau lake system is fed by the following channels: Suyk-Kol (0.3 km, 6 m, 3 m); Tabeken (0.7 km, 12 m, 3 m); Akkoi (1.5 km, 15 m, 5 m); and Aksha-kyz (40 m, 8 m, 2 m).

From 1988 to 1997, many locks on canals were destroyed by spring ice drifts and water backwater from lake systems. Repair and ongoing prevention due to lack of funds were not conducted. The capacity of the canals decreased due to vegetation overgrowing, siltation of the bottom, and collapse of the coastal embankment.

Table 7.5 Bathymetric characteristics of the Aral Sea (Mustafayev and Kozykeeva 2009)

Years	River inflow (km^3/year)				Precipitation (km^3)	Evaporation (km^3)	Water level (m)	Water volume (km^3)	Area of water surface (km^2)	Salinity (g/l)
	Amudarya river		Syrdarya river							
	S	Q	S	Q						
1950	0.47	41.0	0.58	11.9	9.22	66.06	52.90	1058.0	65,607	10.17
1951	0.52	33.4	0.55	13.2	8.07	59.19	52.77	1049.0	64,914	9.74
1952	0.41	55.2	0.46	18.8	8.78	62.62	52.79	1050.0	64,964	10.67
1953	0.41	54.8	0.45	19.5	9.63	64.11	52.94	1059.0	65,706	9.82
1954	0.41	55.1	0.43	21.1	10.87	62.87	53.21	1076.0	67,042	10.21
1955	0.47	41.9	0.49	16.7	9.17	66.13	53.27	1079.0	67,290	10.13
1956	0.44	48.0	0.50	16.4	9.30	67.20	53.32	1082.0	67,537	10.19
1957	0.54	30.9	0.63	9.5	8.51	68.11	53.27	1080.0	67,389	10.01
1958	0.42	52.3	0.60	10.9	7.94	68.93	53.23	1078.0	67,240	10.42
1959	0.45	46.3	0.47	18.3	9.92	70.05	53.39	1086.0	67,884	10.19
1960	0.47	42.0	0.43	21.1	9.41	71.13	53.50	1093.0	68,478	9.93
1961	0.57	31.1	1.14	8.9	6.59	70.43	53.38	1087.0	67,983	9.97
1962	0.51	38.4	1.60	4.0	8.63	70.93	53.07	1067.0	66,350	10.80
1963	0.56	31.8	1.28	7.0	11.56	70.64	52.72	1045.0	64,568	10.58
1964	0.51	39.2	1.10	9.4	8.12	64.04	52.58	1038.0	63,974	10.13
1965	0.62	25.3	1.17	3.2	8.48	66.35	52.40	1026.0	63,308	10.81
1966	0.53	35.6	1.33	6.4	6.64	71.13	51.98	1000.0	62,014	11.81
1967	0.58	29.3	1.38	5.9	7.51	57.82	51.66	980.9	61,060	11.02
1968	0.54	34.4	1.49	4.9	6.03	67.35	51.35	960.7	60,299	11.49
1969	0.36	70.6	1.03	10.6	9.06	52.31	51.39	963.7	60,408	10.91
1970	0.56	32.4	1.32	6.5	7.22	62.03	51.44	971.7	60,692	11.20

(continued)

Table 7.5 (continued)

Years	River inflow (km^3/year)				Precipitation (km^3)	Evaporation (km^3)	Water level (m)	Water volume (km^3)	Area of water surface (km^2)	Salinity (g/l)
	Amudarya river		Syrdarya river							
	S	Q	S	Q						
1971	0.65	20.6	1.04	5.6	5.81	59.83	51.11	949.0	59,885	11.38
1972	0.59	24.2	1.15	4.8	5.78	55.34	50.65	917.8	58,935	11.95
1973	0.40	43.5	0.99	6.0	8.95	56.45	50.32	898.9	58,494	11.95
1974	1.01	6.9	2.19	1.3	4.75	60.18	49.92	874.4	57,924	13.02
1975	0.92	9.2	2.47	0.8	4.43	59.99	49.09	824.2	56,757	13.40
1976	0.85	11.3	2.88	0.3	5.79	51.09	48.36	785.3	55,718	14.57
1977	0.99	7.2	2.98	0.2	5.04	45.75	47.74	749.2	54,792	15.44
1978	0.68	18.9	2.79	0.4	6.42	52.52	47.06	717.6	53,981	14.97
1979	0.87	10.9	1.80	2.1	4.87	52.14	46.45	683.4	52,989	15.09
1980	0.92	9.3	1.96	1.7	9.73	50.24	45.76	648.7	51,743	16.80
1981	1.33	6.9	2.03	1.7	11.92	47.11	45.19	620.0	50,714	17.70
1982	2.75	0.3	2.31	1.3	8.52	38.50	44.39	579.8	49,270	18.80
1983	2.06	2.4	3.20	0.5	4.51	47.59	43.55	537.5	47,753	20.30
1984	1.23	8.0	3.53	0.3	5.99	44.33	42.75	502.7	46,243	21.90
1985	2.11	2.2	3.53	0.3	7.19	42.52	41.95	475.0	44,382	22.90
1986	2.69	0.46	3.73	0.20	0.11	0.98	41.94	448.0	41,047	22.9
1987	1.17	8.68	2.58	1.0	0.10	1.0	41.10	432.0	38,831	23.9
1988	0.72	17.81	1.01	5.0	0.11	0.94	40.29	401.0	37,410	25.0
1989	2.30	1.51	1.42	3.1	0.15	0.97	39.75	380.0	36,562	28.0
1990	1.33	6.89	1.67	2.41	0.70	1.04	39.08	354.0	35,349	30.0
1991	1.33	10.48	1.89	2.58	0.80	1.06	38.24	323.0	33,831	32.0

(continued)

Table 7.5 (continued)

Years	River inflow (km^3/year)				Precipitation (km^3)	Evaporation (km^3)	Water level (m)	Water volume (km^3)	Area of water surface (km^2)	Salinity (g/l)
	Amudarya river		Syrdarya river							
	S	Q	S	Q						
1992	0.78	24.27	1.73	3.34	0.10	0.92	37.56	299.0	32,649	34.0
1993	1.06	15.52	1.17	7.5	0.90	0.83	37.20	286.0	32,017	35.0
1994	0.93	18.72	1.09	8.46	0.12	0.97	36.96	278.0	31,564	36.0
1996	1.87	4.92	1.47	4.89	0.19	0.97	36.11	250.0	29,872	42.0
1997	2.68	0.73	1.64	3.82	0.24	0.93	35.48	230.0	28,530	43.5
1998	0.89	20.07	1.18	7.41	0.17	0.88	34.80	210.0	26,956	49.8
1999	1.97	4.17	1.32	6.03	0.90	1.0	34.24	194.0	25,519	50.6
2000	2.51	1.37	1.83	2.86	0.13	0.96	33.80	181.0	24,266	55.8
2001	2.87	0.09	1.79	3.03	0.16	0.95	33.30	169.0	22,745	58.6

Table 7.6 Average min and max parameters of the Aral Sea level and total river flow (Mustafayev and Kozykeeva 2009)

Years	Sea level (abs m)		River flow (km^3)	
	Min.	Max.	Min.	Max.
1941–1945	52.67	52.97	–	–
1946–1950	52.68	53.03	–	–
1951–1955	52.82	53.13	58.6	76.2
1956–1960	53.13	53.46	40.4	64.4
1961–1965	52.54	52.98	28.5	48.6
1966–1970	51.29	51.69	35.2	81.2
1971–1975	49.81	50.38	8.2	49.5
1976–1979	47.03	47.68	7.4	19.3
1980–1984	42.75	45.75	1.8	10.05
1985–1989	39.08	41.10	0.6	21.8
1990–1994	36.9	38.24	11.41	32.24
1995–2000	33.98	36.5	5.17	28.53

Temporary dams on the canals are often washed away, and water flows back to the Syrdarya river, leading to a disruption of the water–salt regime of the lake systems.

The immediate physical reason for the decline in the level of the Aral Sea is a violation of the water balance in the water body: excess water discharge over the water income.

Beginning in 1961 is an active period of anthropogenic influence on Sea's regime. The sharp increase in the irretrievable seizures of the runoff, which in recent years has reached 70–75 km^3/year, the exhaustion of the compensatory possibilities of the rivers, as well as the natural low water periods of the two decades of 1960–1980 years (92%) have led to a violation of the equilibrium of the water and salt balances in the Sea.

Until 1961, the water balance of the Aral Sea was close to steady; the income and outcome of water flow were the same.

After 1961, the loss of water for evaporation from the surface of the sea began to exceed the income of water (the amount of river runoff and precipitation). This has led to a decrease in the water volume of the water pond and a drop in its level. Exceeding the water flow over the income increased, which caused an accelerated decline in sea volume and a decrease in its level, especially in the 1970s–1980s. Since 1989, the Aral Sea has been divided into the Small and the Big Aral. A one-way water flow through the Berg Strait from Small to the Big Aral was established between these water reservoirs.

In the period from 1988 to 1994, the water balance and consequently the level of the Small Sea were stabilized, in general. However, in the Big Aral Sea, the water volume was continued to decrease and the water level was reduced. The main reason for the change in water balance (volume, surface area, and water level) of the Aral Sea is the reduction of river flow (Tables 7.5, 7.6 and 7.7).

The natural water resources of the Amudarya and Syrdarya basins in the runoff formation zone in the highland areas of the Pamir and the Tian Shan are estimated at

Table 7.7 Average annual values of level, volume, and area of the Aral Sea (Mustafayev and Kozykeeva 2009)

Years	Water level (m)	Volume (km^3)	Area of water mirror (thousand km^2)	Water supply in the Aral Sea region (km^3)		
				From Syrdarya river	From Amudarya river	Total
1960	53.40	1083	68.9	21.1	43.0	64.1
1961	53.29	1079	68.5	–	30.9	–
1962	52.97	1060	65.9	5.8	27.6	33.4
1963	52.61	1038	64.3	10.6	33.1	43.7
1964	52.49	1030	64.8	15.0	38.3	53.3
1965	52.31	1019	63.1	4.7	25.5	30.2
1966	51.89	993	61.7	9.6	33.1	42.7
1967	51.57	974	60.9	8.7	27.0	35.7
1968	51.24	952	60.1	7.3	28.0	35.3
1969	51.29	955	60.2	17.5	55.5	73.0
1970	51.43	964	60.3	9.8	28.0	37.8
1971	51.06	940	59.7	8.2	15.8	24.0
1972	50.54	909	58.9	7.0	13.2	20.2
1973	50.22	891	58.4	8.9	31.2	40.1
1974	49.85	870	57.9	1.9	6.3	8.2
1975	49.01	822	56.7	0.61	10.6	11.2
1976	48.27	779	55.7	0.57	11.13	11.7
1977	47.63	742	54.6	0.48	9.0	9.5
1978	47.06	713	53.9	0.78	21.3	22.1
1979	46.45	680	52.9	3.2	11.1	14.3
1980	45.75	644	51.7	2.5	8.6	11.1
1981	45.18	616	50.7	7.4	6.3	8.7
1982	44.39	574	49.3	1.7	0.54	2.2
1983	43.55	532	47.7	0.94	2.3	3.2
1984	42.75	499	46.2	0.60	8.0	8.6
1985	41.94	466	44.6	0.68	2.4	3.1
1986	41.10	432	42.8	0.51	0.44	0.95
1987	40.29	401	41.1	1.6	8.2	9.8
1988	39.75	380	39.9	6.9	16.4	23.3
1989	39.08	354	38.4	4.4	1.0	5.4
1990	38.24	323	36.4	3.5	9.0	12.5
1991	37.56	299	43.8	4.0	12.5	16.5
1992	37.20	286	33.9	4.6	28.9	33.5
1993	36.95	278	33.2	7.9	18.8	26.7
1994	36.60	266	32.3	8.9	21.7	30.6
1995	36.11	250	31.3	5.2	5.1	10.3

(continued)

Table 7.7 (continued)

Years	Water level (m)	Volume (km^3)	Area of water mirror (thousand km^2)	Water supply in the Aral Sea region (km^3)		
				From Syrdarya river	From Amudarya river	Total
1996	35.48	230	29.7	5.1	7.5	12.6
1997	34.80	210	28.0	4.6	2.2	6.8
1998	34.24	194	25.5	7.6	23.9	31.5
1999	33.80	181	23.7	5.5	6.4	11.9
2000	33.30	169	22.9	2.9	2.6	5.5
2001	32.16	143	21.2	2.8	0.40	3.2
2002	30.90					

approximately 75 and 37 km^3/year, respectively (in total only about 112 km^3/year). To the mouths of rivers, the volume of runoff decreased approximately half in natural conditions, due to losses for evaporation and infiltration; the Aral Sea received on average 50–60 km^3/year (Table 7.8) (Borotnik et al. 1991; Kuksa 1994). During the years 1911–1960, the total flow of the Amudarya and the Syrdarya rivers to the Aral Sea averaged 56 km^3/year (Table 7.9). The Amudarya river accounted for an average of 46 km^3/year, and the Syrdarya river was 10 km^3/year.

Table 7.8 Average annual components of water balance in the Aral Sea (km^3/year) (Shivarev and Smerdov 1996)

Part of the Aral Sea	Year	River flow	Visible evaporation (evaporation minus precipitation)	Water change through Berg Strait	Net balance
Small Aral Sea	1988	4.80	2.34	−2.46	–
	1989	3.04	2.70	−0.34	0
	1990	2.52	2.52	−0.00	0
	1991	2.58	2.58	−0.00	0
	1992	3.18	2.12	−1.06	0
	1993	6.17	2.09	−4.10	0
	1994	7.14	2.34	−4.80	0
Big Aral Sea	1988	9.19	28.12	2.46	−16.47
	1989	0.84	29.96	0.34	−28.78
	1990	3.92	29.60	0.00	−25.68
	1991	5.76	28.24	0.00	−22.48
	1992	11.44	22.06	1.06	−9.56
	1993	9.08	21.31	4.10	−8.13
	1994	11.71	23.38	4.80	−6.87

Table 7.9 Average long-term values of water balance in the Aral Sea

Years	Income (km^3)		Outcome (evaporation, km^3)	Water balance
	River flow	Precipitation		
1911–1960	56	9.1	66.1	−1
1961–1980	30	7.1	59.7	−22.6
1971–1980	16.7	6.2	53.7	−30.8
1981–1990	3.45	7.1	40.4	−29.8
1991–1999	7.55	5.8	28.1	−14.8

This amount of river flow was sufficient to maintain the level of the Aral Sea in the first half of the twentieth century in a relatively stable position at around 53 m abs.

Since the beginning of the 1960s, the situation has radically changed: The river flow to the Aral Sea began to decline rapidly. The decrease in water inflow to the Aral Sea is explained by natural low water (20%) and by 80% by the anthropogenic factor: irretrievable withdrawal of the river runoff for irrigation (Borotnik et al. 1991; Kuksa 1994). Recently, there was another hypothesis of reducing river flow to the Aral Sea: an increase in infiltration losses in the Amudarya and Syrdarya riverbeds caused by earthquakes, especially strong ones in 1981–1988 (Lunezheva et al. 1987).

The magnitude of irretrievable drainage in the Aral Sea basin has steadily increased. By 1980, natural water resources of the Amudarya and Syrdarya rivers in the estuaries were used, respectively, at 65 and 82%. Since 1990, these numbers were projected to increase to 87 and 90%, respectively (Mustafayev and Kozykeeva 2009).

The water flow of rivers into the Aral Sea has steadily declined over the past 25–30 years. The only exception was the very high water year in 1969. In the 1970s–1980s, the natural water shortage of these years contributed to the reduction of runoff to the Aral Sea. Flow in the sea increased somewhat in the relatively high water year in the 1990s.

Since 1982, the water discharge from the Amudarya river into the Aral Sea along the mainstream has been discontinued: A dense bulk dam was built near the Kyzylzhar village. All residual river flow entering the delta began to be directed to irrigation in the left bank and watering of the dried delta's water bodies. Inflow to the sea in 1982, 1983, 1985, and 1986 was completely absent. In the subsequent relatively high water years (for instance, in 1987–1988, 1990–1994), part of the Amudarya river flow still fell into the Aral Sea (Mustafayev and Kozykeeva 2009).

Since 1974, due to the almost complete economic use of water of the Syrdarya river and the overlapping of the main riverbed within the delta by several deaf-filled bulk dams, the river flow into the Sea has almost ceased. An insignificant amount of water discharged into the delta by the Kazaly hydroelectric complex was used to water the delta's water bodies. In 1982–1987, the Syrdarya river flow to the Aral

Sea was completely absent. Water has reached in the Sea into small volumes only in relatively high water years, for instance in 1988–1994 (Mustafayev and Kozykeeva 2009). Consequently, the quality of the river flow has also changed. The increase in the share of highly mineralized waste and drainage waters led to a significant increase in mineralization and deterioration of the sanitary state of the river waters. In the dry years, the average annual mineralization of the Amudarya river entering the sea reaches 0.8–1.6 g/l and in the Syrdarya river is 2.0 g/l. In some seasons, its even higher values are noted. As a result, despite the fact that the average annual river runoff in 1961–1980 decreased by more than 46%, the average annual ion flow for the same period decreased by only 4 million tons or by 18% (Mustafayev and Kozykeeva 2009). In addition, other components of the salt balance have also changed significantly.

1. The water salinity has significantly increased. If before 1961 the average salinity of the Aral Sea was about 10 g/kg (10 days), then by 1990 it had increased to 30 days (Table 7.6) and by 1995 apparently to 33–35 days. The hydrochemical class of the water reservoir was changed, turning from a brackish one (S < 24.7 days) into a salty one (S > 24.7 days) in a short time.
2. The salt composition of the waters has changed. With increasing salinity of water and partial accumulation of calcium carbonates, the relative contents of calcium, magnesium, and sulfates decreased and the relative contents of chloride, sodium, and potassium ions increased markedly.
3. The thermal regime of the water reservoir has changed. In connection with the increase in salinity, the freezing temperature (down to 1.5–2.0 °C) and accordingly the water temperature in winter decreased. Simultaneously with the decrease in the depths and volume of water, the water temperature in spring increased by 1.5 °C and in summer by 0.5 °C.
4. The ice regime has become more severe. The water pond now freezes faster, and a few days earlier, the ice melting begins later and passes for a longer period.
5. The ecosystem of the water reservoir was disturbed. Irreversible changes in the ecosystem of the Aral Sea were manifested in a reduction in the number of microorganisms and changes in their species composition; the biomass of phytoplankton and zooplankton was decreased in 5–6 times, a sharp deterioration in the forage reserve, spawning conditions, habitat and reproduction of fish stocks, and a sharp reduction in fish stocks.

The main consequences of the Aral desiccation are:

- Deterioration in the regime of air humidity and a decrease in moisture content in the lower atmosphere;
- Increase in the frequency of early frosts;
- Increase the frequency of dust storms, and increase the removal of sand and salts from the dried part of the sea;
- Progressive desertification of the Aral Sea area.

The change in the regime of both parts of the Aral Sea, the Small and Big Seas, will depend in the future on the amount of river flow entering them, as well as on the possible local hydrotechnical measures to regulate their regime.

According to various forecasts, the level of the Small Sea can vary within 35–40 m abs when the flow of the Syrdarya river fluctuates in the range 0.5–2 km^3/year (Borotnik et al. 1991; Kuksa 1994; Shivarev and Smerdov 1996). It is possible in both natural and artificial ways to completely isolate the Small Aral Sea from the Big Aral Sea. With a lack of river flow, this water reservoir is expected to progressively dry out and become saline (Table 7.10).

In 1960–1990 as a result of irrational use of water resources for the development and expansion of irrigated land areas, the flow of the Amudarya and Syrdarya rivers has decreased significantly. Consequently, the complex of environmental and socioeconomic problems has formed that has consequences with international character.

An area of 35 thousand m^2 of the Sea bottom was dried up. The former saline Sea bottom is a source of dust and salt storms. The dried bottom of the Aral Sea transports 100 million tons/year of salt by dust and sand storms (Mustafayev and Kozykeeva 2009). Its composition is dominated by suspended particles like aerosols with an admixture of agricultural toxic chemicals, fertilizers, and other harmful components of industrial and domestic sewage. Due to reducing the size of the sea and increasing the evaporation and supply of the drainage-collector waters, the salinity of the water increased significantly, which in 1965 was 9.94 g/l and now about 15 g/l (Mustafayev and Kozykeeva 2009).

The decline in inflows since 1960 is caused by both climatic and anthropogenic causes. In the arid1970s, especially in 1974–1975, the influx from the zones of formation of the Amudarya and Syrdarya flows decreased by approximately 30 km^3/year (27%) compared to the average for the preceding 45 years (Kotliakov and Glazovskii 1991; Shkolomanov 1979; Burlibayev et al. 2001; Mustafayev et al. 2003). Minimal flow was also recorded from 1982 to 1986 (Mustafayev et al. 1997). However, the most important factor in reducing river flow is large volumes

Table 7.10 Development of the ecological crisis in the Aral Sea basin (Mustafayev and Kozykeeva 2009)

Parameters	1966	1976	19961	2000
The area of the "new" salty desert, which arose as a result of the desiccation of the sea (km^2)	No	130,200	38,000	42,000
The physical mass of salt, dust, and waste within the saline desert (million tonnes)	Absent	500	2300	3300
The area covered by the removal of salts and dust (thousand km^2)	Absent	100–150	250–300	400–450
Increase in removal and precipitation of salts and dust (kg/g)	Absent	100–200	500–700	700–1100
The population in the zone affected by the ecological crisis (thousand people)	Absent	500–600	3000–3500	3500–7000

of irretrievable water consumption, mainly for irrigation. The annual river runoff in the zones of formation of these rivers averaged 111 km^3 between the years of 1926 and 1970 (Mustafayev et al. 1997). Under natural conditions, only about half of this volume would flow into the Aral Sea due to losses for evaporation, transpiration, and filtration, as these rivers cross the deserts and form deltas.

7.2.4 Use of Water Resources in the Aral Sea Basin

The use of water resources in the Aral Sea Basin mainly for drinking and irrigation began more than 6000 years ago (Mustafayev and Kozykeeva 2009). Water resources began intensively to be used in the twentieth century, especially after 1960 due to rapid population growth (Fig. 7.4), intensive industry development and mainly irrigation and completely regulation of the river flows in the Aral Sea basin (Mustafayev et al. 1997, 2003).

The water resources of the Aral Sea basin consist of Amudarya and Syrdarya river basins (Fig. 7.5), and their water resources mainly are used by Central Asian countries. The main water consumer in the basin is Uzbekistan and Turkmenistan (Fig. 7.4).

More than 60 water reservoirs have been constructed in the Aral Sea basin with a useful water volume of over 10 million m^3 each. The total volume of water reservoirs is 64.5 km^3, including a useful volume of 46.5 km^3, 20.2 km^3 in the Amudarya river basin, and 26.3 km^3 in the Syrdarya river basin (Mustafayev and Kozykeeva 2009) (Table 7.11).

The potential hydropower resources of Central Asia are estimated at 531 billion kW/h (Table 7.12) (Debol'skii and Osmonbetova 1999; Ibatullin et al. 2005).

There are 45 hydroelectric power stations in Central Asian courtiers with a total capacity of 34.5 GW, each varying from 50 to 2700 MW. The largest hydropower

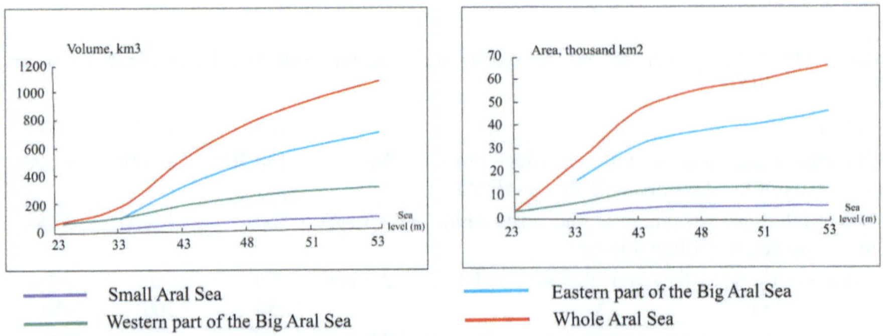

Fig. 7.4 Dynamics in use of water resources in the Aral Sea basin (million m^3)

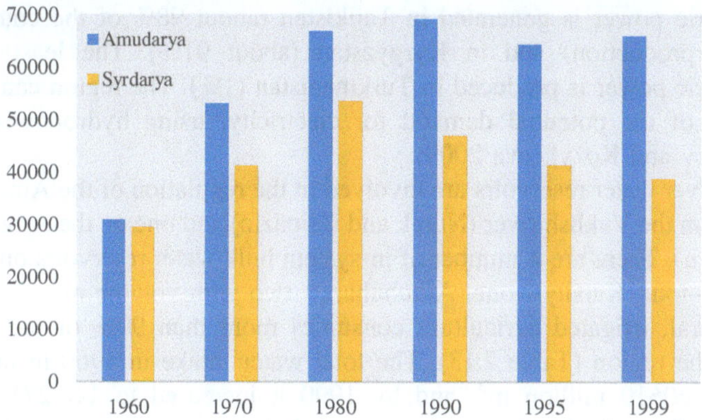

Fig. 7.5 Dynamics in use of water resources from Amudarya and Syrdarya rivers (million m^3)

Table 7.11 Water reservoirs in the Amudarya and Syrdarya river basins

Water reservoir	Useful volume (million m^3)	River
Amudarya river basin		
Nurek	4500	Vagsh
Tuiamoiyn	4500	Amudarya
Iuzhnosurhan	700	Suhandarya
Syrdarya river basin		
Toktagul	14,000	Naryn
Shardara	4700	Syrdarya
Kairakum	2500	Syrdarya
Andizhan	1750	Karadarya

Table 7.12 Potential hydropower resources of Central Asia

Country	Potential resources		Resources possible to use	
	Billion kW/h	%	Billion kW/h	%
Kyrgyzstan	136	26.5	73	53.7
Tajikistan	282	55.0	143	50.7
Turkmenistan	22	4.3	5	22.7
Uzbekistan	73	14.2	25	34.3
Total	513	100	246	48.0

stations are the Nurek (in Tajikistan on the Vakhsh River) with a capacity of 2700 MW and Toktogul (in Kyrgyzstan on the Naryn River) with a capacity of 1200 MW (Mustafayev and Kozykeeva 2009).

Hydropower accounts for 27.3% of the total energy consumption of the Aral Sea basin. However, in some countries, this indicator varies considerably. Most of all hydroelectric power is generated in Tajikistan (about 98% of the country's total electricity production) and in Kyrgyzstan (about 91%). The least amount of hydroelectric power is produced in Turkmenistan (1%). The region can meet more than 71% of the potential demand for electricity, using hydropower resources (Mustafayev and Kozykeeva 2009).

Three river water reservoirs are involved in the regulation of the Amudarya river flow: two on the Vakhsh river (Nurek and Baipazin) and one on the Amudarya river (Tuyamoiyn). There are a number of in-system bulk water reservoirs on the canals: Karakum—four, Karshy—one, Amubuhara—two with volume more than 6 km^3.

In general, irrigated agriculture consumes more than 90% of the total water intake in the region (Table 7.13). The total water intake in 1960 in the Aral Sea basin was 60610 million m^3, and by 1990 it increased to 116,271 million m^3 (Fig. 7.6) (Table 7.14). Over the same period, the population in this territory increased by 2.7 times, irrigation areas by 1.7 times (Table 7.15), consequently agricultural production by 3 times, and gross national product almost by 6 times (Mustafayev and Kozykeeva 2009).

Table 7.13 Dynamics in use of water resources in the Aral Sea basin for irrigation (million m^3)

Years	Aral Sea basin		Amudarya river basin		Syrdarya river basin	
	Total	For irrigation	Total	For irrigation	Total	For irrigation
1960	60.61	56.152	30.97	28.55	29.64	27.602
1970	94.56	86.837	53.22	49.282	41.34	37.555
1980	120.69	106.79	66.95	60.345	53.74	46.445
1990	116.271	106.404	69.247	65.151	47.024	41.253
1995	105.805	96.72	64.392	60.700	41.253	41.413
2000	104.955	94.657	66.079	59.568	38.876	35.089

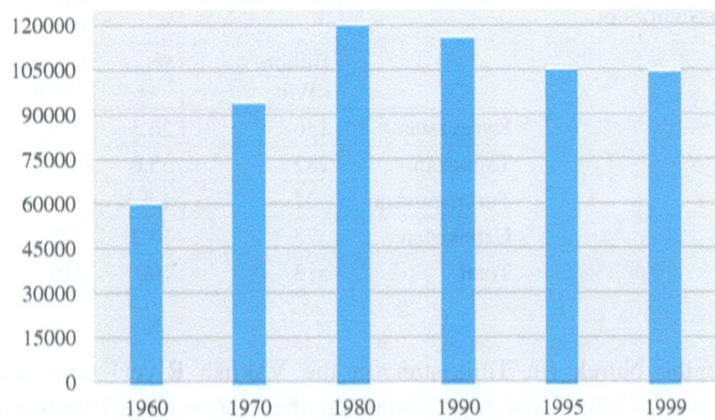

Fig. 7.6 Dynamics of water intake from Aral Sea basin (million m^3)

Table 7.14 Dynamics in use of water resources in the Aral Sea basin (million m^3)

Country	Years					
	1960	1970	1980	1990	1995	1999
Kazakhstan	9750	12,850	14,200	11,320	11,300	8235
Kyrgyzstan	2210	2980	4080	5155	4966	3291
Tajikistan	9800	10,440	10,750	9259	12,089	12,521
Turkmenistan	8070	17,270	23,000	23,338	23,230	18,075
Uzbekistan	30,780	48,060	64,910	63,611	54,220	62,833
Aral Sea basin	60,610	94,560	120,690	116,271	105,805	104,955
Amudarya	30,970	53,220	66,950	69,247	64,392	66,079
Syrdarya	29,640	41,340	53,740	47,024	41,413	38,876

Table 7.15 Main parameters in use of water–land resources in the Aral Sea basin

Parameters	Years				
	1960	1970	1980	1990	2000
Population (million people)	14.1	20.0	26.8	33.6	41.5
Area of irrigated lands (thousand ha)	4510	5150	6920	7600	7990
Irrigated area per capita (ha for person)	0.32	0.27	0.26	0.23	0.19
Total water intake (km^3/year)	60.61	94.56	120.69	116.27	105.0
For irrigation (km^3/year)	56.15	86.84	106.79	106.4	94.66
Specific water intake for 1 ha of irrigated land (m^3/ha)	12,450	16,860	15,430	14,000	11,850
Specific water intake per capita (m^3 for person in year)	4270	4730	4500	3460	2530

After the collapse of the Soviet Union in 1991, the total use of water in the region began to decline due to general economic degradation.

Until recently, the water resources of the Syrdarya river basin were used in natural state without flow control, except for a few small water reservoirs on small rivers in the Fergana Valley. In the 1950s, it was necessary to rapidly increase the cotton production and to expand the area of irrigated land in Central Asia (mainly in Uzbekistan) within the Syrdarya river basin. It was planned to increase the cotton production from 4.3 million tons in 1960 to 10–11 million tons in 1980 (Bobchenko 1991).

This required an extensive program on irrigation construction, including the construction of hydroelectric power stations and water reservoirs. The basins of the Syrdarya and Amudarya rivers are not homogeneous. The conditions for the development of irrigated agriculture are very limited in the mountains. But the bulk of precipitation accumulates in the mountains, and the flow of all rivers is formed there (in the territory of Kyrgyzstan). The valley is the most developed part of the basin, and the main zone of cotton growing begins from the Uch-Kurgan hydroelectric power station and includes: the Fergana Valley, the Golodnaya steppe,

Dal'verzin steppe, Djizak and Farish steppes, Kyzylkum steppe, and the land of the right bank of the Syrdarya River (mainly the territory of Uzbekistan) (Sorokin 2002).

The lower reaches of the Syrdarya river such as Zhanakurgan, Kyzylorda, and Kazaly massifs are the least developed part of the basin. Those massifs are suitable for planting rice and forage crops in natural conditions.

A cascade of power electricity station was designed on Naryn river taking into account irrigation requirements in the Syrdarya river basin and the development of Kyrgyzstan's energy sector (Table 7.16) (Debol'skii and Osmonbetova 1999).

More than 35 water reservoirs were built in the Amudarya river basin with a capacity of more than 10 million m^3 each. The total capacity of these water reservoirs exceeds 29.8 km^3 (Mustafayev and Kozykeeva 2009).

Due to development of irrigation and drainage systems in the region, there is a constant increase in the formation of return waters. The most intensive period is observed in 1960–1990. After 1991, the volume of return water stabilized and even began to decrease somewhat due to a temporary reduction in the used irrigated areas and degradation of drainage systems. For the period 1990–1999, the total volume of return waters ranged from 28.0 to 33.5 km^3 per year. About 13.5–15.5 km^3 of return water was formed annually in the Syrdarya basin and about 16–19 km^3 in the Amudarya basin. More than 51% of the total volume of return water is diverted by collectors into rivers, about 33% in depressions. Only 16% of return water is reused for irrigation, which is due to their contamination. Thus, the amount of return water in the Aral Sea basin only from irrigated lands is 29.55 km^3, that is corresponding to the flow of the Syrdarya river. It characterizes on the one hand and lacks the culture and responsibility in using of water resources, on the other hand; it allows to solve ecological problems of the region.

Table 7.16 Hydropower stations with a capacity of more than 300 thousand kW (Toktogul HPS is not included)

Hydropower stations	Rated water discharge (m^3/s)	Established capacity (thousand kW)	Average long-term production (million kW/h)
On the Naryn river			
Atbashy	240	310	115
Alabugin	370	700	2600
Toguz-Toro	430	540	2150
Kara-Bulak	350	330	1650
Kambarta	600	1000	4400
Kurpsai	800	500	2200
Tash-Kumyr	735	360	1580
On the Kokomeren river			
Susamyr	140	300	1325
Kokomeren	120	600	2610

Intensive development of irrigated agriculture and drainage of lands in Central Asian countries, while increasing water consumption for industrial and domestic needs, has caused an increase in the volume of freshwater intake and discharges to water sources of pollutants together with return waters.

The main sources of pollution are residues of agrochemicals that are washed into drainage systems and mixed with river water. Wastewater from municipal and industrial sewer systems is the source of pollution, and they impact on the quality of water resources.

Return waters are the main source of environmental pollution in the Aral Sea basin. Sulfates, chlorides, and sodium ions predominate in drainage waters, which also contain pesticides, nitrogen, and phosphorus compounds. Up to 25% of nitrogen, 5% of phosphorus, and 4% of pesticides that enter the soil are transferred by drainage collectors from fields (Mustafayev and Kozykeeva 2009). The concentration of these pollutants in drainage water exceeds the maximum permissible concentration by 5–10 times. Similar trends in the change of water quality are also traced in the Syrdarya river basin. The mineralization of water in the upper reaches of this river does not exceed 0.3–0.5 g/l, but even at the outlet from the Fergana Valley it reaches 1.2–1.4 g/l and in the alignment of Kazaly exceeds 1.7–2, 3 g/l (Mustafayev and Kozykeeva 2009)

In all control areas, mineralization of water has increased compared with the period 1960–1970 years. Simultaneously with the increase in the total salinity of river water, there is an increase in the content of such chemical components as magnesium, copper, iron, sulfates, chlorides, and others (Mustafayev and Kozykeeva 2009). As a result surface waters not only in the lower but also in the middle reaches of the Syrdarya river are not suitable for drinking needs. Significant contamination of the river as a source of drinking water supply often leads to an increase in the incidence among the local population. More than 50 million tons/year of salts entered the river; natural runoff brings only about half and the rest is formed by collector drainage runoff.

Salinization and concentration of basic ions are also high; it increases from mountain regions down to plains. Amudarya river area between the Nurek water reservoir and the site of its confluence with the Vakhsh river receives about 8.5 million tons of salt per year and the Panj river to 0.8 million tons. Between the confluence of the Vakhsh and Panj rivers and the Tuyamoiyn water reservoir up to 16.5 million tons of salts are discharged into the Amudarya river (Mustafayev and Kozykeeva 2009). Below the Tuyamoiyn water reservoir, another 1.6 million tons of salts are dumped into the Amudarya river. Data on salinization of water in two main points of the Amudarya river indicates that irrigation has caused salinization, exceeding the permissible norm in the Tuyamoiyn water reservoir which already reached 1.76 g/l in the 1980s.

The balance of river salt and irrigated massifs makes possible to clearly establish salt accumulation zones due to well-being meliorative of irrigated lands where soil fertility either decreases or is kept at the maximum permissible level. Such zones in the Amudarya river basin are the land massifs of the Turkmen coast, Tashauz, and Karakalpakistan.

7.3 Balkash Lake Basin

Balkash Lake is located in the southeast of Kazakhstan, in the lowest western part of the Balkash–Alakol depression, and on the border of three administrative regions of Kazakhstan such as Karagandy, Zhambyl, and Almaty. The lake extends from west to east. It is a drainage basin with tectonic origin.

The drainage basin of the Balkash Lake occupies about 501 thousand km^2, including 400 thousand km^2 within Kazakhstan. The length of the basin from the west to the east is more than 900 km and from the north to the south is 680 km (Philonets 1981). The northern border of the basin is the Karkaraly-Aktau Massif and the Shyngystau ridge; the eastern border is Tarbagatay ridge, the Urkashar, Barlyk and Maily ridge and the Zhetysu (Dzungar) Alatau and the ranges of Borokhoro, Iren-Khabirga, Narat and Khalyntau (in China); the Southern border is eastern spurs of the Terskey and Kungei Alatau ranges and Ile Alatau; the Western border is the Shu–Ile Mountains. The total length of the watershed line is about 4000 km.

According to the nature of the relief in the catchment area, there are several physical–geographical areas distinguished: the Kazakh hummocky terrain, the plains of the Balkash–Alakol depression, the systems of the Kazakh-Dzungarian mountain region, the Shu–Ile mountains, and the eastern part of the Northern Tian Shan. The height of the catchment ranges from 342 to 6995 m (Philonets 1981).

The great amplitude of the centuries-old and secular fluctuations in the water balance and the level of the Balkash lake significantly changes its morphometric and hydrological characteristics. The water and salt balances of the lake is shown in Fig. 7.7. At a water level of 342 m, the area of the water mirror is 18,200 km^2, the length is 614 km, the width is 74 km, the average width is 30 km, the coastline is 2383 km, the maximum depth is 27 m, and the average depth is about 6 m. The volume of water is about 106 km^3 (Philonets 1981).

In the current century, the water mirror of the Balkash lake changed from 23,444 km^2 in 1910 to 15,730 km^2 in 1946 and the volume of water from 163.9 to 82.7 km^3.

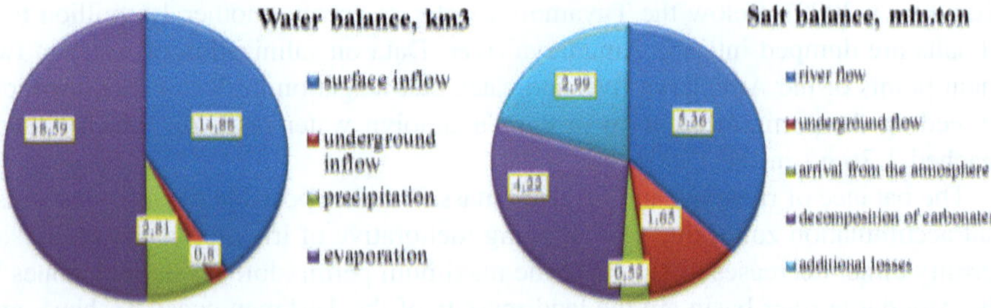

Fig. 7.7 Balkash Lake basin

The lodge of the lake consists of a series of alternating vast hollows. In Western Balkash, there are two such hollows with a depth of up to 7 m: one at the western shore, stretched north from Tasaral Island to Korzhyntubek Cape. The second hollow stretches to the south from Bertys Bay. The Bertys Bay is the deepest place (14 m) in the Western Balkash. In East Balkash, the depression up to 16 m deep is located to the northeast of Korzhyn Island. The same hollow in the depth but more extensive in area is located west of the Shaukar Peninsula. The deepest depression (up to 27 m) is located in the center of the eastern Burlitobe area.

The western and northern shores of the lake are elevated up to 20–30 m above the water's edge and are composed of solid Paleozoic rocks such as porphyry, tuff, granite, shale, limestone. Individual hills reach a height of 40–60 m and even 115 m. There are narrow coastal shoals, covered with fragments of rocks, gravel or pebbles between the rocks of the coastal hills and the lake. There are coastal shafts on some sites.

There are several small plains folded by lake or lacustrine river sand deposits on the northern shore, in the lower reaches of the Tokrau and Ayaguz rivers and in the area of the Karakamys tract. The southeastern shore of the lake from Burlitobe Bay to Karashagan is elevated and composed by bedrocks. The southern coast from the Karashagan bay to the modern delta of the Ile river is low and sandy. It rises only 1–2 m above the water, and there it is covered with low (5–10 m) sandy hills and barkhans. As a result of the low position of the lake bank, it is often covered with reeds, water level fluctuations, and sand mobility, and the coastline is constantly changing its outlines.

Balkash Lake has many large bays, capes, peninsulas. The largest of the bays are Alakol, Karakamys, Kashkantengiz, Sary Shagan, Bertys, Balyktykol, and Karashagan. Korzhyntubek, Karaagash, Bertys, Baigabyl, Shaukar, Kentubek, and Saryisek are the largest peninsulas. The Saryisek Peninsula, which is 21 km into the lake, is divided into two parts: the East and the West Balkash that are connected by the Uzunaral Strait.

There is an archipelago of islands under the common name Ucharal in the southern part of the lake and opposite the mouth of the Ile river. It includes three large islands such as Basaral, Ortaaral, and Ayakaral and many small ones. There is a small shrub bogalish on the north coast; there are the gingil, jantak, teresken, reed thickets, and chin on the southern coast of the lake. Minor thickets of jiggies and tala are found only in the lower reaches of the Ile and Lepsi rivers. Turanga occurs only occasionally on some parts of the northern coast. Large areas in old Bakanas region are occupied by saxaul forests.

7.3.1 Hydrography and Water Regime of the Balkash Lake

The rivers of the Balkash lake basin can be divided into two groups: southern and northern tributaries. In the north, the river network is rare and varies from 0.01 to 0.5 km/km^2; in the south, it ranges from 0.6 to 3 km/km^2. The following rivers flow into the lake from the south: The Ile River is giving 78.2% of the surface inflow, Karatal is 15.1%, Lepsi is 5.4%, and Aksu is 0.43% (Philonets 1981). They play the main role in the water balance of the lake. The Ayaguz, Bakanas, Tokrau, and Mointy rivers are the northern tributaries. These rivers do not give a superficial inflow to the lake.

The average depth of the lake is 6 m. The average depth of West Balkash is 4.8 m. East one is 8.8 m. It contains about 106 km^3 of water (Philonets 1981). The height of the waves in the eastern part reaches 3–3.5 m, and in the west it does not exceed 2.5 m. The currents in the lake are of two types. This is a constant runoff current in the western part of the lake that arises at the mouth of the Ile river under the influence of the river waters, and temporary up-and-over, overtaking and compensatory currents that arise in different parts of the reservoir under the influence of winds (Philonets 1981).

Transparency of water in the lake ranges from 0.2 to 0.4 m in the mouth of the Ile river and up to 10–12 m in the eastern part. The least transparency is in summer and the largest one in winter. The color of water gradually changes, its shades from a dull yellowish in the southwest to emerald green with a bluish tinge in the east. The average water temperature in the western part of the lake is higher than in the eastern part that is 9.9 and 8.5 °C, respectively; however, it is vice versa in winter. The maximum water temperature is observed in July. It is 23.8 °C in the western part of the lake and in the eastern part is 20.1 °C. Thermal stratification is very weak (Philonets 1981).

The freeze-off on the whole Balkash water area occurs at different times, but mainly in the second half of November in the Western Balkash and in December in Eastern Balkash. This is explained by the fact that the average salinity of water in the lake increases from 0.7 to 0.9% in the region of the mouth of the Ile river and up to 4.1–5.5% in the eastern extremity of the lake. Accordingly, the freezing temperature of water along the lake varies from −0.04 to −0.30 °C. At the end of December, the ice thickness in the coastal regions is 35–40 cm and in some years 50–65 cm. Ice buildup ceases in late February to early March. The average thickness of ice is 50–85 cm. The oscillation varies from 30 to 105 cm and sometimes 150 cm. The lake is opened in the first half of April in the southwestern, then in the western, and in the eastern parts of the lake. The average period of clearing the lake from ice is the beginning of May. The period with ice phenomena is 145–160 days (Philonets 1981).

Fluctuations in the water level are intra-annual and perennial. It depends directly on the intra-annual distribution of outflowing river flows and evaporation from the lake surface. Perennial fluctuation is due to the variability of air temperature, the amount of precipitation and other components. The annual amplitude of the level fluctuations is 17–67 cm (Philonets 1981).

The fluctuations in the water level of the lake are cyclical (Fig. 7.8). A high water level was observed in Balkash in 1908 and a low one in 1946. During this period, half a century of average annual water river flow with a total duration of about 37 years passed. Since 1947, the water level in the lake has started to rise—the next century cycle, which is expected to end in 2016–2020. Natural condition of

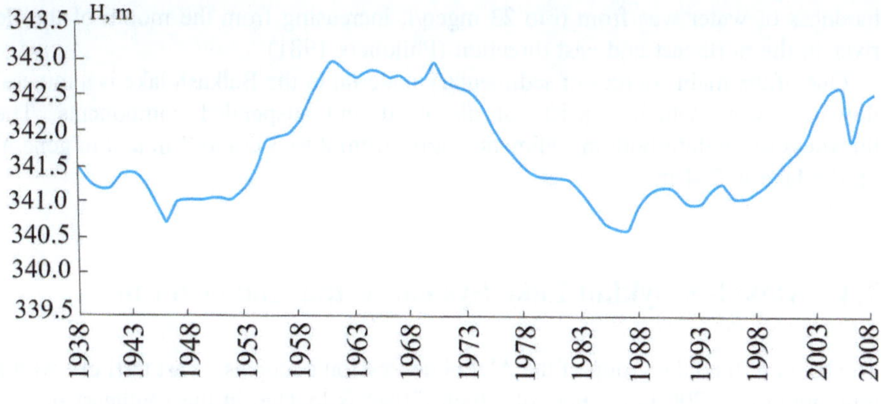

Fig. 7.8 Main tributaries of the Alakol lake system

the hydrological regime on the Balkash lake was disturbed due to the construction of the Kapshagay HPP on the Ile river. The multi-year amplitude of the water level fluctuation is more than 3 m.

Hydrochemistry of the Balkash lake is a great heterogeneity of mineralization and chemical composition of water along the water area (Fig. 7.9). The mineralization of water rises from 0.5 g/kg at the mouth of the Ile river up to 4.4 g/kg to the eastern extremity of the lake. The average mineralization for the whole lake was 2.94 g/kg (Philonets 1981).

The oxygen content in the surface layer of the lake is distributed within 90–100% in the water area. Its vertical stratification is insignificant. The pH of the water varied from 7.6 to 9 and increases from the Ile river delta to the eastern extremity of the lake. Permanganate oxidation (in alkaline medium) varied from 2 to 6 mg O_2/l, increasing northeastward from the southwest coast, and the total

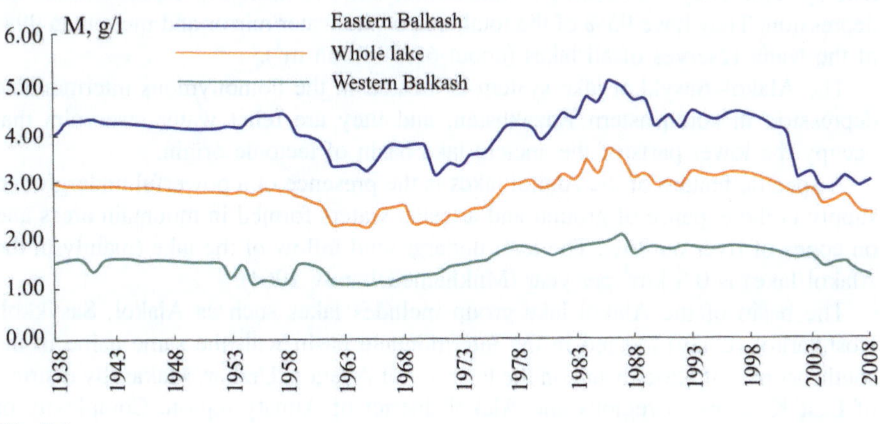

Fig. 7.9 .

hardness of water was from 6 to 23 mgeq/l, increasing from the mouth of the Ile river in the northeast and east direction (Philonets 1981).

One of the main sources of sedimentary material in the Balkash lake is a mineral flow of rivers, which consists of dissolved and suspended components. The thickness of modern bottom sediments varies from 2 to 3 cm to 7 m and in general on the lake is 2–4 m.

7.4 Alakol–Sasykkol Lake System in the Alakol Basin

The basin of the lake group of the Alakol depression occupies a vast territory with a total area of 68,700 km^2, most of which (70%) is located in the southeast part of Kazakhstan and the other one on the adjacent part of China. By orographic condition, the basin can be divided into mountainous that is 48,600 km^2 (44% of the area) and the flat part is 56% (Philonets 1981).

The Alakol depression is an intermontane tectonic deflection filled with a thickness of alluvial and mountain deposits. Its flat part represents the bottom of the existing in the Upper Quaternary, possibly even later Balkash–Alakol water reservoir, and then the Alakol water reservoir. The surface of this bottom was later largely recycled as a result of the activity of rivers and wind. It is in the central part complicated by the latest tectonic uplifts along the fault line of the block structures. These processes of relief formation determined the current location and structure of the lacustrine basins of the Alakol lakes, which are the relics of the former vast water reservoirs. At the same time, modern lake basins are the product of the development and "vital activity" of the very modern lakes, their tributaries, and coastal vegetation.

There are 529 lakes on the territory of the Alakol depression, 513 of them with a mirror area of less than 1 km^2. The Alakol lake system includes four large lakes such as Alakol, Koshkarkol, Sasykkol, and Zhalanashkol, which form a peculiar lake system. They are located in the central and the most depressed zone of the depression. They have 95% of the total area of the water mirror and more than 99% of the water reserves of all lakes (about 61.6 billion m^3).

The Alakol–Sasykkol lake system is located in the homonymous intermontane depression in southeastern Kazakhstan, and they are relict water reservoirs that occupy the lower parts of the ancient lake basin of tectonic origin.

A specific feature of the Alakol lakes is the presence of a powerful underground supply at the expense of ground and artesian waters formed in mountain areas and on cones of river outflow. The total underground inflow of the lake (mainly in the Alakol lake) is 0.8 km^3 per year (Mukhamedzhanov 1964).

The basin of the Alakol lake group includes lakes such as Alakol, Sasykkol, Koshkarkol (Uialy) located in the intermontane basin with the same name in the southeast part of Kazakhstan on the territory of Ayagoz, Urzhar, Makanshy districts of East Kazakhstan regions and Alakol district of Almaty region. Complexity of geology and orography of the considered region and a large amplitude of elevation

(300–4107 m) cause a variety of natural and climatic conditions. The Alakol depression in the natural and climatic relation refers to the semiarid zone of the foothill dry steppe plains and deserts of Kazakhstan.

The deepest part of the basin is occupied by the main lake system that is the deepwater drainless Alakol lake, which concentrates 95% of the total water mass of four large lakes (Fig. 7.1). Higher in the longitudinal axis of the basin, there are sewage (periodically flowing) shallow lakes: to the north—Koshkarkol and Sasykkol, to the south—Zhalanashkol, dropping excess water in the Alakol lake.

Alakol lake is the largest lake in the Alakol lake system. It is located in a semi-desert zone and in the east of the Balkash-Alakol depression, at an altitude of 343 m. It is stretched from the west to the east, with the area (with islands) is 2696 km^2, the length is 104 km, the maximum width is 52 km, the depth is up to 54 m, the water volume is 58.56 km^3. Annual fluctuations of the level are 1.2 m (Fig. 7.10). The lake water has a sodium chloride composition. It is the second largest among the water reservoirs in Kazakhstan and the only deep lake among the drainless lakes in Kazakhstan. The shoreline of the Alakol lake is distinguished by its great ruggedness. It forms numerous peninsulas, capes, spits, bays, and coves. There are several islands on the lake with a total area of 80 km^2. The shores of the lake are high and steep, except the western, eastern, and northeastern ones, where lowland and shallow by reed banks predominate. The gravel and pebble beaches stretch along the precipitous northwestern, southwestern, and southeastern shores. The width and height of the beaches depend on the phase of long-term fluctuations in the water level. The maximum depth of the lake is 54 m, and the average depth is 22 m (Table 7.17). The water and salt balances of the lake is shown in Fig. 7.11.

There are a number of small drying rivers on the southeastern and southern coast of the Alakol lake, as well as the rivers of the spring type formed by groundwater outlets on the line of conjugation of the foothill and lacustrine plains. In addition, numerous springs are observed directly along the coastal cliff on the southwest coast of the lake.

Sasykkol lake occupies the most northerly position among the Alakol lake groups. It is the second largest lake after Alakol lake and is characterized by a

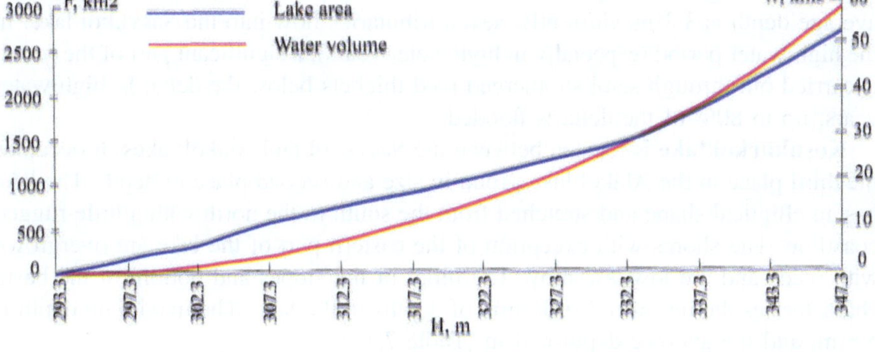

Fig. 7.10 .

Table 7.17 Main characteristics of lakes in the Alakol lake system

Characteristics	Lake's name			
	Alakol	Sasykkol	Koshkarkol	Zhalanashkol
Lake length (km)	104	49.6	18.3	8.8
Max width (km)	52	19.8	9.6	6.3
Average width (km)	25.5	14.8	6.5	4.7
Length of shoreline (km)	384	182	57.3	26.2
Area of water mirror (km^2)	2650	736	120	40.6
Max depth (m)	54	4.7	5.8	3.4
Average depth (m)	22.1	3.3	4.1	2.4
Water volume (million m^3)	58,560	2434	489	99.5

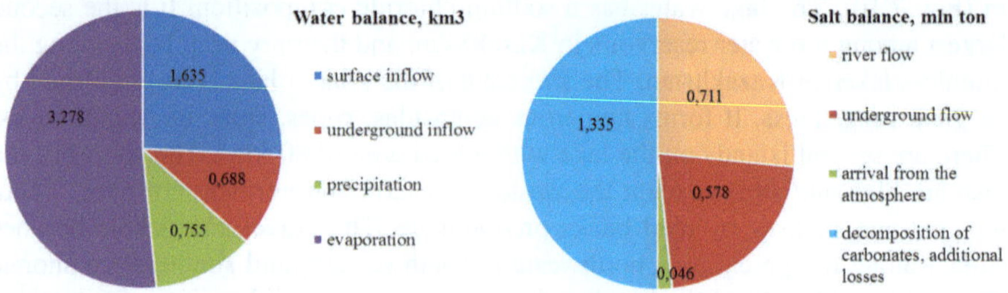

Fig. 7.11 .

winding coastline. In the southeastern part, the Aral-Tobe Peninsula cuts into the lake, forming two gulfs—Bargen and Zhartas. The Aral-Tobe Island is located in the western part of the lake. The shores of the lake are almost all along low, swampy, and overgrown with reeds. The shores are composed by loams and clays with the exception of outcrops of bedrock to the Aral-Tobe Peninsula. Sasykkol lake is a shallow-water reservoir with a monotonously regular basin stretched from east to west. The gentle slopes of the basin almost along the entire coastline caused the presence of large areas with depths of 0–1.5 m and a smooth transition to the zone with the greatest depths. The maximum depth of the lake is 4.7 m, and the average depth is 3.3 m. Currently, seven tributaries flow into the Sasykkol lake. In the high water period (especially in high water years), a significant part of the runoff is carried out through semi-submerged reed thickets below the delta. In high water years, up to 80% of the delta is flooded.

Koshkarkol lake is located between the Sasykkol and Alakol lakes. It occupies the third place in the Alakol lake group in size and second place in depth. The lake has an elliptical shape and stretched from the south to the north with a little-rugged coastline. The shores with exception of the eastern part of the lake are overgrown with reeds and are low, swampy. The relief of the slopes and bottom of the basin characterizes the uniform distribution of depths in the lake. The maximum depth is 5.8 m, and the average depth is 4 m (Table 7.18).

Table 7.18 Long-term average amount of evaporation from water surface of the Alakol–Sasykkol lake system (mm)

Lake	Months											
	January	February	March	April	May	June	July	August	September	October	November	December
Sasykkol	Freezing	Freezing	4.3	97.2	130	158	190	197	132	86	36.5	Freezing
Koshkarkol	Freezing	Freezing	13.5	69.3	121	161	197	177	133	82.7	39.2	Freezing
Alakol	114	Freezing	38.5	39	71	115	154	212	202	159	108	158
Zhalanashkol	Freezing	Freezing	14.8	107	164	184	239	243	181	101	78.9	Freezing

Zhalanashkol lake ("Naked Lake") is located in the southeastern part of Kazakhstan near the Dzungar Gate and to the south of the Alakol lake, at an altitude of 372.5 m above sea level. The area of the lake's water surface was 37.5 km^2, the length of 9 km, the width of 5.8 km, the length of the coastline 23.8 km, and the average depth 2.6 m (the largest is 3.25 m).

Alakol lake groups do not have clearly defined watersheds and catchments, respectively. "Watershed" between the Sasykkol and Koshkarkol lakes passes along the shore accumulative cane shaft of the eastern and southeastern coast of the Sasykkol lake. Flow from the lake is located toward the Koshkarkol and Alakol lakes along permanent and temporary channels on the indicated coast, as well as by filtration and overflow through the coastal shaft. "Watershed" between the Koshkarkol and Alakol lakes passes along the inter-lakes strait, as well as by swamped depression located to the south and north of the Koshkarkol lake, which has lower grades than the strait. Sasykkol and Koshkarkol are surrounded by continuous ring of swampy and semi-submerged reed thickets with a total area of about 770 km^2, and they are located within the horizon with a mark of about 352 m.

The drainage basin of the Alakol lake includes basins of the whole lake groups. Almost 2/3 of the area of this basin falls on the southern exposition of the Western Tarbagatai Ridge, where many rivers flow, but only three of them such as Urzhar, Katynsu, and Emel are spreading their waters to the Alakol lake. The remaining 1/3 of the area is occupied by the northeastern slopes of Zhetysu Alatau, where Tentek, Zhamanty, and Yrgaity rivers originate from and flow into the Alakol lake. These rivers account for the bulk of the water balance of the lake.

The feeding of the lake basin is mainly due to the surface runoff of the rivers Tentek, Zhamanty, Yrgaity (the rivers of the Zhetysu Alatau), and Emel, Katynsu (the rivers of the Tarbagatay and Barlyk ranges) and several others. According to the classification of Schults (1965), the rivers of the Zhetysu Alatau belong to the rivers of snow and glacier feeding; the rivers of the Tarbagatay and Barlyk ranges belong to the rivers of snow feeding. The share of glacial runoff for individual rivers when leaving the mountains is 5–10% of the annual runoff and for the total average runoff in the basin is 1.2% (Korovin 1965). The Yrgaity and Tentek rivers have a fast current on all extent of their riverbeds. The riverbeds of these rivers are composed by self-peeling from gravel deposits in average sizes from 5 to 10 cm. The gravel deposits alternate with deposits of coarse sand in the lower part of the rivers. A gradual migration of sediments is observed in the delta sections of these rivers.

Atmospheric precipitation varies from 250 mm in the central part of the Alakol–Sasykkol system to 800–1000 mm in the watershed zone of the Tarbagatai and Zhetysu Alatau ridges. The average precipitation is 442 mm; it is 595 mm in the mountain (the drain-forming part). The main part of the precipitation falling in the basin is in the mid-mountain zone (1000–2000 m). The entire river runoff of the basin is formed in the mountain and foothill zones. The flat part of the basin occupied 22.6% of the total area and belongs to the drainless region. The average runoff in the basin is 120 mm, in the southern mountainous part (Zhetysu Alatau) is 349 mm, and in the northern mountainous region (Tarbagatay, Barlyk, Maily) is 215 mm.

Seasonal snow plays the main role in feeding the rivers of the northern region, and long-term snow reserves in the mountains and glaciers play a significant role in feeding the rivers of the higher-mountainous and humidified Zhetysu Alatau region. In-season distribution of runoff for the Tentek, Zhamanty, and Yrgaity rivers is characterized by a long spring–summer high water period: from early April to the end of July. Peak of high water period occurs in May. The Katynsu, Karakol, Urzhar, and Emel rivers are characterized by a short springwater period: from March to May, with a maximum in April.

Steady snow cover is observed from 2–3 decades of November to 2–3 decades of March. The greatest height of the snow cover is observed in mid-February and in the south somewhat earlier. The average height of the snow cover is 27 cm. In different years, it varies from 13 to 45 cm. Snow lies on the Alakol lake unevenly, and the height of the snow cover does not exceed 5–10 cm. In the southeastern part of the lake, the snow is almost completely blown away by hurricane southeasterly winds and is retained only on hummocky areas and coastal thickets.

The speed of the wind over the large water reservoirs of the region is much higher than over the land surface. This is due to the lower roughness of the water surface compared to the land surface. As a result, the vertical wind profile in the surface layer undergoes a rearrangement during the flow from land to water and from water to land. Wind speed increases most strongly in the coastal strip. The increase in wind speed slows down as the air flow continues to move depending on the security and roughness, and stops at a distance of 1–2 km from the shore. Based on the data on coastal weather stations, the following values of the maximum wind speed above the water of the Alakol lakes can be taken: Sasykkol—40 to 45 m/s; Koshkarkol—40 to 45 m/s; northwestern and northern areas of the Alakol lake—45 to 50 m/s; southeastern and central region of the Alakol lake—50 to 60 m/s; Zhalanashkol—60–70 m/s.

Temperature and air humidity. The air temperature in the surface layer of the atmosphere changes when the air flow goes from land to the water reservoir. It is due to the difference between the conditions of heating and cooling and respectively, the temperatures of the land surface and water changes. The sign and magnitude of changes in air temperature over the water body change during the day and year depending on weather conditions (wind, cloudiness), seasonal and diurnal variations in air and water temperature.

In summer, when the weather is mild, the temperature of the air over the Alakol lake is close to the water surface temperature and deviates from it by 1–2 °C. At the same time, the temperature difference over the surface of water and land can reach 7–8 °C. During the period of autumn cooling and an increase in the daily amplitude of the air temperature over the land, the difference between water–air temperatures reached 3-4 °C with an increase in the daily amplitude of the air above water to 5–8 °C. During this period, the difference in air temperatures over the water reservoir and land reached 8–10 °C.

Air humidity above the water reservoir is usually greater than over the land, but sometimes an inverse relationship is observed in the spring and autumn period. It is

Table 7.19 Hydrological characteristics of the main rivers in the Alakol lake basin

River's name	Lakes (into falls)	Basin area (km^2)	Glaciation area (km^2)	River length (km)	Mean annual water discharge (m^3/s)	Volume of annual runoff (million m^3)
Tentek	Sasykkol and Uialy	5380	94/20 km^3	183	47.5	1395
Urzhar	Alakol	4550	–	186	1.56	49
Katynsu	Alakol	387	–	71	4.5	213
Emel	Alakol	21,800	–	102	4	142
Yrgaity	Alakol	1660	15.7/ 505 km^3	28	5.7	179.7
Zhamanty	Alakol	667		51	2.5	78.8

associated with the condensation of moisture above the water surface, which to some extent withers the air.

The reverse effect is most pronounced in large deepwater reservoirs that slowly warm up in the spring and cool for a long time in the autumn. Air humidification above the water reservoir is most pronounced in the coastal zone, but it continues during the entire air mass above the water reservoir. The increase in humidity over the water reservoir continues until there is equilibrium between the arrival of steam from the evaporating water surface and its removal to the upper atmosphere.

Evaporation from the surface of the Alakol lake was first studied by TM. Trifonova. According to her calculation, the periods of incomplete freeze-up are December, January, February, and March. Although the area of the water mirror is free of ice, it is relatively small (10–30%); but due to intensive wind activity in winter and a large temperature difference at the water–air boundary, the average evaporation in winter is 210 mm. The amount of evaporation during the transition periods (December–February, March, and April) is 92–150 mm (Table 7.19).

7.4.1 Formation of the Alakol Lake System and Hydrographic River Network of the Basin

The Balkash–Alakol system was formed after the Likhvinsky glaciation, about 1 million years ago. During the glaciation of the Tian Shan and Zhetysu Alatau ridges, the Alakol lake group and Balkash Lake constituted a single water body (Davydova et al. 1966), whose area reached 6250 km^2 in the postglacial period (Muravlev et al. 1969). During this period, the formation of an aboriginal hydrofauna related to the Nagorno-Asian zoogeographic subregion occurred. The Balkash–Alakol sea basin first disintegrated into the Balkash and Alakol–Sasykkol reservoirs, and then it was divided into the Sasykkol, Koshkarkol, and Alakol lakes, while in the first half of the nineteenth century there was a decrease in their levels and sizes (Kurdyukov 1951; Korovin 1965). At the end of the Paleozoic, there was a mountainous country in this territory, which was leveled as a result of intensive erosion–denudation processes and turned into a peneplain in the Mesozoic. The remaining equalization surfaces in Zhetysu Alatau, Barlyk, and Tarbagatai can be evidence. In the Tertiary, this shallow hilly plain with the remains composed of Paleozoic rocks experienced insignificant uplifts that led to the erosion of the accumulated weathering crust and the deposition of red-colored sandy clays in hot and arid conditions. At the end of the Tertiary and Early Quaternary periods, the bottom of the Alakol depression fell down along with the powerful uplift of Zhetysu Alatau and Tarbagatai, and erosion processes led to the formation of alluvial–proluvial trails at the foot of Zhetysu Alatau and Barlyk (Popov 1965; Geology of the USSR 1971).

The hydrographic river network of the basin is represented by rivers flowing from the southern slopes of the Tarbagatai Range, the western slopes of the Barlyk Ridge and the northern slopes of the Zhetysu Alatau. The Tentek, Urzhar, Katynsu, and Emel rivers are the largest rivers of the basin. The Emel, Katynsu, and Urzhar rivers, in contrast to the rivers that flow into the southern part of the lake system, have rather extended sections of riverbeds passing along the flat part. The water flow of these rivers is characterized by lower velocities due to more gentle slopes. The fine sediment fractions move in the suspended state, while the turbidity of water associated with the snow melting processes is high only during the spring flood period. There is a ridge migration of sediments in the form of misfit and collateral in the riverbeds of these rivers. The ridges in the middle part of rivers are composed by fine and medium sand. Fine sand and silt fractions are predominated in the delta areas.

The surface inflow of rivers flowing into the Alakol lake system is the determining factor in the water balance of the lake system (Table 7.19).

Tentek, Urzhar, Katynsu, Emel, and Zhamanty rivers are the main tributaries of the Alakol lake system (Fig. 7.8). Tentek carries about 40% of the total surface inflow into the lakes. In the central part of the Alakol depression, it forms a vast delta common to the three lakes—Sasykkol, Koshkarkol, and Alakol, which also indicates the existence of a single Sasykkol–Alakol water reservoir in the past. The total area of the delta is about 1000 km^2, including the ancient delta, the Shubartubek tract, and the "irrigation fan" of the Presasykkol main irrigation canal, which is currently not operational. The area of the Presasykkol part of the delta is 295 km^2. The top of the modern delta is 4.5 km below the Aktogay railway bridge, where the Tentek river is divided into three branches Tentek (left), Karatentek (middle), and Borgan (right) and forms the three main branches of the estuary system of the delta.

The *Tentek River* is the largest river of the basin, and it originates from the glaciers of Zhetysu Alatau. The catchment area is 5390 km^2, and the length is 200 km. In the area of Usharal City, from the left bank, the river takes in the tributary of the Shynzhyly river. The Tentek river bears about 40% of the surface influx of lakes and forms a delta area of 295 km^2 at its confluence into the Sasykkol lake.

The *Urzhar river* is formed from the confluence of two mountain rivers in the central and the most elevated parts of the Tarbagatai Range. The catchment area of the river is 5280 km^2, and the length is 206 km. The largest tributaries of the river are: Kusak on the left side and Eginsu river on the right side. The Urzhar river forms a swamped delta in the lower course at the confluence of the Alakol lake.

The *Katynsu river* coast began on the southern slopes of the Tarbagatai ridge. The catchment area is 2650 km^2. The river has one tributary on the left that is Kokterek river. The flow of the Katynsu river falls into the Alakol lake.

The *Emel river* originates on the slopes of the Tarbagatai Range outside of our country. The catchment area is 21,800 km^2. For 102 km within Kazakhstan, the river crosses a sandy valley and flows into the Alakol lake. The Emel river on the right side takes the tributary of the Karabuta river.

The rivers such as Karakol and Ai flowing from the northwestern spurs of the Tarbagatai Range do not transmit their waters to the Sasykkol lake in the mid- and low water years. They are lost in swamped spills near the coast of the lake. The Karakol river forms a delta, when approaching the lake. The delta appeared, likely in the past and more humid period. In the modern period, this delta operates only in the years of maximums of the perennial and intra-century cyclicity of river flow. A significant part of the delta of the Karacol river as well as the delta of the Tentek river currently flooded Sasykkol lake.

Currently, the Alakol lake is experiencing a phase of transgression. Fluctuations in the lake level, which reach 5–6 m in the long-term section, are accompanied by significant changes in the position of its shoreline. Koktuma village which is located on the western shore of the lake was blurred, and entire street within the range of 350–400 m was destroyed. The Rybachye village was postponed for 20 years twice, and delta of the Urzhar river was flooded.

According to the long-term data of the hydrometeorological service in Kazakhstan, an underground inflow into the Alakol lake is 323 mm or 770 million m^3 per year (Table 7.20). The arrival of water due to atmospheric precipitation is 274 mm or 653 million m^3 per year, and evaporation is 1194 mm.

Table 7.20 Water balance of the Alakol lake (Burlibayev 2007)

Income	mm	million m^3	%	Outcome	mm	million m^3	%
Surface inflow	664	1579	53	Evaporation	1194	2844	95
Underground inflow	323	770	26	Surface outflow	10	30	1
Precipitation	274	653	21	Changes of water reserve	54	128	4
Total	1261	3002	100	Total	1261	3002	100

7.4.2 Modern State of Water Resources in the Alakol Lake System

The current state of water bodies shows that large water period has come in the region. Because the rivers flowing into the Alakol lake system are characterized by increased water content, the water level in the lakes is increasing (Burlibayev 2007). Raising the water level in lakes is an ambiguous factor in the sustainable development of aquatic ecosystems. In addition, the area of river deltas increases, and new shallow-water lakes are formed in the main wetland areas. All this contributes to the successful development of river biota, and consequently feeding base for birds and animals increases. The rise in water level in delta areas leads to the flooding of traditional nesting sites for birds. In addition, its mineralization decreases due to the increase in the amount of water. Consequently, this affects the changes in the quantitative and qualitative composition of many microorganisms.

The Alakol system is the largest reserve of nesting waterbirds in Kazakhstan, the site of their mass molting and migratory stops, through which hundreds of thousands of waterfowl and waterbirds migrate annually. To preserve them, it is necessary to create specially protected areas, taking into account the historically developed landscape complex and existing socioeconomic conditions, which should be the standards of the protected environment and have an extensive protected area.

The rise in the water level of the Alakol lake also led to negative consequences for the population of coastal villages, where as a result of flooding a part of the population was forced to move or abandon their homes (Burlibayev 2007). Increasing the water level of the Zhalanashkol lake leads to a repeated erosion of the road embankment. Strengthening the banks of the road embankment and the arrangement of additional culverts will require considerable investment.

The structure of the Yrgaity and Zhamanty river valleys, as well as the granulometric composition of their bottom sediments, makes it possible to conclude that considerable mudflows have passed through these rivers in the past. The danger on mudflow of these rivers in the present period becomes especially important because of the crossing of their riverbeds by the oil pipeline. A possible accident on the pipeline can cause irreparable harm to the entire ecology of the region. Therefore, the investigation of these rivers for mudflow danger is a very urgent task.

A particular danger to the ecology of the Alakol lake recently represents a transboundary Emel river, where industrial enterprises are located in the upper reaches of the territory of China.

Large and unjustified losses of water for irrigation are now smoothed by the increased water content of the rivers. With a subsequent decrease in the water content of rivers, the problem of the irrevocable volume of water for irrigation will begin to lead to the inhibition of their delta areas. As a result, the area of wetlands will be reduced. Some of the shallow-water lakes will dry up. Therefore, it is necessary to organize comprehensive observations on elements (parameters) of the water balance of the lake.

Besides above-mentioned factors, the climate of the region is affected by the global warming. The increase in the water content of rivers and as a consequence the water levels in the lakes of the Alakol lake system are primarily related to the intensive melting of glaciers. The subsequent decrease of glaciation in the area will affect the climatic characteristics of the region. It will lead to a decrease in the water flow to rivers and as a result to a water shortage. To develop preventive measures and measures for lack of water, it is necessary to organize monitoring of the main climatic characteristics, including actinometric observations in the region.

7.5 Tengiz–Korgalzhyn Lake System

The Tengiz–Korgalzhyn lake system is located in the southeast of the northwestern part of the Central Kazakhstan hilly region (melkosopochnik) in the Akmola region. It is the remnant of an ancient mountainous country that lined with processes of denudation and weathering and largely buried under loose products of destruction of bedrock. The Tengiz–Korgalzhyn lake system is one of the most important wetlands for transitional birds in Kazakhstan and the whole Central Asia. It is located at the intersection of the two main flyways of birds' migratory: the Central Asian-Indian and Siberian-Black Sea-East African ways and it is a place of concentration of a huge number of waterfowl and waterbirds during seasonal migration. The lake system occupies an area of 2600 km^2, and it feeds from the Nura river (Burlibayev 2007). The Nura and Kulanotpes Rivers are the main source of water for the lake system. The greatest width of the Nura river up to 80 m and the depth of the river up to 4.5 m prevail on the plain, where the speed of the current is often barely caught by floats. The average depth of the river is 0.4–0.5 m. The banks of the river are high and steep, and in some areas they are washed and merged with the ledge of the terrace, rising to a height of 10 m at steepness of 80–90°. The river banks are usually composed by dense loam, and sand and sandy-pebble soils are found in the upper part of the riverbed.

The Korgalzhyn group of lakes: Birtaban, Sholakshalkar, Shalkar, Zhanybekshalkar, Uialyshalkar, Zhandyshalkar, and Korgalzhyn are located in the lower part of the valley of the Nura river (Posokhov 1948, 1949, 1955). The fresh lakes of the Tengiz–Korgalzhyn depression based on morphological, geological, and hydrobiological conditions can be divided into two main isolated groups. The first group is the Besshkalar group of lakes—in the lower reaches of Nura river. It starts from Lake Birtaban and ends in the northeast of Lake Shiyshalkar.

The second group is the southeast or Uzynkol–Zharlykol lakes which are located southward, and begins from the Touzas lake and ends with the Uzynsor lake.

The lakes are separated by narrow stripes of land. Between the Besshalkar and Uzynkol–Zharlykol groups are Alkarsor lakes, genetically related to the first group.

All lakes are large, medium, and small with number about 50. Lakes Uzynkol-Zharlykol lake group and Alkarsor are mainly solonetsous. The area of lakes is from 9.4 km^2 (Uzynkol) to 37 km^2 (Sholak-Shalkar), and the depth rarely exceeds 1.5–2.5 m (Posokhov 1948, 1955). Most lakes have an oval–oblong shape with different orientations to their main axis. The predominant form of the basins of modern lakes is saucer-like. On the coasts of large lakes, there are small lakes, formed by separating the bays with sand spits at the confluence of streams into the lakes.

The third part of the territory under consideration belongs to the Delta River Nura. It is a straggly wavy plain with many lakes of various sizes and different salinities. The absolute heights of this part of the depression are 300–350 m.

The positive forms of the relief of the lake system are so weak that even the watershed of the Nura and Esil rivers near Astana rises only 3–4 m above the low water level of these rivers. The lowered relief forms are expressed by shallow depressions occupied by lakes, most of which are located along the left bank of the lake system. According to the degree of mineralization, water in the lakes is found from highly mineralized to freshwater.

The surface of the plains is mostly flat and almost uncrossed; however, in some places it is disturbed by undulating ridges that rise to a height of up to 10 m. The hilly region is composed mainly by loamy soils covered with superficial carbonate chestnut soils. In the plains, dense loams and tertiary clays often contain sandy–loamy dark chestnut solonetsous complex soils. The riverbed part of the Nura river is mainly occupied by sandy–pebbly sediments of the river covered with a thin layer of humus. Flooding occurs annually on average at the end of the second decade of April. The depth of flooding in low places up to 2–3 m, at elevated parts, does not reach 1 m. The duration of flooding is 15–20 days (Burlibayev 2007).

Due to significant fluctuations in the water level in the river, both over the natural climatic and anthropogenic (regulation of the riverbed and the retention of water in the reservoirs), the phenomena of overflow and drying of the lakes of the system periodically recur, which negatively affects the status of the most important habitats of waterfowl and waterbirds. To maintain the optimal water level of the Korgalzhyn Lake created artificial dams, which due to imperfections of their structures often break through. This makes an ecological catastrophe for the Korgalzhyn lakes (sharp shallowing of reservoirs and fish freezes), and the islands on the Teniz lake which are nesting places of colonial species (in particular flamingos) are flooded.

Great damage to the fauna of the reserve is also caused by the person as a result of the descent of water in late May–early June (just in the period of intensive breeding of birds) from the reservoirs. Water floods the floodplain of Nura, fills the lake behind the lake, floods the coastal areas and low islands, and forms huge spills in places of relief lowering. For most waterbirds, this is a real disaster: Thousands of nests perish.

References

Abuduwaili J, Liu Dongwei, Guangyang Wu (2010) Saline dust storms and their ecological impacts in arid regions. J Arid Land 2(2):144–150

Bobchenko VI (1991) Water loses for cotton and salt washing. J Amelioration Water Manag 3:22–27 (in Russian)

Borotnik VN, Kuksa VN, Tsytsapne AG (1991) Modern condition and Aral Sea in the future. Izvestia Acad Sci USSSR, Geogr Ser 4:62–68 (in Russian)

Bortnik VN, Chistyaevaya SP (1990) Hydrometeorology and hydrochemistry of the seas of the USSR. Volume VIII: Aral Sea. Gidrometeoizdat, Leningrad (in Russian)

Burlibayev MZh (eds) (2007) Globally significant wetlands of Kazakhstan (Alakol-Sasykkol lake system). Astana, p 271

Burlibayev MK, Dostai ZhD, Tursunov AA (2001) Aral-Syrdarya basin: hydroecological problems, issues of water allocation. Almaty, Dauir, p 180 (in Russian)

Chub VY (1998) Regional problems of estimating the influence of aerosols (of anthropogenic and natural occurrence) on the climatic characteristics of the Central Asiatic region. In: Materials for the regional scientific seminar aeolian transport of salt in the Aral Sea basin, pp 3–4 (in Russian)

Chub VY (2002) Climate change and its influence on the natural resource potential of the Republic of Uzbekistan. Glavgidromet, Tashkent (in Russian)

Debol'skii VK, Osmonbetova DK (1999) Concept on use of water resources in the neighboring countries of Central Asia. Izvestia Acad Sci Geograph Ser 6:63–67 (in Russian)

Dukhovnyi VA, Razakov PM (1988) Aral Sea. J Amelioration Water Manag 9:27–32 (in Russian)

Environmental Programm of United Nations World Bank (1998) Aral Sea basin program (Kazakhstan, Kyrgyz Republic, Tajikistan, Turkmenistan, and Uzbekistan). Water Environment Management Project, Washington, DC

Faizov KSh, Tapalova AS (2003) Ecology of crisis areas in the Aral Sea region. Problems and solutions. Almaty, p 110 (in Russian)

Fayanas E (2013) El mar Aral: Historia de la desaparicion de un mar

Gerasimov IP, Kuznetsov NT, Kes' AS, Gorodetskaya ME (1983) Problems on the Aral Sea and anthropogenic desertification in the Aral Sea Region. J Prob Desert Develop 6:22–23 (in Russian)

Gills TE (1996) Aeolian sediments generated by anthropogenic disturbance of playas: human impacts on the geomorphic system and geomorphic impacts on the human system. Geomorphology 17:207–228

Glazovskiy NF (1990) The Aral crisis: causative factors and means of solution. Nauka, Moscow (in Russian)

Groll M, Opp C, Aslanov I (2012) Spatial and temporal distribution of the dust deposition in Central Asia—results from a long term monitoring data. J Aeolian Res. doi:http://dx.doi.org/10.1016/j.aeolia.2012.08.002

Ibatullin SR, Mustafayev ZhS, Koibagarova KB (2005) The balanced use of water resources of the transboundary rivers. Taraz, p 111 (in Russian)

Issanova G, Abuduwaili J (2017) Aeolian processes as dust storms in the deserts of central Asia and Kazakhstan. Springer, Singapore, p 133

Issanova G, Abuduwaili J, Semenov O (2014) Deflation processes and their role in desertification of the southern Pre-Balkhash deserts. Arab J Geosci 7:4513–4521. https://doi.org/10.1007/s12517-013-1106-z

Issanova G, Abuduwaili J, Kaldybayev A, Semenov O, Dedova T (2015a) Dust storms in Kazakhstan: frequency and division. J Geol Soc India 85(3):348–358

Issanova G, Abuduwaili J, Galayeva O, Semenov O, Bazarbayeva T (2015b) Aeolian transportation of sand and dust in the Aral Sea region. Int J Environ Sci Technol 12:3213–3224. https://doi.org/10.1007/s13762-015-0753-x

Kes' AS, Andrianov BV, Itina MA (1980) Dynamics of hydrographic network and changes in water level of the Aral Sea. Moscow, Nauka, pp 185–197 (in Russian)
Khvorog GV (1992) Ecological map of the near Aral region. Alma-Ata, Kazakhstan (in Russian)
Kotliakov VM, Glazovskii NPh (1991) The main positions of the concepts of Restoration of the Aral Sea and Normalization of situation in the Aral Sea Region. J Amelioration Water Manag 12:10–14 (in Russian)
Kudekov TK (2002) Modern ecological condition of Balkhash Lake basin. Almaty, p 388 (in Russian)
Kuksa VI (1994) Southern seas (Aral, Caspian, Azov and Black seas) under anthropogenic stress. Hydrometeoizdat, p 319 (in Russian)
L'vovich MI (1971) Rivers of USSSR. Moscow, Mysl', p 231 (in Russian)
Lunezheva MS, Kiatkin AK, Polishuk VP (1987) Middle Asia and Kazakhstan—ancient region of irrigated agriculture. J Hydrotechnika Amelioration 10:65–68 (in Russian)
Mal'kovskii IM (2003) Geographical bases of water supply and ecological sustainability of the natural-economic systems in Kazakhstan. Abstract of the doctoral thesis, p 50 (in Russian)
Mal'tsev EA (1969) Land-water resources of Middle Asia and their agricultural use. Frunze, Ilim, p 246 (in Russian)
Medoev A (1981) Geochronology of the Paleolithic of Kazakhstan. Nauka, Alma-Ata, p 64
Meedecins sans Frontieres (2000) MS Fin Uzbekistan and Turkmenistan. Act Report, 1999–2000. Tashkent, Uzbekistan
Micklin P (1990) Shrinking of the Aral Sea: water-economic catastrophe. J Amelioration Water Manag 5:16–19
Micklin PP (1991) The water management Crisis in Soviet Central Asia: The Carl Beck papers in Russian and East European, Stud. No. 905. Pittsburgh: Cent. Russ./East Eur. Stud
Micklin PP (1992) The Aral crisis: introduction to the special issue. Post-Sov Geogr 33(5):269–283
Micklin PP (2004) The Aral Sea crisis. See Nihoul et al. 2004, pp 99–125
Micklin PP (2007) The Aral Sea disaster. Annu Rev Earth Planet Sci 35:47–72
Mustafayev ZhS and Kozykeeva AT (2009) Aral Sea basin: past, present and future. Taraz, p 318 (in Russian)
Mustafayev ZhS, Pulatov K, Kozykeeva AT, Mustafayeva LZh (1997) The ways in improvement of the natural-ecological situation in the Aral Sea Basin (Analitical review). Taraz, p 80 (in Russian)
Mustafayev ZhS, Kozykeeva AT, Mustafayeva LZh (2003) The Aral Sea Basin: past, present and future. J Hydrometeorol Ecol 1:115–129 (in Russian)
Novikova N (1996) The Tugai of the Aral Sea Region is dying: can it be restored? Russ Conserv News 6:22–23
Novikova N (1997) Principles of preserving the botanical diversity of the deltaic plains of the Turan. Scientific Report for the Doctoral Degree in Geography, Institute of Water Problems, Moscow (in Russian)
O'Hara SL, Wiggs GF, Mamedov B, Davidson G, Hubbard RB (2000) Exposure to airborne dust contaminated with pesticide in the Aral Sea region. Lance 355(9204):627–628
Palvaniyazov M (1989) The influence of dust storms on the habitats of certain mammals of the coastal zone of the Aral Sea. Probl Osvoyeniya Pustyn 1:55–59 (in Russian)
Philonets PP (1981) Essays on the geography of inland waters of the Central, Southern and Eastern Kazakhstan (lakes, water reservoirs and glaciers). Nauka, KazUSSR, Alma-Ata, 283 p (in Russian)
Posokhov EV (1948) About self-planted lakes of the Tengiz-Korgalzhyn basin. Vestnik of the Academy of Sciences of the Kazakh SSR (in Russian)
Posokhov EV (1949) Salt lakes of the Tengiz-Korgalzhyn basin. Izvestia of the Academy of Sciences of the Kazakh SSR. Geol Ser 10(72) (in Russian)
Posokhov EV (1955) Salt lakes of Kazakhstan. Moscow (in Russian)

Prospero J, Ginoux P, Torres O et al (2002) Environmental characterization of global sources of atmospheric soil dust identified with the Nimbus 7 Total Ozone Mapping Spectrometer (TOMS) absorbing aerosol product. Rev Geophys 40(1):1–31

Severskiy I, Chervanyov I, Ponamorenko Y et al (2005) Global International Waters Assessment (GIWA) 24, Aral Sea. University of Kalmar, Sweden

Shivarev SP, Smerdov BA (1996) About modelling of levels of the Aral Sea and its parts. J Hydrometeorol Ecol 3:96–115 (in Russian)

Shkolomanov IA (1979) Anthropogenic changes in water of rivers. Hydrometeoizdat, Leningrad, p 240 (in Russian)

Shults VL (1965) Rivers of middle Asia. Hydrometeoizdat, Leningrad, p 129 (in Russian)

Sorokin AG (2002) Modeling of the processes in management of water resources of the transboundary rivers of the Syrdarya and Amudarya basins. J Amelioration Water Manag 1:48–53 (in Russian)

Spivak L, Terechov A, Vitkovskaya I, Batyrbayeva M, Orlovsky L (2012) Dynamics of dust Transfer from the desiccated aral sea bottom analysed by remote sensing. In: Breckle S.-W et al. (eds) Aralkum—a man-made desert: The desiccated floor of the aral sea (Central Asia). Chapter 7, pp 97–106

Thompson (2008) "The Aral Sea Crisis". Columbia Edu

Walters P (2010) Aral Sea recovery? National Geogr Daily News

Wucherer W (2013) Uzbekistan: Rehabilitation de las tierras degradadas por la actividad humana: La estabilizacion de los suelos desecados de la cuenca del Mar de Aral. Universidad de Bielefeld

Zenkevich LA (1963) Biology of seas in USSSR. Academy of Sciences of USSSR, Moscow, p 453 (in Russian)

Chapter 8
Water Resources and Lakes in Kyrgyzstan

8.1 Water Resources and Hydrographic System in Kyrgyzstan

The basis of the hydrographic system in Kyrgyzstan is the river systems and water catchments that are separated by watershed mountain ranges from each other (Fig. 8.1). The hydrographic system of river basins is made up of lakes, glaciers, groundwaters, swamps, and wetlands. All of them are products of orography, relief/topography, underlying surface, and climate. They have an interaction and mutual influence and take a direct part in the formation of the water balance of the river flow regime. Area of 198.5 thousand km^2 in Kyrgyzstan is occupied by river flow. They consist more than 2047 watercourses (rivers and tributaries) longer than 10 km with the total length about 35,000 km (Fig. 8.2; Table 8.1).

According to the nature of the water balance, the territory of Kyrgyzstan is divided into areas of formation and dispersion of river flow (Alamanov et al. 2013; Bol'shakov 1974). The area of runoff formation is estimated at 171,800 km^2 or 86.5% of the country's territory. They cover mountain slopes and high-mountainous uplifts where a significant part of the atmospheric precipitation is spent for the formation of surface, soil, and riverbed flows. The area of runoff dispersion is confined to the foothill alluvial and deluvial sloping slopes and plains where mainly irrigated agriculture is concentrated. The area of the irrigated agriculture in Kyrgyzstan is insignificant: 26,700 km^2 or 13.5% of the territory.

According to the regional hydrography scale, the river systems in Kyrgyzstan refer to the Aral Sea basin (76.5% of the territory), to the Tarim river basin (12.4%), to the inner basin of the Issyk-Kul lake (10.8%) and 0.3% is to the Balkash basin. In case of Kyrgyzstan, it is advisable to consider its hydrographic systems at the level of large watersheds separated by orography that differing geographically and hydrologically from each other. Accordingly, there are nine hydrographic systems identified: the Naryn river basin; watersheds of the rivers in the mountain range of the Fergana Valley; Chatkal river basin; the basin of the Issyk-Kul Lake; the basin

J. Abuduwaili et al., *Hydrology and Limnology of Central Asia*, Water Resources Development and Management, https://doi.org/10.1007/978-981-13-0929-8_8

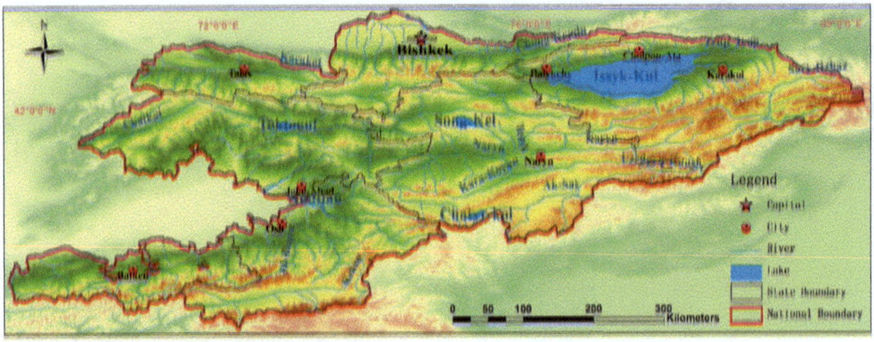

Fig. 8.1 Hydrographic system and the main lakes in Kyrgyzstan

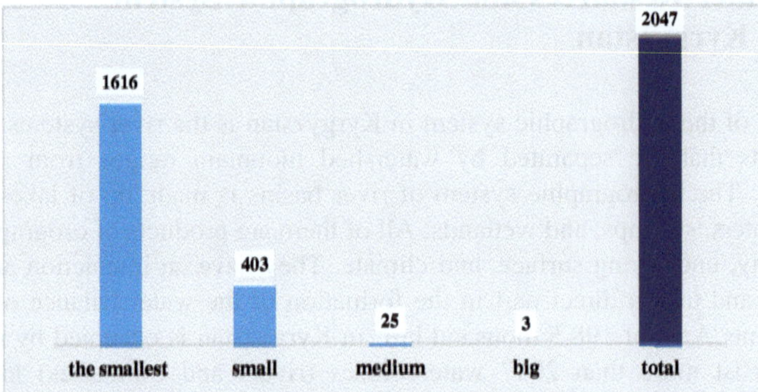

Fig. 8.2 Number of rivers in Kyrgyzstan

Table 8.1 Distribution of watercourses (rivers and tributaries) in Kyrgyzstan (Alamanov 2016)

Gradation of watercourses	Length (km)	Number of watercourses		Total length	
		Number	%	km	%
The smallest	10–25	1616	78.9	12,117	34.7
Small	26–50	321	15.7	10,916	31.2
	51–100	82	4	6061	17.3
Medium	101–200	24	1.2	3216	9.2
	201–300	1	0.05	253	0.8
	301–500				
Big	501–1000	2	0.1	1239	3.5
	Above 1000	1	0.05	1186	3.3
Total		2047	100	34,988	100

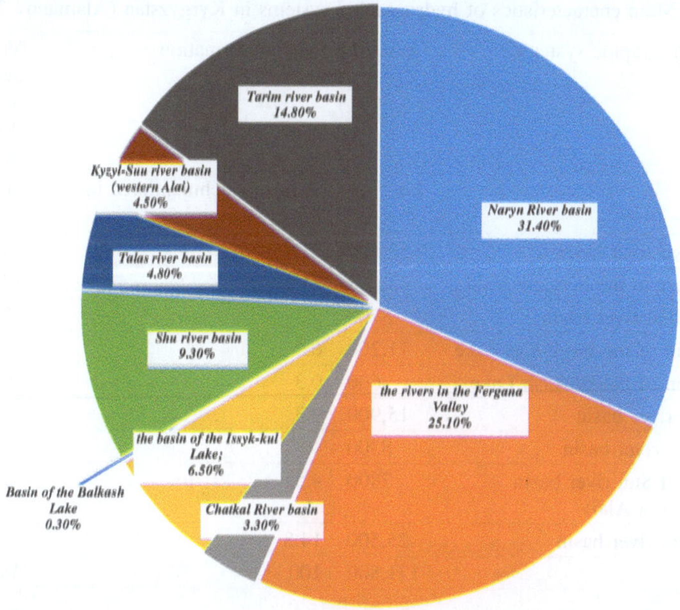

Fig. 8.3 A portion of catchment area of different river basins in Kyrgyzstan

of the Balkash Lake; the Shu river basin; the Talas river basin; Kyzyl-Suu river basin (western Alai); and the Tarim river basin (Table 8.3). The biggest river basin is Naryn which takes 53,700 km^2 in area of river flow and 31.4% from catchment area in Kyrgyzstan. The rivers in the Fergana Valley takes 43,100 km^2 (25.1% from catchment area), Tarim river basin is 25,500 km^2 (14.8%), and 28.7% of catchment area is belong to the rest of the river basins (Fig. 8.3).

The average annual total water flow in the rivers of Kyrgyzstan is 1543 m^3/s.

The total amount of river flow in the Kyrgyzstan is estimated 48.6 km^3 (Table 8.2). The resources of the river flow are distributed within the country extremely unevenly. They are concentrated mainly in the not yet lived-up, economically underdeveloped areas. There are 258,000 m^3 of water/year per 1 km^2 in Kyrgyzstan.

The Zhalal-Abad region (oblast) is more affluent with an average of 386 thousand m^3 of river flow/1 km^2, 272 and 246 thousand m^3 in Naryn and Talas regions, respectively. In Issyk-Kul region is 244 thousand m^3 of river flow/1 km^2. In the Issyk-Kul basin, where practically the entire population of the region lives, the resources of the river flow are 250 thousand m^3 (Fig. 8.4). The total amount of river flow resources is only 21.3% in Shu Valley and Osh region where 62% of the country's population is concentrated (Table 8.3).

Table 8.2 Main characteristics of hydrographic systems in Kyrgyzstan (Alamanov 2016)

No.	Hydrographic systems	Area of river flow formation		Volume of average multi-year (long-term) flow	
		km^2	% from catchment area in Kyrgyzstan	km^3	%
1	Naryn river basin	53,700	31.4	14.6	30
2	Rivers in the Fergana Valley	43,100	25.1	12.4	25.5
3	Chatkal river basin	5700	3.3	2.74	5.65
4	Basin of the Issyk-Kul Lake	11,200	6.5	3.96	8.15
5	Basin of the Balkash Lake	600	0.3	0.37	0.76
6	Shu river basin	15,900	9.3	3.84	7.90
7	Talas river basin	8300	4.8	1.72	3.54
8	Kyzyl-Suu river basin (western Alai)	7800	4.5	1.98	4.10
9	Tarim river basin	25,500	14.8	6.99	14.4
Total		171,800	100	48.6	100

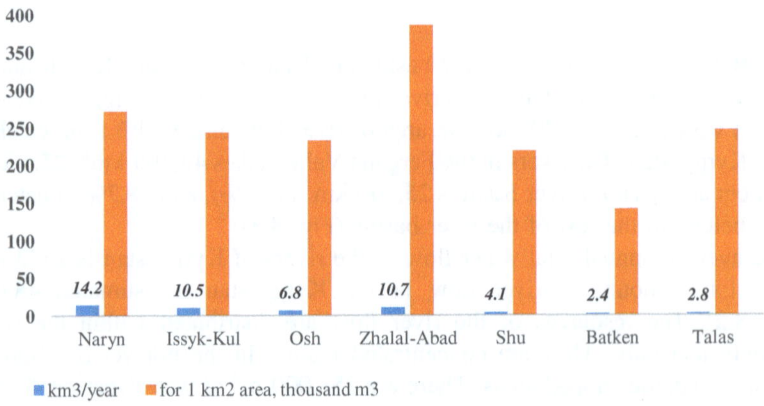

Fig. 8.4 Distribution of river flow in different regions of Kyrgyzstan

According to the distribution of river water resources per capita is 54.2 thousand m^3/year per person in Naryn region, the Issyk-Kul region is 23.3 thousand m^3/year and 10.2 thousand m^3/year in Zhalal-Abad region. In economically developed areas such as Shu valley and Osh region are 2.3 and 6.0 thousand m^3/year, respectively.

Table 8.3 Distribution of resources of river flow in different regions of Kyrgyzstan (Alamanov 2016)

No.	Region	Area (thousand km^2)	Resources of river flow			
			km^3/year	%	For 1 km^2 area (thousand m^3)	Per capita (thousand m^3/year)
1	Naryn	52.2	14.2	27.7	272	54,2 (65,7)
2	Issyk-Kul	43.1	10.5	20.5	244	23,3 (24,3)
3	Osh	29.22	6.8	13.3	233	6 (6,6)
4	Zhalal-Abad	26.9	10.7	20.3	386	10,2 (14,3)
5	Shu	18.7	4.1	8	219	2,3 (2,9)
6	Batken	16.98	2.4	4.7	141	5,55 (5,5)
7	Talas	11.4	2.8	5.5	246	11,7 (10,7)
Total for Kyrgyzstan		198.5	51.5	100	258	8,5 (11,2)

8.2 Water Balance of Natural Belts in Kyrgyzstan and Artificial Hydrological Network

There are four natural belts such as lowland-desert, low mountain-steppe, meadow-steppe landscapes; middle mountain-steppe, meadow-steppe, and forest-meadow-steppe landscapes; high-mountain meadow and meadow-steppe subalpine, and alpine landscapes; glacial–nival belt in the territory of Kyrgyzstan (Ergeshov et al. 1992).

In the belt of lowland-desert, low mountain-steppe, meadow-steppe landscapes, an average of 400 mm of atmospheric precipitation falls out and 85% of which evaporates (Table 8.4). 8–10% of precipitation spends for surface runoff and is only 5% for underground. The gross moistening of the territory is relatively large (80–90%), but only 6–10% of it goes to the formation of underground runoff, and the rest is spent for evaporation. The ratio of the surface and subterranean components is 67 and 33% of the river runoff, respectively.

The structure of the water balance in the belt of the middle mountain-steppe, meadow-steppe, and forest-meadow-steppe landscapes is as follows: Atmospheric precipitation is 510 mm, of which 290 mm or 56.8% evaporates; 17% of precipitation goes to the formation of underground runoff. More than 70% of precipitation is the gross moistening of the territory, and 22% of which is spent on the formation of underground runoff. 26% of the precipitation falls off superficially. Surface runoff is almost 60% of the total runoff.

A considerable amount of precipitation (600 mm) falls during the year in the belt of high-mountain meadow and meadow-steppe subalpine and alpine landscapes. The relative value of atmospheric precipitation evaporation is 38%, surface runoff is 35%. Underground runoff is 43% of the total runoff.

Table 8.4 Structure of the water balance in the high-altitude natural belts of Kyrgyzstan (Alamanov et al. 2013)

No.	High-altitude belts	Area (thousand km^2)	Precipitation (mm)	Runoff (mm)			Evaporation (mm)	Total humidification	Runoff coefficient
				Total	Surface	Underground			
1	Lowland-desert, low mountain-steppe, meadow-steppe landscapes (below 2000 m)	55.4	400	60	40	20	340	360	0.15
2	Middle mountain-steppe, meadow-steppe, and forest-meadow-steppe landscapes (2000–3000 m)	62.6	510	220	130	90	290	380	0.43
3	High-mountain meadow and meadow-steppe subalpine and alpine landscapes (3000–3500 m)	36.2	600	370	210	160	230	390	0.61
4	Glacial–nival belt (above 3500 m)	45.3	640	480	290	190	160	350	0.75

Almost more than 40% of the total humidification of the territory goes to the formation of underground runoff. The relative magnitude of underground runoff from precipitation is small (26%).

Most precipitation falls in the glacial–nival belt (640 mm) where 160 mm or 25% of the precipitation evaporates. 65% of atmospheric precipitation goes to the total humidification of the territory and 41% of which goes to the formation of underground runoff. An essential part of the total river flow of mountain rivers is formed here, which accounts for almost 42% of the runoff formed at the catchments (Table 8.4).

The structure of the water balance of each natural belt is generally preserved irrespective of the position of ridge in the orographic system of Kyrgyzstan. Because the absolute values of their elements of water balance may differ. Thus, in the low mountain-steppe belt, more than 80% of the precipitation evaporates, 40% of the rainfall is spent for the formation of a complete river runoff in the low mountain-middle, mountain-forest, shrub-meadow belt, and 60% for evaporation, and 75–80% of precipitation forms a complete river runoff in the high-mountain-alpine belt and 20–25% for evaporation (Alamanov et al. 2013).

The underground and surface components of the flow for Kyrgyzstan are practically equivalent in economic significance. Underground drain does not need special regulation, and mountain rivers provide water throughout the year. The value of surface runoff is to provide water to irrigated agriculture. However, it needs regulation for more complete use; crops: tobacco, cotton, vegetables, and melons, as well as pastures, hayfields, perennial grasses, orchards of perennial plantings are watered from March to April. Watering is produced depending on the culture, from 5 to 9 times during the entire growing season.

Total humidification of the territory of Kyrgyzstan is 78.1 km^3 per year, and 28% of this amount is spent for groundwater and 72% for evaporation. Therefore, the coefficient of supply of rivers by groundwater for Kyrgyzstan is 0.28, and evaporation—0.72 (Alamanov et al. 2013).

An assessment of water resources in natural belts/zones of Kyrgyzstan showed that humidification is 34.7 km^3 in the highland and glacial–nival belt that is more than half the resources of the river flow in Kyrgyzstan. Only 3.2 km^3 in the plain-low mountain belt, and in middle mountain belt is 19.3 km^3.

The regions located in the peripheral parts of the mountain ranges, i.e., in the most humid areas, are the most provided regions with water resources (Table 8.5). The flow here is 35–40% of the amount of precipitation.

The resources of the river flow are distributed extremely unevenly, and they are concentrated mainly in still uninhabited, economically underdeveloped regions. There are 258,000 m^3 of water per year per 1 km^2 in Kyrgyzstan (Alamanov et al. 2013).

The more affluent is the Zhalal-Abad region where an average of 386,000 m^3 of river flow per 1 km^2, in Naryn and Talas regions are 272 and 246 thousand m^3, respectively (Table 8.5). There are 244 thousand m^3 of river flow per 1 km^2 in the Issyk-Kul region. In addition, the resources of the river flow make up only 102 thousand m^3 in the Issyk-Kul depression where the entire population of the region

Table 8.5 Distribution of resources of the river flow in different regions of Kyrgyzstan (Alamanov et al. 2013)

No.	Regions	Area (thousand km^2)	Resources of the river flow			
			km^3/year	%	Per 1 km^2 area (thousand km^3)	Per capita (thousand km^3/year)
1	Osh	29.22	6.8	13.3	233	6.6
2	Batken	16.98	2.4	4.7	141	5.5
3	Zhalal-Abad	26.9	10.7	20.3	386	14.3
4	Issyk-Kul	43.1	10.5	20.5	244	24.3
5	Talas	11.4	2.8	5.5	246	10.7
6	Naryn	52.2	14.2	27.7	272	65.7
7	Chui	18.7	4.1	8.0	219	2.9
Total for Kyrgyzstan		198.5	51.2	100	258	11.3

lives. The total amount of river flow resources is only 25.9% in the Chui Valley and Osh regions where 62% of the country's population is concentrated (Alamanov et al. 2013).

The Naryn region is distinguished by high water resources per capita with 65.7 thousand m^3/year, then the Zhalal-Abad region is 14.3 thousand m^3/year, in Issyk-Kul region is 24.3 thousand m^3/year, while in economically developed areas of Chui valley and Osh region are 2.9 and 6.6 thousand m^3/year, respectively.

The artificial hydrographic network of Kyrgyzstan in terms of size, significance, and impact on the environment has become comparable to the natural hydrographic system. According to the Department of Water Resources and the Irrigation Institute of Kyrgyzstan, the total length of irrigation canals is 30,836 km or 88.1% of the length (34,988 km) of all rivers in Kyrgyzstan, including the smallest (10–25 km); and 6200 km of those length belong to inter-farm channels and the cross section is designed to pass 2528 m^3/s of water. This is almost for 1000 m^3/s higher than the runoff of all rivers per year with average water content (1543 m^3/s). Channels of on-farm irrigation systems with a length of 19,200 km are formed the basis of an artificial hydrographic network. The collector-drainage network with a total length of 5436 km is sufficiently branched (Alamanov et al. 2013). In addition, there are more than 620 artificial water reservoirs that can regulate river runoff for one day to two days, and there are 2200 wells that bring groundwaters for irrigation.

8.3 Water Reservoirs, Fresh Groundwaters, and Wetlands in Kyrgyzstan

Reservoirs have become an integral part of the hydrographic system of Kyrgyzstan. Their construction was in demand by the need to develop two basic branches of economy: irrigated agriculture and hydropower. It was necessary to optimize the use of river flow in order to obtain a stable high yield and uninterrupted supply of electricity to the country. The most acceptable way to implement such a requirement was the construction of water reservoirs that allow to regulate the regime of river flow and to redistribute them across the territory and in time.

A small Kara-Suu–Kara-Balta water reservoir was the first built in the Chui Valley on the Kara-Balta river in 1937–39 years. In 1946 on the Ak-Suu river was built water reservoir with a volume of 200 thousand m^3 in order to increase the supply of irrigation for sugar beet crops and expand their areas in the Chui Valley. In 1951, after the completion of the construction of the Sokuluk water reservoir, a large-scale work began on the development of a network of water reservoirs of various purposes and different sizes (Table 8.6).

At present, in Kyrgyzstan, there are 18 water reservoirs with total volumes from 13 to 19,500 million m^3 of water and more than 200 artificial reservoirs of decadal and seasonal flow regulation with a total volume of about 105 million m^3, and about 400 basins regulating day and decade fluctuations have been constructed and are operated (Alamanov and Akmatov 2006; Bilik 2001).

According to the generally accepted hydrology classification in Kyrgyzstan, only one reservoir, the Toktogul water reservoir, with a surface area of 284.3 km^2 belongs to large reservoirs.

Its total volume is 19,500 km^3, i.e., 81.5% of the total water mass that can accumulate in artificial reservoirs in Kyrgyzstan. The large artificial reservoirs include the Kempir-Abad reservoir. The vast majority of reservoirs in terms of water accumulation are shallow (42%), small (25%), and medium (25%). There are six mountain reservoirs located within the heights of 1200–2000 m. In addition, there are 11 reservoirs located in the piedmont at the 700–1200 m. The remaining reservoirs located at lower altitudes are conditionally assigned to the plain (flat) reservoirs. The amplitude of level fluctuation of reservoir is the main indicator characterizing the water regime of the reservoir. Its greatest changes reaching 63 m are observed in the Toktogul water reservoir. The range of level fluctuations is 59.5 m in the Kempir-Abad water reservoir, in Papan is 57 m, in Kara-Buuri is 53.4 m, in Orto-Tokoi is 31, and 24 m in Kiurp-Sai. The level fluctuation of the main reservoirs is 5–10 m. The water reservoirs of Kyrgyzstan differ from each other by the intensity of water exchange which is defined as the ratio of its total volume to the average annual flow of water through the dam. Thus, the longest time that equal to 1.72 years (1 year 9 months) is required for a complete change in the water mass of the Toktogul water reservoir. This indicator is 5–7 months for the Kempir-Abad, Kara-Buuri, Papan, and Orto-Tokoi water reservoirs (Alamanov and Akmatov 2006).

It should be noted that individual reservoirs of Kyrgyzstan were built on transboundary rivers. They are used to provide irrigation interests of neighboring countries. For instance, waters of the Toktogul reservoir are used for irrigation of the fields in Uzbekistan (720 thousand ha), Tajikistan (60 thousand ha), and Kazakhstan (140 thousand ha). In addition, Uzbekistan uses water resources accumulated in the Kempir-Abad, Kerkidon, Papan, and Kasan-Say water reservoirs. Kazakhstan

Table 8.6 Main characteristics of water reservoirs in Kyrgyzstan (Alamanov et al. 2013)

	Water reservoir name, river	Total volume (mln m^3)	Useful volume (mln m^3)	Surface area at total volume (km^2)	Dam height (m)	Year of exploitation
1	Toktogul, Naryn river	19,500	14,000	284.3	215	1976
2	Kempir-Abad, Kara-Darya river	1950	1930	50	110	1975
3	Kara-Buuri, Talas river	550	540	28.6	83.7	1975
4	Orto-Togoi, Shu river	470	450	24	52	1960
5	Kiurp-Sai, Naryn river	370	35	12	113	1982
6	Papan, Ak-Buura river	200	180	7.1	100	1983
7	Kasansai, Kasansai river	160	150	–	–	–
8	Tash-Komir, Naryn river	140	16	6	73	1986
9	Tortkol, Isphara river	90	75	6.6	34	1972
10	Kerkidon, Isphairam	90	86	–	–	–
11	Kambar-Ata2, Naryn river	61	31	3.3	70	–
12	Uch-Korgon, Naryn river	53	20	74	29	1962
13	Ala-Archa, Ala-Archa river	51	48	6.3	24.5	1967
14	Naiman, Abshyr river	40	38	4.2	40.5	1966
15	Shamaldy-Sai, Naryn river	39.3	5.7	5	37	1995
16	Bazar-Korgon, Kara-Unkir river	22	20	2.7	25	1962
17	Spartak, Ak-Suu river	22	21	1.8	11	1967
18	Sokuluk, Sokuluk river	13	11	1.8	22.5	1954
19	At-Bashy, At-Bashy river	9.6	3	0.46	79	1970
20	Kara-Balta, Kara-Balta river	4.3	4.2	0.46	11.7	1964

receives water from the Orto-Tokoi and Kara-Buuri water reservoirs, and Tajikistan receives water from the Toktogul water reservoir (Alamanov et al. 2013).

The constantly growing needs of various branches of the economy in the adjusted flow determine the necessity of building an even larger number of water reservoirs. In the future was planned to build another 18 large and about 100 small artificial water reservoirs in Kyrgyzstan. Each such facility will occupy a certain territory on which there already existed peculiar landscapes and socioeconomic infrastructure that will be destroyed. Studies of the impact of already existing water reservoirs show that along with the benefits that reservoirs bring, they are also a number of negative processes and phenomena in the socioeconomic sphere and the environment (Alamanov and Akmatov 2006). Due to water resources accumulated in the water reservoirs of Kyrgyzstan, the supply of irrigation water to the farmland in Kyrgyzstan and neighboring countries has been increased. Based on the waters of the Toktogul water reservoir, 400 thousand ha of land has been developed and 920 thousand ha of fields in Uzbekistan, Tajikistan, and Kazakhstan is being irrigated. Settlements with the appropriate time and sociocultural infrastructure are built in the territories adjoining the water reservoir, and industrial and agro-industrial enterprises are being created. Rain-fed and conditionally irrigated lands and pasture are provided by irrigation systems.

Water reservoirs quite actively change the physical–geographical appearance of the territories. The most significant among them is the flooding of various types of established man-made and natural landscapes. Thus, 47 thousand ha of agricultural land in Kyrgyzstan was under water during the construction of water reservoirs. There were flooded forests, mineral deposits, archaeological and historical sites, and cemeteries. There were places of accommodation of 32 settlements with objects of infrastructure: power lines, communications, roads, social institutions under the water.

The hydrological, chemical, and thermal regime of the river and groundwater significantly changes below and above the dam that damages the aquatic flora and fauna. In general, it damages the landscapes of the territory surrounding the water reservoir bowl and the river bed. The level of groundwater rises in the adjacent areas. Thus, as a result of this phenomenon caused by the action of the Toktogul and Kempir-Abad water reservoirs, about 900 ha of crop rotation and the quality of 3000 ha of arable land in the Toktogul, Ozgon, and Kara-Suu regions are deteriorated; more than 200 residential buildings came into emergency condition (Alamanov et al. 2013).

Changes in the regime and magnitude of the climate elements have occurred in the adjacent areas to the water reservoirs. So, after the construction of the Toktogul water reservoir, the average annual air temperature has increased by 1.8 °C compared with the previous multi-year period, according to the Toktogul coastal meteorological station data calculated for the period 1979–2002. The same tendency in smaller values is observed at meteorological stations adjacent to relatively large water bodies. The average annual air temperature has increased by 0.4 °C according to the data from the Kochkor station which is under the influence of the Orto-Tokoi water reservoir; Kirov station by 0.4 °C (Kara-Buuri water reservoir).

Besides, the cooling effect of artificial water reservoirs on the air temperature is observed in the warm months of the year. There is a decrease in the amount of precipitation in winter from 4.3 to 37.5% at adjacent stations of different water reservoirs and an increase in the spring rains from 3.6 to 9.8% (Alamanov and Akmatov 2006).

It is established that the construction of large reservoirs in mountainous conditions with a dense network of tectonic faults and different-scale cracks in the earth's crust contributes to the disturbance of its unstable equilibrium condition. For instance, after the filling of the Toktogul water reservoir, the situation in the adjoining hollow has changed significantly, and the number of small and medium earthquakes has increased significantly (Alamanov et al. 2013).

Fresh groundwaters. Fresh groundwater is widely distributed on the territory of Kyrgyzstan. Their significant part is unloaded into riverbeds, accounting for 30–70% of the average long-term (perennial) annual flow. In fact, the flow of rivers is formed by groundwater during the period of November–March. A high degree of subdivision of the relief (topography), a dense network of cracks and faults in the upper stratum of the earth's crust, and the presence of powerful deposits of loose sediments in intermontane depressions and foothills (cones of river outflow) determine the widespread wedging of springs. In this regard, springs are found everywhere in the plains and riverbeds of river valleys, and on steep mountain slopes. Depending on the hydrogeological conditions, groundwaters either wedge out in the form of single springs or come out to the earth's surface with a system of linear and point currents (flows) moisturizing and swamping large areas.

According to the estimates of the geological service of Kyrgyzstan, the static reserves of fresh groundwater are estimated at 650 km^3 in the country. A significant part of these reserves is concentrated in Chui (300 km^3), Talas (75 km^3), Issyk-Kul (58 km^3), and Ak-Say (50 km^3) valleys (Alamanov et al. 2013). This explains the abundance of springs in these regions. It should be noted that there has not been a targeted study on the geography of springs. They have been studied in other studies as companion materials for hydrology, hydrogeology, and water management. The information below is based on reports of scientific and production expeditions conducted by the laboratories of the Institute of Geology of the Academy of Sciences of Kyrgyzstan and the Department of Geology of Kyrgyzstan. The expeditions were conducted in the 1960–1990s and covered about 2–3% of all springs in Kyrgyzstan (Alamanov et al. 2013).

Renewable resources are one of the main indicators characterizing the dynamics of the groundwater regime and determining the water availability of underground sources. They represent the flow rate of groundwater that flowing in the hydrogeological structure along the pores and cracks in the rock formations. This stream is fed by the infiltration of underground parts of precipitation, rivers, canals, lakes, and water reservoirs. The renewable resources are distributed within the hydrogeological regions in Kyrgyzstan. The Chui hydrogeological region is the richest region in renewable resources (Fig. 8.5) and then Issyk-Kul hydrogeological region with 71.3 km^3/s of renewable resources.

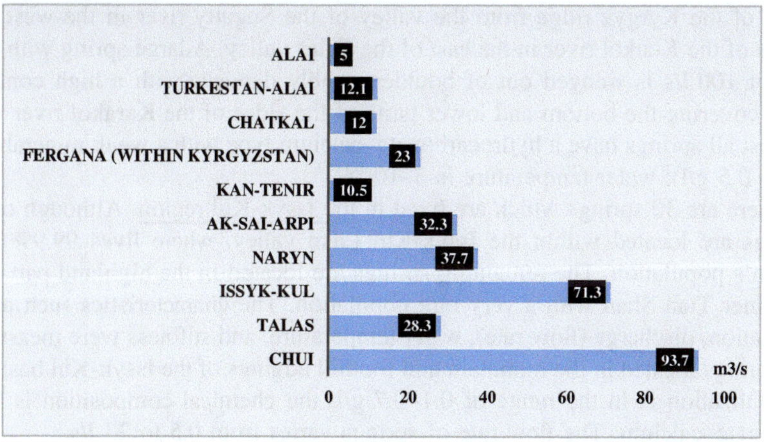

Fig. 8.5 Renewable resources of the fresh groundwaters in the hydrogeological regions within Kyrgyzstan

The geological service of Kyrgyzstan recorded 27 springs in the Chui region. Their mineralization, discharge, temperature, stiffness, and chemical composition were determined. Springs with small flow rates (0.5–1.5 l/s) are typical for the mid-mountain and high-mountainous parts of the region. They are mainly hydrocarbonate–calcium type with a mineralization in the range of 0.1–0.5 g/l. A spring that wedges off from Nijnia Serafimovka on the right bank of the Noorus river is an exception. Its water has a chloride–sodium composition and mineralization of 295.4 g/l. The springs of the Chui region are found in the altitude range of 2000–3000 m in the mountainous parts of the river basins such as Zhardy-Kaindy, Kara-Balta, Ak-Suu, Sokuluk, Ala-Archa, Alamedin, Issyk-Ata, and Kegeti. The springs in the Kichi-Kemin and Chon-Kemin river basins are distinguished by more waterborne and their flow rates reach 4–10 l/s, water temperatures within 3–11 °C. A substantial part of the underground waters of the Chui Valley is discharged into ravines that embedded in the earth surface to a depth of 5–10 m. They form small rivers and rivers with a constant runoff which are called "Kara-Suu." Their flow rates reach from 100 to 1000 l/s, and on this basis ponds are formed on the negative forms of surface.

There are 19 springs which have been recorded in the Talas region. Their mineralization, discharge, temperature, stiffness, and chemical composition have been determined. According to the lithological composition of the rocks in which springs are formed, the springs can be divided into two groups.

The first group of springs are sources that form in non-metamorphosed rocks: granites and porphyrites. They are characterized by relatively small flow rates of 1.5–6 l/s. They are confined to the middle and upper steep slopes of the river valleys such as Uch-Koshoi, Chon-Koshoi, Karakol, Besh-Tash, and Keksu. The second group of springs with a flow rate of 30–100 l/s is formed in marbles, marble limestones, and limestones. They wedged into the logs and ravines of the southern slope of the Kyrgyz ridge from the valley of the Sugatty river in the west to the valley of the Kenkol river in the east of the Talas valley. A large spring with a flow rate of 100 l/s is wedged out of boulder–pebbly deposits with a high content of loam covering the bottom and lower parts of the sides of the Karakol river valley. Almost all springs have a hydrocarbonate–calcium type with a weak mineralization (0.02–0.5 g/l); water temperature is 3–10 °C.

There are 30 springs which are fixed in the Issyk-Kul region. Although only 11 springs are located within the Issyk-Kul Lake valley, where lives 99.9% of the region's population. The remaining springs are located in the highland part within the Inner Tian Shan with a very rare population. The characteristics such as mineralization, discharge (flow rate), water temperature, and stiffness were measured in the springs located in the mountain and foothill edgings of the Issyk-Kul basin. The mineralization is in the range of 0.1–0.7 g/l; the chemical composition is hydrocarbonate–calcium. The flow rate of springs varies from 0.5 to 23 l/s.

The outcrops of the springs in the basin are concentrated in river valleys located to the east of the Boster village on the northern shore and from the Bokonbaevo village on the southern shore.

There are outcrops of springs in the river valleys such as Sasykbulak, Karagaibulak, Chon-Ak-Suu, Kichi-Ak-Suu, Baisoorun, Kuturgu, Tyup, Zhyrgalan, Karakol, Chon-Kyzyl-Suu, Zheti-Oguz, Zhuuku, Barskoon, and Akterek with a flow rate of 10–15 l/s which are not included in the main list of springs (Alamanov et al. 2013).

Wetlands in Kyrgyzstan. The formation of wetlands and swamping lands in Kyrgyzstan is determined by the hydrological, geological, geographical conditions and the impact of human activities. The wetlands and swamping lands occupy only 600 km^2 or 0.3% of the territory of Kyrgyzstan (Issaev 1956). They are distributed as small arrays and patches over a wide range of heights from flat valleys to high-mountainous uplands. Basically, they are confined to areas with a close occurrence of the day surface of waterproof layers and the support of water from large water bodies and rivers. Such wetlands are common for the coast of the Issyk-Kul, Son-Kul, Chatyr-Kul lakes and floodplain of the Chon-Naryn, Kichi-Naryk, and Kara-Darya rivers.

Waterlogging occurs in the case of wedging a weak stream to a terrain with a small slope, making it difficult to drain. This phenomenon is common in high-mountainous highlands with an even surface. The waterlogging is associated with permafrost and small deviations in the trough valleys and karas.

The development of irrigated agriculture has led to anthropogenic waterlogging that associated with poor quality of both irrigation technology and the poor state of the irrigation network. On the one hand, canals and collector-drainage systems which have mainly earth channels supply water to underground horizons that contribute to the increase of water level. On the other hand, they overgrown with hydrophilic plants and filled with soil that maintain streams of water and also contributing to flooding of lands. Such lands gradually turning into swamps and wetlands. They are widespread in the lower parts of Chui, Talas, and Ferghana valleys and occupy areas of hundreds of hectares (Alamanov et al. 2013).

8.4 Use of Water Resources in Kyrgyzstan

Water resources of the Kyrgyzstan are a natural factor affecting both the development of the country and the formation of international relations in the Central Asian countries. The total average annual volume of river flow equals to 52 km^3/year and a significant amount that is 75–80% goes to contiguous countries such as Kazakhstan, China, Tajikistan, Turkmenistan, and Uzbekistan. The quota of water intake within Kyrgyzstan in the Syrdarya the river basin was 4.03 km^3 from 29.8 km^3 (14%), in the Amudarya river basin was 0.2 km^3 from 1.98 km^3 (5.2%) flow forming in the territory of Kyrgyzstan. The Shu (3.84 km^3) and Talas (1.72 km^3) rivers were almost equally divided between Kyrgyzstan and Kazakhstan. The Karkyra river (0.37 km^3) flows into Kazakhstan. The runoff of the Tarim Basin (6.99 km^3) where the Sary-Zhaz, Uzengu-Kuush, Ak-Sai, and Kyzyl-Suu (eastern) territories are located in the Kyrgyzstan is completely drained to China.

Now complex water relations are formed in the region connected with the initiatives of Kyrgyzstan which suggests neighboring countries to change existing water allocation and adapt it to new conditions taking into account the national interests of sovereign states. Such initiatives are supported by only Tajikistan, not any other Central Asian countries located in the lower reaches of rivers (Alamanov 2016).

At present, the water supply for users is provided by the artificial hydrographic network of Kyrgyzstan. This artificial hydrographic network in terms of size, significance, and environmental impact has become comparable to the natural hydrographic system. According to the Department of Water Resources and the Irrigation Institute in Kyrgyzstan, the total length of irrigation canals is 30.836 km or 88.1% of the length of all rivers in Kyrgyzstan (34.988 km), including the smallest (10–25 km) and 6.200 km is for inter-farm channels; the cross section is designed to pass 2528 m^3/s of water. This is more almost 1000 m^3/s than the runoff of all rivers per year in average water content (1543 m^3/s). The basis of an artificial hydrographic network is formed by the channels for farm irrigation systems with a length of 19,200 km. The collector-drainage network with a total length of 5436 km is sufficiently branched. The network is complemented by artificial reservoirs (more than 620 water reservoirs) that capable of regulating river runoff for a period of days to 2 years. There are 18 water reservoirs with total volumes from 13 to 19.5 million m^3 of water; more than 200 artificial reservoirs of decadal and seasonal flow regulation with a total volume of about 105 million m^3 and about 400 basins of daily and decade regulation (Mamatkanov et al. 2006; Alamanov 2016).

The river flow is regulated by 23.5 km^3 or 47% of the available surface water resources. There are 2.200 wells supply groundwater for domestic and drinking purposes, and irrigation of farmland.

Since 1983, water management has reached the greatest development. By 2015, the withdrawal of water from water bodies has decreased by 40%. The water consumption for irrigation is decreased by 40% in 2010, and by 70% in 2015 due to the indicator in 1983. Irrigation is the main water user in the country (Fig. 8.6).

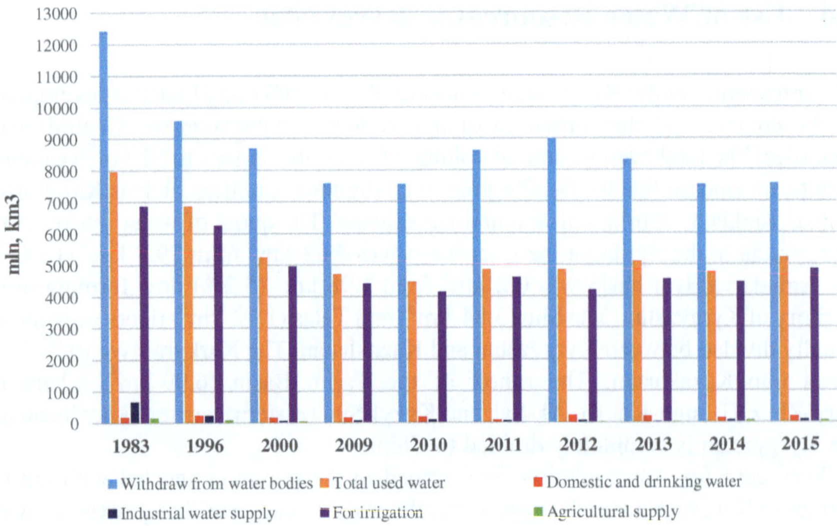

Fig. 8.6 Dynamics of water use in Kyrgyzstan, mln m^3

Table 8.7 Dynamics of water use in Kyrgyzstan, mln m^3 (Alamanov 2016)

Years	Withdraw from water bodies	Total used water	Domestic and drinking water	Industrial water supply	For irrigation	Agricultural supply	Water loss during transportation (%)
1983	12,428	7979	190	681	6884	159	–
1996	9596	6878	254	236	6278	107	1990/20
2000	8715	5262	183	58	4972	48	1962/22
2009	7600	4729	180	79	4417	8	1862/24
2010	7562	4478	206	90	4153	10	1768/23
2011	8634	4864	106	78	4620	14	1877/21
2012	9006	4863	243	82	4198	28	1955/21
2013	8327	5114	206	40	4544	28	1699/20
2014	7539	4768	143	81	4452	79	2005/26
2015	7569	5224	194	87	4853	70	2092/27

Notice Water resources are spent on replenishment of water reservoirs and transfer outside the Kyrgyzstan

The water consumption of the production sector in individual years is ranged from 5% (2013) to 12% (2015). The amount of water used for household and drinking water supply reflecting the condition of urban water utilities is also varied in a wide range of from 254 million m^3 (1996) to 106 million m^3 (2011) (Table 8.7).

A reduction in the water consumption of the industrial sector is associated with the decline in production at industrial enterprises. Deterioration of the condition in the water supply system and urban irrigation networks affected the significant decrease in water consumption in cities and villages of the country. Loss of water during transportation ranges 20–27% of its intake from sources (Table 8.7).

There is a significant degradation of agricultural water supply which is understood in the country as the operation of water supply systems transferred to local self-government. Its volume in 1983 amounted to 159 million m^3, in 2009 fell to 8 million m^3 (by 95%) (Table 8.7). However, in recent years there has been a tendency to restore this important social component of water use. Its volume has increased to 70–79 million m^3 (Table 8.7).

Changes in the volume of water usage were observed in recent decades that is due to the radical restructuring of political and economic systems in the country which has not yet been completed (Alamanov 2016).

Land reform that crushing farmland massifs especially grazing lands was the reason not to use significant areas reaching hundreds of thousands of hectares a year. For example, according to the information of the Ministry of Agriculture of Kyrgyzstan, in 2013 100.4 thousand ha of arable land out of 1170.4 thousand ha was not used. The situation is continued in subsequent years (Fig. 8.7).

Use of huge energy reserves of river flow of large and small rivers is a perspective direction in Kyrgyzstan. Their potential is estimated at 142.5 billion kWh. Hydroelectric resources are estimated at 55 billion kWh. Currently, only 8–9.5% of hydroresources is used which is 90% of the energy produced in the country. At the same time, only 3% of hydropower resources of small rivers has been developed. At present, the Toktogul HPS with a capacity of 1200 MW, Kurpsay 800,000 kW, Tash-Kumyr is 450,000 kW, and Shamaldy-Sai is 240,000 kW, Uch-Kurgan HPS is 180,000 kW, Kambar-Ata2 is 120,000 kW, and At-Bashy is 40,000 kW are built and are operating (Fig. 8.8) (Alamanov 2016).

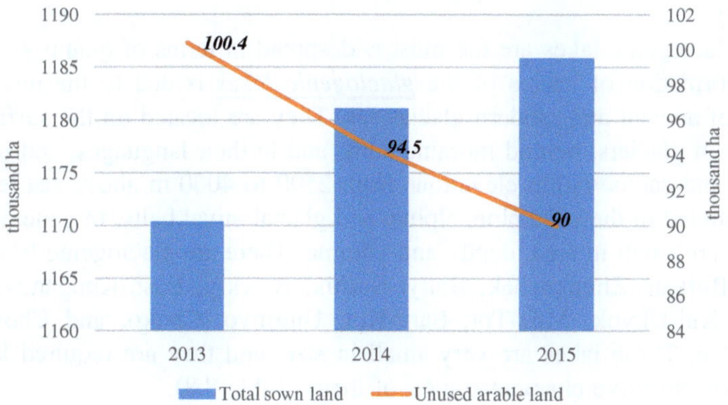

Fig. 8.7 Relationship between the total sown area and unused arable land in Kyrgyzstan for recent years

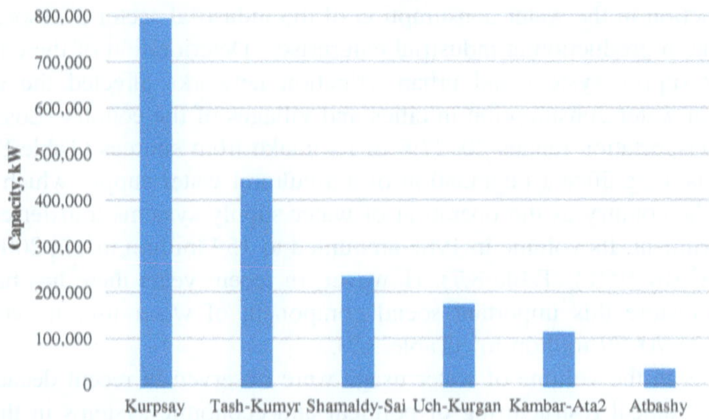

Fig. 8.8 Operating hydropower stations (HPS) in Kyrgyzstan

8.5 Lakes in Kyrgyzstan

According to the studies of the Hydrometeorological Service in Kyrgyzstan and the Tian Shan Highland Physical Geographic Station of the National Academy of Sciences, there are 1923 lakes in the territory of Kyrgyzstan with a total area of 6847.2 km^2 and lake volume of 1760 billion m^3 (Bol'shakov 1974). The average percentage of lakes is 3.4. It varies from 0.02% in the Shu Valley and to 30% in the Issyk-Kul Basin. 84% of the country's lakes are distributed in the mountains at an altitude of 3000–4000 m. Most mountain lakes are concentrated in modern glaciers and alpine zones. The spatial distribution of lakes is the alpine line (Wu et al. 2011).

Geological structure, relief, and climatic conditions are the main factors influencing the formation of lakes in the Kyrgyzstan. The peculiarities of their combination determine the formation of various genetic types of lakes, which are subdivided into tectonic, glaciogenic, hydrogenic, and drift (accumulated) ones (Fig. 8.9).

The glaciogenic lakes are the most widespread in terms of quantity.

The formation of basins of the *glaciogenic lakes* is due to the results of the activity of ancient and modern glaciations. They are located on the surface and in the body of glaciers, behind moraine dams and in their languages, and in moraine deposits and carats within elevations from 2500 to 4000 m above sea level. They are distributed in the subalpine, alpine, and glacial–nival belts. In general, lakes of this type are small in area, depth, and volume. There are glaciogenic lakes such as Arabel, Burkan, Zhuukuchak, Baltyr-Beshik, Balykty, Bash-Kuugandy, Zhuuku, Kenkol, Kul-Ukyok, Muz-Tor, Sary-Kul, Uiurmyo, Choko, and Chon-Tash in Kyrgyzstan. These lakes are very small in size, and they are required large-scale studies for objective characterization of them (Table 8.8).

Tectonic lakes formed as a result of the filling of inter-mountain depressions with water. They occupy the main share (portion) of both surface areas and water

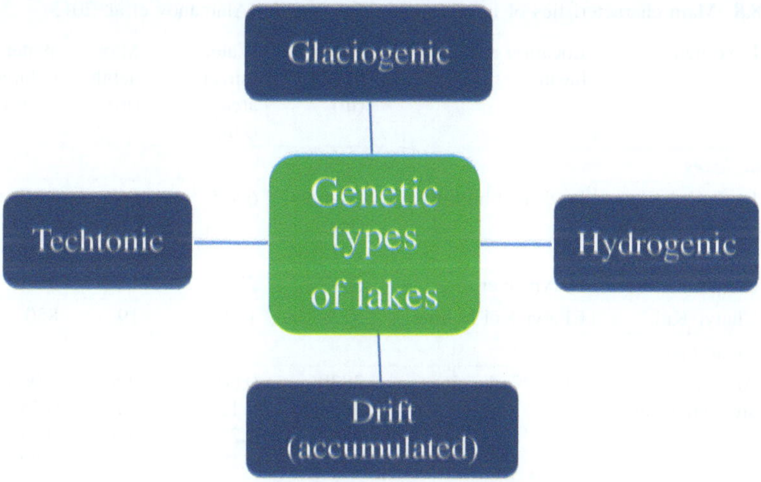

Fig. 8.9 Genetic types of lakes in Kyrgyzstan Notice: A capacity of the Toktogul HPS is 1200 MW

volumes of the lakes in Kyrgyzstan, and are characterized by great depths (Table 8.8). The total area of the water mirror of the tectonic lakes such as Issyk-Kul, Son-Kol, and Chatyr-Kol is 6670 km^2, i.e., they account for 97.4% of the surface of all lakes in the country. The area of the Chatyr-Kol lake alone is 153 km^2, i.e., twice the total area of all high-mountainous lakes.

Drift (accumulated) lakes are formed as a result of the partitioning of riverbeds and valleys by rockfalls or landslides. They are usually associated with the geological fault and discharge zones. They have formed in the periods of tectonic activity. This group of lakes is few in comparison with glaciogenic and hydrogenic lakes, but they are relatively large in size. The Sary-Chelek lake is the largest among this lake group and then followed by Kara-Suu, Kulun, Kul-Ukyok, Kara-Toko. Small lakes such as Ai-Kul, Bash-Kuugandy, Kapka-Tash, Kul-Suu, Kylaa-Kul, and Uiurmyo are belonging to this lake group.

The basins of *hydrogenic lakes* are formed from the activities of river waters and groundwater. Hole (depression) lakes are typical representatives of this group. The depressions of the earth's surface are filled by waters. They are formed under the influence of karstic or thermokarstic processes. They are typical for the permafrost zone. This lake group includes Enilchek, Okirgen-Kul, Zhashyl-Kul, Tash-Bulak, Kul-Tor, and Kogala lakes.

Issyk-Kul lake is the largest lake in Kyrgyzstan (Fig. 8.1) and one of the largest lakes in the world. The Issyk-Kul lake is located in a tectonic basin framed by the mountain ranges of Kungei Alatau from the north and Teriskei Alatau from the south. According to hydrographic and bathymetric studies by scientists from the Tian Shan High-altitude Physiographic Station, Moscow State University, and Kyrgyz Hydrometeoservice, the length of the lake is 178 km, the maximum width

Table 8.8 Main characteristics of large lakes in Kyrgyzstan (Alamanov et al. 2013)

No.	Lake name	Location and river basin	Water height (m)	Water mirror area (km^2)	Max depth (m)	Water volume (mln m^3)
Tectonic lakes						
1	Issyk-Kul	Issyk-Kul hollow	1606.99 (for 2012)	6247	668	1738×10^3
2	Son-Kul	Naryn river	3013	270	13.2	2640
3	Chatyr-Kul	Chatyr-Kol hollow	3530	153	19	850
Glaciogenic lakes						
4	Arabel (there are four lakes)	Chon-Naryn river	3830 3820 3800 3910	0.007 0.12 0.25 0.19	2.5 18.5 4.5 5	0.006 0.75 0.37 0.82
5	Burkan	Kichi-Naryn river	3822	0.49	10	1.7
6	Zhuukuchak	Naryn river, Arabel river	3766	1.5	–	8
7	Baltyr-Beshik	Sary-Zhaz river, Ak-shyirak river	3700	0.12	4	–
8	Balykty	Southern slope of Talas ridge, Balykty-Suu river	3620	0.07	14	0.41
9	Bash-Kuugandy	Zhumgal river	2860	0.26	22	1.72
10	Zhuuku	Zhuuku river	3790	0.073	9.5	0.20
11	Kul-Ukyok	Shu river, Kol-Ukiok river	3048	1.6	17	34
12	Muz-Tor	Naryn river, Itagar river	3010	0.20	23	1.82
13	Sary-Kul	Naryn river, Zhaman-Echki river	3477	1.2	–	20
14	Uiurmyo	At-Bashy river	3800	0.21	8	0.50
15	Choko (Ak-kol)	Zhuuku river	3755	0.84	10.6	2.75
16	Chon-Tash	Zhumgal river	2590	0.23	33	3.95
17	Kashka-Suu	Naryn river, Kashka-Suu river	3882	1	–	48
18	Besh-Tash	Talas river	2994	0.54	28	6
19	Ala-Kul	Issyk-Kol, Karakol river	3532	1.3	–	43
20	Mertsbakher (Enilchek)	Sary-Zhaz river, Enilchek river	3304	4.5	60–70	129

(continued)

Table 8.8 (continued)

No.	Lake name	Location and river basin	Water height (m)	Water mirror area (km^2)	Max depth (m)	Water volume (mln m^3)
Hydrogenic lakes						
21	Okirgen-Kol	Naryn river, Taragai river	3628	1.2	–	6
22	Zhashyl-Kol	Naryn river, Arabel river	3779	1	15.5	5
23	Tash-Bulak	Chatyr-Kol hollow	3540	2.5	–	13
24	Kol-Tor	Shu river, Kegeti river	2725	0.20	–	1.8
25	Kogala	Chatkal ridge, Kaba-Sai river	2606	0.84	100–130	–
Drift (accumulated) lakes						
26	Ai-Kul	Ak-Suu river, Leilek region	2937	1	–	57
27	Kara-Toko	Chatkal, Kara-Toko river	2876	1.1	111	49
28	Sary-Chelek	Naryn river, Kara-Suu (right)	1873	4.9	234	483
29	Kara-Suu	Naryn river, Kara-Suu (left)	2022	4.2	153	223
30	Kapka-Tash	Naryn river, Kara-Suu (left)	2303	1	50	21
31	Kulun	Kara-Darya river, Kulun river	2856	3.3	91	118
32	Kul-Suu	Ak-Sai river	3514	4.5	20	338
33	Kylaa-Kul	Naryn river, Kara-Suu (right)	1847	0.30	20	2.7

is 60.1 km, the coastline is 688 km, the average depth is 278.4 m, and the maximum depth confined to the southern shore in the central part of the lake basin is 668 m. The area of the lake's mirror was determined to be 6236 km^2, and the water volume is 1738 km^3. The catchment area is 22,080 km^2.

Significant fluctuations in level of the drainless Issyk-Kul lake are its peculiarity. During the time of continuous instrumental observations which began in 1927, the next level course is traced. The level decreased by 3.00 m from 1927 to 1986, i.e., its fall was 5.1 cm per year. However, the level was raised to 30 cm in 1956–60. From 1986 to 1998, the level fell by 31 cm; and from this year to 2011, the level increased by 84 cm and reached to 1707.02 m. Naturally, such level fluctuations lead to changes in hydrographic and bathymetric characteristics. So, the area of the Issyk-Kul lake was reduced by 50 km^2 due to decreasing of the water level in the Issyk-Kul lake over the period from 1968 to 1986 and the volume of water decreased by 7 km^3 (Romanovskii 1990). According to the data from the Institute of Water Problems and Hydropower of the National Academy of Sciences of Kyrgyzstan, the water level of the Issyk-Kul lake was 1606.98 m and the area of the lake's mirror was 6247 km^2 in 2006.

However, there are a number of assumptions about the influencing factors, and the causes of fluctuations in the water level of the lake have not yet been established reliably. It has been subjected to the most experimental balance studies, both on the lake and in the catchment area (Alamanov et al. 2013).

According to the degree of mineralization, the water of the lake refers to brackish, chloride–sulfate, sodium–magnesium with an ion concentration of 6 mg/l and with a cation of 1.952 g/kg, anions of 4.016 g/kg. The transparency of the water is high. It reaches 15–16 m in the middle of the lake. Due to the huge water mass that accumulating a large heat reserve, the lake does not freeze even in the coldest winters. The waters of the shallow coastal strip have a temperature of 4.2–5.0 °C in January–February; in July–August, they warm up to 20–24 °C and sometimes reach a maximum of 29 °C. Night and day breezes are typical for the coast of the lake due to the significant difference in the temperatures of the water surface and the surrounding land. The lake and its coastal zone are characterized by high recreational value.

Son-Kul Lake is the second largest lake in Kyrgyzstan (Fig. 8.1). It is located in a tectonic hollow between the mountain ranges of Moldo-Too, Son-Kul, and Booralbas. The lake's water height is at 3013 m of high. The lake length is 29 km, the largest width is 18 km, and it is in the western part of the lake. The length of the coastline is 102 km. The maximum depth is 13.2 m with an average value of 9.2 m. The area of the mirror is 270 km^2, and the volume of water in the lake is 2.64 km^3. The catchment area covers an area of 1120 km^2. Only four rivers of the 45 watercourses of various sizes and types (erosion cuts, rivers and streams, streams, etc.) that bring water to the lake have a steady flow. These are the Kum-Bel, Ak-Tash, Tash-Tobe, and Kara-Keche rivers. From the Son-Kul follows the only river is Kazhyrty. The widespread waterlogged coast indicates a significant underground inflow into the lake.

The water in the lake is fresh. Mineralization is hydrocarbonate–sulfate–calcium and varies in the range from 249 to 446 mg/l. Transparency of water in the lake reaches 7 m.

In summer, the water temperature in the lake is 11–12 °C; in winter, it is not above—3 °C. Ice cover begins in the second half of October and completes in May. The maximum ice thickness reaches 70–110 cm in January–February (Alamanov et al. 2013).

The natural conditions of the Son-Kul valley have great potential for the development of mountain tourism in the warm season.

Chatyr-Kul is the third largest lake in Kyrgyzstan (Fig. 8.1). It occupies the lowest part in the intermontane fold between the mountains of At-Bashy and Torugart-Too in the Inner Tian Shan. The lake is drainless, and the water level is at an altitude of 3530 m. The length of the lake is 23 km, and the maximum width is 11 km. The maximum depth is estimated at 16.5 m with an average depth of 3.8 m.

The area of the mirror is 153 km^2, and the water volume is 0.85 km^3. The length of the coastline of the lake is 58.8 km, and the catchment area is 1050 km^2. The mineralization of water is weak (2%). The transparency reaches up to 4 m. The water warms up to 10 °C in summer and sharply decreasing to 4.4 °C at a depth of 0.6 m. Ice cover begins in October and lasts until the end of April. The maximum ice thickness reaches 1.25–1.50 m in the central part of the lake.

The basin is in conditions of an arid and sharply continental climate of uplands in the zone of permafrost development. Atmospheric precipitation falls within the limits of 200–270 mm; in summer the air warms up to +24 °C; in winter the minimum air temperature drops to −50 °C. Due to this, thermokarst formations are widely distributed in the region.

Sary-Chelek is one of the largest lakes in Kyrgyzstan. It is located at the junction of the Chatkal and At-Oynok ranges. The basin of the lake is formed in the valley of the Toskol river as a result of the collapse of its slope during the earthquake. The water level is at 1873 m. The lake has an elongated shape and repeating the outline of a flooded mountain valley. The length is 7.5 km with a maximum width of 2.28 km and a surface area of 4.9 km^2. It is the second deepest lake in Kyrgyzstan. The greatest depth is 234 m with an average value of 98 m. Very small intra-annual fluctuations in the water level is a feature of the lake that distinguishes it from other drift lakes. The intra-annual fluctuations are only 40–50 cm per year. The water of the lake is distinguished by a high degree of transparency that reaches 16 m. The mineralization of water is carbonate–sulfate–calcium–magnesium, and the value increases from 200 mg/l on the surface to 500 mg/l at the bottom. Maximum water is heated in August (up to 19.8 °C). The minimum temperature is observed in February (0 °C). The lake is flowing. The Sary-Chelek river flows into it and from it flows Toscool that is the tributary of the Kozho-Ata river. The lake is the core of a biosphere reserve and created to preserve a unique geosystem representative of the South-Western Tian Shan (Alamanov et al. 2013).

The Kara-Suu—Kapka-Tash system located on the Ferghana Fault between the Kankol and Taktalyk mountains (the south–western spurs of the Ferghana Ridge) in the basin of the Kara-Suu River is a characteristic for the mountainous conditions of Kyrgyzstan. The dams of these lakes are formed by ancient landslides of tectonic origin descended on the line of the Talas-Fergana fault. Water mirror of the Kapka-Tash is at an absolute altitude of 2305 m and occupied an area of 0.96 km^2. The maximum accumulated water in the lake is 26 million m^3 (with an average volume of 21 million m^3). Its maximum depth of the dam reaches 50 m. The intra-annual amplitude of the level fluctuation in the lake reaches 10–12 m. The water that seeps through this dam and flows out through its top drains into the Kara-Suu lake. The Kara-Suu lake is located below at a level of 2022 m. The area of the lake's mirror is 4.2 km^2 with a water volume of 223.5 million m^3. Its maximum depth reaches 153 m. The intra-annual amplitude of the level fluctuation is 20–30 m (Alamanov et al. 2013).

The flow from Lake Kara-Suu is carried out only by filtration from under the dam. This lake can serve as an example of the negative impact of human economic activity on the environment. In recent times, the water is polluted by the waste. This affected negatively the environment conditions in the catchment area in high degree.

The system of glaciogenic Arabel lakes in the upper reaches of the Chon-Naryn that consisting of 3 reservoirs is an example of a cascade of high-mountain lakes. The uppermost of them is located at an altitude of 3830 m and not far from the Arabel Pass. During the period of its filling, the water flowing through the top of the dam flows into the lower lake which is at an altitude of 3820 m. The water flowing from this lake along the riverbed which was formed on the northeastern part of the dam fills the next lake located 700 m to the east, at a level of 3800 m. The water transparency of the lake is low; it is 2.5–3 m. The moraine lake is lying 2 km to the south at an altitude of 2910 m and having no hydrological connection with the mentioned lakes also enters to the Arabel lakes system.

Lakes and weakly resistant dams take a special place among highland lakes of the country. They can be destroyed under the influence of seismic, hydrological, or anthropogenic factors. There are 200 such kind of lakes, which are included in the type of explosive in Kyrgyzstan. They are under a systematic or periodic monitoring by the services of the Ministry of Emergencies in Kyrgyzstan. Since 1952, over the observation period, there have been more than 70 cases of breakthrough of glacial and moraine–glacial dams of mountain lakes. Annually in the summer–autumn period, the glacial dam of Merzbacker lake breaks and forms catastrophic floods on the Sary-Zhaz river. Five cases of breakthrough of the high-mountain lakes were recorded in the Issyk-Kul basin; three in the Chui valley; three in the Osh and Batken regions; two in the Talas region; two in the Naryn region; and one in the Zhal-Abad region. The following lakes are among the most explosive ones: Kara-Suu (left), Kapka-Tash: Kara-Suu (right), Sary-Chelek, Kara-Toko, Kulun, Ak-kul, Tiuz-Ashuu, Top-Karagai, Teke-Tor, Kul-Tor, Ala-Kul, Choktal, and others.

References

Alamanov SK (2016) Water resources in Kyrgyzstan and their use. J Water Manage Kaz 3(71):17–25 (Almaty) (in Russian)

Alamanov SK, Akmatov RT (2006) Water reservoirs of Kyrgyzstan. Bishkek, 224 p (in Kyrgyz)

Alamanov SK, Sakiev K, Akhmedov S et al (2013) Hydrology of Kyrgyzstan. Physical geography of Kyrgyzstan. Bishkek, 211–274 pp (in Russian)

Bilik O (2001) Water management. Kyrgyzstsan: Encyclopedia. Bishkek, 218–221 pp (in Russian)

Bol'shakov MN (1974) Water resources of rivers of the Soviet Tian-Shan and calculation methods. Frunze, 307 p (in Russian)

Ergeshov AA, Tsigel'naya ID, Muzakeev MA (1992) Water balance of Kyrgyzstan. Ilim, Bishkek, 152 p (in Russian)

Issaev DI (1956) Wetlands of Northern Kirgizia. Kirstateuniversity, Frunze, 59 p (in Russian)

Mamatkanov DM, Bazhanova LV, Romanovskii VV (2006) Water resources of Kyrgyzstan in modern stage. Ilim, Bishkek, 266 pp (in Russian)

Romanovskii VV (1990) Issyk-Kul lake as a natural complex. Ilim, Frunze, 168 p (in Russian)

Wu L, Li X, Liu S et al. (2011). Remote sensing based glacial lake inventory in the Hindu Kush-Himalaya region. International Conference on Remote Sensing, pp. 1451–1454.

Chapter 9
Hydrographical and Physical–Geographical Characteristics of the Issyk-Kul Lake Basin and Use of Water Resources of the Basin, and Impact of Climate Change on It

9.1 Physical–Geographical Characteristics of the Issyk-Kul Lake Basin

The Issyk-Kul lake basin is located in a vast intermontane basin. Its width in the meridian direction near the Tosor village is 75 km. From here to the east and west, the basin narrows. Its total length from Chaar-Zhon to Kyzyl-Ompol is 240 km (Chupakhin 1964). The area of the lake basin is 22,080 km^2. The Issyk-Kul basin is surrounded by high mountain ranges: Kungei Ala-Too from the north and Terskey Ala-Too from the south. The lake basin is surrounded by the mountain ranges and forms around it an inter-temporal space closed in all directions that stretched in a latitudinal direction. The Issyk-Kul basin is connected with the Karkara River (the basin of the Balkash Lake) through the Santash Pass in the east. The basin in the west is closed by the low Kara-Koo and Kyzyl-Ompol mountains. On the westernmost part of the basin, the Chu river flows to the valley of the same name through the Boom Gorge (length 40 km) (Kadyrov 1986).

The relief of the Issyk-Kul lake basin is a complex. It distinguishes three genetically related complexes: mountain, foothill-adyrny, and plains (Issaev 1959, 1962; Rantzman 1959). The mountain complex is represented mainly by the slopes of the ridges framing the hollow. The relief of the ridge is generally rocky with crystalline rocks.

The average height of Kungei Ala-Too is 3800–4000 m above the sea level. The ridge splits into a number of spurs in the area of Boom Gorge. The total length of the Kungei Ala-Too is 280 km. The highest point of the ridge is in the upper reaches of the Chok-Tal River (4770 m) (Kadyrov 1986).

Terskey Ala-Too rises more than 5000 m above sea level. It begins in the east from the Semenov Peak and bends around the Issyk-Kul Lake. Its total length is about 350 km. High peaks are in the upper reaches of the Chon-Kyzyl-Suu, Zheti-Oguz, Karakol, and Ak-Suu rivers. One peak reaches 6216 m between the upper reaches of the Karakol and Zheti-Oguz rivers. The absolute elevations of the

J. Abuduwaili et al., *Hydrology and Limnology of Central Asia*, Water Resources Development and Management, https://doi.org/10.1007/978-981-13-0929-8_9

ridge are descended in the west direction. In the hypsometric ratio, the Terskey Ala-Too is much higher than the Kungei Ala-Too and there are more glaciers and firn fields on it (Kadyrov 1986).

The climate of the Issyk-Kul region is continental. It is affected by relief, significant altitude above sea level, large volume of lake water, and its area. The large diurnal and annual amplitudes of temperature fluctuations are common for Central Asia. The temperature decreases in the Issyk-Kul lake basin sharply under the influence of a significant water mass. Constant evaporation from the water surface creates an increased humidity in the coastal zone.

The Issyk-Kul basin, as well as the whole of Central Asia, is characterized by the invasion of the cold Arctic (Boom Gorge and the Santash Pass) and moderate air masses that causing a drastic change in the weather. Sometimes, tropical cyclones invade the basin from the southwest.

The average annual pressure at the surface of the lake is 840 mb, which is 173 mb less than the normal pressure at the ocean level. It is 746 mb at an altitude of 2550 m (Chon-Kyzyl-Suu). The pressure drop on the slopes in the Issyk-Kul lake is 10 mb (7.5 mm) for every 100 m of the uplift.

The western airflow over the lake prevails throughout the year in the upper layers of the atmosphere. The mountain–valley winds blow in the valleys, which are directed from the top down at night and from the bottom upwards in the daytime.

There are breezes arising in the coastal strip of Issyk-Kul lake. Breezes are under the influence of difference in air temperature over the lake and land. At night, their direction coincides with the direction of mountain winds blowing perpendicular to the extent of the mountain ranges; in daytime, they are directed from the central part of the lake to the coast.

The mountain wind prevails over the lake within a day in the whole basin. The speed of mountain–valley winds and breezes on the northern and southern coasts is about 2 m/s, on the eastern is 3 m/s, and on the western is 4 m/s. The speed of the western winds reaches 20–24 m/s, sometimes 30–40 m/s (Kadyrov 1986).

The lake in winter has a mitigating effect on the climate. The northern and southern coasts of Issyk-Kul lake in winter are warmer than in other intermontane valleys of Northern Kyrgyzstan. The average air temperature in January is 2–3 °C, and the duration of winter with frequent thaws is 3 months (Kadyrov 1986).

The summer on the coast of the Issyk-Kul lake is moderately warm and lasts about 5 months. The average daily temperature varies from 10 to 20 °C. The average temperature in July and August is about 17 °C. In the high-mountain part of the basin such as Zheti-Ogys, Chon-Kyzyl-Suu, the average temperature in January is −8 to 10 °C; in July respectively 14–10 °C.

The Issyk-Kul lake basin has a continental–marine climate in the eastern part and "sea" in all the rest part (Kaigorodov 1955).

Absolute maximum and average monthly temperature of summer along the entire coast are the same and reach 30–32 °C. Frosts in the lower zone of the basin cease in early May and resume in October. However, in certain years there may be deviations.

The distribution of precipitation in the Issyk-Kul lake basin is uneven. It is due to the general circulation of the atmosphere and the emergence of descending air currents in the west and the ascending currents in the east. Only 115 mm of precipitation falls on the west coast of the lake (Fig. 9.1). The precipitation increases in the direction from west to east. Precipitation falls mainly during the warm season. May, June, July, and August account for 50% of annual amount of precipitation. In December, January, February, it is 10%. In the western part of the basin, maximum precipitation is observed in June and July (50%), a minimum one from December to February (1% of the annual precipitation) (Fig. 9.1). In the eastern part of the hollow, the snow cover remains from the middle of November to March. However, in certain years there are deviations from these terms. At an altitude of 2500 m, the snow lies from October to April.

There are up to 40 lightning days during the year, mainly in June, July, and August. Relative air humidity in the Issyk-Kul lake basin always is higher in summer and winter compared to other intermontane valleys of Kyrgyzstan due to greater altitude and evaporation from the lake. The average monthly relative air humidity in the western and southern coasts is 54–64%, in the northern is 62–73%, and on the slopes of the basin is 60–74% (Kadyrov 1986).

The minimum cloud cover over the basin would be in September and the maximum in winter. Fogs in the coastal zone of the basin are rarely observed (no more than 1–2 days per year), mainly in the eastern part. Fogs are frequent in the highlands.

The number of hours with sunshine in the lower zone of the basin is about 2600 per year. The maximum is observed in July (300 h), the minimum is in December and January (135 h per month) (Perova 1960).

The climatic features of the Issyk-Kul lake basin affect the distribution of soil cover, vegetation, the formation of glaciers in the highland zone, the water content of rivers on which the chemical composition of the water depends primarily.

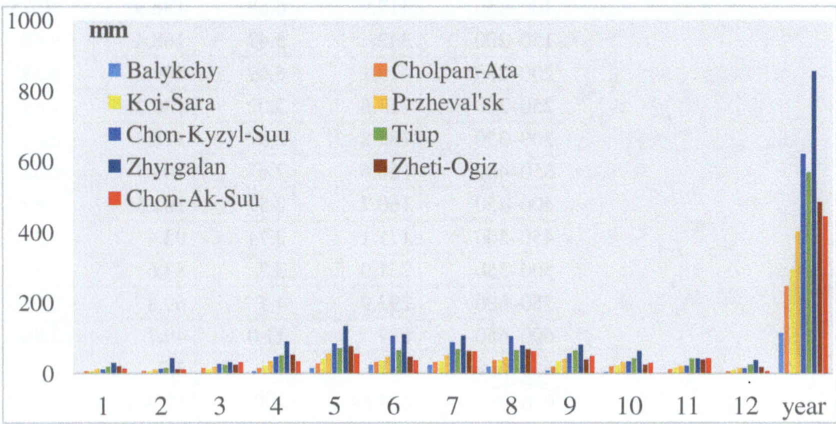

Fig. 9.1 Averaged monthly and annual precipitation in the Issyk-Kul lake basin

9.2 Hydrography of the Issyk-Kul Lake

The Issyk-Kul Lake is one of the largest and most interesting high-mountain reservoirs in the world. It is located at an altitude of 1608 m above sea level. The Issyk-Kul Lake is the largest mountainous lake in CA with area of 6247 km^2. It is the second highland lakes of the globe after Lake Titicaca in Peru and Bolivia (6900 km^2) (Kadyrov 1986). The area of the Issyk-Kul lake is only 2.5 times less than the entire catchment area (15,685 km^2). This ratio of the feeding area and the lake area is characteristic mainly for the inner seas. Its ratio is 7.0 for the Aral Sea and 7.5 for the Balkash Lake (Strakhov 1954). Water volume is 1738 km^3. It is length is 178 km, width is 60.1 km, and length of the coastline is 688 km. Average depth is 278.4 m, and maximum depth is 668 m (Table 9.1). The maximum depth of the Issyk-Kul is 109 m more than the maximum depth of the Baltic Sea and 328 m more than of the White Sea. The average depth of Issyk-Kul is three times higher than the average depth of the Baltic Sea (Romanovsky et al. 2014).

The area of the basin of the Issyk-Kul Lake is 22,080 km^2. It stretched 252 km in latitudinal direction and in the meridional at 146 km. The lake area is 6247 km^2 and 3092 km^2 is the foothill plain, which is a zone for the dispersion of river flow. The remaining part of the basin (12,741 km^2) is occupied by mountain areas, which are a zone for the formation of river runoff (Romanovsky et al. 2014).

Table 9.1 Morphometric parameters of the Issyk-Kul lake (Romanovsky et al. 2014)

Layer (m)	Area		Volume	
	km^2	%	km^3	%
0–10	436.5	6.99	60.2	3.46
10–20	409.4	6.55	56.1	3.23
20–30	292.2	4.68	52.4	3.02
30–50	620.7	9.93	96.6	5.55
50–100	535.2	8.56	209.1	12.02
100–150	417.9	6.69	186.9	10.75
150–200	342	5.47	168.4	9.68
200–250	418.1	6.69	149.2	8.58
250–300	313.6	5.02	130.2	7.49
300–350	185.8	2.97	118.4	6.81
350–400	166.6	2.67	109.7	6.31
400–450	160.7	2.57	101.5	5.84
450–500	171.1	2.74	93.4	5.37
500–550	231.0	3.7	83.6	4.81
550–600	293.9	4.7	69.8	4.02
600–650	687	11.0	49.7	2.86
650–668	567	9.1	3.7	0.21
0–668	6249.4	100	1739.1	100

Hydrography of Issyk-Kul lake basin as well as other areas is caused by climatic factors and landscape conditions. The water content of rivers is determined mainly by the distribution of atmospheric precipitation in certain areas of the basin. For this basin, there is a significant predominance of summer precipitation over winter one. Its amount increases from west to east and as well as intensive evaporation from the surface of the Lake.

Up to 80 large and small rivers from the mountain ranges of Kungey and Terskey Ala-Too flow to the Issyk-Kul lake basin. They are turbulent steep currents with significant energy and erosive power in the upper reaches. A large number of suspended particles are carried with water.

According to the hydrological division of Central Asia, there are region of formation and the area of flow dispersion which are characteristic for the Issyk-Kul lake basin (Shul'ts 1965; Bolshakov and Shpak 1960; Bolshakov 1974). The first region includes the mountainous southern slopes of the Kungei and the northern ones of the Terskey Ala-Too. Here, the moisture consumption is less than the arrival. The second region is characterized by a significant predominance of moisture consumption over its inflow. The flow is redistributed to surface and underground flows in this region. It extends a narrow strip (1–15 km) along the coast of the lake. The area extends to 30–40 km in the eastern part of the basin and in the lower reaches of the Tyup and Zhyrgalan rivers. Its upper boundary is located at an altitude of 1800–1900 m. In the western part of the basin, it rises to 1900–2000 m. Here, the hydrographic network is relatively poorly developed. Due to the dry climate, the large loss of water for filtration and the expense of irrigation most rivers do not reach the lake by surface way. The river network in this part of the basin is represented mainly by temporarily watercourses.

Toward to the east, the hydrographic network considerably expands and reaches its maximum development in the outermost part of the basin where the largest rivers form.

The Issyk-Kul lake basin hydrologically is divided into five orographic regions (Bolshakov 1974).

The first region includes the southern slope of Kungei Ala-Too from the western tip of the lake to the Kabirga river. In this region, the hydrographic network is extremely poorly developed due to the small amount of precipitation and the absence of large glaciers. The feeding of rivers is mainly due to snow. The river runoff is almost completely disassembled for irrigation in the summer.

The second region consists of the rivers of the middle part of the southern slope of the Kungei Ala-Too: from the Chok-Tal river to the Chon-Ak-Suu river. The river network is well developed here and the rivers are more abundant in comparison with the first region due to meltwater supply of alpine snow and glaciers. The Kungei Ala-Too reaches an altitude of 4500–5200 m in this region. The rivers have highly developed cones of removal and water filtered through them and partly reappear on the surface in the form of separate keys and often form wetlands or "saz."

The third region includes the rivers of the eastern part of Kungei Ala-Too from Kichi-Ak-Suu river to the Tiup river. Here, the precipitation falls more than in the first two regions. Kungei Ala-Too has a significant height within the region.

Therefore, thawed snow and summer rains is played main role in the formation of river flow. Rivers are also characterized by highly developed cones of removal forming a relatively wide foothill plain where "saza" meet.

The fourth region is represented by the rivers of the northwestern slopes of the northeastern part of Terskey Ala-Too from the Tiup to the Kichi-Kyzyl-Suu rivers. There is a lot of precipitation here. Terskey Ala-Too reaches maximum height. Its slopes serve as a condenser of moisture and are characterized by a developed glaciation. The rivers of the region are the most abundant (Tiup, Zhyrgalan, etc.). The Zhyrgalan river is the largest in the Issyk-Kul lake Basin. It has left tributaries such as Turgen-Ak-Suu and Ak-Suu-Arashan. The total area of its basin is 2085 km^2. The second largest river is Tiup. The total area of the basin is 1130 km^2.

The fifth region unites the rivers of the western part of the northern squab of Terskey Ala-Too from the Dzhuuku river to the Ulakhol river. The amount of precipitation increases from west to east, which significantly affects the development of the hydrographic network. The eastern part of the range of the Terskey Ala-Too has a significant height (4000–4700 m) within the region. The region is rich in glaciers and eternal snows, which are of great importance for feeding rivers. The largest rivers in this part are Dzuuku and Barskoon. Their basins have a well-developed hydrographic network. A chain of low mountains stretches between the ridge and the lake that forming small intermontane valleys with slopes. Mountain streams after passage of these valleys are sorted for irrigation and infiltrated through alluvial deposits. Therefore, their water content is relatively insignificant. The largest rivers are Tamga, Tosor, Ton, Ak-Say, and Ulakhol.

Groundwaters are wedged out on the surface and form a saz in the lower parts of the intermontane depressions. Konurulen Lake was formed from such waters between the valleys of the Ak-Terek and Ulakhol rivers.

The flow regime of the Issyk-Kul lake basin depends on the sources of their supply. Most of the rivers of the basin have mainly glacial-snow feeding. The importance of seasonal snow in the general feeding of rivers increases to the west. Rainwater plays an important role in the feeding of rivers in the eastern part of the Issyk-Kul lake basin (Bolshakov 1960, Bolshakov 1974).

There are glacial-snow feeding rivers are 64% and snow-glacial rivers are 36% of all the rivers of the basin. The total share of runoff from meltwater of high-mountain snow and glaciers is 30–35% in the total annual runoff of the Issyk-Kul lake basin. The average long-term annual runoff of rivers is estimated at 118.2 m^3/s (Bolshakov 1974).

The beginning and maximum water discharge of the rivers depends on the altitude of their basins and the type of feeding. Spring-summer floods of rivers in the Issyk-Kul lake basin are characterized by a considerable duration, which is due to the heterogeneity of the melting of snow and glaciers. The beginning of the flood of rivers coincides with the snow melting located in various altitudinal zones. Thus, in April and early May, it is due to the melting of snow in the lower zone of the basin. In May and June, snowmelt occurs in the middle high-altitude zone of the basin. Most of the rivers located in the northwestern part of the basin and feeding on mostly melted snow waters reach a maximum at this time. In July and the first half of August, glaciers and snow in the highland zone begin to melt. As a result, the water discharge of glacier feeding reaches the greatest value. The high water during for the indicated period is typical for the rivers on the eastern part of the Issyk-Kul lake basin. The most prolonged high water period (flood) is typical for the rivers such as Chon-Ak-Suu, Karakol, Chon-Kyzyl-Suu, Ton, Ulakhol; the shortest water period for Toru-Aygyr, Uital, and Kuturga rivers. The low water period comes in the cold part of the year (October–March), when the melting processes fade. The season of minimum water discharge depends on the river feeding. The minimum comes in December and February for rivers that feed mainly by seasonal snows; for glacier–snow rivers in March and April.

The formation of the ice cover of rivers is due to their vertical zoning under certain temperature conditions as well as the geomorphological features of individual parts of the river basin. The periods of ice formations and their duration are different for different rivers. Some of the rivers in the lower reaches do not freeze at all. In general, the formation of the ice cover of the Issyk-Kul lake basin reaches at the end of November and the beginning of December. It releases in March.

The winter regime of some rivers such as Ak-Suu-Arashan, Zheti-Oguz, and Karakol is significantly influenced by thermal springs.

The average duration of the general ice formations is 90 days on the rivers of the Issyk-Kul lake basin. River sycophancy is observed only 16 days/year, and about 40% of the rivers are not slushy.

The hydrological features of the rivers in the Issyk-Kul lake basin have a significant effect on the formation of the chemical composition of surface waters.

9.3 Hydrochemistry of the Issyk-Kul Lake Basin

The Issyk-Kul lake is one of the largest mountain lakes of continental origin—it differs in depth, because it can contain a huge amount of water, more than the Aral Sea. But the lake's feature is surprising homogeneity of water both in area and in depth, as indicated by relative constancy, specific gravity, salinity, and chemical composition (Kadyrov 1986).

The water resources of the basin of Lake Issyk-Kul are represented by the inland lake, numerous rivers flowing into it and underground waters. Natural waters are a dynamic chemical system containing in its composition complex of gases, mineral, and organic substances in the form of solutions, as well as suspensions and colloids, which depends not only on environmental conditions, but also on various processes that occur both outside and directly in the water body. Knowledge of the chemical composition of water is of great practical importance for the development of various part of the economy. The chemical composition must be taken into account when using natural waters for all types of water supply and water use (drinking, domestic, industrial, agricultural, and transport). The study of qualitative and quantitative chemical composition of natural waters (hydrochemistry) allows possibility of practical application with the aim of raising the level of productivity of organisms.

The value of the hydrochemical studies increases in connection with increasing pollution of water bodies and anthropogenic activities.

Numerous scientists and researchers (Schmidt 1882; Matveev 1930, 1932, 1935; Berg 1930; Kadyrov 1986) carried out studies on the conditions and formation of chemical composition within the Issyk-Kul Lake Basin. However, the results of hydrochemical studies differed from each other. The quality and use of lake water and the Issyk-Kul Lake Basin were also studied (Kawabata et al. 2014; Alymkulova et al. 2016). Some studies have been carried out for climate change on water level fluctuation of Issyk-Kul Lake (Abdyzhapar uulu et al. 2014).

The content of the main ions of Issyk-Kul water in its open (deep-sea) and coastal zones, as well as in the wrecks, is not the same. However, the water of the open part of the lake is characterized by a uniform distribution of the main ions both in the water area and in the depth. This can be explained by distinctive hydrochemical features of the Issyk-Kul lake. In addition, the lake has a good vertical and horizontal water circulation, constantly acting multi-directional winds and lake drainage, an extremely small amount of river runoff relative to the volume of the water mass and other factors (Kadyrov 1986).

The chemical composition of the waters in the coastal zone of the lake is not homogeneous. The degree of desalination of lake water depends on the amount of river flow, but the influence of the amount of river flow on the change in its mineralization affects a certain distance from the shore and with depth.

The results of the previous studies (Matveev 1932; Kadyrov 1986) show that cations such as Na^+ and Mg^+ and anions such as Cl^- and SO_4^{2-} are predominate for various mineralization values in the central and coastal zones of the lake. Consequently, the main chemical type of lake water is Cl–SO_4–Na–Mg that does not change. The lake water belongs to the sulfate class (Levchenko 1946, 1953). The amount of sulfates and chlorides is approximately the same. This also distinguishes Issyk-Kul lake from other large water bodies (Levchenko 1946, 1953).

The total water alkalinity is mainly due to the content of HCO_3^- and partly to CO_3^- ions. The concentration of the other ions that affect the alkalinity of water is very small. The concentration of HCO_3^- and CO_3^- ions depends on the mineralization of water and CO_2. The value of the total alkalinity averages 5.20 meq/l. Thus, the water of the lake is characterized by high alkalinity compared with other large water reservoirs of the CIS countries.

The content of all dissolved substances, including salts in water, is called salinity of water or mineralization. Mineralization is more often used to characterize the amount of dissolved substances in freshwater. The average mineralization of the lake water is 6.01 g/kg. It is one of the salty lakes in the world (Table 9.2). This mineralization says that the drainage water reservoir such as Issyk-Kul lake has a young age. This is confirmed by paleogeographical, geomorphological, and archaeological data (Korolev 1978).

Chemical composition of lake water. Water of the Issyk-Kul lake is refer to chloride-sulfate-sodium group by ionic composition. The anions $C1^-$ is 24.1% and SO_4^{2-} is 23.8%, and the cations Na^+ is 33% and to a lesser extent Mg^{2+} is 13% (Table 9.3) (Romanovsky et al. 2014). Consequently, the lake belongs to salty water bodies.

Table 9.2 Salinity of the some seas and large lakes in the world

Water body	Mineralization (g/kg)	Water body	Mineralization (g/kg)
Elton lake	265	Issyk-Kul lake	6.01
Big Salty lake (USA)	230.4	Balkash lake	3
Dead sea	205.8	Sevan lake	0.66
Iletskoye sea	155.2	Tanganika	0.438
World Ocean	35.1	Erie lake	0.134
Van lake	17.2	Chud lake	0.106
Kukunor (Qinghai)	13	Huron lake	0.105
Caspian sea	12.9	Baikal lake	0.09
Aral sea	10.5–12.5	Onega lake	0.049

Table 9.3 Average chemical composition of water in the Issyk-Kul lake

Ions	g/kg	Mg/equ	% Equ
$C1^-$	1.585	44.70	24.1
SO^{2-}	2.115	44.03	23.8
HCO^-	0.240	3.93	2.1
Sum of anions	3.940	92.66	50.0
Ca^+	0.114	5.69	3.1
Mg^+	0.294	24.18	13.0
K^+	0.068	1.74	0.9
Na^+	1.407	61.04	33.0
Sum of cations	1.883	92.65	50.0
Total	5.823	185.31	100

The average long-term chemical composition of the water of the open part of the Issyk-Kul Lake is shown in Fig. 9.2. According to the results of the study, the lake refers to brackish water reservoirs. The salinity of Issyk-Kul lake is 6.22 g/l. The largest saline lake in the Caspian Sea is salinity of up to 13 g/l, the Aral Sea, 12.5 g/l; the eastern Balkash Lake, 5.2 g/l; the World Ocean, 35.1 g/l (Romanovsky et al. 2014). In the waters of the lake, sulfates and chlorides are contained in large concentration (Fig. 9.2).

The water of the lake belongs to the sulfate class; the sulfates slightly predominate over the chlorides. The water of the open part of the lake is characterized by a uniform distribution of the main ions both in the water area and in depth, which is one of the distinctive hydrochemical features of the lake. This is because of the vertical and horizontal circulation of water, constantly acting in different directions; the drainage of the lake; the small river runoff relative to the volume of the water mass and other factors. This distinguishes Issyk-Kul from other large reservoirs. They are more or less evenly distributed. As a result, the decrease in activity of the concentration of the sulfate ions pH is higher. The ion content shows that Na^+ and Mg^+ from Cl^- and SO_4^{2-} anions predominate at various mineralization

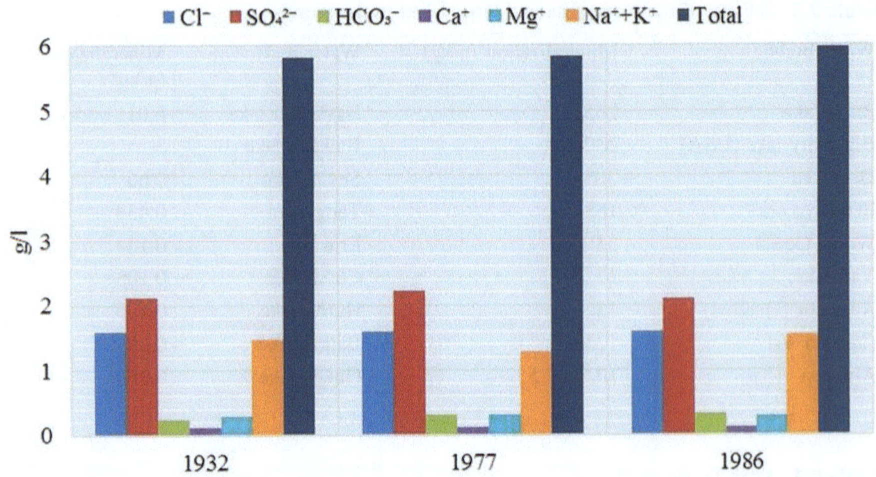

Fig. 9.2 Average long-term chemical composition of lake water

values in the coastal zones and bays of cations, and the pH of the water varied from 8.69 to 8.75. The water in the lake has an alkaline reaction, and the pH of the water was within 7.95–8.82 in 2015. The total alkalinity is mainly due to the content of HCO_3^- and partially CO_3^{2-} ions. The lake water has an alkaline reaction (pH is 8.0–8.6). It is characterized by a high content of dissolved oxygen due to good aeration and mixing of the water mass (Korolev 1978). The mineralization of water is homogeneous in the horizontal and vertical directions down to the most significant depths. This feature is due to the good mixing of the water mass.

The concentration of other ions affecting the alkalinity of the water (H_2BO_3, HPO_4^{2-} $H_2PO_{4-,}$ $HSiO_3$) is very small (Kadyrov 1986). The concentration of HCO^{3-} and CO_3^{2-} ions depends on the mineralization of water and CO_2. The content of the main ions of sodium and potassium in the waters of the lake in comparison with 1932 and 1986 shows a slight increase. A slight change in the content of ions of chlorine, sulfate, and other ions, as well as mineralization, over the water area and depth in the open part of the lake, indicates the presence of good water exchange between its individual sections and the propagation of vertical circulation of water masses to the very bottom.

Hydrochemistry of the coast. The chemical composition of the water in the lake's coastal zone is not homogeneous (Fig. 9.3). The isolation of some of them, a peculiar water and thermal regime, as well as a combination of physical, geographical, and hydrological features affect the chemical composition of water.

On the coasts, rivers flow into the lakes and carry a large number of suspended sediments. As a result, the mineralization is markedly reduced. Thus, the mineralization of water in the bay and the coast from the mouth of the river toward the open part of the lake is gradually increasing. That is, in the coasts of the lake the mineralization of water is low than in the open part of the lake. For example, the

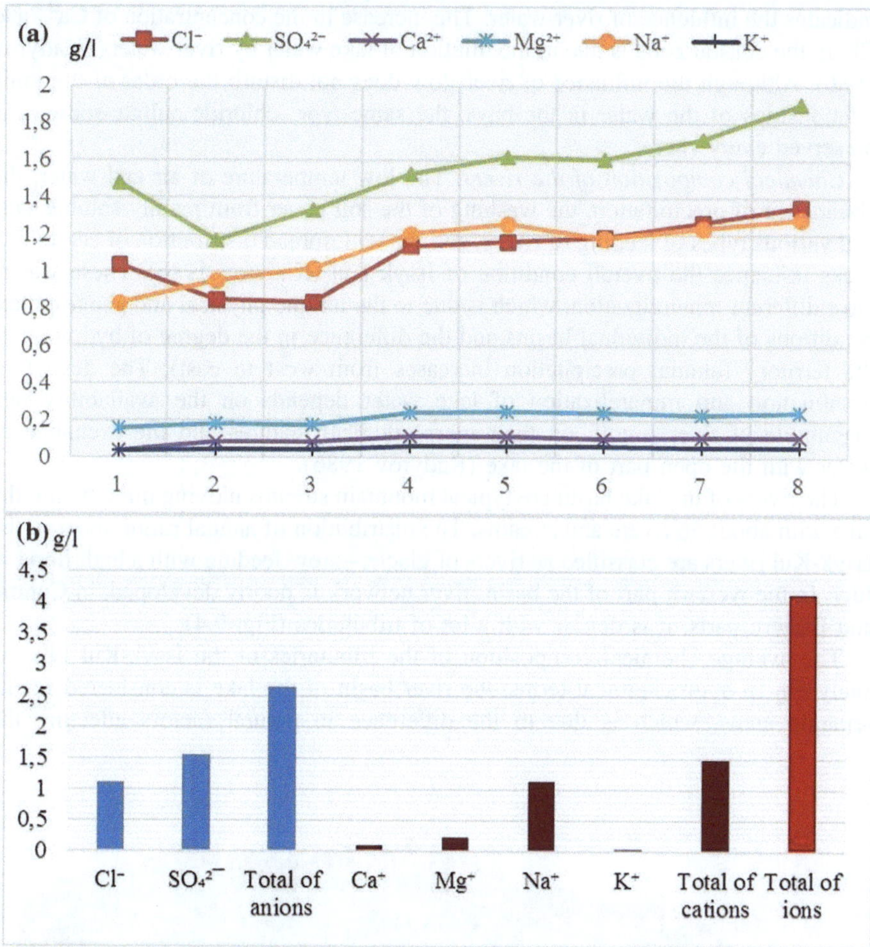

Fig. 9.3 **a** Chemical composition of the lake coast water; **b** the average value (2014)

chemical composition of the water in the open part of the lake is Cl^- 1.5 g/l and SO_4^{2-} 2.1 g/l, $Na^+ + K^+$ 1.5 g/L, while Cl^- 1.1 g/l and SO_4^{2-}—1.5 g/l, $Na^+ + K^+$—1.2 g/l in the coast of the lake. The value of mineralization of the northern coast (2.3 points), unlike other coasts, is lower than other bays, and coasts and mineralization are close to the open part of the lake (Fig. 9.3a). However, the eastern and southern coasts are characterized by high mineralization. This is due to the development of agriculture, primarily with agricultural land improvement in the coastal zone of the lake, and the entry of groundwater with a high content of chloride, sulfate, and sodium ions into the lakes (Kutseva 1980). For water in the coastal zone, the average values of the chlorine coefficients are always greater than in the open part of the lake, especially for calcium and bicarbonate ions. The content of which sharply decreases with increasing mineralization of water, which indicates the influence of river water. The increase in the concentration of Ca^{2+} and Cl^{-} in the coastal zone is due to the dilution of lake water by river waters (Kadyrov 1986). Although the influence of river flow does not disturb the order of the ionic relationships of the water in the bays, the same type, chloride sulfate-sodium, is preserved everywhere.

Chemical composition of the rivers. The low temperature of air and water, the abundance of precipitation, the washing of the soil cover from readily soluble salts and various types of feeding of rivers, and the widespread distribution of crystalline rocks influence the overall condition of Issyk-Kul (Kadyrov 1986). These waters have different mineralization, which is due to the unique physical and geographical conditions of the individual basins and the difference in the degree of hydration of the territory (annual precipitation increases from west to east). The degree of desalination and mineralization of lake water depends on the availability and magnitude of river runoff, on their morphological features and the exchange of water with the open part of the lake (Kadyrov 1986).

The rivers of the lake basin are typical mountain streams moving directly into the lake with about 56 rivers and streams. The distribution of annual runoff most of the Issyk-Kul rivers are classified as rivers of glacier–snow feeding with a high flood in July. In the western part of the basin, river network is poorly developed; in Central and Eastern parts, it is denser with a lot of tributaries (Fig. 9.4).

The average chemical composition of the tributaries of the Issyk-Kul Lake is analyzed. In hydrochemical terms, the river basin of the lake is considered for its separate areas, which is due to the difference in natural factors affecting the

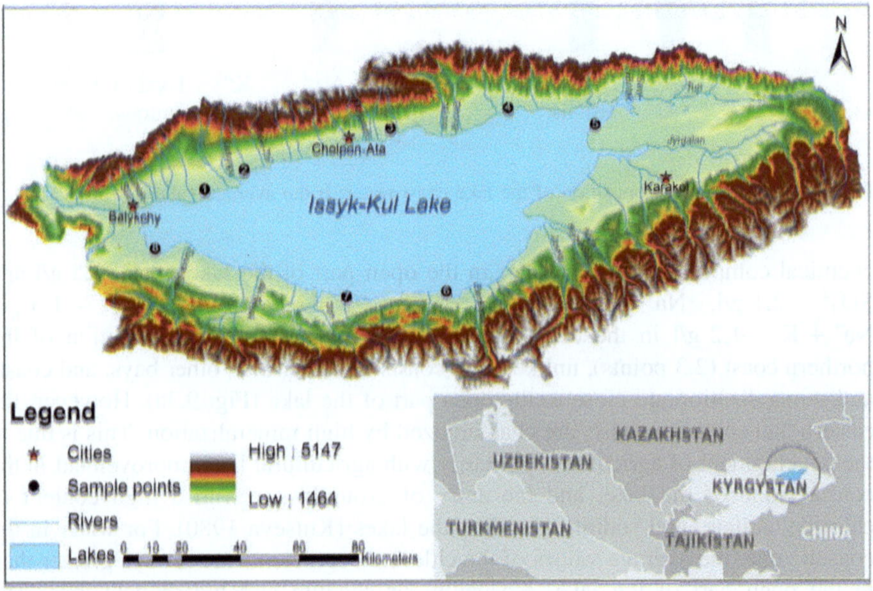

Fig. 9.4 Issyk-Kul Lake basin

formation of the chemical composition of water. The northern tributaries of the lake are rivers from the river Kabyrga up to the river Sary-Bulak, and the Chon-Ak-Suu river is the largest in the region. The greatest mineralization of the northern tributaries of the lake is characterized by the water of the Chetki-Dolonotu river, the smallest Orto-Koi-Suu (Fig. 9.5a). And in the chemical composition of the water of rivers, calcium prevails, and the concentration of the remaining ions is small.

Among the southern tributaries, the Ak-Terek, Ak-Sai, Kichi-Zhargylchak, and Sary-Bulak rivers have the highest salinity (Fig. 9.5b). The rivers Zhuuka, Ak-Terek, Barskoon are the largest rivers of the southern tributary of the lake. The chemical composition of the water of rivers is represented mainly by calcium ions.

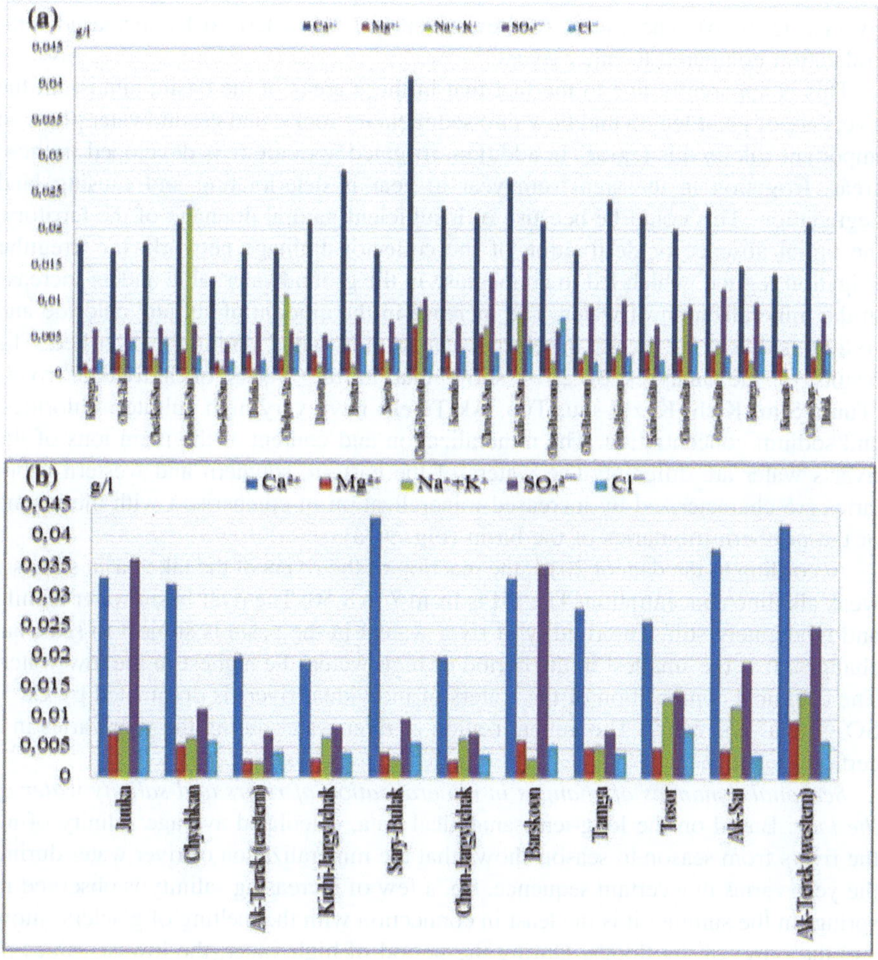

Fig. 9.5 Average chemical composition of water is Issyk-Kul lake: **a** northern tributaries, **b** southern tributaries

The concentration of the remaining ions is small. In some cases, in different seasons of the year, the water ions of rivers change.

The mineralization of the western tributaries of the lake increases in comparison with the rivers of the northern and southern tributaries. This is due to the arid climate in this area of the basin. The most mineralized water is the Tuura-Suu river, and the smallest river is Taldy-Bulak (Fig. 9.6b). The chemical composition of the water of the eastern tributaries is much higher than that of other areas of the lake (Fig. 9.6a). In the eastern region, the largest rivers of the Issyk-Kul Lake basin are concentrated from the Tup river to the Kichi-Kyzyl-Suu river. The mineralization and content of the main ions of the river's water are different. However, in the chemical composition of the river's water, Ca^{2+} and SO_4^{2-} predominate (Fig. 9.6). The Ak-Suu, Jyrgalan, Juuka, Chon-Zhargylchak, and Tup rivers have the highest concentration of ions, and other SO_4^{2-}, Cl^-, Na^+ contents are higher than other rivers (Fig. 9.7b). The waters of these rivers are characterized by increased mineralization compared to other rivers.

This is apparently due to the fact that in these areas of the basin, almost all the rivers are of great length and flow into sedimentary rocks, and groundwater plays an important role in this region. In addition, irrigated agriculture is developed in these areas. Irrigation in the area from year to year is deteriorating and causing land degradation. This could be because of insufficient natural drainage of the territory, the initial absence or destruction of the collector-drainage network, the irregular irrigation regime, which led to an increase in the groundwater table and an increase in the mineralization of water. The increase in the amount of sulfate, chloride and sodium is due to the receipt of return water from irrigated land in the river bed. The results of the analyses of 2014 show that in the waters of individual rivers (Tuura-Suu, Kichi-Kyzyl-suu, Tup, Ak-Terek) have very high sulfates, chlorides, and sodium concentration. The mineralization and content of the main ions of the river's water are different. The waters of the eastern, southern and western tributaries are characterized by increased mineralization in comparison with the waters of the northern tributaries of the basin (Fig. 9.7a).

According to the data of 2014, the reaction of the rivers of the lake basin shows a weak alkaline concentration. The pH is from 7 to 8.50. The river basin water is mild and moderately stiff. The rigidity of river waters in the basin is subject to seasonal changes. It is the smallest in the period of high water, the highest in the low water. The chemical composition of the waters of individual rivers is dominated by Ca^{2+}, SO_4^{2-}, Na^+ and Mg^{2+}. The mineralization of river water during the year varies in a certain sequence.

Seasonal dynamics of changes in mineralization of rivers and salinity water of the lake. Based on the long-term analytical data, calculated average salinity of all the rivers from season to season shows that the mineralization of river water during the year varies in a certain sequence. So, a few of increasing salinity is observed in spring. In the summer it is the least in connection with the melting of glaciers, snow and the onset of the floods. During the period of high water, the lowest values of mineralization and concentration of the main ions are observed. In autumn and winter, a minimum water runoff is observed on the rivers, formed mainly by

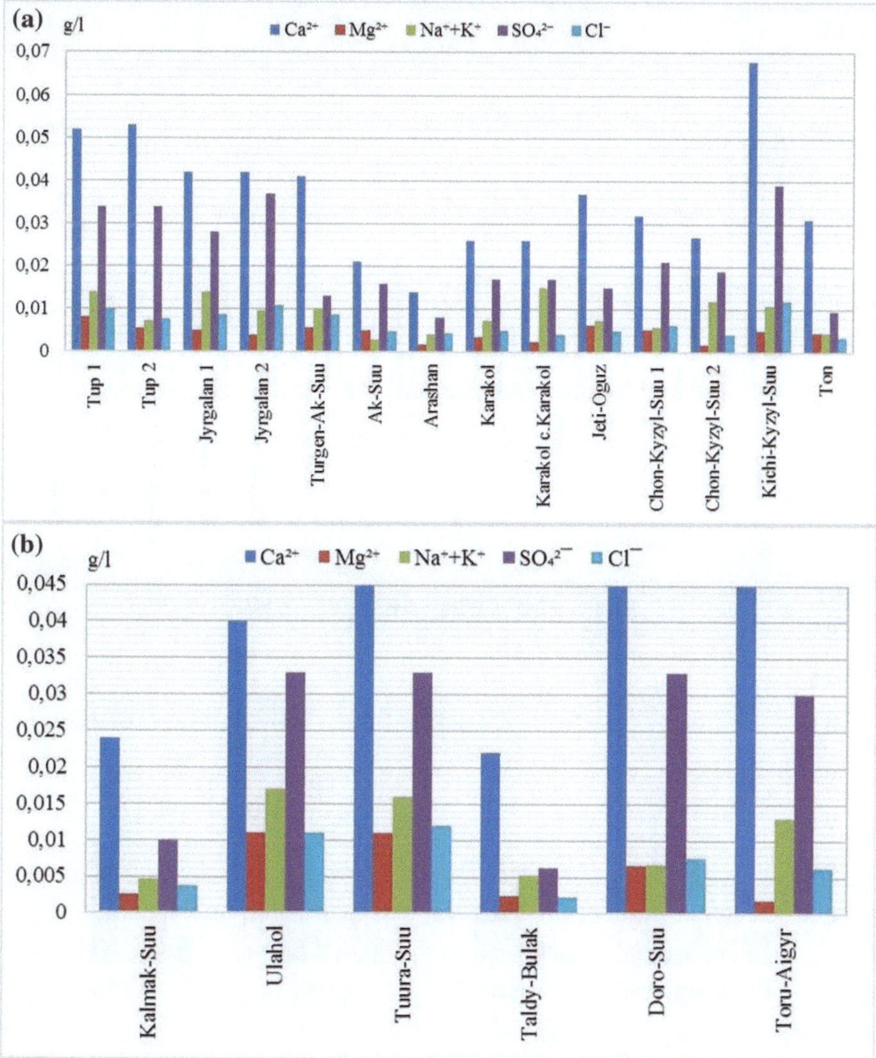

Fig. 9.6 Average chemical composition of water is Issyk-Kul lake: **a** eastern tributaries, **b** western tributaries

mineralized groundwater (Fig. 9.8). In the period of minimum expenditure, a sharp increase in mineralization and the main ions is observed. The saline balance of the lake is formed by salts coming from river flow, atmospheric precipitation, groundwater, which as a result precipitates. The main quantity of salts enters the lake with groundwater 72%, with river waters 22% and precipitation 6% (Romanovsky et al. 2014). Half of the incoming salts (49.8%) replenishes the salt

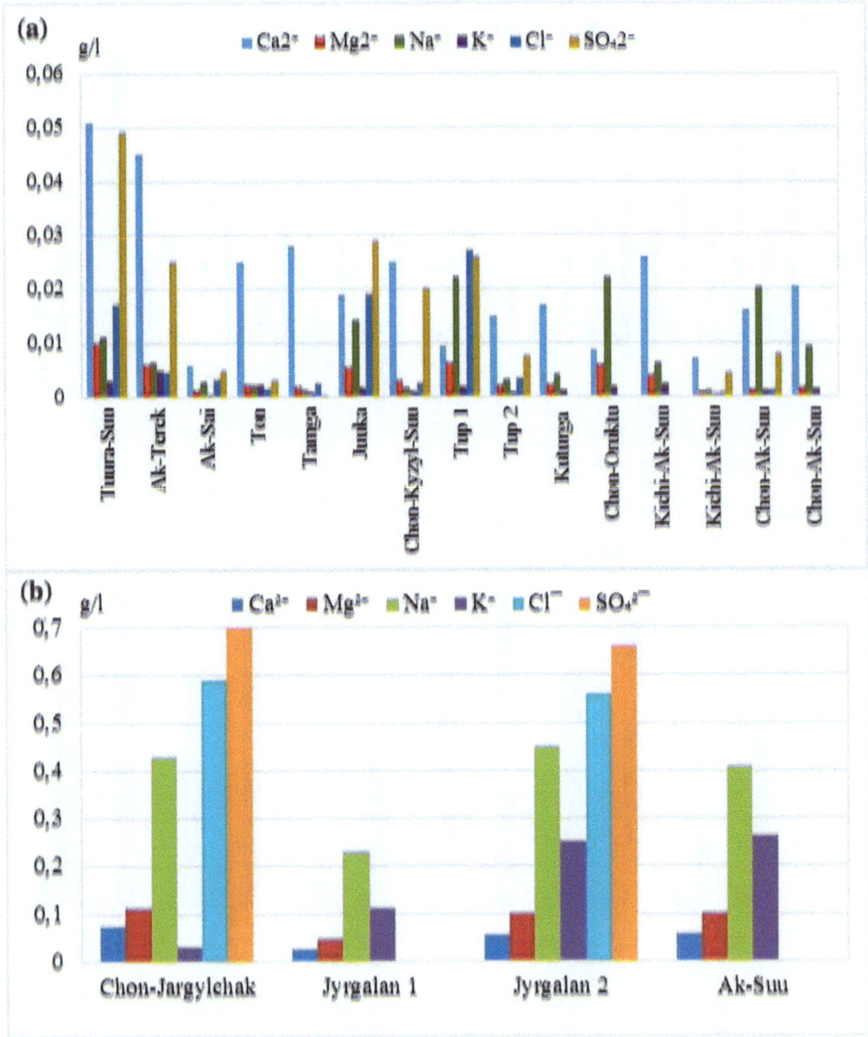

Fig. 9.7 Chemical composition of river waters (2014)

reserve of the lake, the remaining part (50.2%) is then precipitated. The content in water for all dissolved substances, and not only salts is called salinity.

The salinity of the lake water from 1928 to 2014 increased from 5.823 to 6.22 g/l respectively (Fig. 9.9). The concentration of salts in the Issyk-Kul Lake increases. The increase in salinity of water in the lake is determined by the fact that the incoming part of the water balance is much less than the evaporation. Since the removal of salts during evaporation is insignificant. Consequently, there is an accumulation of salts in the lake, which is expressed in an increase of the

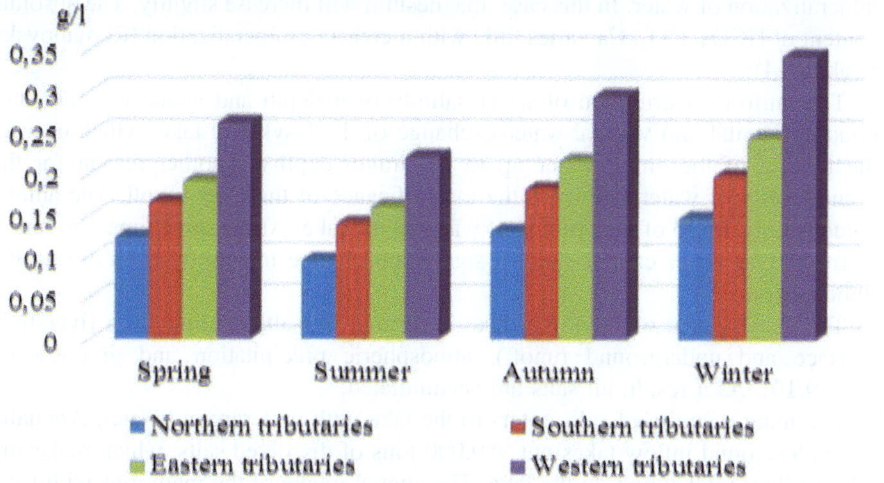

Fig. 9.8 Seasonal dynamics of changes in mineralization of river waters

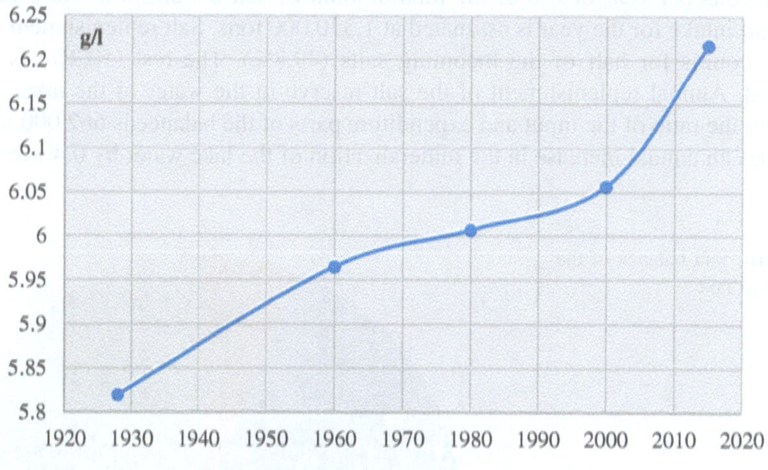

Fig. 9.9 Salinity of water in the Issyk-Kul lake

mineralization of water. In the lake, like evaporation, the removal of salts is insignificant, accumulation of salts occurs, which is expressed as an increase in the mineralization of water.

The concentration of individual chemicals and salinity of water in Issyk-Kul lake is distributed uniformly in different parts of the open lake and at different depths. The lowest values of salinity are observed at the mouths of the largest rivers such as Zhyrgalan and Tyup, where it decreases to 0.6 g/kg. The concentration of SO_4, Cl and alkali metals increases sharply in the coastal zone while increasing mineralization of water. In this case, magnesium will increase slightly. The absolute content of HCO_3, CO_3, Ca varies little with increasing mineralization (Romanovsky et al. 2014).

The uniform distribution of water salinity over depth and water area indicates good horizontal and vertical water exchange of the Issyk-Kul lake, which ensures the mixing of the entire water up to maximum depths. Another reason for the homogeneity of water salinity is the insignificance of the river runoff. The annual value is only 1/530 of the volume of water in the lake. At the same time, the runoff of fresh river water cannot cause a significant change in salinity even during the flood period.

The salt balance of Issyk-Kul lake is formed by salts coming from river flow (surface and underground runoff), atmospheric precipitation and groundwater (Fig. 9.10). As a result, all salts are accumulated.

The main quantity of salts enters to the lake with underground water. Annually the underground inflow takes out 959,000 tons of dissolved salts, which makes up 72% of their total supply to the lake. The annual intake of the main ions with river waters (including irretrievable water loss for irrigation) accounts for 296,000 tons or 22% of the inflow amount. The flow of salts from the atmospheric precipitation is 75,000 tons per year or 6% of the total amount of salt coming into the lake. The total salt intake for the year is estimated at 1,330,000 tons. Salt replenishment of the lake accounts for half of the incoming salts (49.8%). The rest (50.2%) is accumulated. Annual replenishment of the salt reserve in the water of the lake, calculated by the ratio of the input and expenditure parts of the balance is 662,000 tons. It provides an annual increase in the mineralization of the lake water by 0.4 mg/l. The

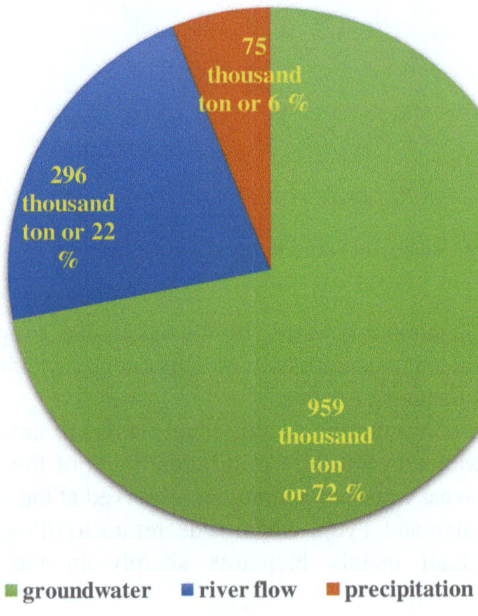

Fig. 9.10 Salt balance of the Issyk-Kul lake

amount of accumulation of salts in the lake determined by the change in the reserve of substances for the period 1928–1978 is 1,110,000 tons, which corresponds to an annual increase in mineralization of water by 0.6 mg/l (Romanovsky et al. 2014).

The change in the chemical composition of natural waters depends on the time and depending on the chemical, physical, and biological processes and anthropogenic factors (intensive tourism and agriculture). The Issyk-Kul Lake is brackish. The salinity of Issyk-Kul lake is 6.22 g/l. The predominant ions of lake water are sulfates, chlorides, sodium, and magnesium. In the chemical composition of water, sulfates predominate over chlorides, which is a feature of the continental origin of the salt composition of the lake fed by high-mountain rivers. The cations are dominated by sodium and magnesium ions. Thus, the water in the lake refers to the chloride sulfate-sodium-magnesium type. The water in the lake has an alkaline reaction, and the pH of the water is within 7.95–8.82. The hydrochemistry of the coast is not uniform. This is due to the water and thermal regimes, physical–geographical and hydrological features of the lake coast. And so the mineralization of the coast water gradually increases toward the open part of the lake, that is, the mineralization of water is low on the coasts than in the open part of the lake. This is due to suspended sediment transported by rivers in large quantities to the lake. The eastern and southern coasts are characterized by high mineralization. This is because to the development of agriculture and the introduction of lakes of groundwater with a high content of chloride, sulfate, and sodium ions. It is established that the formation of the chemical composition of the water of the rivers of the basin of Issyk-Kul lake is subject to vertical zones, and their hydrochemical regime depends on the change of the water regime. The western tributaries of the lake have a high mineralization in comparison with the northern and southern tributaries. This is due to the arid climate in this area of the basin. The chemical composition of the water of the Eastern tributaries is much higher than other tributaries of the lake. The greatest concentration of ions is in the rivers Ak-Suu, Jyrgalan, Juuka, Chon-Jargylchak, Tup owning to the fact that almost all these rivers are of great length and flow into sedimentary rocks. Also, irrigated agriculture is developed in these areas. And in the chemical composition of the rivers water, Ca^{2+} and SO_4^{2-} predominate, and the concentration of the remaining ions is small. The mineralization of river water during the year varies in a certain sequence. In autumn and winter, there is a sharp increase in mineralization due to minimal water flow. And in the summer, the smallest mineralization is observed due to the melting of glaciers, snow and the onset of floods. In the spring, the mineralization slightly increases.

9.4 Use of Water Resources of the Issyk-Kul Lake Basin and Sustainable Water Management

Agricultural irrigation practices, rapid urbanization, and climate change increase pressure on water resources. The world population is projected to be 9.7 billion by 2050, an increase of about 2 billion from today (7 billion); this world population may increase its pressure on both water quality and quantity (Wilson et al. 2005; Vörösmarty et al. 2000). Worldwide water utilities experience seasonal fluctuations in their aggregate consumption levels and these variations generally cause specialized maintenance and administrative schedules to develop individual utilities as a mean of optimizing the natural resources (Fullerton and Elias 2004). Water from rivers, lakes, and aquifers is mostly used for irrigation, bathing, washing, and other human needs. Increasing exploitation of natural resources, inappropriate land-use practices, and uncoordinated sectoral policies and development activities in lake basins impair various important functions of water. Therefore, appropriate approaches for water management are urgently needed (Rost et al. 2008; Avramoski 2004; Koop and van Leeuwen 2015). Natural forces and human activities have been reported to damage water resources. Despite increasing public awareness on the role of water in human health and development, water scarcity and pollution are experienced due to lack of consistency and consensus of economic and political willingness on water resources management (Kundzewicz et al. 2008; Piao et al. 2010). Water plays an important role in the productive process, since this natural resource can become either a limiting factor for development or the driving force behind economic growth. Therefore, it is necessary to know the situation regarding water resources and its relationship with the regional productive process (Velazquez 2006). Water size and salinity fluctuate whenever the balance between hydrological inputs (precipitation, surface runoff, and groundwater inflows) and outputs (evaporation and seepage losses) changes due to seasonal and climatic variation or anthropogenic activities (Jellison et al. 2016; Ramana et al. 2013). While the impact of water diversions for irrigated agriculture and urban consumption on several large salt lakes (Aral Sea, Dead Sea, Mono Lake) has been widely publicized, the global extent and rapidity with which saline lakes are being impacted throughout the world has yet to be fully appreciated (Polunin 2008; Atadjanov et al. 2012).

Kyrgyzstan is a Central Asian country with the advantage of having water resources fully formed in its own territory. Kyrgyzstan has significant resources of underground and surface waters, stocks of which are in the rivers, glaciers, and eternal snow arrays. The country has more than 3500 rivers and streams, which belong to the main basins; the Syrdarya river, the Amudarya, Chu, Talas, Tarim, and Issyk-Kul lake (Klerx and Imanackunov 2002; Mamatkanov et al. 2006; Asykulov 2002). On its territory is the Issyk-Kul lake basin formed by a plurality of streams, of which about 123 are used for irrigation purposes. Surface water is characterized by sediment, salts, and fine sediments transported by the lake surrounding rivers. Mineralization of surface waters in general is less than 1 g/l, which makes it suitable for irrigation (Giralt et al. 2002; Vollmer et al. 2002a, b). The

Issyk-Kul Lake has been subject to changes in its water level, increasing sediments, and other anthropogenic activities which changed its physical-chemical properties (Gavshin et al. 2004). Previous studies conducted at Issyk-Kul Lake (Vollmer et al. 2002a, b; Wang et al. 2006; Tsigelnaya 1995) have identified its main problems related to water pollution and its increasing water demand, which in turn, threaten sustainable use of the lake, fisheries, glacial retreat, agriculture, water diversions, biodiversity, tourism, and its biosphere reserve. This shows that Issyk-Kul Lake Basin is under increasing pressure which needs accurate and sustainable management for its continuous use. Therefore, the objectives of this study are to (1) consider historical water consumption and its main users and (2) provide suggestions, based on the results, regarding how the lake water can be sustainably productive and usable.

Issyk-Kul Lake has been subject to several water consumers, including but not limited to construction, irrigation, and recreation. These water uses led to variation in the lake's water level being headed by agricultural irrigation water demand, which was 43% in 1998 of the total lake tributaries (Gavshin et al. 2004), against today's increase, which is more than 70% compared to other lake. The lake registered little seasonal variation of water levels from 1606.73 m in 1980 to 1606.29 m in 1999, while from 2000, the water level slightly increased until 2011 (1607.02 m) and ended in 2012 with small decreasing numbers (1606.99 m). These alterations of water level can result from the lake water uses and the system under which the water is used and/or climate change, which is reported to affect water resources.

The main water consumers are agricultural irrigation, manufacture, and household. The agriculture is still the highest water consumer compared to manufacture and household. Long-term dynamics of water consumption reveal decreasing number (Table 9.4). This reduction in water consumption could be a result of the collapse of the Soviet Union in 1989–1992, where water users decreased and the remaining were deployed in different places, meaning that they used other water sources (Propastin 2013; Vollmer et al. 2002a, b). However, as the population in the lake basin grows, consequently water demand rises, and leads to a water level decrease which was detected and controlled by government policies through using other lakes and basins, managing the lake's streams, soil erosion control measures, rainwater harvesting, etc. (Giralt et al. 2002; Gavshin et al. 2004; Jailoobayev et al. 2009; Fereres and Soriano 2007). In addition, agriculture is among the world's most significant water consumers and its water consumption is projected to keep increasing as long as the human population grows, where adaptive policies are to be regarded. Moreover, the population located near Issyk-Kul Lake grew from 441.3 to 458.5 thousand in 2010 and 2014, respectively (Antwi-Agyei et al. 2012). This expresses increasing food demand, household and other development activities which require availability and access to water. For the problem to be dealt with, it is proposed to interchangeably use the lake and rainfall water (Pereira Filho et al. 2010; Viala 2008). Despite the fact that integrated water resources management action is being taken, it is still mandatory to consider climate change, global population growth, and other increasing water demands for its sustainable

Table 9.4 Lake water consumers for period of 1980–2014 (million m^3) (Alymkulova et al. 2016)

Years	Agriculture	Manufacture	Household
1980–1989	2029.42	17.74	27.02
1990–1999	981	9.46	18.06
2000–2014	461.76	8.74	16.55

management (Angelakis and Zheng 2015). Furthermore, even though the results of this study revealed decreasing water demand (Table 9.4), to mitigate and adapt to the impact of climate change on water resources, particularly at Issyk-Kul lake basin, it is good to consider the hydrological weather-related changes and adapt accordingly.

Although agriculture water demand showed decreasing numbers (Table 9.4), its water demand has been reported (Piao et al. 2010; Pereira et al. 2002) to change over time with great likelihood of increasing water demand depending on the type of crops grown. In addition, the types of irrigation like flooding or sprinkler cause high volume water consumption, because they are associated with environmental effects like water salinization, while, drip irrigation is suggested for both quantity and quality of water management, as it does not consume much water but helps to reach crops with minimum total water use (Afolayan et al. 2014; Lamek Nahayo and Zhao 2016). Therefore, it is good to consider the best irrigation practices which favor water management at Issyk-Kul Lake Basin.

The Issyk-Kul lake has been subject to changes in its water level, increasing sediments and other anthropogenic activities such as expansion of agricultural irrigated lands and rise of recreational, household, and industrial water demand (Vollmer et al. 2002a, b; Wang et al. 2006; Tsigelnaya 1995). These threaten the lake fisheries, glacial, water quality, biodiversity, biosphere and other socio-economic and environmental services and call for appropriate measures for its sustainable management and use. Water demand at Issyk-Kul Lake Basin reveals decreasing numbers, which could be a result of the region's background and the lake water management measures adopted (Propastin 2013; Vollmer et al. 2002a, b). However, as the basin is used for many other purposes, such as ecological or socio-economic (Wang et al. 2006; Fereres and Soriano 2007). This exposes the lake to natural and anthropogenic forces, meaning that whatever measures are taken for its management, there is great need to consider each and every demand and/or use to sustainably manage the lake water due to its role in the sustainable socio-economic development of the region.

Sustainable development, socio-economic development, healthy ecosystem, and human survival are rooted in water availability. Water helps in hosting and preserving development activities upon which human depend in daily life. It also serves as the link between the climate system, environment, and human society (Sapkota et al. 2014; Mancosu et al. 2015). Water, if efficiently and equitably managed, can help to achieve sustainable development. However, it has been reported that water scarcity is among the most significant development challenges worldwide, and access to water is not equal (Braden et al. 2013; Griggs et al. 2013).

The United Nation's (UN) overarching goal is to secure sustainable water for all, due to the role of water in ensuring people's health, prosperity, resilient communities, equitable societies, and protected ecosystems, which in turn are the basis of sustainable development (Vonk and Shackelford 2012). Nevertheless, this cannot be realized without universal access to water, its sustainable management and effective governance (Libert et al. 2008). Water management is also a serious problem in Eastern Asia, where a number of problems, compounding the issue include but not limited to floods, increased salinization and a loss of freshwater reserves due to agriculture along with pollution from sewage and industry have severely damaged the water supply with dangerous effects on the health of local people (Törnqvist et al. 2011).

For sustainable water management of Issyk-Kul Lake Basin, whose water is used for many purposes. There is great need to take into consideration: variation of water resources; the changing climate which causes fluctuations in intensity and frequency of precipitation, one of the sources of the lake's water; managing the changes (increase/decrease) in the demand and uses of the lake water along with decision makers and common community understanding on the use, conservation and protection of the lake.

9.5 Impact of Climate Change on the Issyk-Kul Lake Basin

Lake level changes occurred as a result of regional climate change and recent human activities (Mason et al. 1994; Salamat et al. 2015). Lakes are the main components in any hydrological cycle and interact with many aspects of ecosystems and human activities in a catchment area. Since lakes are very sensitive to changes in nature, they serve as an important indication of global climate changes and regional environment variations (Bai et al. 2011). The water level of Issyk-Kul Lake was not stable in previous geologic periods as regressings and transgressings were noted (Berdovskaya and Egorov 1986). The radiocarbon dating method showed that 500–700 years ago the water level of Issyk-Kul Lake was 2–5 m lower than the present level (Trofimov and Grigina 1979). Even in the first half of the nineteenth century, the water level of the lake was close to its maximum (i.e., 1620 m asl) and was about 13 m higher than at present. Many studies about lake variations in Central Asia have been reported for a while (Savvaitova and Petr 1992; Aladin and Plotnikov 1993; Stanev et al. 2004; Lehner and D ll 2004; Kezer and Matsuyama 2006; Ma et al. 2007, 2011, 2012), and the causes of water level fluctuations of Issyk-Kul Lake have been discussed by many scientists in the past (Shnitnikov 1980; Tsigelnaya 1995; Klerx and Imanackunov 2002; Romanovsky 2002a; Peeters et al. 2003; Hofmann et al. 2008; Wang et al. 2010; Bai et al. 2011).

There are two categories of hydrodynamic processes which lead to water level changes: (1) short-term fluctuations lasting from seconds to hours that are caused by oscillations and traveling surface waves, and (2) long-term fluctuations lasting from days to years (Hofmann et al. 2008). The long-term fluctuations of the water level of the lake due to climate change are mostly considered and analyzed (Salamat et al. 2015). According to the research results, the causes and processes for the decrease in water level from the mid of twentieth century were identified as well as the cause of this change was due to long-term changes in air temperature or precipitation or the result of the human activities was studied (Salamat et al. 2015). There are various factors that affect the decrease of water level of the Lake. The increase in human population and the expansion of agricultural lands—and consequently the enlargement of the irrigation system and its channels in the basin—caused an increase in the water discharged into irrigated plains, thus allowing less water to flow into the lakes (Sehring 2007; Taft et al. 2011). The relationship between the total runoff and the total number of tributaries of a river showed that the correlation coefficient was more than 0.84. At the present time, the trend of total surface runoff that the lake receives is positive, which means that the water level of Issyk-Kul Lake is rising (Salamat et al. 2015).

Lake Issyk-Kul is drain less, salty (5.90‰), and does not ice cover during winter. The number of all inflows to Issyk-Kul Lake is 110, with includes 80 large rivers. The rivers in the Issyk-Kul catchment are fed by snow and glaciers, with the exception of the Tiup River which is the second major river of the basin. This river fed by snow in as much as its catchment occupies the most humid east area of the lake (Kadyrov 1986).

The coastal area has a temperate maritime climate. The Issyk-Kul Lake has a moderate effect on all watershed basins. The western part of the Issyk-Kul Basin is dry with a low amount of precipitation (115 mm per year), especially during winter, when regular and strong winds are common (Gronskaya 1983; Heinicke 2003; Taft et al. 2011). In contrast, the eastern part of the basin has an average precipitation of about 600–800 mm (Mamatkanov et al. 2006). In spite of the natural patterns of weather conditions, which are typical for the entire region of Central Asia, there is a local atmospheric circulation in Issyk-Kul Lake Basin that has a significant impact on the precipitation distribution in this basin (Gronskaya 1983). However, in recent years, the amount of rainfall has increased, which can be a result of the impact of global warming that leads to changes in the local climate (Huntington 2006). The high mountains that surround Issyk-Kul Lake Basin create barriers to the free flow of air masses that would bring moisture and cold air from the west, northwest, and north, especially during the cold period. Consequently, this physical–geographical feature of the basin has an influence on the spatial distribution of precipitation in this basin (Aizen et al. 1997).

Data of air temperature and precipitation for the periods of 1880–2010 and 1930–2010 from the weather stations (WSs) such as Karakol (Prezheval'sk), Balykchy (Rybachie), Cholpon-Ata and Tian Shan located in the lake basin were analyzed to ascertain the fact of the changing climate. In addition, in order to find out the correlation among climatic factors: air temperature and precipitation and runoff, the Gray correlation analysis and Pearson cross-correlation test was used in the study. Those correlations showed the relationship between two or more variables aimed to express the correlation degrees and properties among variables (Salamat et al. 2015; Xu et al. 2006; Ling et al. 2011; Li et al. 2012). As well as nonparametric Mann–Kendall monotonic test was used in trend analyses, and abrupt changes were determined by a Mann–Whitney test. These tests have been widely used in hydrologic studies (Salamat et al. 2015).

According to the study analysis and results, the increasing trend in air temperature and in the quantity of precipitation was observed in above mentioned WSs. The average annual air temperature was rised. Local climate warming has an impact on increasing air temperatures, due to winter warming and summer dehydration, and a significant increase in average annual air temperatures and a negligible increase in the amount of precipitation accompany a deficiency of moisture in foothills (Morrison et al. 2002). If such changes continue into the future, the water content of the rivers will be reduced (Siegfried et al. 2012).

Database analyses of WSs located at different altitudes around Issyk-Kul Lake indicate rising air temperatures for the last 80 years. The temperature trend in lower zone showed a rise of 1–1.5 °C/80 years (Figs. 9.11a and 9.12a) (Salamat et al. 2015).

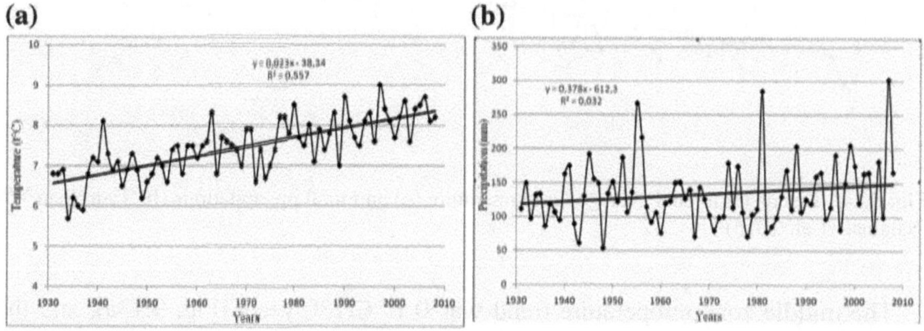

Fig. 9.11 Trends for annual average air temperature (**a**) and total precipitations (**b**) Balykchy WS (Salamat et al. 2015)

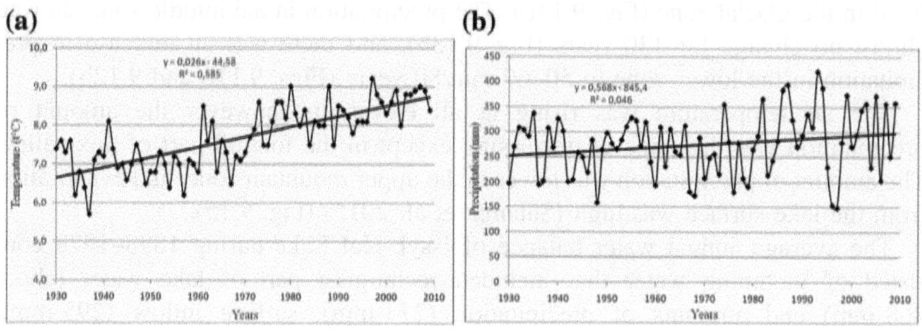

Fig. 9.12 Trends for annual average air temperature (**a**) and total precipitations (**b**) Cholpon-Ata WS (Salamat et al. 2015)

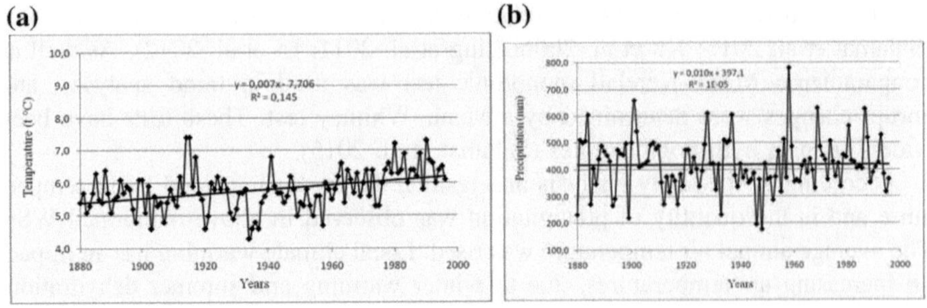

Fig. 9.13 Trends for annual average air temperature (**a**) and total precipitations (**b**) Karakol WS (Salamat et al. 2015)

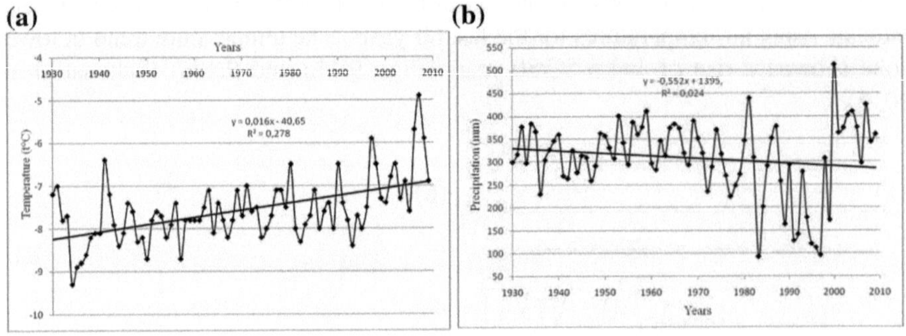

Fig. 9.14 Trends for annual average air temperature (**a**) and total precipitations (**b**) Tian Shan WS (Salamat et al. 2015)

The middle zone temperature trend was 0.7 °C/130 years (Fig. 9.13a), and the lowest trend of rising temperatures was approximately 1–1.2 °C/80 years in the glacial–nival zone (Fig. 9.14a). The air temperature increased more rapidly in the foothills than other places, but the precipitation trends were totally opposite: A decrease of precipitation to 45 mm/80 years over the entire monitoring period was fixed in the glacial zone (Fig. 9.14b). The precipitation in the middle zone showed almost no change for 130 years (Fig. 9.13b), and there was an increase in precipitation in the lower zone to 50–60 mm/80 years (Figs. 9.11b and 9.12b).

The air temperature was rising at all elevations, however the amount of precipitation was the same or decreasing except in the foothill part of this valley. The amount of precipitation was lower in the upper mountain zone, and evaporation from the lake surface was high (Salamat et al. 2015) (Fig. 9.15).

The average annual water balance of Issyk-Kul Lake during 1898–1978 consisted of incoming water that included recharging part of lake water (about 868 mm) and amounts of precipitation (274 mm), surface inflow (295 mm), underground tributary discharge (299 mm), and water consumption (914 mm), including evaporation (836 mm) and irrigation (78 mm). In this case, the rate of

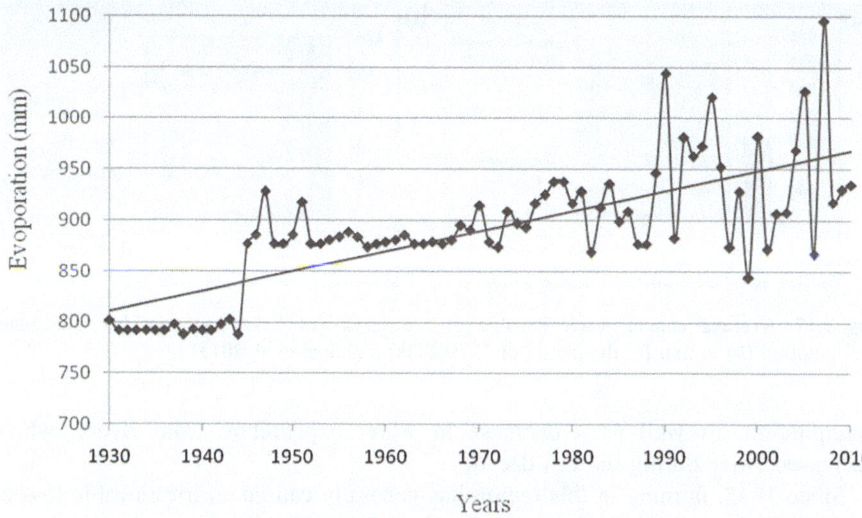

Fig. 9.15 Evaporation from Issyk-Kul Lake surface (1930–2010) (Salamat et al. 2015)

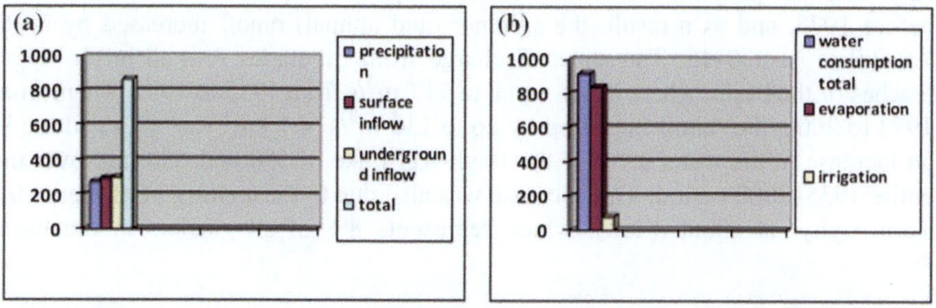

Fig. 9.16 Average annual water balance of the Issyk-Kul Lake: recharge (**a**) and water consumption (**b**) in mm for the period of 1898–1978 (Salamat et al. 2015)

decrease in water level was about 50 mm per year (Fig. 9.16). The water balance of Issyk-Kul Lake from 1979 to 2000 was 941 mm, including precipitation (301 mm) and a combined amount of surface influx and groundwater tributary inflow of 640 mm. But, the lake's water discharge during this time was 996 mm, which included irrevocable water use for irrigation from rivers that flow into the lake (47 mm) and evaporation (949 mm). In this case, the water level of Issyk-Kul Lake had an average declining water level of 25 mm year^{-1} from 1980 to 2000 and, for the whole 20-year period, a total decreasing water level of 30 cm (Fig. 9.17). Within the study period, there were five periods when a positive ratio of water balance was observed: 1981, 1987–1989, 1993, 1994, and 1999–2005. This positive increase involved both an increase in the water contribution from rivers and

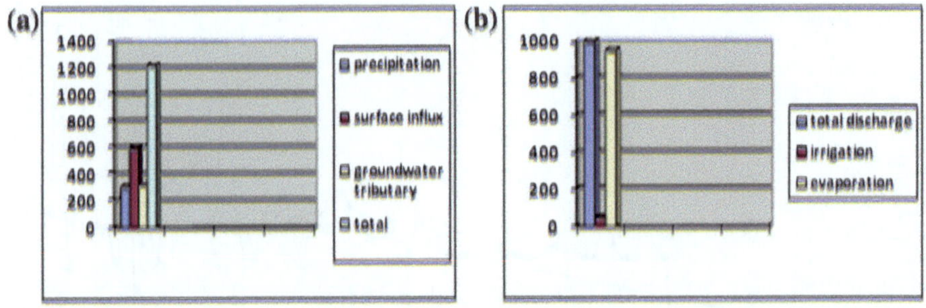

Fig. 9.17 Average annual water balance of the Issyk-Kul Lake: recharge (**a**) and water consumption (**b**) in mm for the period of 1979–2000 (Salamat et al. 2015)

precipitation, as well as a decrease in water expenditure from rivers, which decreased twice during the last decade.

Since 1945, farming in this region has generally caused an irretrievable loss of water from rivers and ultimately the lake (Fig. 9.18). Temperature and precipitation in the region began to change after 1972 (Figs. 9.11, 9.12, 9.13, and 9.14). For example in the ice-snow-fed Juuku and Ak-suu, river discharge during the month of July over the period 1973–2000 increased by up to 2.2–3.0 m^3/s compared to before 1973, and as a result, the summer (and annual) runoff increased by up to 1.0–3.0 m^3/s or 6–15. The water discharge from tributaries formed in the upper reaches of the basin, where it was equal to 117 m^3/s from 1935 to 1972. Then, from 1973 to 2000, the runoff increased by up to 130 m^3/s (4.1 km^3 $year^{-1}$), resulting in an increase in the water level of the Issyk-Kul Lake of 590 and 650 mm over the entire 1935–2000 period. This increase was also due to the melting of glaciers, like Toru-Aighyr at about 4 km^3, which represents 8% of all glaciers in the basin

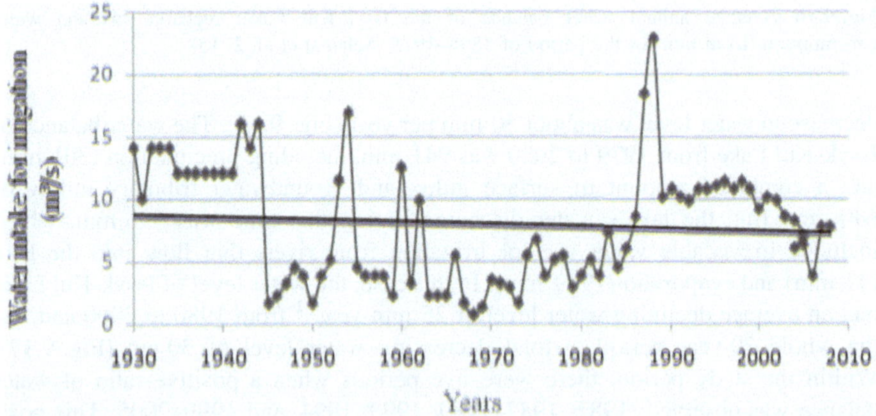

Fig. 9.18 Total summary water withdrawal from rivers for irrigation in Issyk-Kul Lake Basin (Salamat et al. 2015)

(Mamatkanov et al. 2006). The warming climate has changed the regime of river runoff (Mamatkanov et al. 2006), and at present, the intensity of glacier melt has led to an increase in runoff into the Issyk-Kul Basin (Dikikh 2000; Romanovsky 2002a). However, continuing warming can lead to a considerable shrinkage of the glaciers and consequently to a decrease of river runoff during the summer period (Kazimir et al. 2011; Tynybekov 2011). In the Tup River, the maximum water from river discharge shifted from May to April and reached 2.7 m^3/s, while May saw a decrease to 6.0 m^3/s. This shift was also accompanied by a decline in the annual average river discharge. In the Toru-Aighyr river basin, there are no longer any glaciers (Mamatkanov et al. 2006). This river is fed by snow and ice, but nowadays, the participation of glaciers in the Toru-Aighyr River runoff has been reduced to a negligible amount. The increase in river runoff into IKB is due to the melting of ancient glaciers, but this process has only a temporary nature. Ultimately, the continual warming of the local climate around IKB will lead to the shrinkage of glacial resources, and as a result, a decrease in the annual summer runoff in the future. The basin of the Toru-Aighyr River will show the true results of climate warming with the temporary increase in water discharge followed by a reduction in river volume (Dikikh 2000). The threat from decreasing summer runoff is highest for rivers found on the sunny southern slopes of the Kungei Ala-Too Range. Except for the glaciers feeding the Toru-Aighyr River, which are located on south slopes of the Kungei Ala-Too Range, there are several glaciers that are located at same position and at the same level. Unfortunately, the process of glacial shrinkage was also very fast in the Teskei Ala-Too Range, despite the fact that these glaciers generally have a northern exposure (Aizen et al. 1997; Bolch et al. 2011). In the future, the decrease in the amount of river runoff from increased air temperatures, as well as an increase in lake evaporation, will have a negative influence on Issyk-Kul Lake and will affect the agriculture in the basin that is based on irrigation. The melting of glaciers will decrease in volume, which, in turn, will have a negative influence on their annual average runoff as a whole (Wang et al. 2010). And as the temperature continues to rise, this reduction in water flow will impact the Toru-Aighyr, Tiup, Zhyrgalan, and other rivers in this region at an alarming rate. Climate change increases temperature and precipitation, thus accelerating the glacial/snow melting process and the hydrological cycle in the region (Ling et al. 2011) The cycle of water level fluctuations of the Issyk-Kul Lake were negligible during individual years; however, over a 10-year period, these fluctuations were significant, as in the water level increase noted in 1900–1910 (Shnitnikov 1979). Through geologic history of the region, the water level of Issyk-Kul Lake has not been stable, showing periodic times of regression and transgression. Since the middle of the nineteenth century, the water level of IKL has declined, although through the milieu of this steady fall, there were also periodic occurrences of increases, as in 1900–1910 when the water level increased up to 1.5 m, in 1956–1960 it rose up to 32 cm, and in 1998–2003 the level rose to 56 cm (Fig. 9.19). Through observations of data gathered from hydrologic stations located around the lake, the water level of Issyk-Kul was shown to decrease about 2.75 m from 1927 to 1999, while in recent years, the level has been increasing (Fig. 9.19).

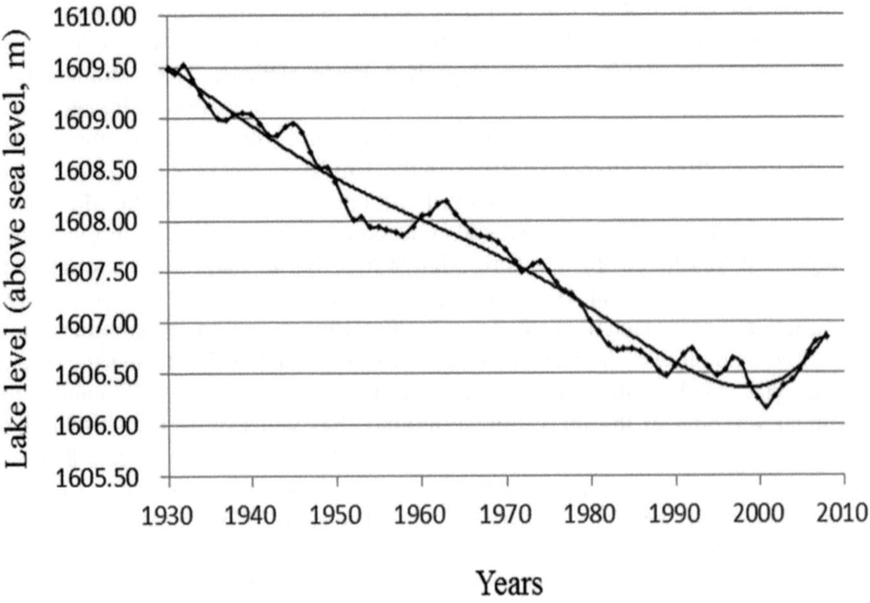

Fig. 9.19 Changing of annual water level of Issyk-Kul Lake (1930–2010) (Salamat et al. 2015)

From the analysis of the Karakol WS database from 1880 to 1996 and data from other WS for the years 1997 to 2010, we developed an estimation of index dynamics for the Karakol WS. Within the general context of increasing air temperatures in the century-long weather indexes, there were climate variations in IKB. Data from the Karakol WS showed that there were cycles of declining air temperatures that lasted around 28–29 years. The beginning of these periods air temperature increased abruptly and then gradually decreased to the end of each period. These periods occurred between 1879–1911, 1912–1941, 1942–1972, and 1973–2005 and are separated by abrupt changes in air temperature during July, the hottest month of the year, with an increasing trend line of up to 1–1.5 °C at the end of each cycle in 1912, 1942, and 1973 (Fig. 9.20).

The decreasing in air temperature at the end of each period was accompanied by an increase in precipitation and relative humidity, a decrease in evaporation, and a rise in the water level of Issyk-Kul Lake. At the end of the last cycle (1973–2005), the rise in the average annual precipitation reached about 60 mm. Regardless of these cycles, the general temperature within our study area has increased. We explained this phenomenon as part of the climate dynamics of this region. The relationships between the lake level variations and the average air temperature within the Issyk-Kul Basin (1930–2010) are shown in Fig. 9.21.

The changes in the average annual precipitation, evaporation rate, and air temperature were closely correlated with the water level changes of Issyk-Kul Lake.

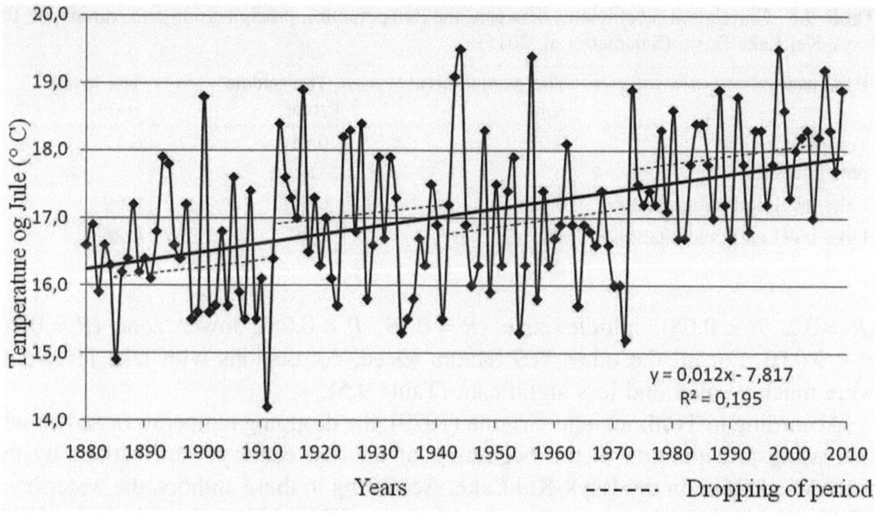

Fig. 9.20 Cycle of dropping temperature at July long-term air t^{0}C of Karakol WS (Salamat et al. 2015)

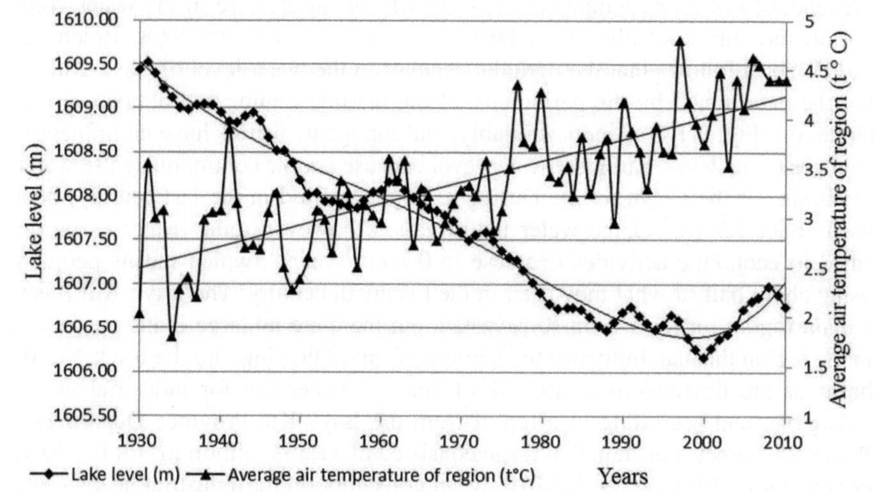

Fig. 9.21 Relationship between lake level variation and average air temperature of Issyk-Kul Lake Basin (1930–2010) (Salamat et al. 2015)

The Pearson's correlation coefficient values showed that the temperature in the Issyk-Kul Basin had a strong and significant correlation with water level of Issyk-Kul Lake [glacial–nival zone ($R = 0.41$, $P < 0.05$), middle zone ($R = 0.48$, $P < 0.05$), lower zone ($R = 0.69$, $P < 0.05$)], and the precipitation that fell in the Issyk-Kul Lake Basin also correlated with the lake level [glacial–nival zone

Table 9.5 Correlation coefficient between the temperature, precipitation, and runoff of the Issyk-Kul Lake Basin (Salamat et al. 2015)

Parameters	The glacial–nival zone	The middle zone	The lower zone
Temperature and precipitation	0.03	0.01	0.17
Lake level and temperature	0.41	0.48	0.69
Lake level and precipitation	0.2	0.19	0.26

($R = 0.2$, $P < 0.05$), middle zone ($R = 0.19$, $P < 0.05$), lower zone ($R = 0.26$, $P < 0.05$)]. For all the other WS factors tested, correlations with lake level data were much weaker and less significant (Table 9.5).

According to Trofimov and Grigina (1979), the dropping temperature and slowly increasing precipitation at the beginning of the last century were caused by the water level rising in the Issyk-Kul Lake. According to these authors, the water level of this lake increased by up to 2 m over the years 1900–1910, 12 cm in 1941–1942, and less than 8 cm during 1970–1971. In recent years, we found other significant rises in water level, and as a result, the water level reached about 1607 m asl in July 2004. In case of a continuing rise in the dynamics of the climate in the region, with the tendency of rising temperatures, there will be an increase in the water volume due to the shrinking glaciers (Aizen et al. 1997; Amanaliev 2008; Bolch et al. 2011). So, it follows that the periodic changes in the water level of Issyk-Kul Lake will be determined by the periodicity of the thermic regime and atmospheric precipitation (Fig. 9.17b). Unquestionably, anthropogenic factors have an influence on changes to the Issyk-Kul Lake water level because people continuously use water in the basin for their own needs. During a 70-year period in the last century, people used about 1.5 km^3 of the water for irrigation; however, today water usage dedicated to economic activities decrease to 0.7 km^3 $year^{-1}$, which means people are using about half of what they used in the twentieth century. The Issyk-Kul Basin is a main region for tourism in Kyrgyzstan, but there are no large industrial factories in the region that can influence the amount of runoff flowing into the Issyk-Kul river basin or the fluctuations in the lake level. So, water use for industrial needs is negligible, and according to a report from the Issyk-Kul Province Department of Water Resources, the annual average consisted of 11.393 million m^3 for the 30-year period from 1980 to 2009. In 2010, water consumption for industrial needs was 8.5 million m^3, which was considerably less than in 1980, when industrial water consumption reached 20.15 million m^3. The average annual water supply for public use was 20.422 million m^3 from 1980 to 2010. The annual average withdrawal of water for irrigation was equal to 1132.180 million m^3, and the irrigated area for the entire basin was 155.728 ha in 2010. During 1980, the water intake for irrigation was 2122 million m^3, whereas in 2010 the water consumption for irrigation decreased to 568.2 million m^3 (Fig. 9.16b). As you can see, water use by humans has dropped by at least four times since 1980, which means that human activity does not have a significant impact on the water level fluctuations in Issyk-Kul Lake over the last few decades (Fig. 9.18). With this information, it seems obvious that climate change is the main factor contributing to the water level fluctuations in Issyk-Kul Lake.

Investigations of the Issyk-Kul Basin confirmed that air temperatures have been rising in all mountain elevations, and precipitation amounts have been dropping in the lower elevations of the basin. Precipitation was high and evaporation was low at high mountain elevations. And because of the increase in glacier melt, the volume of river runoff flowing into Issyk-Kul Lake increased, thereby raising the water level of the lake. These fluctuations in water level are caused more by the climate changing in the basin (rising moisture due to an increase in air temperature in the catchment area of the lake) than by human activity (diversion of water for irrigation). So, the periodicity of fluctuations in the water level of Issyk-Kul Lake can be determined by the periodicity of the rate of precipitation and the thermal regime of the region with a general tendency toward rising temperatures.

The modern ecological problems in challenging Issyk-Kul Lake today were determined by three factors: (1) abrupt changes in water level, (2) regional warming of climate, and (3) anthropogenic factors (which were found to have a much lower effect).

9.6 System Dynamics Modeling of Water-Level Variations of the Issyk-Kul Lake

Water is considered the most precious natural resource, and its availability is directly threatened by human activities (Vorosmarty et al. 2010; Meybeck 2003). In arid regions, water scarcity is already today of the highest significance while ongoing socioeconomic growth and the climate change will only intensify the challenges related to managing the increasingly limited water resources. Climate change leads to an increase of the air temperature and more variable rainfall regimes, with severe consequences for the frequency and magnitude of droughts and flood events, and an accelerated meltdown of glaciers, which can increase the river runoff in the short term but ultimately alters the discharge regimes in the long term, reducing the amount of available water (Setegn et al. 2011; Abbaspour et al. 2009). Combined with projected population growth, expanding irrigation schemes, increased evapotranspiration rates, and the growing water demand for domestic and industrial purposes, these climatic changes will have a significant impact on the water balance in the near future (Smiatek et al. 2014). This thematic issue is particularly serious in Central Asia (Ibatullin 2009), a region characterized by intense water-related issues and heavily affected by the climate change (White et al. 2014; Yessekin and Bogachev 2004; Ososkova et al. 2000).

Most of Central Asia is determined by an arid and semiarid climate. Therefore, areas close to shallow groundwater, rivers, and lakes are characterized by unique water-dependent ecosystems and human societies, which are closely related to the limited water resources and have been developed for thousands of years (Karthe et al. 2015). At the beginning of the twenty-first century, global climate change, population growth, high agricultural water use, rising levels of pollution, river flow regulation, and dropping lake levels exert multiple pressures on the Central Asian water resources (Deng et al. 2015; Dukhovny et al. 2013; Makhmudov et al. 2008). It is evident that in many lake catchments the agriculture development is associated with increasing water withdrawal from lake tributaries, which causes serious lake water-level drops (Badescu and Schuiling 2010; Hwang et al. 2011; Salamat et al. 2015).

Endorheic lakes located in arid and semi-arid environments are highly sensitive to climate oscillation since minor changes in their environmental conditions can cause dramatic changes of the lake level (Song et al. 2014). The sensitivity of this lake is even higher when they are located in high-altitude mountain environments (Caruso et al. 2017). In addition, many researchers indicated that the notable climate warming and decrease in precipitation since the nineteenth century is associated with an abrupt lake water-level drop in arid regions worldwide (Matsuyama and Kezer 2009; Bai et al. 2011; Romanovsky 2002b). Lake Issyk-Kul is a prime example of these endorheic mountain lakes in arid Central Asia, and the variations of its water level reflect the global climate processes very well (Shabunin and Shabunin 2002).

Lake Issyk-Kul is a closed lake, located in the Tian-Shan mountain belt within the Republic of Kyrgyzstan, in arid Central Asia (Vermeesch et al. 2004). From 1973 to 1999, the average monthly temperature in the lake basin increased, resulting in an acceleration of glaciers melting and an increased river runoff during the summer period. However, the lake level has decreased by about 13 m since in the mid-19th century, while 3 m of lake level decrease has been recorded from 1927 to 1966 (Romanovsky 2002b). In recent years, Lake Issyk-Kul became a subject of special concern because of the continuous water level decrease and its contamination related to human activities (Karmanchuk 2002). The variations of the lake water level have impacts on the agriculture, the local population, economic activities, and the environment (Yuan et al. 2015; Jalili et al. 2012; Dusini et al. 2009). Studies conducted on the Issyk-Kul during the last decades covered topics ranging from geological characteristics and earthquakes in the Northern Tian-Shan mountain range (De Grave et al. 2013); the reconstruction of the paleoclimate based on sediment cores and isotopes (Macaulay et al. 2016; Ricketts et al. 2001); the lake basin glaciation and the impacts of the climate change (Bolch 2007); the physicochemical characterization of the basins water resources in general (Vollmer et al. 2002a, b; Lyons et al. 2001) and the influence of uranium mining in particular (Uralbekov et al. 2011; Gavshin et al. 2005), to the potential for a biodiversity-friendly fishing industry in the lake (Alamanov and Mikkola 2011). Specific components of the Issyk-Kul water balance have also been studied by several projects. The main focus of those studies has been on the groundwater resources and the deep-water renewal as a part of the Issyk-Kul water balance (Peeters et al. 2003; Hofer et al. 2002; Vollmer et al. 2002a, b; Waugh et al. 2002; Mandychev 2002), while only very few analyses of the lake-level variability have been published (Salamat et al. 2015; Romanovsky 2002b; Qin and Yu 1998; Romanovsky et al. 2013; Alymkulova et al. 2016). These previous studies dealing with the water balance are too limited in their focus, as they only analyze a single aspect of it (groundwater, climate change impact on glaciers, or the physico-chemical properties of the lake's water) of the complex hydrological lake catchment interactions. They neglect equally important aspects like the land cover and its dynamic, human activities or the variability in precipitation discharge. As well, that the few studies focusing on the lake-level variations, as an integral of the water balance, cover a temporal scale that is more related to paleoclimatic research and the influence of the last glaciation then on the present-day water resources problems.

In order to close this knowledge gap and to allow a better understanding of the lake-level variations and its main driving forces, a complex, holistic, and dynamic simulation model of the lake and its whole catchment is needed. However, these requirements seem difficult to be met as the system is highly complex and its individual parameters are controlled by numerous cause and effect relations, with feedback mechanisms and nonlinear characteristics (Winz et al. 2009; Kotir et al. 2016; Forrester 1971). One method suitable for this ambitious task is the system dynamics (SD) model, which was originally developed by Forrester in 1961, and is an approach for understanding the interactions among driving factors and inter-connected subsystems that drive the dynamic behavior of a system (Ahmad and Prashar 2010; Ahmad and Simonovic 2001). Over the years, a number of SD models have been developed for water balance simulation and have been used to evaluate various water-related solutions (Qaiser et al. 2011, 2013; Rusuli et al. 2015), such as water resource planning models (Kalra et al. 2013; Shrestha et al. 2012; Dawadi and Ahmad 2013; Butler and Adamowski 2015), hydrologic extremes models (Choubin et al. 2014), agriculture water management models (Yaeger et al. 2014; Valipour 2015), and water balance models, which have been developed to test water-related and environmental issues in developing countries where the data availability is lacking (Abadi et al. 2015). With this background, the SD model satisfies the requirements for a complex analysis of the Issyk-Kul water-level fluctuations and its driving factors.

The novelty of the present study is the holistic approach which incorporates the whole lake catchment and a wide range of parameters influencing the water balance, including land use, the population, the economy, water supply and demand, and how these parameters change over time and interact with each other. To achieve this goal, a dynamic water balance model for the period 1980–2012 was developed in a first step, and the relations between the relevant factors were formulated based on mathematical equations. The second step was the calibration and validation of the model using historical data. The third and final step was to evaluate the main factors impacting the lake water level via a sensitivity analysis and various water resource scenarios.

Model Structure. System Dynamics Model

The event-oriented view of the world and linear thinking cannot sufficiently address complex issues. The SD model, based on the nonlinear causal thinking, has been developed to expand the dynamic simulation model (Abadi et al. 2015). In this study, based on the SD principles and a series of casual loops and mathematical equations, a novel stock–flow diagram has been developed for the Issyk-Kul lake basin, considering the hydrological and socioeconomic drivers of the lake water-level change and their interactions (Fig. 9.22). This model uses the concepts of water balance and rainfall-runoff transformation and estimates the water availability for the uses of agricultural, domestic, and industrial demands. Based on land use changes, population growth, and industrial productivity progression, the model generates water demand data for agricultural, domestic, and industrial use, respectively. The SD model also improves the understanding of the internal working of a complex socio-hydrological system with its various direct and indirect interconnections.

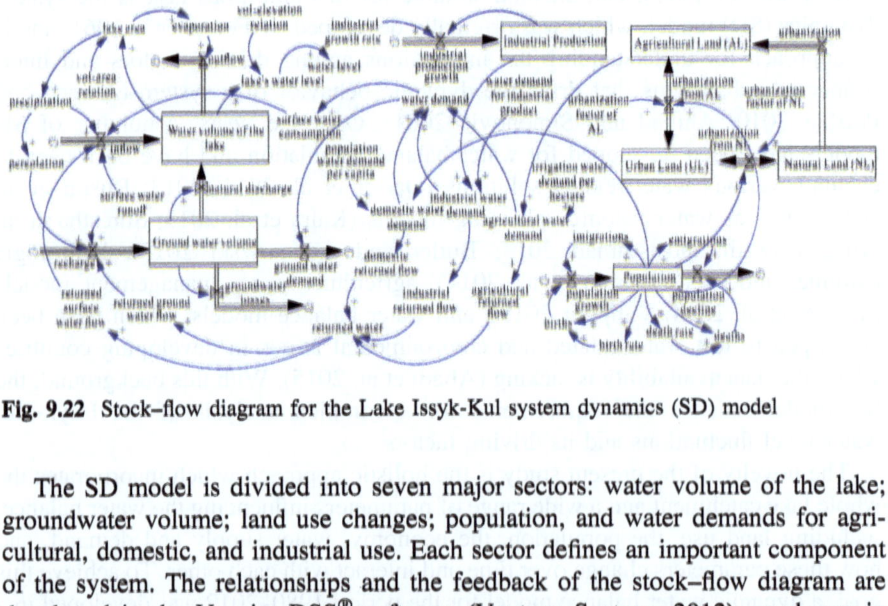

Fig. 9.22 Stock–flow diagram for the Lake Issyk-Kul system dynamics (SD) model

The SD model is divided into seven major sectors: water volume of the lake; groundwater volume; land use changes; population, and water demands for agricultural, domestic, and industrial use. Each sector defines an important component of the system. The relationships and the feedback of the stock–flow diagram are developed in the Vensim DSS® software (Ventana Systems 2012), as shown in Figure 9.22 to characterize the system processes. The "+" marks close to the arrows show an increase of the variable at the head of the arrow, while the "–" marks indicate a decrease of that variable. The time horizon for the simulation model extends from 1980 to 2040, and the model operates on an annual time step. The period from 1980 to 2012 is used for the model calibration and the remaining period (2013–2040) for the simulation of possible future developments.

Model Equations. In the equations of the SD model, the variables are either stocks or flows. Stocks are accumulations and are used to characterize the state of the system and create the information upon which decisions are made. Flow variables are used to define rates which can change the stock variables (Ragni et al. 2011). Stocks accumulate the difference between its inflow and outflow. Therefore, a stock with one inflow and one outflow is formulated as follows:

$$\text{Stock}(t) = \int_{t_0}^{t} [\text{Inflow}(t) - \text{Outflow}(t)]\text{d}t + \text{Stock}(t_0) \tag{9.1}$$

where Stock(t) = Amount of stock at time t, Inflow(t) = Inflow at time t, Outflow (t) = Outflow at time t, and Stock(t_0) = Stock in time t_0. According to the Equation (9.1), other mathematical equations of the major processes of inflow and outflow represented in the following sections from Sects. 2.2.1–2.2.5.

Lake Water Volume. The Issyk-Kul water volume (storage) is the main stock of the model. Annual changes of the lake's water volume are controlled by the in- and outflow of water. The inflow to the lake includes precipitation, surface, and groundwater inflow. As the Issyk-Kul is an endorheic lake, the outflow is limited to evaporation. The amount of water in the Issyk-Kul at a time t is calculated as follows:

$$V(t) = V(t_0) + \int_{t_0}^{t} [\text{GW}(t) + (\text{SW}(t) - W(t)) + P(t) - E(t)]\text{d}t, \tag{9.2}$$

where $V(t)$ = Water volume of the lake at time t, $V(t_0)$ = Water volume of the lake at the time t_0, GW(t) = Groundwater into the lake at time t, SW(t) = Surface water flow at time t, $P(t)$ = Precipitation at time t, and $E(t)$ = Evaporation from the lake surface at time t. $W(t)$ = Water consumption from lake tributaries at time t, which is divided into Q_1, Q_2, and Q_3—is the amount of water used for domestic, agricultural, and industrial, respectively. Surface water inflow = ((SW) − $W(t)$).

Groundwater. Groundwater inflow is an important component in the lake water budget and thus was first considered as a separate component of the Issyk-Kul water balance by Kaplinsky and Timchenko (1997), who showed that the groundwater component provides about 30% of the income. The groundwater aquifers are located in permeable coarse Meso-Cenozoic sediments [(cemented) pebbles, sand, and clay as well as siltstone conglomerates] of up to 5 km in depth. The youngest groundwater layer can be found in up to 300-m-thick Quaternary sediments (consolidated coarse clastics) with a high permeability (0.04–4.0 m/h) and a total storage capacity of 58 km^3. This Quaternary groundwater discharges into the lake through littoral and submarine vents and provides most of the groundwater inflow (>70%), though deeper groundwater aquifers (e.g., the Pliocene-quaternary Sharpyldak or the Paleogene-Neogene Issyk-Kul aquifer) also contributes to the lake's water balance (Mandychev 2002). The special topography and geological structure of the lake Issyk-Kul basin, as well as the groundwater gradient maps, show no evidence for groundwater outflow from the lake and its adjacent aquifers in the region (Mandychev 2002). Therefore, only the groundwater inflow of into the lake was considered in this study. The groundwater inflow of the lake was calculated with this equation:

$$\mathrm{GW}(t) = \mathrm{GW}(t_0) + \int_{t_0}^{t} [R(t) - W(t) - \mathrm{GW_L})]\mathrm{d}t \quad (9.3)$$

where GW(t) = Groundwater into the lake at time t, GW(t_0) = Groundwater into the lake at time t_0, $R(t)$ = Recharge at time t, $W(t)$ = Water demand/consumption at time t (it is the amount of water used for domestic, agricultural and industrial), and $\mathrm{GW_L}$ = Groundwater losses at time t (evaporation from the groundwater surface (Mandychev 2002).

The recharge primarily comes from the infiltration from surface water bodies, recharge from domestic, agricultural, and industrial water use, and percolation of precipitation. The recharge into the groundwater is calculated as:

$$R(t) = R(t_0) + \int_{t_0}^{t} [C_\mathrm{I} \times P(t) + (R_1 + R_2 + R_3) + I_\mathrm{swb}]\mathrm{d}t \quad (9.4)$$

where $R(t)$ = Recharge at time t, $R(t_0)$ = Recharge at time t_0, C_I = Coefficient of infiltration, $P(t)$ = Precipitation at time t, R_1 = Recharge from domestic water demand, R_2 = Recharge from agricultural water demand, R_3 = Recharge from industrial water demand, I_swb = Infiltration from surface water bodies.

Water Demand. The GIP is a measure of the productivity of the population and hence is associated with the water consumption. In this model, water demand includes the domestic water demand, the agricultural water demand, and the industrial water demand. These can be calculated as:

$$T_\mathrm{wd} = D_\mathrm{wd} + A_\mathrm{wd} + I_\mathrm{wd} \quad (9.5)$$

where T_wd = Total water demand (in million m^3/year), D_wd = Domestic water demand (in million m^3/year), A_wd = Agricultural water demand (in million m^3/year), and I_wd = Industrial water demand (in million m^3/year).

The domestic water demand is the quantity of water used in people's daily life. It can be estimated by the size of the population and the water demand per capita per day. The domestic water demand was formulated as follows:

$$D_{wd} = p(t) \times D_{pwd} \times 365/10^3/10^8 \tag{9.6}$$

where $p(t)$ = Population, and D_{pwd} = Population water demand per capita (L/person·day).

The agricultural water demand is the water used in irrigation. It can be calculated by the agricultural area and the irrigation water demand per hectare calculated as follows:

$$A_{wd} = A_{pwt} \times L_{agr}/10^8 \tag{9.7}$$

where A_{pwd} = Irrigation water demand per hectare in cubic meters (m^3/ha), and L_{agr} = Agricultural land area (ha).

The industrial water demand includes the water used in the process of production. Water use in industry varies significantly; the gross industrial product was used to calculate the industrial water demand. It is calculated as the gross industrial product multiplied by the average water demand for one unit of industrial product, formulated as:

$$I_{wd} = I_{pwd} \times \mathrm{GIP}/10^8 \tag{9.8}$$

where I_{pwd} = Water demand for the industrial production (Kyrgyz Som (Ks)/m^3), and GIP = Gross industrial product (Ks).

Population. The inflows to the population stock consist of births and immigration; the outflows consist of deaths and emigration. The population and net population change in the lake basin are formulated as:

$$p(t) = p(t_0) + p(t_0) \times (\mathrm{Br} - \mathrm{Dr} + \mathrm{Ir} - \mathrm{Er}) \tag{9.9}$$

$$np(t) = p_g \times p(t) \tag{9.10}$$

where $p(t)$ = Population at time t, $p(t_0)$ = Population at time t_0, Br = Birth rate, Dr = death rate, Ir = Immigration rate, Er = Emigration rate, $np(t)$ = net population change, and p_g = population growth rate.

Land Use Changes. Land use patterns in the study area are divided into three categories—agricultural, natural, and urban land. Land use changes are a complex process, managed by several factors including population growth, economics, and legislation. In the SD model, it is not possible to foresee and simulate the policy changes, but the population growth and the resulting land use dynamics can be simulated. For this, it was assumed that based on the population dynamics, natural land is converted into agricultural land, and natural land and agricultural land are both transformed into urban land. The urban land was determined by cumulating the changes in the agricultural land and natural land over time which is calculated as follows:

$$L_{\mathrm{urb}}(t) = L_{\mathrm{urb}}(t_0) + \int_{t_0}^{t} \left[R_{\mathrm{agr}} + R_{\mathrm{nat}}\right] \mathrm{d}t \tag{9.11}$$

$$L_{\mathrm{agr}}(t) = L_{\mathrm{agr}}(t_0) + C_{\mathrm{agr}} - \int_{t_0}^{t} R_{\mathrm{agr}} \mathrm{d}t \tag{9.12}$$

$$L_{\mathrm{nat}}(t) = L_{\mathrm{nat}}(t_0) - \int_{t_0}^{t} R_{\mathrm{nat}} \mathrm{d}t \tag{9.13}$$

The urbanization rates for agricultural land and natural land were used to estimate the total land use change into urban land. The urbanization rate is a function of land patterns, which is computed by the population growth and urbanization factors of natural land and agricultural land. These parameters were determined by analyzing the change in the total population and land use in the lake basin. The urbanization rates for natural and agricultural land in the lake basin were calculated using:

$$R_{\mathrm{nat}}(t) = F_{\mathrm{nat}} \times np(t) \tag{9.14}$$

$$R_{\mathrm{agr}}(t) = F_{\mathrm{agr}} \times np(t) \tag{9.15}$$

where L_{agr} = Agricultural land (ha), L_{urb} = Urban land (ha), L_{nat} = Natural land (ha), C_{agr} = Natural land converted into agricultural land (ha), R_{agr} = Urbanization rates for agricultural land (ha/year),

R_{nat} = Urbanization rates for natural land (ha/year), F_{agr} = Urbanization factor of agricultural land (ha/person), and F_{nat} = Urbanization factor of natural land (ha/person).

The Issyk-Kul is an endorheic mountain lake, located on the northern slopes of the Tian-Shan mountains, Republic of Kyrgyzstan, in arid Central Asia (42° 25′ N, 77° 15′ E) (Fig. 9.23). It is situated at an altitude 1607 m above sea level, covering an area of 6236 km^2. The maximum depth of the lake is 668 m, and the total water volume is 1736 km^3. It is surrounded by high mountain ranges: the Kungey Ala-Too Range in the north with the highest peaks reaching 4770 m, and the Teskey Ala-Too Range in the south with peaks exceeding 5200 m (Salamat et al. 2015). More than 118 rivers feed into the lake. The main tributaries are the Pzhergalan River and the Tyup River, and their runoff regime is dominated by meltwater from glaciers and snow (Vermeesch et al. 2004; Bolshakov et al. 1986). At present, the lake has no outlet. Up to the late Pleistocene, the Chu River was flowing into and out of the lake at its western extremity. As a result of a tectonic shift, the river changed its course less than five kilometers before reaching the lake and now flows in southwestern direction into the Boom Canyon and into the Naryn

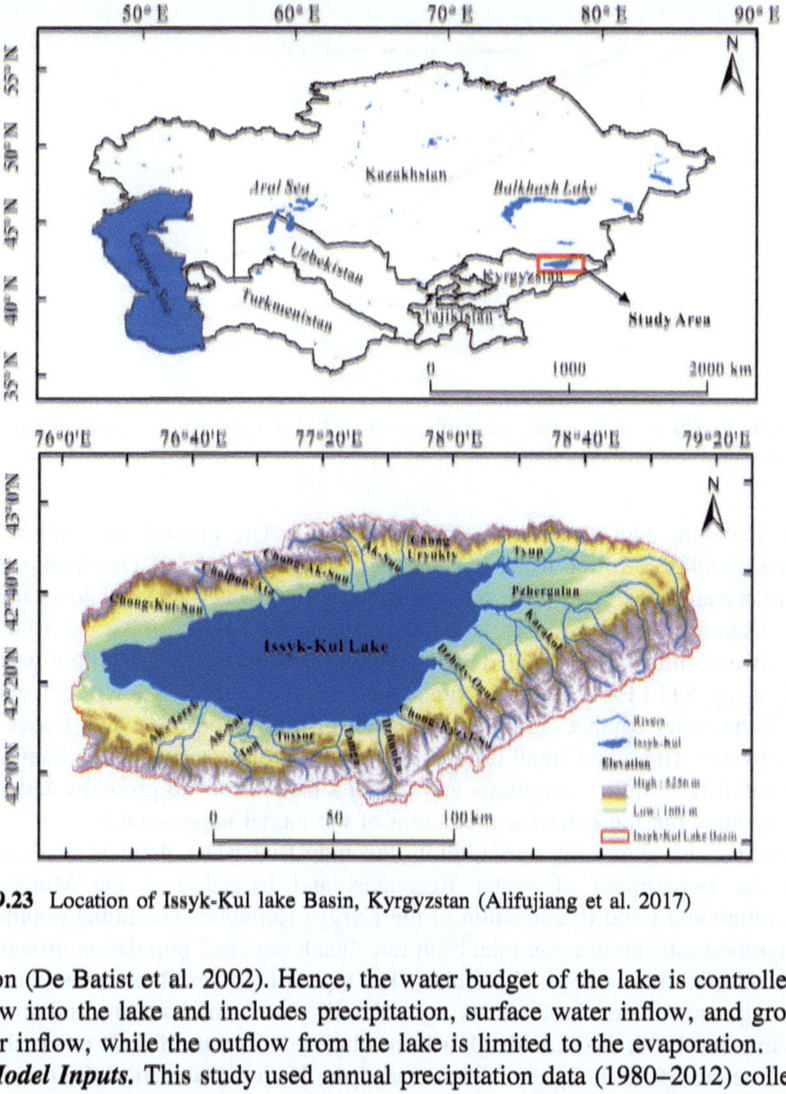

Fig. 9.23 Location of Issyk-Kul lake Basin, Kyrgyzstan (Alifujiang et al. 2017)

region (De Batist et al. 2002). Hence, the water budget of the lake is controlled by inflow into the lake and includes precipitation, surface water inflow, and groundwater inflow, while the outflow from the lake is limited to the evaporation.

Model Inputs. This study used annual precipitation data (1980–2012) collected from four meteorological stations (Balykchy, Karakol, Kyzyl-Suu, and Cholpan-Ata) located around the lake. The average of these stations was considered as the precipitation data inflow into the lake. Annual surface runoff from 16 hydrometric stations (Ak-Terek, Ak-sai, Ton, Tossor, Tamga, Dzhuuku, Chong-Kyzyl-Su, DZhety-Oguz, Karakol, Pzhergalan, Tyup, Chong-Uryukty, Ak-Suu, Chong-Ak-Suu, Cholpon-Ata, and Chong-Koi-Suu) (Fig. 9.23) which are the main rivers leading up to the Issyk-Kul was used to determine the surface water flow. For the evaporation term, estimates of the mean annual energy budget at the water surface and evaporation from the lake surface were obtained from Salamat et al. (2015). The lake volume-area and volume-depth curve data were extracted

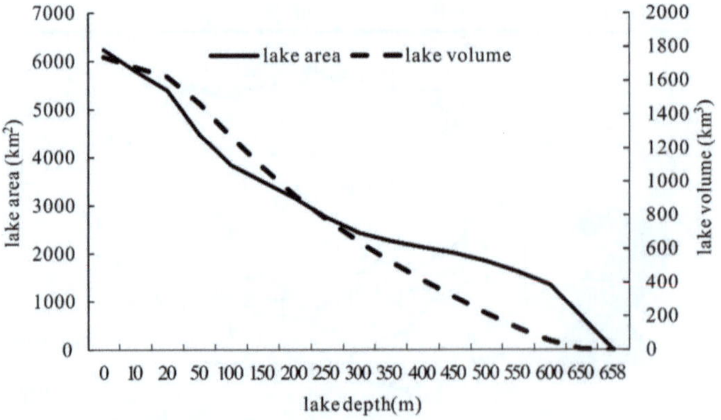

Fig. 9.24 Variations of area and volume of Lake Issyk-Kul at different water depths (data sources from Shabunin and Shabunin 2002)

from Shabunin and Shabunin (2002) (Fig. 9.24). The groundwater inflow contributes about 30% of the water balance income (Kaplinsky and Timchenko 1997), but no research has been done on how much groundwater actually flows into the lake. Therefore, in this paper, according to the Equations (9.3) and (9.4), the groundwater inflow is calculated. The resulting total groundwater inflow was then fitted, using STELLA's graphical functions to estimate the unknown relevant factors of the water balance, like the coefficient of infiltration and water losses from groundwater. There is a small difference between previous research estimations of the groundwater influx variations and model's results for the groundwater inflow into the lake, but the behavior prediction of the model is reasonable.

The data for domestic, agricultural, and industrial water demand was derived from the Department of Water Resources and Irrigation at the Ministry of Agriculture and Land Reclamation of the Kyrgyz Republic. The initial population, immigration rate, emigration rate, birth rate, death rate, and population growth rate were collected for the Issyk-Kul Oblast. The types of land use/land cover change in the study area were classified as agricultural, natural and urban, and the data available for 1990, 2000, and 2010 were obtained from the Data-Center Sharing Infrastructure of Earth System Science, and the National Statistical Committee of the Kyrgyz Republic. Other important parameters like the urban and rural population water demand per capita, net irrigation water demand per hectare, and water demand for industrial products were derived from the CAWATER Info-Indicators of Sustainable Development for Central Asia Countries (Kyrgyzstan water resources, Socioeconomic indices 1990–2010).

Model Calibration and Validation. Model calibration is a process by which certain model variables are adjusted to obtain a match between model output and historical data (Rykiel 1996). The model must be validated in terms of its structure, behavior, and in order to check its applicability and accuracy. First, a behavioral replication was used as a verification method to test whether the model can reproduce both qualitatively and quantitatively the behavior of key parameters (Mirchi et al. 2012). The linear correlation coefficient (CF), Nash-Sutcliff efficiency coefficient (NSE), multiplicative bias (MBias), values of the mean absolute error (MAE), and root mean square error (RMSE) were calculated based on the difference between measured historical data and the simulated data (Jiang et al. 2012; Sidike et al. 2016):

$$\mathrm{CF} = \frac{\sum_{i=1}^{n}[(V_s - \bar{V}_s)(V_0 - \bar{V}_0)]}{\sqrt{\sum_{i=1}^{n}\left[(V_s - \bar{V}_s)^2\right]\sum_{i=1}^{n}\left[(V_0 - \bar{V}_0)^2\right]}} \tag{9.16}$$

$$\mathrm{NSE} = 1 - \frac{\sum_{i=1}^{n}(V_s - V_0)^2}{\sum_{i=1}^{n}(V_0 - \bar{V}_0)^2} \tag{9.17}$$

$$\mathrm{MBias} = \frac{\sum_{i=1}^{n} V_s}{\sum_{i=1}^{n} V_0}, \tag{9.18}$$

$$\mathrm{MAE} = \frac{1}{n}\sum_{i=1}^{n}|(V_s - V_0)|, \tag{9.19}$$

$$\mathrm{RMSE} = \sqrt{\frac{1}{n}\sum_{i=1}^{n}(V_s - V_0)^2} \tag{9.20}$$

where V_0 = Observed value of the variable, V_s = Simulated value of the variable, $\bar{V}_0$ = Mean of the observed value of the variable, and $\bar{V}_s$ = Mean of the simulated value of the variable.

The CF is used to assess the agreement between the simulated data and observed data. The range of CF values is between −1 and +1. A CF value of exactly +1 indicates a perfect positive fit, while a value of exactly −1 indicates a perfect negative fit. The values of NSE coefficient range roughly from −∞ to 1. A perfect estimation would result in an MBias value of 1. Underestimation will lead to values smaller than 1 and overestimation to values greater than 1. MAE and RSME are used to assign the average magnitude of the error. The optimal values of the MAE and RSME are 0 (Moazami et al. 2013).

In the second step, using a sensitivity analysis, the most important variables have to be identified for the developed model or system development. This means that the sensitivity analysis can identify the main links between observations, model inputs, and predictions. Thus, the variable selection is very important as they are the basis for the regulation strategy (Omlin et al. 2001). Calculating model outputs by changing one variable while keeping other parameters unchanged, the calculation formula is as follows:

$$S_q = \left| \frac{\Delta q(t)}{q(t)} \times \frac{y(t)}{\Delta y(t)} \right|, \tag{9.21}$$

where t is time, $q(t)$ indicates the system's state at time t, $y(t)$ denotes the system parameters that affect the system's state at time t, S_q is the sensitivity degree of state q toward parameter y while $\Delta q(t)$ and $\Delta y(t)$ represent increments of state q and parameter y at time t, respectively. The results obtained can help to identify the parameters that maximally affect the system's behavior, and thus, they can potentially help guide the analysis of simulation scenarios.

Water-Level Variations of Issyk-Kul Lake

The annual average water level of Lake Issyk-Kul has fluctuated since 1980 (Fig. 9.25), showing a downward trend throughout the 1980s and 1990s. The water level decreased from 1606.73 m in 1980–1606.18 m in 1998, dropping 0.57 m in 18 years (−3.17 cm/year), while from 1999 to 2011, the water level increased. The year 2012 was characterized by another small decrease with a water level of 1606.99 m at the end of that year (+0.81 m in 14 years; +5.79 cm/year). The

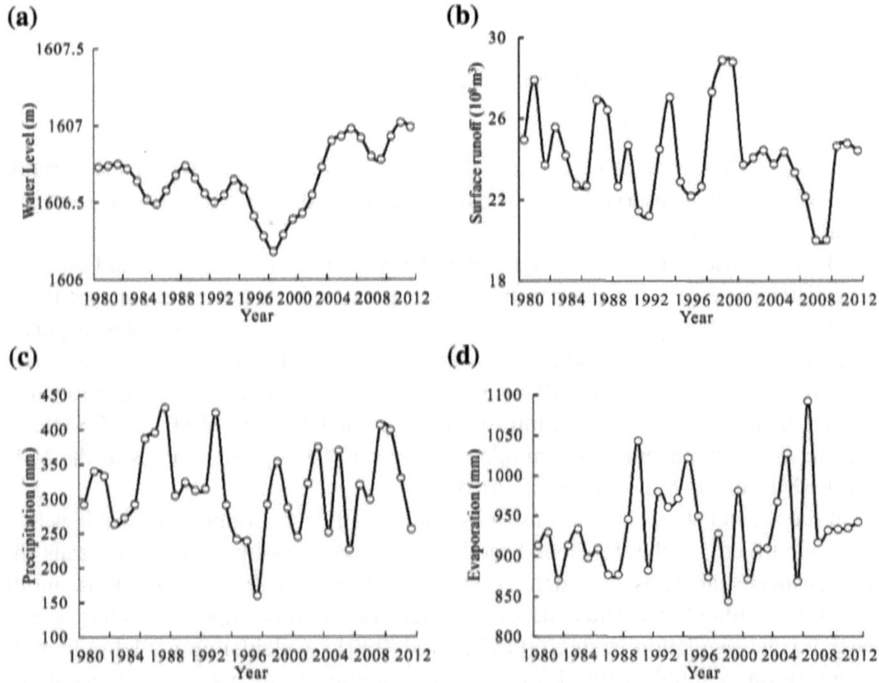

Fig. 9.25 Annual variability of the lake water level, surface runoff, precipitation, and evaporation based on the historical data from 1980 to 2012. **a** The annual water level of Lake Issyk-Kul; **b** surface runoff averaged across 16 hydrometric stations; **c** annual precipitation averaged across four meteorological stations; **d** annual evaporation at the Karakol station (Alifujiang et al. 2017)

highest water level during the study period was detected in 2011 (1607.02 m), and the lowest water level was recorded in 1998 (1606.18 m), while the years from 1998 up to 2011 registered an increasing lake-level amplitude of 0.84 m, indicating a high temporal dynamic. These fluctuations of the Issyk-Kul water level are the result of the climate change and the anthropogenic water use in the catchment.

Factors Influencing the Water Balance. Climate–Hydrologic Change

According to the water balance equations, the lake water level of Lake Issyk-Kul is closely related to climatic hydrological factors, including the precipitation within its basin, evaporation over the lake's surface area, and surface water runoff. The annual precipitation, evaporation, and natural runoff variability during the period of 1980–2012 are shown in Figure 9.25. The trends of surface runoff, precipitation, and evaporation variations are basically consistent with the changing trend of the lake water level and show the same pattern of decrease, increase, and decrease again. Over the period from 1980 to 2012, the mean annual natural runoff was 2421 million m^3, and the annual maximum and minimum surface runoff were 2888 and 1998 million m^3, respectively. The mean annual precipitation during the study period was 313.46 mm. The annual maximum and minimum precipitation were 396 and 238.5 mm, respectively. And the mean annual evaporation from lake surface area was 933.96 mm.

Anthropogenic Activity

In the present study, human activities are characterized by the total population, land use changes, and the gross industrial product. Kyrgyzstan's economy is growing fast, and an impact on the hydrology of the lake basin can be expected. The population of the Issyk-Kul Oblast as a proportion of the total population of Kyrgyzstan has declined slightly over the study period from 10.05% in 1980 to 8% in 2012, but the total population increased from 355,000 in 1980 to 448,000 in 2012. This equals an annual population growth rate of +0.712% for recent 33 years. This increase in the population could lead to an increasing use of water as the population change is the main driving factor behind the amount of water consumed in the lake basin. It not only determines the domestic water demand but also industrial water demand and the area of agricultural land, which in turn affects the agricultural water demand.

Land use patterns in the study area are divided into three categories, namely agricultural land, natural land, and urban land. The distribution of land use in 1990 was 9.9% agricultural, 88.7% natural, and 1.4% urban. In 2012, this changed to 10.5% agricultural, 87.9% natural, and 1.6% urban. Overall, the changes during this period were only small, with only minor changes in the water consumption. The analysis of the land use changes shown a process of ongoing urbanization in the Issyk-Kul catchment. Natural land and agricultural lands were found to be converted into urbanized areas by 0.012 and 0.009 ha per person. But the land use distribution alone stands for only one part of the anthropogenic influence on the water balance as the land use intensity is also a major factor. The annual water consumption for agricultural irrigation, domestic and industrial purposes has decreased from 1120, 16.95, and 10.59 million m^3 in 1980 to 380, 6, and 6.8 million m^3 in 2012, respectively. This is a drop in water consumption of 66% for the agrarian sector, 64% for the domestic sector, and 36% for the industrial water use. This is the result of the disintegration of the Soviet Union between 1989 and 1992, which led to a decline of the agrarian and industrial water use, in combination with adopted lake water management measures. However, with the growth of the population in the lake basin, the local water demand has been rising, leading to a water level decline. This is being detected and controlled through various government policies, such as the use of other lakes and basins, management of lake streams, soil erosion prevention measures, and rainwater collection (Propastin 2013; Vollmer et al. 2002a, b). Although the overall demand for agricultural water has been reduced, the water consumption of this sector is linked to the crop types grown in the basin, and there is a great likelihood of an increasing water demand as more irrigation intensive crop types (e.g., cotton) are cultivated. In addition to the crop types, the applied irrigation techniques such as flood, furrow, sprinkler, or drip irrigation have a significant impact on the water use efficiency and are related to processes like secondary salinization, which in turn contributes to an increase of the water demand due to leaching. Therefore, considering the best irrigation practices is conducive to the water management in the Issyk-Kul lake basin (Farre and Faci 2009; Fereres and Soriano 2007).

The GIP arises from industrial activity which requires water, though, in the case of the industry, much of the water is not "consumed" but recycled into the river network as wastewater. It could be argued that as economic development proceeds, such recycling will increase, as will the quality of the recycled water (Alcamo et al. 2003). After Kyrgyzstan's independence, the annual industrial effluent discharge fell from 674 million m^3 in 1991 to 466 million m^3 in 2000 (a decrease of 208 million m^3 or 31%) due to the collapse of the industrial sector. The amount of treated industrial wastewater remained stable at approximately 150 million m^3, which means that only 32% of the industrial wastewater was channeled through wastewater treatment plants. Other sewage discharges, mainly municipal wastewater and farmland drainage water, increased from 2.90 million m^3 (1991) to 3.80 million m^3 (2000). Only about 56% of Kyrgyzstan's cities and towns are connected to a centralized sewage treatment system. More than half of the small- and medium-sized cities, villages, and towns have no centralized wastewater treatment, but are responsible for 27% of the total wastewater, which is discharged untreated.

Calibration and Validation

In this study, calibrations of the model provided new estimations of the groundwater inflow of the lake. The estimated coefficient of infiltration and water losses from groundwater using SELLA's graphical function for calibration between 0.32 and 0.54, and 0.13 to 0.21, respectively. The model was validated using multiple tests, as described by Bayer (Bayer 2004), including structure assessment, dimensional consistency, extreme condition test, and behavior reproduction tests.

The structure assessment test compares the model structure with the real system (Dawadi and Ahmad 2013). The dimensional consistency test ensures uniformity in the units of measurement of all the variables in the model (Ahmad and Prashar 2010). An extreme condition test evaluates the model's responses under extreme input values. The behavior reproduction test checks the model's ability to replicate the behavior of the real system (see Sect. 4.3.2). The model performed satisfactorily under all these tests. The model was validated by comparing the model output with the observed data for the lake water level, the agricultural water demand, and the total population in the Issyk-Kul Oblast between 1980 and 2012.

Extreme Condition Test

The structure of a system dynamics model should permit extreme combinations of levels (state variables) being properly represented in the system. By examining the model structure for extreme conditions, the confidence in the model's ability to behave plausibly for a wide range of conditions is developed, and thereby, the model's usefulness to explore policies that move the system outside of historical ranges of behavior is enhanced (Ji et al. 2006). The extreme condition used for this test in this study was to reduce the evaporation from the lake surface (the only outflow of the lake) to zero. As a result of the absence of evaporation, the lake's volume should rise progressively, which the model properly represented (Fig. 9.26).

Behavior Reproduction Test

For the behavior reproduction test, the generated model behavior is judged against the recorded historical behavior. According to Equations (9.16)–(9.20), the model was validated by comparing the simulation results with the historical data collected from 1980 to 2012. Based on the research objectives, the lake water level, agricultural water demand, and the total population in the Issyk-Kul Oblast were selected as significant variables for the validation of the SD model. The results of

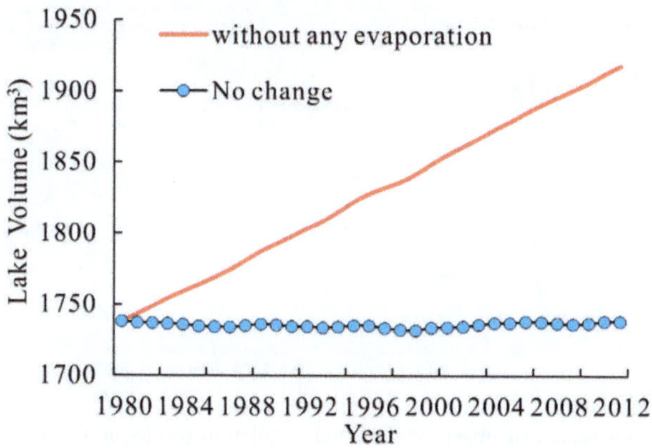

Fig. 9.26 Change of Issyk-Kul lake volume with and without evaporation (Alifujiang et al. 2017)

this model validation are shown in Table 9.6 and in Figure 9.27. They show the comparison of the simulation results and the observed data of the lake water level.

The validation results of the observed and simulated trends of the lake's water level, agricultural water demand, and total population were found to be acceptable for a complex model. This suggests that the developed SD model is capable of reproducing the behavior of different parameters within the system, and using this model is appropriate in this research.

Sensitivity Degree Analysis

In this study, to assess the degree of sensitivity for each state variable, selected parameters were increased or decreased by about ±10% for the study period from 1980 to 2012. Based on Equation (9.21), nine important model parameters were selected for a sensitivity analysis of the lake: precipitation (*P*); surface water flow (SW); groundwater inflow (GW); and evaporation (*E*) represented the natural factors; and population (*p*); agricultural water demand (A_{wd}); industrial water demand (I_{wd}); domestic water demand (D_{wd}); and agricultural land area (L_{agr}) were used as anthropogenic factors. The water level of the lake was selected as the target variables for the sensitivity analysis. Table 9.7 shows the results of this model analysis and the influence of variations in the selected parameters on the lake level.

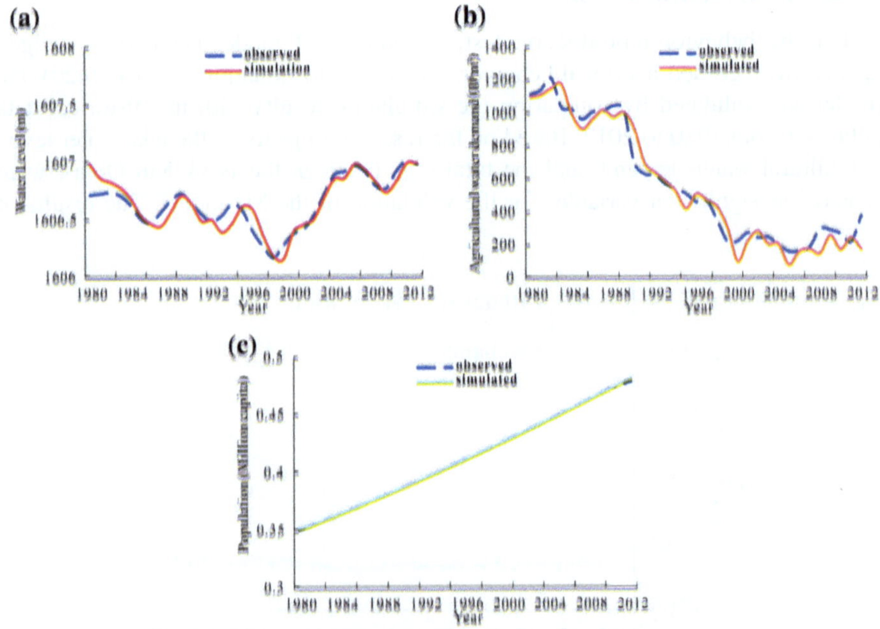

Fig. 9.27 The comparison of observed data and simulation results. **a** The annual water level of Lake Issyk-Kul; **b** The agricultural water demand of Issyk-Ku basin; **c** The total population in the Issyk-Kul Oblast (Alifujiang et al. 2017)

Table 9.6 Analysis of the model validation for simulated and observed values of variables

Variables	CF	MAE	RMSE	MBias	NSE
Lake level	0.928	0.0667	0.0906	1.000	0.9132
Total population	0.999	0.0002	0.0006	1.000	0.9998
Agricultural water demand	0.966	0.5816	0.7761	0.988	0.9327

Alifujiang et al. (2017)

The variation rankings for each selected parameter are as follows: $SW > E > GW > P > A_{wd} > L_{agr} > p > D_{wd} > I_{wd}$.

Overall, the variation in the values of the agricultural area, population, domestic water demand, and industrial water demand had little impact on the lake water level. On the other hand, the water level was much more sensitive to the dynamics in the surface water flow, evaporation, groundwater inflow, precipitation, and the agricultural water demand. This attests that the inflow and outflow of water plays an important role for the water balance, and the impact of natural factor variability is still stronger than the impact of the human activities on the lake water level.

Simulation Scenarios and Results

The simulation includes a base run and four scenarios. The base run was used to estimate the lake water level from 2013 to 2040, and to assess the dynamics of water level vulnerability using the average values of the observation period between 1980 and 2012. For this simulation run, it was assumed that the averages of the observed values will remain unchanged in the future. Besides this base run, four scenarios were evaluated, each of them focusing on a different combination of the main driving forces for the lake water-level variability (Table 9.8). Figure 9.28 indicates the behavior of the water-level variations under the extreme scenarios in the simulation period. The future scenarios for estimating the water level cover a

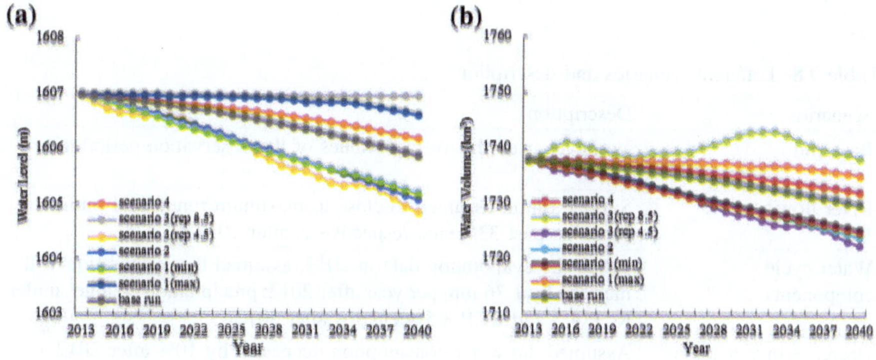

Fig. 9.28 Impact of scenario analysis on lake water level and volume in the period of 2013–2040 (Alifujiang et al. 2017)

Table 9.7 Sensitivity analysis of the model (%)

Parameter change (%)	SW		GW		*E*		*P*			
	+10	−10	+10	−10	+10	−10	+10	−10		
Lake level	1.47	1.09	0.57	0.63	−1.02	−0.53	0.41	−0.08		
Parameter change (%)	*p*		A_{wd}		I_{wd}		D_{wd}		L_{agr}	
	+10	−10	+10	−10	+10	−10	+10	−10	+10	−10
Lake level	0.045	−0.040	0.16	0.08	0.006	0.007	0.007	−0.007	0.073	0.10

Alifujiang et al. (2017)

period from 2013 to 2040; analyzing the relationships between the corresponding parameters is expressed quantitatively.

Scenario 1: Impact of the river runoff on the lake level:

The first scenario considers the impact of the river discharge into Lake Issyk-Kul based on historical data for 16 rivers. The historical discharge data series for the 16 main tributaries in the Issyk-Kul basin show that those rivers account for a total annual average runoff of 2421 million m^3, with a maximum runoff of 2888 million m^3 and minimum runoff of 2091 million m^3 between 1980 to 2012 (Fig. 9.25). The first scenario was run under the assumption that the river runoff in the future remains close to its maximum runoff and minimum runoff during the past 33 years.

Scenarios 2 and 3: Impact of the water cycle components on the lake level:

The historical annual average evaporation from the Lake Issyk-Kul surface from 1930 to 1950, 1951 to 1970, 1971 to 1990, and 1991 to 2010 was 820.31, 884.81, 907.05, and 941.51 mm, respectively. The evaporation increased from 800.34 mm in 1930 to 942.76 mm in 2012, increasing 142.42 mm in 83 years (+1.716 mm/year) (Salamat et al. 2015). The second scenario was run under the assumption that this positive trend continues for the period 2013–2040.

Scenario 3: The third scenario uses the annual evaporation increase from scenario 2 in combination with predicted precipitation changes. These predictions were based on the representative concentration pathway (RCP) 4.5 and RCP 8.5 climate change scenarios and were extracted from the 20-km Weather Research and

Table 9.8 Different scenarios and description

Scenario	Description
Base run	Variables used the average values of the observation period after 2012
River flow	Surface runoff assumed to close to maximum runoff and minimum runoff in past 33 years, respectively, after 2012
Water cycle components	Based the evaporation data in 2012, assumed the evaporation will increase by 1.76 mm per year after 2012; precipitation changes under RCP 4.5 and RCP 8.5 scenarios after 2012
Decrease in water consumption	Assumed the water consumption decreased by 10% after 2012

Alifujiang et al. (2017)

Forecasting (WRF) model (downscaled and bilinear interpolated) (Yuan 2017) for four meteorological stations (Balykchy, Karakol, Kyzyl-Suu, and Cholpan-Ata) for the period 2013–2040.

Scenario 4: Impact of the water consumption on the lake level:

The annual water consumption for agricultural irrigation, domestic and industrial purposes has decreased from 1120, 16.95, and 10.59 million m^3 in 1980 to 380 million m^3 (−66%), 6 million m^3 (−64%), and 6.8 million m^3 (−36%) in 2012, respectively. This is the result of the disintegration of the Soviet Union between 1989 and 1992, which led to a decline of the agrarian and industrial water use, in combination with adopted lake water management measures which have increased the water use efficiency. This fourth scenario was run under the assumption that the water consumption will be decreased by 10% until 2040.

The simulated results from these four scenarios indicate that the water-level fluctuations vary based on the applied water resource scenario. The scenario simulations (Fig. 9.28) demonstrate that water cycle components and river inflow to the lake are the crucial factors. The behavior of the lake water level in response to changes in the model inputs indicate that

1. The behavior of the lake water level under the scenarios decrease in water consumption (scenarios 4), the annual lake water level will not change much; the annual average lake water level would increase from 1606.532 to 1606.688 m (+0.156 m).
2. The river flow and water cycle components scenarios (1, 2, and 3) have a greater influence on the water level. As the river runoff (scenario 1) close to its maximum runoff and minimum runoff during the past years, the annual average lake water level drops from 1606.532 to 1606.105 m (−0.427 m), or increase to 1606.913 m (+0.381 m). Moreover, as the evaporation increases (scenario 2) by 1.76 mm during the future, the annual average lake water level drops from 1606.532 to 1606.109 m (−0.423 m). Scenario 2, in combination with predicted precipitation changes (RCP 4.5 and RCP 8.5), as scenario 3, have a significant impact on the water level. This impact was greater than for scenario 2 and scenario 4, as the annual average lake water level would decrease from 1606.532 to 1606.013 m (+0.519 m), or increase to 1607 m (+0.468 m).
3. The analysis determines that the river inflow and the water cycle components scenarios significantly impact the lake water level. They are the key strategies for managing the lake water level. Reducing the water consumption in the catchment is indeed leading to an increase of the surface water inflow to the lake and in turn to a rising water level.

The results presented here indicate that changes of the river discharge and the water cycle components are the main driving factors for the Issyk-Kul water-level variations. The issue of water cycle components changes in response to the lake's water level has been discussed for this region. The water balance equation analyzed the factors of the water balance in the Issyk-Kul catchment and concluded that the correlation between these factors and the meteorological conditions is statistically significant. This assessment confirms the results provided by the SD model and is presented here. Salamat et al. (2015) studied data from hydro-meteorological stations around the Issyk-Kul and used correlation analyses to determine the fluctuations of the lake water level that are caused by surface runoff, precipitation, and evaporation. Burul et al. (2016) collected data on water consumption to identify the main water consumers and their likely impact on Lake Issyk-Kul. The results of those two studies are based on linear correlation analyses and thus tend to provide less accurate insights than the results derived from the nonlinear SD approach used here. This is mainly because of SD model's ability to provide proper insights into potential consequences of system perturbation, which are dependent on efficiently recognizing the main constituents and feedback loops among them. The study presented here takes the two main driving forces—water cycle component factors and anthropogenic factors—into account and considers the interactions between them and their influence on the Issyk-Kul water level. These nonlinear correlations and quantitative analyses show that the water cycle components are an essential parameter for the dynamic of the lake water level.

Many researchers have developed SD models for lake water balance assessments in arid regions and have demonstrated that the surface water runoff in arid regions varies both temporally and spatially and is the largest term in the lake's water balance. Among various socio-hydrologic factors, surface water runoff is most commonly related to the lake water level. In a complex water balance system, the SD model is a great way to solve these complex problems as it provides a unique framework for integrating different physical and socioeconomic systems that are critical to the watershed management. Several studies in lake basins have reported how the different social, economic, and environmental subsystems interact (Rusuli et al. 2015; Guo et al. 2001; Li et al. 2010; Hassanzadeh et al. 2012). They developed SD models for the environmental planning and the management of lake basins, and also to simulate the impact of climate change and variability.

According to the Issyk-Kul SD model (Fig. 9.22), some important points have been achieved. The model used in this study includes, as a novel approach, hydrologic and socioeconomic sub-models, implementing the previous research for water-level fluctuation of the Issyk-Kul, and finalized versions of other simple models from other regions. The results show a complex response between variables, such as the connections between the population growth and the positive response of the agricultural area, as farming is the main economic activity in the Issyk-Kul basin. A continued increase of these factors may have a dramatic negative effect on the lake water level. On the other hand, according to the four simulation scenarios, certain changes, like decreases in the water consumption (scenario 4), would stabilize the Issyk-Kul water level. The scenarios showed that the lake system is highly vulnerable to water cycle components (scenarios 2 and 3) and how much inflow from its tributaries is required for a stable water balance (scenario 1).

Overall, the increase of the surface water inflow into the lake or the decrease of water consumption for irrigation alone cannot guarantee a sustainable development for the Issyk-Kul basin. In addition, maintaining a higher or lower lake level has an impact on the social and economic development of this region. Thus, in order to reach a socioeconomically sustainable development, the anthropogenic water consumption, the ecological water use and the proportion of the industrial water must be managed holistically, for example through the promotion of water-saving practices throughout society or the improvement of water use efficiency.

A dynamic water balance model of the Issyk-Kul lake basin was developed for the period from 1980 to 2012, and the relations between all relevant and available environmental and socioeconomic factors were formulated by mathematical equations. The test run with historical data validated this model (CF, NSE, and MBias > 0.91; RMSE and MAE < 0.78). In a second step, the main factors impacting the lake's water level were evaluated via sensitivity analysis and various water resource scenarios. Results based on the sensitivity analysis indicate that the socio-hydrologic factors had different impacts on the lake water level change (SW > E > GW > P > A_{wd} > L_{agr} > p > D_{wd} > I_{wd}), with the main influence coming from the water inflow dynamic, namely the increasing decreasing water withdrawal from lake tributaries. Land use changes, population increase, and water demand decrease were also important factors for the lake water-level variations. Results of four scenario analyses demonstrated that water cycle components changes as the evaporation, and the precipitation and the variations of river runoff into the lake are essential parameters for the dynamic of the lake water level. Overall, the water balance of the Issyk-Kul is highly complex, and any attempt for a sustainable water resource management needs to address a wide range of relevant parameters, from land use change and water consumption and population dynamics.

References

Abadi LSK, Shamsai A, Goharnejad H (2015) An analysis of the sustainability of basin water resources using vensim model. KSCE J Civ Eng 19:1941–1949

Abbaspour KC, Faramarzi M, Ghasemi SS, Yang H (2009) Assessing the impact of climate change on water resources in Iran. Water Resour Res 45. https://doi.org/10.1029/2008wr007615

Abdyzhapar uulu SAU, Abuduwaili J, Shaidyldaeva N (2015) Impact of climate change on water level fluctuation of Issyk-Kul Lake. Arab J Geosci 8: 5361–537

Afolayan SO, Ogedengbe K, Lateef SA, Akintola OA, Oladele OJ (2014) Response of tomato (Lycopersicium lycopersicun, CV UC82B) to drip irrigation and planting conditions. Afr J Agric Res 9:1543–1549

Ahmad S, Prashar D (2010) Evaluating municipal water conservation policies using a dynamic simulation model. Water Resour Manage 24:3371–3395

Ahmad S, Simonovic SP (2001) Integration of heuristic knowledge with analytical tools for the selection of flood damage reduction measures. Can J Civ Eng 28:208–221

Aizen VB, Aizen EM, Melack JN, Dozier J (1997) Climatic and hydrologic changes in the Tien Shan, Central Asia. J Clim 10:1393–1403

Aladin NV, Plotnikov IS (1993) Large saline lakes of former USSR: a summary review. Hydrobiology 267:1–12 (in Russian)

Alamanov A, Mikkola H (2011) Is biodiversity friendly fisheries management possible on Issyk-Kul Lake in the Kyrgyz Republic? AMBIO 40:479–495

Alcamo J, Doll P, Henrichs T, Kaspar F, Lehner B, Rosch T, Siebert S (2003) Global estimates of water withdrawals and availability under current and future "business-as-usual" conditions. Hydrol Sci J 48:339–348

Alymkulova B, Abuduwaili J, Issanova G, Nahayo L (2016) Consideration of water uses for its sustainable management, the case of Issyk-Kul Lake, Kyrgyzstan. Water 8:298

Amanaliev M (2008) Water resources management system of Kyrgyzstan. In: Sustainable use and development of watersheds, pp 61–80

Angelakis AN, Zheng XY (2015) Evolution of water supply, sanitation, wastewater, and stormwater technologies globally. Water 7:455–463

Antwi-Agyei P, Fraser ED, Dougill AJ, Stringer LC, Simelton E (2012) Mapping the vulnerability of crop production to drought in Ghana using rainfall, yield and socioeconomic data. Appl Geogr 32:324–334

Asykulov T (2002) The socio-economics and natural environment of eastern Kyrgyzstan and development issues of biosphere reserve Issyk-Kul. Dissertation, Graysvald (in Russian)

Atadjanov S, Tulegabylov N, Bekkulova D, Baidakova N, Grebnev V (2012) The national report on the state of the environment of the Kyrgyz Republic for 2006–2011. United Nations Development Programme (UNDP), New York, NY, USA

Avramoski O (2004) The role of public participation and citizen involvement in lake basin management. Available online: http://www.worldlakes.org/uploads/Thematic_Paper_PP_16Feb04.pdf. Accessed 14 Apr 2004

Badescu V, Schuiling RD (2010) Aral Sea; irretrievable loss or irtysh imports? Water Resour Manage 24:597–616

Bai J, Chen X, Li J, Yang L, Fang H (2011) Changes in the area of inland lakes in arid regions of Central Asia during the past 30 years. Environ Monit Assess 178:247–256. https://doi.org/10.1007/s10661-010-1686-9

Bayer S (2004) Business dynamics: systems thinking and modeling for a complex world. Interfaces 34:324–326

Berdovskaya NI, Egorov AN (1986) Reconstruction of nature condition between Pleistocene and Holocene. Issyk-Kul Lake and its natural development tendency. Nauka, Leningrad, pp 165–179 (in Russian)

Berg LS (1930) Hydrologic research in Issyk-Kul Lake in 1928. In: Proceedings of the State Hydrological Institute. No 28, pp 9–24

Bolch T (2007) Climate change and glacier retreat in northern Tien Shan (Kazakhstan/Kyrgyzstan) using remote sensing data. Global Planet Change 56:1–12

Bolch T, Peters J, Egorov A, Pradhan B, Buchroithner M, Blagoveshensky V (2011) Identification of potentially dangerous glacial lakes in the northern Tien Shan. Nat Hazards 59:1691–1714. https://doi.org/10.1007/s11069-011-9860-2

Bolshakov NM (1974) Water resources of rivers of the Soviet Tian Shan and methods of their calculation. Ilim, Frunze, 305 p (in Russian)

Bolshakov MN, Shpak VG (1960) Water and energy resources of Kyrgyz SSR. Academy of Sciences of Kyrgyz SSR, Frunze, 248 p (in Russian)

Bolshakov VA, Virina EI, Faustov SS, Chernova LM (1986) Magnetic-properties of Issyk-Kul Lake bottom sediments. Izv Akad Nauk SSSR Fiz Zemli 1:99–105

Braden JB, Jolejole-Foreman MC, Schneider DW (2013) Humans and the water environment: the need for coordinated data collection. Water 6:1–16

Butler C, Adamowski J (2015) Empowering marginalized communities in water resources management: addressing inequitable practices in participatory model building. J Environ Manage 153:153–162

Caruso B, Newton S, King R, Zammit C (2017) Modelling climate change impacts on hydropower lake inflows and braided rivers in a mountain basin. Hydrol Sci J 62:928–946

Choubin B, Khalighi-Sigaroodi S, Malekian A, Ahmad S, Attarod P (2014) Drought forecasting in a semi-arid watershed using climate signals: a neuro-fuzzy modeling approach. J Mt Sci 11:1593–1605

Chupakhin VM (1964) Physical geography of Tian Shan. Academy of Sciences, Alma-Ata, 373 p (in Russian)

Dawadi S, Ahmad S (2013) Evaluating the impact of demand-side management on water resources under changing climatic conditions and increasing population. J Environ Manage 114:261–275

De Batist M, Imbo Y, Vermeesch P, Klerkx J, Giralt S, Delvaux D, Lignier V, Beck C, Kalugin I, Abdrakhmatov KE (2002) Bathymetry and sedimentary environments of Lake Issyk-Kul, Kyrgyz Republic (Central Asia): a large, high-altitude, tectonic lake. Lake Issyk-Kul Its Nat Environ 13:101–123

De Grave J, Glorie S, Buslov MM, Stockli DF, McWilliams MO, Batalev VY, Van den Haute P (2013) Thermo-tectonic history of the Issyk-Kul basement (Kyrgyz northern Tien Shan, Central Asia). Gondwana Res 23:998–1020

Deng HJ, Chen YN, Wang HJ, Zhang SH (2015) Climate change with elevation and its potential impact on water resources in the Tianshan Mountains, Central Asia. Glob Planet Change 135:28–37

Dikikh AO (2000) Glaciations in the Issyk-Kul Basin: its role as a flow source. In: Nature and people of Kyrgyzstan. Special edition. Bishkek

Dukhovny VA, Sokolov VI, Ziganshina DR (2013) Integrated water resources management in central Asia, as a way of survival in conditions of water scarcity. Quat Int 311:181–188

Dusini DS, Foster DL, Shore JA, Merry C (2009) The effect of Lake Erie water level variations on sediment resuspension. J Gt Lakes Res 35:1–12

Farre I, Faci JM (2009) Deficit irrigation in maize for reducing agricultural water use in a mediterranean environment. Agric Water Manage 96:383–394

Fereres E, Soriano MA (2007) Deficit irrigation for reducing agricultural water use. J Exp Bot 58:147–159

Forrester JW (1971) Counterintuitive behavior of social systems. Theor Decis 2:109–144

Fullerton TM, Elias A (2004) Short-term water consumption dynamics in El Paso, Texas. Water Resour Res 40(8)

Gavshin VM, Sukhorukov FV, Bobrov VA, Melgunov MS, Miroshnichenko LV, Klerkx J, Kovalev SI, Romashkin PA (2004) Chemical composition of the uranium tail storages at Kadji-Sai (southern shore of Issyk-Kul lake, Kyrgyzstan). Water Air Soil Pollut 154:71–83

Gavshin VM, Melgunov MS, Sukhorukov FV, Bobrov VA, Kalugin IA, Klerkx J (2005) Disequilibrium between uranium and its progeny in the Lake Issyk-Kul system (Kyrgyzstan) under a combined effect of natural and manmade processes. J Environ Radioact 83:61–74

Giralt S, Klerkx J, Riera S, Julia R, Lignier V, Beck C, De Batist M, Kalugin I (2002) Recent paleoenvironmental evolution of lake Issyk-Kul, in lake Issyk-Kul: its natural environment. Springer, Berlin, pp 125–145

Griggs D, Stafford-Smith M, Gaffney O, Rockström J, Öhman MC, Shyamsundar P, Steffen W, Glaser G, Kanie N, Noble I (2013) Policy: sustainable development goals for people and planet. Nature 495:305–307

Gronskaya TP (1983) Water balance and expecting water level of Issyk-Kul lake. Dissertation, State Hydrology Inst, Leningrad (in Russian)

Guo HC, Liu L, Huang GH, Fuller GA, Zou R, Yin YY (2001) A system dynamics approach for regional environmental planning and management: a study for the Lake Erhai Basin. J Environ Manage 61:93–111

Hassanzadeh E, Zarghami M, Hassanzadeh Y (2012) Determining the main factors in declining the Urmia Lake level by using system dynamics modeling. Water Resour Manage 26:129–145

Heinicke T (2003) Mires within the dry steppe zone of the Issyk-Kul basin (Kyrgyzstan)—part 1: soils, stratigraphy and hydrology. Telma 33:35–58

Hofer M, Peeters F, Aeschbach-Hertig W, Brennwald M, Holocher J, Livingstone DM, Romanovski V, Kipfer R (2002) Rapid deep-water renewal in Lake Issyk-Kul (Kyrgyzstan) indicated by transient tracers. Limnol Oceanogr 47:1210–1216

Hofmann H, Lorke A, Peeters F (2008) Temporal scales of water-level fluctuations in lakes and their ecological implications. Hydrobiologia 613:85–96. https://doi.org/10.1007/s10750-008-9474-1

Huntington TG (2006) Evidence for intensification of the global water cycle: review and synthesis. J Hydrol 319(1–4):83–95. https://doi.org/10.1016/j.jhydrol.2005.07.003

Hwang C, Kao YC, Tangdamrongsub N (2011) Preliminary analysis of lake level and water storage changes over lakes Baikal and Balkhash from satellite altimetry and gravimetry. Terr Atmos Ocean Sci 22:97–108

Ibatullin S (2009) Sector report: the impact of climate change on water resources in central Asia. Eurasian Development Bank, Almaty, Kazakhstan

Issaev DI (1959) The features of relief of the Issyk-Kul hollow. In: Questions on physical geography of the Tian Shan. Frunze, 40–46 p (in Russian)

Issaev DI (1962) Relief. In: Nature of Kyrgyzia. Frunze. 72–86 p (in Russian)

Jailoobayev A, Neronova T, Nikolayenko A, Mirkhashimov I (2009) Water quality standards and norms in Kyrgyz Republic. Regional Environmental Centre for Central Asia (CAREC), Almaty, Kazakhstan

Jalili S, Kirchner I, Livingstone DM, Morid S (2012) The influence of large-scale atmospheric circulation weather types on variations in the water level of Lake Urmia, Iran. Int J Climatol 32:1990–1996

Jellison R, Zadereev YS, DasSarma PA, Melack JM, Rosen MR, Degermendzhy AG, DasSarma S, Zambrana G (2016) Conservation and management challenges of saline lakes: a review of five experience briefs. Lake basin management initiative: thematic paper. Available online: http://citeseerx.ist.psu.edu/viewdoc/download?doi=10.1.1.552.9109&rep=rep1&type=pdf. Accessed on 14 July 2016

Ji J, Yang R, Sun ZH, Yuan HY (2006) Validation and evaluation of large eddy simulation based field-zone model for smoke movement in building fires. Prog Saf Sci Technol 6:721–724

Jiang SH, Ren LL, Hong Y, Yong B, Yang XL, Yuan F, Ma MW (2012) Comprehensive evaluation of multi-satellite precipitation products with a dense rain gauge network and optimally merging their simulated hydrological flows using the bayesian model averaging method. J Hydrol 452:213–225

Kadyrov VK (1986) Hydrochemistry of Issyk-Kul and its basin. Ilim, Frunze (in Russian)

Kaigorodov AI (1955) The natural zonal classification of the climate in Earth. Academy of Sciences, Moscow, 119 p (in Russian)

Kalra A, Ahmad S, Nayak A (2013) Increasing streamflow forecast lead time for snowmelt-driven catchment based on large-scale climate patterns. Adv Water Resour 53:150–162

Kaplinsky MI, Timchenko LS (1997) Water balance of Lake Issyk-Kul over many years. Bull SARNIGMI iss 50:87–101 (in Russian)

Karmanchuk AS (2002) Water chemistry and ecology of Lake Issyk-Kul. Lake Issyk-Kul Its Nat Environ 13:13–26

Karthe D, Chalov S, Borchardt D (2015) Water resources and their management in central Asia in the early twenty first century: status, challenges and future prospects. Environ Earth Sci 73:487–499

Kawabata Y, Kurita T, Nagai M, Aparin V, Onwona-Agyeman S, Yamada M, Fujii Y, Katayama Y (2014) Water quality in the Issyk-Kul and the river flowing into it. J Arid Land Stud 24(1):105–108

Kazimir A, Karimov KA, Gainutdinova RD (2011) Regional climate changes in Kyrgyzstan: impact of natural and anthropogenic factors. In: Baba A et al (eds) Climate change and its effects on water resources. NATO Science for peace and security series C: environ security 3. https://doi.org/10.1007/978-94-007-1143-3_15

Kezer K, Matsuyama H (2006) Decrease of river runoff in the Lake Balkhash basin in Central Asia. Hydrol Process 20:1407–1423

Klerx J, Imanackunov B (2002) Lake Issyk-Kul: its natural environment, vol 13. Springer Science & Business Media, Berlin

Koop SH, van Leeuwen CJ (2015) Assessment of the sustainability of water resources management: a critical review of the city blueprint approach. Water Resour Manage 29:5649–5670

Korolev VG (1978) Issyk-Kul lake. Ilim, Frunze, 209 p (in Russian)

Kotir JH, Smith C, Brown G, Marshall N, Johnstone R (2016) A system dynamics simulation model for sustainable water resources management and agricultural development in the Volta River Basin, Ghana. Sci Total Environ 573:444–457

Kundzewicz ZW, Mata LJ, Arnell NW, Döll P, Jimenez B, Miller K, Oki T, Şen Z, Shiklomanov I (2008) The implications of projected climate change for freshwater resources and their management. Hydrol Sci J 53:3–10

Kutseva PP (1980) Elements of the balance of chemical substances in the Issyk-Kul Lake. Collection of articles: studies of water balance, thermal and hydrochemical regime of Lake Issyk-Kul. Hydrometeorological Publishing, Leningrad

Lamek Nahayo LL, Zhao X (2016) Consideration of precipitation variability under climate change at Kaidu river watershed, China. East Afr J Sci Technol 6:16–23

Lehner B, D ll P (2004) Development and validation of a global database of lakes, reservoirs and wetlands. J Hydrol 296:1–22

Levchenko VM (1946) Mineral waters from a geochemical point of view, vol 4–5. Izvestia of Kyrgyz Department of the Academy of Sciences of USSR, 101–105 pp (in Russian)

Levchenko VM (1953) Classification of the natural waters. Materials of hydrochemistry. Part 21, 86–96 pp (in Russian)

Li YH, Jin HL, Zhao SW (2010) Based on system dynamics model of evolvement of water resources and social economic system in Xinjiang Aibi Lake area. Model Simul 4:122–125

Li B, Chen Y, Chen Z, Li W (2012) Trends in runoff versus climate change in typical rivers in the arid region of northwest China. Quat Int 282:87–95

Libert B, Orolbaev E, Steklov Y (2008) Water and energy crisis in Central Asia. In: China and Eurasia forum quarterly, vol 6. Central Asia-Caucasus Institute & Silk Road Studies Program, Bishkek, Kyrgyzstan, pp 9–20

Ling H, Xu H, Shi W, Zhang Q (2011) Regional climate change and its effects on the runoff of Manas river, Xinjiang, China. Environ Earth Sci 64:2203–2213. https://doi.org/10.1007/s12665-011-1048-2

Lyons WB, Welch KA, Bonzongo JC, Graham EY, Shabunin G, Gaudette HE, Poreda RJ (2001) A preliminary assessment of the geochemical dynamics of Issyk-Kul Lake, Kirghizstan. Limnol Oceanogr 46:713–718

Ma M, Wang X, Veroustraete F, Dong L (2007) Change in area of Ebinur Lake during the 1998–2005 period. Int J Remote Sens 28(24):5523–5533

Ma L, Wu J, Abuduwaili J (2011) The climatic and hydrological changes and environmental responses recorded in lake sediments of Xinjiang,China. J Arid Land 3(1):1–8

Ma L, Wu J, Abuduwaili J (2012) Climate and environmental changes over the past 150 years inferred from the sediments of Chaiwopu Lake, Central Tianshan Mountains, Northwest China. Int J Earth Sci. https://doi.org/10.1007/s00531-012-0838-4

Macaulay EA, Sobel ER, Mikolaichuk A, Wack M, Gilder SA, Mulch A, Fortuna AB, Hynek S, Apayarov F (2016) The sedimentary record of the Issyk Kul basin, Kyrgyzstan: climatic and tectonic inferences. Basin Res 28:57–80

Makhmudov EJ, Makhmudov IE, Sherfedinov LZ (2008) Problems of water resource management in central Asia. In: Moerlins JE, Khankhasayev MK, Leitman SF, Makhmudov EJ (eds) Transboundary water resources: a foundation for regional stability in central Asia. Springer, Dordrecht, pp 11–28

Mamatkanov DM, Bajanova LV, Romanovskiy VV (2006) Water resources of Kyrgyzstan in modern days. Ilim, Bishkek (in Russian)

Mancosu N, Snyder RL, Kyriakakis G, Spano D (2015) Water scarcity and future challenges for food production. Water 7:975–992

Mandychev AN (2002) Groundwater in the Issyk Kul basin. Lake Issyk-Kul Its Nat Environ 13:71–76

Mason IM, Guzkowska MA, Rapley CG (1994) The response of lake levels and areas to climate change. Clim Change 27:161–197

Matsuyama H, Kezer K (2009) Long-term variation of precipitation around Lake Balkhash in central Asia from the end of the 19th century. Sola 5:73–76

Matveev VP (1930) Hydrological and hydrochemical researches in Issyk-Kul in 1928. Materials of the Commission for Expeditionary Research of the USSR Academy of Sciences. Issyk Kul expedition of 1928. Leningrad, pp 71–110

Matveev VP (1932). About water salinity of the Issyk-Kul lake, vol 6. West State Hydrological Institute,80–90 pp (in Russian)

Matveev VP (1935) Hydrological researches in Issyk-Kul in 1932. Lake Issyk-Kul. Materials on hydrology, ichthyology and fisheries, Issue 2. Publishing House of the USSR Academy of Sciences, Moscow, pp 7–56

Meybeck M (2003) Global analysis of river systems: from earth system controls to anthropocene syndromes. Philos Trans R Soc B 358:1935–1955

Mirchi A, Madani K, Watkins D, Ahmad S (2012) Synthesis of system dynamics tools for holistic conceptualization of water resources problems. Water Resour Manage 26:2421–2442

Moazami S, Golian S, Kavianpour MR, Hong Y (2013) Comparison of PERSIANN and V7 TRMM Multi-satellite Precipitation Analysis (TMPA) products with rain gauge data over Iran. Int J Remote Sens 34:8156–8171

Morrison J, Quick MC, Foreman MGG (2002) Climate change in the Fraser river watershed: flow and temperature projections. J Hydrol 263:230–244

Omlin M, Brun R, Reichert P (2001) Biogeochemical model of Lake Zurich: sensitivity, identifiability and uncertainty analysis. Ecol Model 141:105–123

Ososkova T, Gorelkin N, Chub V (2000) Water resources of central Asia and adaptation measures for climate change. Environ Monit Assess 61:161–166

Peeters FD, Hofer M, Brennwald DM, Livingstone Kipfer R (2003) Deep-water renewal in Lake Issyk-Kul driven by differential cooling. Limnol Oceanogr 48:1419–1431

Pereira LS, Oweis T, Zairi A (2002) Irrigation management under water scarcity. Agric Water Manage 57:175–206

Pereira Filho AJ, Carbone RE, Janowiak JE, Arkin P, Joyce R, Hallak R, Ramos CG (2010) Satellite rainfall estimates over South America-possible applicability to the water management of largewatersheds. Wiley, Hoboken

Perova OA (1960) Energy resources of the Sun in Kyrgyzia. All Union Geographical Community Issue 2, 43–57 p (in Russian)

Piao S, Ciais P, Huang Y, Shen Z, Peng S, Li J, Zhou L, Liu H, Ma Y, Ding Y et al (2010) The impacts of climate change on water resources and agriculture in China. Nature 467:43–51

Polunin NVC (2008) Aquatic ecosystems: trends and global prospects. Cambridge University Press, Cambridge, pp 94–110

Propastin P (2013) Assessment of climate and human induced disaster risk over shared water resources in the Balkhash Lake drainage basin. Springer, Berlin, pp 41–54

Qaiser K, Ahmad S, Johnson W, Batista J (2011) Evaluating the impact of water conservation on fate of outdoor water use: a study in an arid region. J Environ Manage 92:2061–2068

Qaiser K, Ahmad S, Johnson W, Batista JR (2013) Evaluating water conservation and reuse policies using a dynamic water balance model. Environ Manage 51:449–458

Qin BQ, Yu G (1998) Implications of lake level variations at 6 ka and 18 ka in mainland Asia. Glob Planet Change 18:59–72

Ragni M, Steffenhagen F, Klein A (2011) Generalized dynamic stock and flow systems: an AI approach. Cogn Syst Res 12:309–320

Ramana RV, Krishna B, Kumar SR, Pandey NG (2013) Monthly rainfall prediction using wavelet neural network analysis. Water Resour Manage 27:3697–3711

Rantzman EY (1959) Geomorphology of the Issyk-Kul hollow and its Mountain frame. Academy of Sciences of the USSR, Moscow, 88 p (in Russian)

Ricketts RD, Johnson TC, Brown ET, Rasmussen KA, Romanovsky VV (2001) The Holocene paleolimnology of Lake Issyk-Kul, Kyrgyzstan: trace element and stable isotope composition of ostracodes. Palaeogeogr Palaeoclimatol Palaeoecol 176:207–227

Romanovsky VV (2002a) Water level variations and water balance of Lake Issyk-Kul. In: Klerkx J, Imanackunov B (eds) Lake Issyk-Kul: its natural environment, NATO science series. Kluwer Academic, The Netherlands, pp 45–57

Romanovsky VV (2002b) Water level variations and water balance of Lake Issyk-Kul. Lake Issyk-Kul Its Nat Environ 13:45–57

Romanovsky VV, Tashbaeva S, Crétaux J-F, Calmant S, Drolon V (2013) The closed Lake Issyk-Kul as an indicator of global warming in Tien-Shan. Nat Sci 5:608–623

Romanovsky V, Mamatkanov D, Kuz'michenok V, Podrezov O (2014) All about the Issyk-Kul Lake. Bishkek, 444 pp (in Russian)

Rost S, Gerten D, Bondeau A, Lucht W, Rohwer J, Schaphoff S (2008) Agricultural green and blue water consumption and its influence on the global water system. Water Resour Res 44(9)

Rusuli Y, Li L, Ahmad S, Zhao X (2015) Dynamics model to simulate water and salt balance of bosten lake in Xinjiang, China. Environ Earth Sci 74:2499–2510

Rykiel EJ (1996) Testing ecological models: the meaning of validation. Ecol Model 90:229–244

Salamat AU, Abuduwaili J, Shaidyldaeva N (2015) Impact of climate change on water level fluctuation of Issyk-Kul Lake. Arab J Geosci 8:5361–5371

Sapkota M, Arora M, Malano H, Moglia M, Sharma A, George B, Pamminger F (2014) An overview of hybrid water supply systems in the context of urban water management: challenges and opportunities. Water 7:153–174

Savvaitova K, Petr T (1992) Lake Issyk-Kul, Kirgizia. Int J Salt Lake Res 1(2):21–46

Schmidt K (1882) Research on the waters of Issyk-Kul Lake. Pharmacist J 878–884

Sehring J (2007) Irrigation reform in Kyrgyzstan and Tajikistan. Irrig Drain Syst 21:277–290. https://doi.org/10.1007/s10795-007-9036-0

Setegn SG, Rayner D, Melesse AM, Dargahi B, Srinivasan R (2011) Impact of climate change on the hydroclimatology of Lake Tana Basin, Ethiopia. Water Resour Res 47. https://doi.org/10.1029/2010wr009248

Shabunin GD, Shabunin AG (2002) Climate and physical properties of water in Lake Issyk-Kul. Lake Issyk-Kul Its Nat Environ 13:3–11

Shnitnikov AV (1979) Issyk-Kul. Nature, protection and perspective of lake use. Academy of Science of Kirghiz SSR, Kirghiz Society of Geography. Ilim, Frunze (in Russian)

Shnitnikov AV (1980) Lakes of Tien Shan and their history. Nauka, Leningrad (in Russian)

Shrestha E, Ahmad S, Johnson W, Batista JR (2012) The carbon footprint of water management policy options. Energy Policy 42:201–212

Shul'ts VL (1965) Rivers in Central Asia. Hydrometeoizdat, Leningrad, 691 p

Sidike A, Chen X, Liu T, Durdiev K, Huang Y (2016) Investigating alternative climate data sources for hydrological simulations in the upstream of the Amu Darya River. Water 8:441

Siegfried T, Bernauer T, Guiennet R, Sellars S, Robertson AW, Mankin J, Bauer-Gottwein P, Yakovlev A (2012) Will climate change exacerbate water stress in Central Asia? Clim Change 112:881–899. https://doi.org/10.1007/s10584-011-0253-z

Smiatek G, Kunstmann H, Heckl A (2014) High-resolution climate change impact analysis on expected future water availability in the Upper Jordan catchment and the Middle East. J Hydrometeorol 15:1517–1531

Song CQ, Huang B, Ke LH, Richards KS (2014) Seasonal and abrupt changes in the water level of closed lakes on the Tibetan Plateau and implications for climate impacts. J Hydrol 514:131–144

Stanev EV, Peneva EL, Mercier F (2004) Temporal and spatial patterns of sea level in inland basins: recent events in the Aral Sea. Geophys Res 31(5505). https://doi.org/10.1029/2004gl020478

Taft BJ, Philippe LR, Dietrich CH, Robertson KR (2011) Grassland composition, structure, and diversity patterns along major environmental gradients in the Central Tien Shan. Plant Ecol 212:1349–1361. https://doi.org/10.1007/s11258-011-9911-5
Törnqvist R, Jarsjö J, Karimov B (2011) Health risks from large-scale water pollution: trends in Central Asia. Environ Int 37:435–442
Trofimov AK, Grigina OM (1979) Paleogeography of Issyk-Kul lake//Littoral zone of Issyk-Kul lake. Ilim, Frunze (in Russian)
Tsigelnaya ID (1995) Issyk-Kul Lake. Enclosed seas and large lakes of Eastern Europe and Middle Asia, vol 199. SPB Academic Publishing, Amsterdam, p 229
Tynybekov A (2011) Glacier degradation from GIS and remote sensing data. In: Kogan F et al. (eds) Use of satellite and in-situ data to improve sustainability. NATO Science for peace and security series C: environ security. Springer, Berlin, pp 159–274. https://doi.org/10.1007/978-90-481-9618-0_18
Uralbekov BM, Smodis B, Burkitbayev M (2011) Uranium in natural waters sampled within former uranium mining sites in Kazakhstan and Kyrgyzstan. J Radioanal Nucl Chem 289:805–810
Valipour M (2015) Future of agricultural water management in Africa. Arch Agron Soil Sci 61:907–927
Velazquez E (2006) An input–output model of water consumption: analysing intersectoral water relationships in Andalusia. Ecol Econ 56:226–240
Ventana Systems (2012) Vensim DSS. 6.0b ed. Ventana systems, Inc., Harvard, MA. Available online: http://vensim.com/vensim-software/. Accessed on 18 Dec 2017
Vermeesch P, Poort J, Duchkov AD, Klerkx J, de Batist M (2004) Lake Issyk-Kul (Tien Shan): unusually low heat flow in an active intermontane basin. Geol Geofiz 45:616–625
Viala E (2008) Water for food, water for life a comprehensive assessment of water management in agriculture. Irrig Drain Syst 22:127–129
Vollmer MK, Weiss RF, Schlosser P, Williams RT (2002a) Deep-water renewal in Lake Issyk-Kul. Geophys Res Lett 29(8). https://doi.org/10.1029/2002gl014763
Vollmer MK, Weiss RF, Williams RT, Falkner KK, Qiu X, Ralph EA, Romanovsky VV (2002b) Physical and chemical properties of the waters of saline lakes and their importance for deep-water renewal: Lake Issyk-Kul, Kyrgyzstan. Geochim Cosmochim Acta 66:4235–4246
Vonk J, Shackelford TK (2012) The oxford handbook of comparative evolutionary psychology. Oxford University Press, New York, p 574
Vörösmarty CJ, Green P, Salisbury J, Lammers RB (2000) Global water resources: vulnerability from climate change and population growth. Science 289:284–288
Vorosmarty CJ, McIntyre PB, Gessner MO, Dudgeon D, Prusevich A, Green P, Glidden S, Bunn SE, Sullivan CA, Liermann CR et al (2010) Global threats to human water security and river biodiversity. Nature 467:555–561
Wang GY, Shen YP, Qin DH (2006) Issyk-Kul Lake level fluctuation during 1860–2005 and its relation with regional climatic and hydrological changes. J Glaciol Geocryol 28:855–860
Wang GY, Shen YP, Wang NL, Wu QB (2010) The effects of climate change and human activities on the lake level of the Issyk-Kul during the past 100 years. Glaciol Geocryol 32:1097–1105
Waugh DW, Vollmer MK, Weiss RF, Haine TWN, Hall TM (2002) Transit time distributions in Lake Issyk-Kul. Geophys Res Lett 29. https://doi.org/10.1029/2002gl016201
White CJ, Tanton TW, Rycroft DW (2014) The impact of climate change on the water resources of the Amu Darya basin in central Asia. Water Resour Manage 28:5267–5281
Wilson JR, Holst N, Rees M (2005) Determinants and patterns of population growth in water hyacinth. Aquat Bot 81:51–67
Winz I, Brierley G, Trowsdale S (2009) The use of system dynamics simulation in water resources management. Water Resour Manage 23:1301–1323
Xu CC, Chen YY, Li WH, Chen YP (2006) Climate change and hydrologic process response in the Tarim river basin over the past 50 years. Chin Sci Bull 51(Supp. I):25–36. https://doi.org/10.1007/s11434-006-8204-1

Yaeger MA, Housh M, Cai XM, Sivapalan M (2014) An integrated modeling framework for exploring flow regime and water quality changes with increasing biofuel crop production in the US Corn Belt. Water Resour Res 50:9385–9404

Yessekin B, Bogachev V (2004) Water resources and sustainable development in central Asia: roles of risk assessment and risk management. In: Teaf CM, Yessekin BK, Khankhasayev MK (eds) Risk assessment as a tool for water resources decision-making in central Asia, vol 34. Springer, Dordrecht, pp 197–212

Yuan Q (2017) Precipitation in central Asia in the mid- and late-21st century. J Clim (manuscript submitted for publication)

Yuan YJ, Zeng GM, Liang J, Huang L, Hua SS, Li F, Zhu Y, Wu HP, Liu JY, He XX et al (2015) Variation of water level in dongting lake over a 50-year period: implications for the impacts of anthropogenic and climatic factors. J Hydrol 525:450–456

Zabirov RD (1963) Issyk-Kul. Academic Science of Kirghiz SSR, Kyrgyz Republic

Chapter 10
Water Resources and Lakes in Uzbekistan

10.1 The Main River Basins in Uzbekistan

There are two river basins in Uzbekistan that form the Aral Sea Basin: The Amudarya and Syrdarya river basins.

The Amudarya river basin covers 81.5% of the country's territory (FAO 2012). There are three sections of the river: (1) the upper course flowing along the border with Afghanistan and Tajikistan, where most of the water flow is formed; (2) the middle course, which first passes along the border of Uzbekistan and Afghanistan and then enters Turkmenistan, (3) as well as the lower course passing through the territory from Uzbekistan to the confluence of the Aral Sea. The Surkhandarya river, Sherabad, Kashkadarya, and Zeravshan rivers are the main tributaries of the Amudarya river in Uzbekistan. The Surkhandarya and Zeravshan rivers originate in Tajikistan. The Zeravshan river was the largest tributary of the Amudarya river before using its waters for irrigation. After this, the remaining part of the runoff evaporates in the Kyzylkum desert near the city of Bukhara. The total volume of runoff formed in the basin of the Amudarya river is estimated at an average of 78.46 km^3/year. It is calculated by summing up internal renewable surface water resources of the basin in different countries as follows: in Tajikistan 59.45 km^3/year, in Kyrgyzstan 1.93 km^3/year, in Afghanistan 11.70 km^3/year, in Uzbekistan 4.70 km^3/year, and in Turkmenistan 0.68 km^3/year, while 5 and 95% of the probability of runoff are estimated at 108.4 and 46.9 km^3/year, respectively (FAO 2012). For the period from April to September, it is 77–80%, and for the period from December to February is 10–13% of the annual runoff. Such intra-annual distribution of flow is very favorable for irrigated agriculture. Due to the significant loss of water of the river flowing in the desert, as well as the selection of most of the water by agriculture, the remaining flow reaching the Aral Sea is less than 10% in the most arid years. About 4.7 km^3/year or 6% of the average total volume of surface water resources in the basin of the Amudarya river are formed on the territory of Uzbekistan.

J. Abuduwaili et al., *Hydrology and Limnology of Central Asia*, Water Resources Development and Management, https://doi.org/10.1007/978-981-13-0929-8_10

The Syrdarya river basin. This basin covers 13.5% of the country's territory. The main course of the Syrdarya river can also be divided into three parts: (1) The upper stream in Kyrgyzstan, where most of the water flow is formed; (2) the middle flow in Uzbekistan and Tajikistan, as well as (3) the lower flow in Kazakhstan, before the confluence of the Aral Sea. The main tributaries of the Syrdarya river in Uzbekistan are the rivers of Chirchik and Akhangaran, which originate in Kyrgyzstan. The total volume of flow produced in the Syrdarya river basin is estimated at 36.57 km^3/year and calculated by summing up the internal renewable surface water resources of the basin in different countries as follows: in Kyrgyzstan 27.42 km^3/year, in Tajikistan 1.01 km^3/year, in Uzbekistan 4.84 km^3/year, and in Kazakhstan 3.3 km^3/year, with a 5 and 95% probability of flow, respectively, at 54.1 and 21.4 km^3/year (FAO 2012). Due to the significant loss of water of the river flowing in the desert, as well as most of the water used by agriculture, and the remaining flow reaching the Aral Sea is less than 10% in the most arid years. About 4.84 km^3/year or 13% of the average volume of surface water resources in the Syrdarya river basin are formed in the territory of Uzbekistan.

There are thousands of small streams in Uzbekistan that disappear in the desert. Many of them are fully used for irrigation.

The total river flow in Uzbekistan is estimated at 9.54 km^3/year, of which 49% of the Amudarya river basin and 51% of the Syrdarya river basin. The share of surface water resources used by Uzbekistan is calculated every year, depending on the climatic conditions of the year and water discharge of rivers (FAO 2012).

10.2 Lakes and Water Reservoirs as a Source of Manufacture/Production

In Uzbekistan, there are few lakes (slightly more than 500), and the largest number of lakes is located in the mountainous area at an altitude from 2000 to 3000 m above sea level. The area of the vast mountain lakes does not exceed 1.0 km^2. The lakes on the plains are mainly river like, and they are located in floodplains and river deltas. There are 32 such lakes with an area of more than 10 km^2 in Uzbekistan (Table 10.1). Lakes formed as a result of water discharge from irrigation systems are also widespread. These lakes depend on the reclamation (amelioration) state of irrigated lands. The lakes of the Arnasay system in the Djizak region, as well as a chain of lakes located along the periphery of the Khorezm oasis, are a typical representative of such lakes. In addition, new artificial lakes–water reservoirs appeared on the territory of Uzbekistan in the twentieth century. They carry out seasonal regulation of the water flow and refer to irrigation and integrated water reservoirs. The size and volume of accumulated water in the water reservoirs of Uzbekistan exceed the same values in natural lakes (Chen Xi et al. 2013).

There are at least 50 water reservoirs in Uzbekistan, with a total capacity of more than 22 km^3 (FAO 2012). The largest water reservoirs are multi-purpose and used for irrigation, flood water regulation, and power generation. The Charvak and

Table 10.1 Characteristics of the main lakes in Uzbekistan

No.	Lake name	Location		Area/km²	Perimeter/km
		N (°)	E (°)		
1	Aral Sea	45.13	60.08	7 803.8	598.3
2	Lake Idar	40.6	66.5	3 223.2	385.5
3	Sodoch Lake	43.44	58.49	330.0	124.2
4	Tudor Lake	34.80	64.80	195.2	62.3
5	Lake Kumzar	39.85	64.83	140.0	53.2
6	Kennedy Geez Lake	39.12	64.18	101.5	74.4
7	Chimkulgan Lake	38.95	66.40	34.0	27.8
8	Ulon Sokur Lake	41.23	60.50	26.0	33.1
9	Lake Tildock	43.67	59.19	25.4	31.9
10	Bashi Achima Lake	40.63	64.55	19.0	19.1
11	Lake Kurt	43.25	60.38	18.8	24.3
12	Lake Saul	40.33	64.86	18.8	16.5
13	Lake Ribataski Lake	43.58	58.89	18.2	23.2
14	Kushzamm Lake	41.25	60.74	16.4	39.3
15	Tashkent Reservoir	40.96	69.33	11.9	21.9
16	Lake Gizark	40.06	67.94	11.6	17.1
17	Lake Kumzur	39.87	64.71	11.2	17.4
18	Lake Araca	41.87	60.85	10.0	18.9
19	Lake Uzil	37.36	67.23	8.1	16.5
20	Lake Krakam	41.25	69.07	7.6	21.1
21	Lake Anderson Lake	40.44	67.72	7.4	17.9
22	Lake Ahan Garland	41.06	70.23	4.9	13.5

Andijan water reservoirs are the largest reservoirs of the Syrdarya river basin. The Charvak reservoir is one of the largest hydroelectric power stations in Central Asia, which is located on the Chirchik river, near the Tashkent city. The reservoir capacity is 1.99 km^3, and the power output is 600 MW. The Andijan water reservoir is located on the Karadarya river in the Ferghana Valley and has a capacity of 1.9 km^3. The largest water reservoir in the Amudarya river basin is the Tuyamuyun in the Khorezm region with a capacity of 7.8 km^3 (FAO 2012). It consists of four separate water reservoirs. One of the reservoirs of this system (Kaparas) is designed to provide drinking water to Karakalpakstan, which is experiencing serious environmental problems as a result of the drying out of the Aral Sea. Most reservoirs were built over 25 years back. During the operation, almost all the reservoirs were silted, which led to a decrease in the useful volume of almost 20–25% (FAO 2012).

The total theoretical hydropower potential of the country is estimated at 88,000 GWh/year, and the economically possible is 15,000 GWh/year. The total installed capacity was 1.7 GW in 1993, which provided about 12% of the country's electricity in 1995 (FAO 2012).

10.3 Use of Water Resources in Uzbekistan

Uzbekistan is an agrarian-industrial country with a predominant share of irrigated agriculture in the economy. The total irrigated area is 4.3 million ha in the country. Cotton, cereals, vegetables, potatoes, fruits, grapes, and rice are grown on irrigated lands.

Arid climate of Uzbekistan requires water in all kinds of economic and production activities. Water resources in Uzbekistan are mainly used for agriculture, i.e., more than 90% of all available water is used for irrigation.

The Amudarya and Syrdarya transboundary rivers are the main sources of water resources in the Central Asian countries. These transboundary rivers flow through the territories of all the countries of the region.

The average long-term consumption of water resources in the basins of the Amudarya and Syrdarya rivers is 114.4 km^3, of which 78.34 km^3 along the Amudarya river basin, and 36.06 km^3 along the Syrdarya river basin. The average long-term water consumption in the Central Asian countries located within the Amudarya river basin is as follows: Uzbekistan—38.94 km^3 (49.8%), Turkmenistan—21.76 km^3 (27.7%), Tajikistan—9.8 km^3 (12.5%), Afghanistan—7.44 km^3 (9.5%) and Kyrgyzstan—0.4 km^3 (0.5%). In the Syrdarya river basin: Uzbekistan—17.28 km^3 (48%), Kazakhstan—12.29 km^3 (34%), Tajikistan—2.46 km^3 (7%) and Kyrgyzstan—4.03 km^3 (11%) (Khamrayev and Rahimov 2016) (Fig. 10.1).

Uzbekistan is the major water consumer in the Aral Sea Basin. The total water use of the Amudarya and Syrdarya rivers accounts for 52% in Uzbekistan, in Turkmenistan for 20%, 11% in Tajikistan, 7% in Kyrgyzstan, and 10% in Kazakhstan (Kulmatov 2014). More than 60% of return waters of the region is formed in the territory of Uzbekistan. About 6% of the Amudarya and 13% of the Syrdarya rivers flows are formed in the territory of Uzbekistan (Kulmatov 2014).

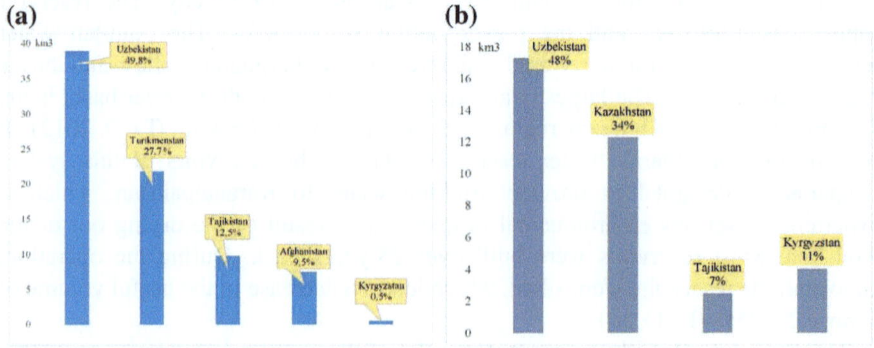

Fig. 10.1 Long-term water consumption: **a** Amudarya RB; **b** Syrdarya RB

The limit of the use of the water resources of the Amudarya and Syrdarya rivers in Uzbekistan as a whole for river basins is 63.02 km^3/year, including 37.53 km^3/year for the Amudarya river basin and 25.49 km^3/year for the Syrdarya river basin (Khamrayev and Rahimov 2016).

The water resources in Uzbekistan consist of external water resources flowing along the rivers from Tajikistan and Kyrgyzstan and as well as inland water resources forming on the territory of Uzbekistan. About 80% of water resources in Uzbekistan come from neighboring countries such as Kyrgyzstan and Tajikistan; only about 20% of water resources are generated in the territory of Uzbekistan.

Water in arid Uzbekistan is a very valuable resource. The sustainable development of the country largely depends on the degree of providing drinking water to almost 30 million people and irrigation water more than 4.3 million ha of agricultural land (Khamrayev and Rahimov 2016).

At present, the total annual use of water resources in the sectors of economy in Uzbekistan is about 51–52 km^3, of which about 46.8 m^3 (90%) is used in agriculture, energy is 3.12 km^3 (6%), communal-domestic is 1.6 km^3 (3%), industry is 0.4 km^3 (0.4%), and fisheries are 0.1 km^3 (0.2%) (Khamrayev and Rahimov 2016) (Fig. 10.2).

Agriculture, energy, and communal-domestic (households) sectors are the main consumers of water resources in the country. The total amount of water consumed for households is equal to 6.4 mln.m^3 per day, 38% of which comes from the surface, and 62% from underground sources. Total annual volume of water required for economic—drinking purposes is 2.33 km^3 (Kulmatov 2014).

The centralized water supply captures 89.4% of city and 73.1% of rural population of the country. The major consumers of surface water sources are irrigated cropland and industry. The greatest uses of water from water reservoirs for

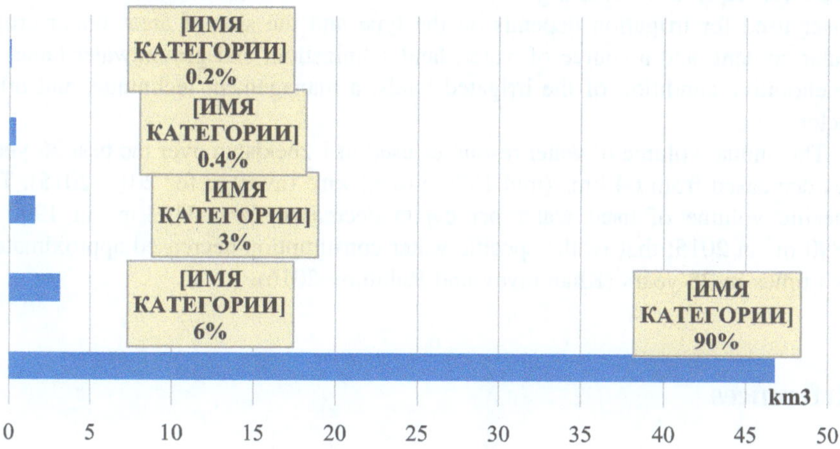

Fig. 10.2 Average use of water resources in the sectors of economy in Uzbekistan

irrigation and industry were recorded in 2005 (59,476.4 mln m^3), in 2006 (58,716.7 mln m^3), in 2004 58,457.3 mln m^3, and in 2003 (56,499.6 mln m^3). The least volumes of water intake for the above-stated needs were recorded in 2008 (43,869.8 mln m^3) and in 2001 (43,869.8 mln m^3). The volume of water used for irrigation has increased from 38,835.7 mln m^3 in 2008 to 44,718.4 mln m^3 in 2009 (15.1%). Out of total amount of water taken from water reservoirs, share of irrigation was 92.82% in 2003, 92.71% in 2001, 88.52% in 2008, 89.03% in 2009, whereas the rest of water was used for household, energy industry needs.

The increase in the water use is partly connected to industrial needs. Growth of volume of water use for industrial purposes begins after 2003. In 2003, the water used was 4054.0 mln m^3 and 4865.0 mln m^3 in 2004. Water use has increased by 811 mln m^3 (20%). In 2009, in comparison with 2000, the volume of water used for industrial needs has increased by 1843.8 mln m^3 (50.33%). Annually, the total amount of water releases from surface reservoirs was between 43,870.0 mln m^3 and 59,476.0 mln m^3. Water use in irrigation was between 38,835.0 mln m^3 and 54,403.0 mln m^3, whereas for industrial needs, it was between 3647.0 mln m^3 and 5506.0 mln m^3 (Kulmatov 2014).

Water use per irrigated area has decreased over time from 1985 to 2010. In 1985, use of water per irrigated hectare (ha) of irrigated area accounted for 22.4 mln m^3/ha; and in 2005, use of water per irrigated hectare of irrigated area accounted for 16.0 m^3/ha. Beyond the analyzed period, overall decrease in water use per hectare of irrigated land can be observed in a longer perspective.

In addition, there is a trend of decreasing water-intensive agricultural products, in the overall composition of agricultural products, the share of cotton decreased down to 30% by 2010, while it used to cover half of the irrigated lands in 1990 (Kulmatov 2014). The situation with water supply in Uzbekistan might worsen due to the expected reduction of existing water resources with the most acute consequences occurring in the Aral Sea area. On the whole, the volume of water used for irrigation depends on the type and the size of areas under crops, water content, and a source of water, land salinization, and groundwater table, an ameliorative condition of the irrigated lands, a management technique, and other factors.

The annual volume of water resources used in Uzbekistan over the past 25 years has decreased from 64 km^3 (mid-1980s) to 51 km^3 (average for 2011–2015). The specific volume of used water per capita decreased from 3193 m^3 in 1990 to 1890 m^3 in 2015; that is, the specific water consumption decreased approximately two times in 25 years (Khamrayev and Rahimov 2016).

References

FAO (2012) Irrigation in Central Asia in numbers. Review AQUASTAT

Khamrayev ShR, Rahimov Shkh (2016) Water management in Uzbekistan in order to increase their productivity. J Water Econ Kazakhstan 3(71):40–48 (in Russian)

Kulmatov R (2014) Problems of sustainable use and management of water and land resources in Uzbekistan. J Water Resour Protect 6:35–42

Xi C, Abuduwaili J, Rakhimov S, Makhmudov E (2013) Water resources and water use in Uzbekistan. Tashkent, 340 p (in Russian)

Chapter 11
Water Resources in Tajikistan

11.1 Lakes and Water Reservoirs in Tajikistan

There are 1300 natural lakes in Tajikistan, with a total water surface area of 705 km^2 and a total volume of about 50 km^3. About 78% of all lakes are located in the mountainous area at an altitude of over 3500 m a.s.l. The largest lake in the country is the Karakul lake on northeast of the country. It is a drainless lake located in the northern part of the Pamirs and to the south of the Zaalaysky Range. The lake is located in a flat mountain valley at an altitude of 3914 m and surrounded by deserted rocky ridges. The surface area of the lake is 380 km^2 and volume is 26.5 km^3. The greatest length of the lake is 33 km, width is 24 km, and the maximum depth is 236 m (FAO 2012).

The lake water is in bright blue color. The water in the lake is bitter–salty, and the transparency is up to 9 m. There are a few small fish, loaches in the mouth of the rivers, where the water is more fresh. In winter, the lake is covered with ice (from the end of November to April), and the maximum thickness of ice in February exceeds 1 m; in summer, the water warms up to 12 °C (Prokhorov 1978).

The Karakul lake region is very dry, with very little rainfall. On the shores of the lake occasionally grow teresken, wormwood, and feather grass, alternating with spaces covered with salt. Several rivers flow into the lake, and the main ones are Karazhylga, Karart, Akzhylga, and Muzkol.

Sarez Lake is the second largest lake in the country with a surface area of 86.5 km^2 and a volume of 17.5 km^3. It is located in Pamir (FAO 2012) (Table 11.1). The length of the lake is about 70 km, the maximum depth is about 500 m, the water level is about 3255 m a.s.l, and the water volume is more than 17 km^3.

The Sarez Lake belongs to the littered or pond lakes, which arose as a result of the catastrophic overlapping of the Bartang River, which occurred on February 18 (March 3), 1911. The intensive filling of the lake was completed in 1926; since 1942, the level of the lake varies vibrationally. The lake poses a danger to the

J. Abuduwaili et al., *Hydrology and Limnology of Central Asia*, Water Resources Development and Management, https://doi.org/10.1007/978-981-13-0929-8_11

Table 11.1 Characteristics of main lakes and water reservoirs in Tajikistan

No.	Name	Location		Area/km^2	Perimeter/km
		N(°)	E(°)		
1	Keirakum Reservoir	40.32	70.06	395.2	116.7
2	Lake Khaluli	39.00	73.49	386.9	172.1
3	Lake Sarez	38.23	72.66	100.0	156.0
4	Nurek Reservoir	38.36	69.43	27.6	53.9
5	Lake Iskul	37.76	72.86	36.9	51.1
6	Lake Zorkul	37.44	73.67	36.1	42.3
7	Lake Langkuli	38.48	74.27	15.7	21.4
8	Lake Tulu Kool Lake Tú lǔ mǎ kù ěr hú Lake Tulu Kool Lake	37.48	72.53	9.2	15.4
9	Chagan Kul Lake	37.40	73.82	8.8	16.5
10	Lake Saskat	37.69	73.18	6.7	10.6
11	Lake Caracur	37.93	72.52	6.6	11.7
12	Sorker Lake	38.45	74.15	6.0	10.1
13	Canakur Lake	37.45	73.43	6.0	17.7
14	Kakaggit Lake	37.45	73.93	3.7	7.3
15	Lake Brenkur	37.72	72.96	3.6	8.7
16	Lake Ischanda	39.07	68.37	3.4	8.4
17	Lake Tuts	37.70	73.12	2.6	9.9
18	Akkala Lake	37.60	72.81	2.4	6.1
19	Muminabad reservoir	38.09	69.99	2.2	6.1
20	Lake Dirham	37.28	72.09	1.9	6.3
21	Purcell Brucer Reservoir	37.86	69.53	1.8	5.9
22	Lake Curikara	37.45	74.01	1.8	5.3
23	Kaguskuri Lake	37.45	73.07	1.6	5.0
24	Lake Chukkul	37.54	73.11	1.3	4.4

settlements located downstream of Bartang, Panj, and Amudarya rivers; as in the case of a breakthrough, a huge mass of water by the mudflow will pass almost to the Aral Sea.

The volume of the lake is approximately 17 km^3 (Makievsky and Mukhabbatov 2016). The total area of the Bartang basin above the Usoi dam is 15,775 km^2, while about 50% of the river's runoff is lost in the highland deserts of the Pamirs (Poslavsky 1968). The lake feeds mainly due to the flow of the rivers such as Murghab, Liangar, and Mardzhanai, which have a constant glacial feeding. The remaining rivers have a source near the lake and feed on the melting snow on the nearby mountains. The water level is at an altitude of 3255 m a.s.l. The upper layer of water is about 50 m, and it is ultra-fresh and high purity. Oxygen content is 93%. A stagnant and highly mineralized layer appeared below, where the oxygen content decreased to 21% (Ashurova 1973). The most transparent water in the western part of the lake and visibility is 15.6 m, and the lowest water transparent is in the east, where the visibility is only 5.8 m (Akulov 1948). Daily changes in temperature are recorded at a depth of 20 m, seasonal changes up to 120 m (Poslavsky 1968). Color of water is light-lac (Akulov 1948).

Regulation of water resources is carried out with the help of hundreds of hydraulic control structures, canals, pumping stations, and water reservoirs. Water plays a major role in regulating water resources for Tajikistan and downstream countries.

In 2010, there were 17 water reservoirs in Tajikistan, of which 4 in the Syrdarya river basin and 13 in the Amudarya river basin (8 on the Vakhsh river, 2 on the Pyanj river, and 3 on the Kafirnigan river. The total capacity of the water reservoirs is about 29.5 km^3 (FAO 2012).

The capacity of ten reservoirs in Tajikistan is more than 10 million km^3 each, and their total volume is 29 km^3. The largest reservoirs are Nurek on the Vakhsh river (10.5 km^3), Kairakkum on the Syrdarya river (4.16 km^3), Farhod on the Syrdarya river (350 million km^3), Boigozi on the Vakhsh river (125 million km^3), Kattasoy on the Kattasoy river (55 million km^3), Muminobod on the Obi Surkh river (31 million km^3), Dakhanasoy on the Dakhanasoy river (28 million km^3), and Sangtudin 1 on the Vakhsh river (25 million km^3). Sangtudin 2 water reservoir on the Vakhsh river (5 million km^3) began work in 2011. The Nurek water reservoir has a power plant with a capacity of 3000 MW. The Nurek and Kairakkum reservoirs store up water for irrigation purposes in Uzbekistan, Turkmenistan, and Kazakhstan. Rogun water reservoir is built on the Vakhsh river (13.3 km^3).

Theoretically, Tajikistan's total hydropower potential is estimated at 527,000 GW/year, of which about half is economically feasible. In 1994, the total installed capacity was about 4 GWh, generating about 98% of electricity in the country. In 1999, Tajikistan was the third largest hydroelectric power development country in the world after the USA and the Russian Federation (FAO 2012).

11.2 The Main River Basins and Rivers in Tajikistan

There are four groups of major river basins in Tajikistan.

1. The Amudarya river basin. About 76% of the Amudarya river flow is formed in Tajikistan. Pyanj river is the largest tributary of the Amudarya river, originates in the mountain ranges of the Pamirs, and forms the border between Tajikistan and Afghanistan almost along the entire length, flowing from east to west. Bartang river is the first large tributary of the Pyanj river. Average annual flow of the Pyanj river is 33.4 km^3/year before merging with the Vakhsh river. The Vakhsh river crosses the country from the northeast to the southwest and is the largest river in Tajikistan. It originates in Kyrgyzstan, where the river is called Kyzyl-Suu, and

then enters Tajikistan, where it is called Surkhob river. The Vakhsh river is formed after the confluence of the Surkhob and Obikhingob rivers.
Its drainage basin is located in the highest part of Tajikistan, at an altitude of more than 3500 m. The Amudarya river formed after the confluence of the Vakhsh and Pyanj rivers on the border with Afghanistan. Kafirnigan river is another major tributary of the Amudarya river. It originates in Tajikistan and it is the border between Tajikistan and Uzbekistan for several tens of kilometers, and then again flows into Tajikistan, and then flows into the Amudarya river in the territory of about 36 km downstream from the confluence of the Panj and Vakhsh rivers on the border between Tajikistan and Afghanistan. On the territory of about 65 km downstream, the Amudarya river leaves the borders of Tajikistan and becomes the border between Afghanistan and Uzbekistan.
Surkhandarya river also originates in Tajikistan, then enters Uzbekistan, and flows into the Amudarya river on the border between Uzbekistan and Afghanistan. Zeravshan river originates in Tajikistan between Zeravshan and Hissar ranges. The total flow of this river, which is formed in Tajikistan, is estimated at 3.09 km^3/year. Then it enters Uzbekistan, previously falling into the Amudarya river on the border between Uzbekistan and Turkmenistan. Zeravshan river was once the largest tributary of the Amudarya river before using water mainly for irrigation in Uzbekistan, after which reduced the flow of the river no longer reached Bukhara city. The total volume of water formed within Tajikistan in the Amudarya river basin is estimated at 59.45 km^3/year (FAO 2012).

2. The Syrdarya river basin. Part of the Syrdarya river basin is in the northwest of the country. Only 1% of the total flow of the Syrdarya river is formed on the territory of Tajikistan by the shallow rivers of Khodjabakirgan, Isfara, and Isfana with a total flow of 1.01 km^3/year (FAO 2012).
3. The Markansu river basin. It is a small river in the extreme northeast of the country, and it comes to China. Data on the flow of this river is not available.
4. Small closed basins: There are several small closed basins formed by small rivers, such as the Kattasu and Basmandasu rivers. However, the annual runoff of these rivers is insignificant compared to the total renewable runoff generated in Tajikistan (FAO 2012).

Rivers. In Tajikistan, a dense river network with glacier–snow and rain types of feeding is developed. There are 947 such rivers. The river length more than 10 km and a total length of more than 28,500 km. The average annual surface runoff reaches 30–45 l/s from 1 km^2 in the central mountain regions of the country, and less than 1 l/s in desert low-lying and high mountain areas. The greatest water discharge in the rivers is observed in June–August—during the maximum melting of snow and ice reserves in the mountains, which very well coincides with the vegetative period of agricultural crops, especially with a period of maximum water consumption (Muhabbatov 2016).

The main river flow is formed in the basins of the Pyanj and Vakhsh rivers, the confluence of which forms the largest river of Central Asia—the Amu Darya. The largest rivers of Tajikistan are: Panj, Vakhsh, Syrdarya, Zeravshan, and Kafirnigan. Thus, we give the main characteristics of the major rivers of Tajikistan.

The Pyanj River is located on the border of the Republic of Tajikistan and is the state border between the CIS and the Islamic Republic of Afghanistan. The length is 921 km, the catchment area is 114,000 km^2, and the drainage volume is 32.1 km^3 of the river. The Pyanj river is the largest on the territory of the republic. It has 494 tributaries with a total length of 11,590 km. The source of the Pyanj river is considered to be the place of the confluence of the Pamir (right) and Vakhandarya (left) rivers. It is glacier–snow powered.

11.3 Glaciers in Tajikistan

Tajikistan is the mountainous country. The mountainous part occupies 93% of its territory and is rich in water resources. It is the main supplier of water in the Aral Sea basin. Its water reserves rank second place among in the CIS, after the Russia. About 64 m^3 water from mountainous zone of Central Asia are formed on the territory of Tajikistan. The main runoff is provided by the Panj, Vakhsh, Kofarnihon, and Zeravshan rivers. Water resources of Central Asia are formed on only 11.2% of the territories Tajikistan; however, 65% of the water resources of the Aral Sea basin are formed on the territory. The water surface is 7% of the territory of country which is 9853 km^2. The water surface was considered without water flow (watercourse).

Water resources have a great importance in the development of various sectors of the national economy of the country, including agriculture, industry, energy, water supply, fisheries. They can be a key importance in the development of the economy not only of Tajikistan but also of neighboring countries located in the Aral Sea basin.

Water resources of Tajikistan are formed on glaciers, river runoff, lakes, water reservoirs, and underground sources.

Tajikistan is a major center of the modern glaciation of Central Asia. It is associated with its peculiarities of orography and climate.

Glaciers are a huge wealth of Tajikistan. They store a huge supply of clean drinking water. Glaciers and eternal snow occupy 8.4,000 km^2, which is 6% of the country's territory. They are regulators of main rivers flow and climate of Central Asia or the Aral Sea basin. They are concentrated mainly in the Pamir.

Glaciers are the main sources of water resources. More than 81% of water resources are formed at 9475 glaciers with a total area of about 9000 km^2.

The Fedchenko Glacier is the largest glacier in Tajikistan and the whole of Central Asia. Its length exceeds 70 km, the average width is 2 km, and the maximum thickness of ice is 1 km. The volume of the glacier with tributaries is about 140 km. It begins at an altitude of 6200 m above sea level. Its language is at an altitude of 2910 m above sea level. According to modern estimates, there are 8000 glaciers in Tajikistan; seven of them have a length of >20 km.

Fig. 11.1 Main glaciers in Tajikistan (by area)

The main glaciers in Tajikistan are located on the Panj, Vakhsh, Kofarnihon, and Zeravshan river basins (Fig. 11.1).

The largest glaciers are Fedchenko, Grumm-Grzhimailo, Garmo glaciers (Fig. 11.1). They are mainly located in the large two river basins such as Panj and Vakhsh in Tajikistan. There are 1225 glaciers in the Zeravshan river basin, and glaciation area is 575 km^2. In the Kofarnihon river basin, 380 glaciers are with glaciation area of 85 km^2. In the basins of the Karakul and Markansu lakes, the number of glacier is 575 with glaciation area of 555 km^2 (Fig. 11.2).

Management and distribution of water resources are carried out by the river network. The river network of Tajikistan is divided into two water basins: the Amudarya and Syrdarya. The Amudarya water basin includes the Panj, Vakhsh, and Kofarnihon rivers with large and small tributaries; the Syrdarya river with all its tributaries belong to the Syrdarya water basin.

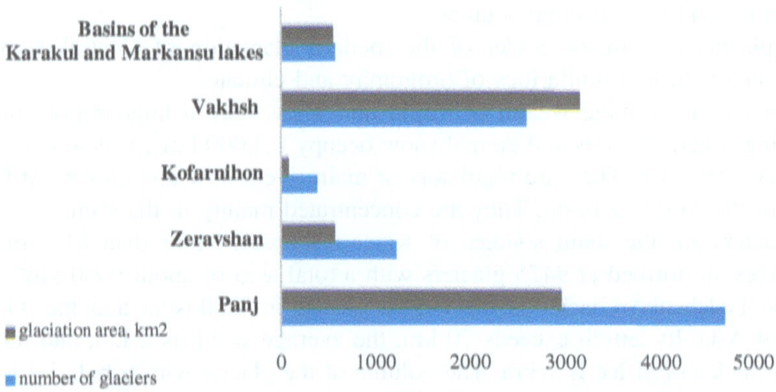

Fig. 11.2 Distribution of glaciation area and glacier number within river basins in Tajikistan

For several years already, the waters of the Zeravshan river do not reach the former mouths of the Amudarya river, since it is fully spent for irrigation of lands, mainly Samarkand and Bukhara oases.

References

Akulov VV (1948) Some observations of the state of Lake Sarez in 1946. J News All-Union Geogr Soc Part 80(3):246–258 (Moscow) (in Russian)
Ashurova M (1973) Parasites of fish in the Sarez Lake (Pamir). J Parasitol Pavlovsky Inst Zool Parasitol (2):164–167 (Academy of Sciences of Tajik USSR) (in Russian)
FAO (2012) Irrigation in Central Asia in numbers. Review AQUASTAT
Makievsky P, Mukhabbatov Kh (2016) Tajikistan: Sarez Lake: geodynamic, technical and social aspects of the problem. CA & CC Press, Sweden
Muhabbatov KHM (2016) The problems of formation and use of water resources in Tajikistan. Water resources of Kazakhstan and their use. J. Water Econ Kaz 3(71):26–30 (in Russian)
Prokhorov AM (eds) (1978) Great soviet encyclopedia, 3rd edn Part 30. Soviet Encyclopedia, Moscow (in Russian)
Poslavsky V (1968) About one catastrophe in the Pamirs (History of Lake Sarez). J Hydrotechnics Reclam (3) Moscow (in Russian)

Chapter 12
Lakes in Arid Regions of Northwest China

12.1 Distribution and Types of Lakes in Arid Regions of Northwest China

There are about more than 700 lakes in arid areas of northwest of China, and they are mainly distributed in Xinjiang (Tables 12.1 and 12.2). There are 29 lakes with size more than 10 km^2, of which three lakes are distributed in both inside and outside the boundary of Inner Mongolia, and rest 26 lakes all located in Xinjiang (Table 12.2) (Abuduwaili 2012).

The large lakes in arid areas are mainly supplied by rivers. They are the trails of rivers and are an important part of the water cycle in arid areas. As the last link in the system of natural and human economic activities in arid areas, its ecological system and environment was first to respond to the interference of human activities.

Bosten Lake is the largest inland freshwater lake, which is located in the south of Tian Shan mountain, made up of two large and small lakes. It is feeding by the Kaidu River, which is the Rump Lake of Kaidu River and source of Konqi River, and is the only throughput Lake in arid areas of China (Xia et al. 2003). The water level of the lake is 1047.5 m, with surface area of 974.5 km^2, the volume of 7.59 billion m^3, the largest depth is 16.8 m, and the average water depth is 7.5 m.

The small lake covers an area of 363.9 km^2, with 16 lakes connected with the old river road of Konqi River, area of 44.5 km^2 of water level. The area of Reed Lake is 802 km^2; area is 39.3 km^2 of pasture and beaches. Ecological detections are settled in the interval of Lake dike hose (brake, culvert), two-way penetration. It is needed to be emphasized that since 2000, the water environment of Bosten Lake has changed. The Bosten Lake sustains a healthy state. The fishery production of Lake was up to 6500 ton in 2005, while was 2620 ton (2.48 times) in 1990. The Bosten Lake owes special position in the local and national economy, ecology and environment in Xinjiang of northwest China.

J. Abuduwaili et al., *Hydrology and Limnology of Central Asia*, Water Resources Development and Management, https://doi.org/10.1007/978-981-13-0929-8_12

Table 12.1 Main lakes in arid and semi-arid areas of China

Province	Name of Lake	Location		Altitude/m	Length/km	Width/km		Area/km²	Remarks
		N	E			Max	Mean		
Xinjiang	Bosten Lake	42° 22′	87° 04′	1,051	60.5	35.4	15.9	960	Inland Playas
Xinjiang	Brenda's Sea	47° 13′	87° 18′	479	41.8	25.3	17.6	736	Outflow Playas
Xinjiang	Aragu kuehl	37° 33′	89° 20′	3,880	48.1	17.6	11.9	570	Inland Playas
Xinjiang	Ebinur	44° 56′	82° 52′	194	54.8	23.9	9.5	522	Inland Playas
Xinjiang	Sailimu Lake	44° 36′	81° 11′	2,074	30.2	24.9	15.0	454	Inland Playas
Xinjiang	Asagkul	37° 04′	88° 22′	4,256	25.8	19.3	13.4	345	Inland Playas
Xinjiang	Whale Lake	36° 20′	89° 25′	4,718	37.0	10.3	7.1	267	Inland Playas
Xinjiang	Jili Lake	46° 53′	87° 27′	1,685	16.8	6.3	9.5	160	Outflow Freshwater Lake
Xinjiang	Lake Aksayi	35° 12′	79° 50′	4,938	19.2	12.1	8.2	158	Inland Playas
Xinjiang	Aiximan Lake	40° 42′	80° 12′	1,051	–	–	–	150	Inland Playas
Xinjiang	Barkol Lake	43° 40′	92° 44′	1,582	–	–	–	88	Inland Playas
Xinjiang	Saligial kollakkul	34° 42′	79° 41′	5,294	10.8	6.0	4.0	43	Inland Playas
Xinjiang	Halakkule	48° 48′	87° 04′	1,392	24.3	2.5	1.9	45	Outflow Freshwater Lake
Xinjiang	Alice Lake	48° 54′	85° 47′	1,445	12.4	4.2	3.5	43	Inland Playas
Xinjiang	Chai Wo-pu Lake	43° 30′	87° 54′	1,091	6.7	5.6	4.3	29	Inland Lake
Xinjiang	Aydingkol Lake	42° 40′	89° 16′	154	8.9	4.3	2.6	23	Inland Lake
Xinjiang	Kumukuli	37° 06′	90° 31′	4,311	14.9	2.5	1.5	22	Inland Lake
Xinjiang	Bera Clarke	36° 22′	90° 44′	4,945	7.7	4.2	2.9	21	Inland Lake
Xinjiang	Kinkik Kuller	36° 58′	90° 44′	4,478	9.5	3.1	1.9	19	Inland Lake
Xinjiang	Unknown Lake	41° 14′	83° 34′	947	6.8	4.9	2.6	18	Inland Lake
Xinjiang	Changhonghu area	36° 03′	86° 00′	4,975	13.9	1.9	1.2	17	Inland Lake

(continued)

Table 12.1 (continued)

Province	Name of Lake	Location		Altitude/m	Length/km	Width/km		Area/km^2	Remarks
		N	E			Max	Mean		
Xinjiang	Unknown Lake	40° 34′	87° 38′	849	12.1	2.6	1.4	17	Inland Lake
Xinjiang	Urukkule	35° 40′	81° 37′	5,639	7.0	3.6	2.1	15	Inland Lake
Xinjiang	Salt Lake	43° 27′	88° 02′	1,091	7.9	4.3	1.9	15	Inland Lake
Xinjiang	Unknown Lake	37° 28′	90° 19′	3,640	9.2	4.3	1.6	15	Inland Lake
Xinjiang	Ashkul	35° 44′	81° 43′	5,566	5.4	2.9	2.2	12	Inland Lake
Xinjiang	Ashman	40° 50′	80° 04′	1,083	6.0	2.0	1.8	11	Inland Lake
Xinjiang	Gulimi Kutler	40° 46′	86° 54′	872	7.9	2.2	1.4	11	Inland Lake
Xinjiang	Aksukule	36° 36′	84° 27′	4,409	6.0	4.1	1.9	11	Inland Lake
Xinjiang	Peninsula Lake	35° 54′	85° 16′	4,875	8.7	2.5	1.3	11	Inland Lake
Xinjiang	Sai Yitekule	40° 52′	86° 44′	876	9.4	1.6	1.1	10	Inland Lake
Xinjiang	Inkur haizi	40° 47′	86° 50′	879	17.6	2.2	0.6	10	Inland Lake
Xinjiang	Lydger Lake	34° 54′	79° 21′	5,199	7.3	2.1	1.3	9.3	Inland Lake
Xinjiang	Akekule	49° 02′	87° 34′	6.8	–	4.3	1.3	9.1	Outflow Freshwater Lake
Xinjiang	Chatea Kartaga North Lake	42° 14′	90° 44′	883	–	–	–	9	Inland Playas
Xinjiang	Arisi Sui Maba	38° 39′	77° 19′	1,196	4.5	2.5	2	8.8	Inland Lake
Xinjiang	Cuolulecuo	35° 52′	78° 34′	4,710	4.3	2.8	2.2	8.7	Inland Lake
Xinjiang	Little Erika Lake	45° 46′	85° 35′	272	9.5	2	0.9	8.7	Inland Lake
Xinjiang	Unknown Lake	40° 54′	80° 04′	1,069	6.5	3.4	1.3	8.2	Inland Lake
Xinjiang	Quartz beach north Lake	41° 33′	93° 40′	1,240	–	–	–	8	Inland Playas
Xinjiang	White jade Lake	41° 10′	93° 25′	1,135	–	–	–	8	Inland Playas
Xinjiang	Tanyi Sua Maba	38° 32′	77° 19′	1,201	4.9	2.2	1.5	7.5	Inland Lake

(continued)

Table 12.1 (continued)

Province	Name of Lake	Location		Altitude/m	Length/km	Width/km		Area/km^2	Remarks
		N	E			Max	Mean		
Xinjiang	Eight fort Lake	41° 04′	92° 03′	1,056	–	–	–	7	Inland Playas
Xinjiang	Chatea Kartaga North Lake	42° 05′	90° 46′	1,104	–	–	–	7	Inland Playas
Xinjiang	Unknown Lake	37° 10′	90° 10′	3,920	3.4	2.7	1.9	6.4	Inland Lake
Xinjiang	Unknown Lake	40° 06′	77° 18′	1,571	3.7	2.6	1.6	6.1	Inland Lake
Xinjiang	Sergio Alba	38° 38′	77° 19′	1,207	4.5	2.4	1.3	6	Inland Lake
Xinjiang	Haizi	47° 10′	89° 44′	1,172	3.9	2.5	1.5	6	Inland Lake
Xinjiang	Naomao Lake	43° 57′	94° 35′	306	323	–	–	6	Inland Playa
Xinjiang	Aguosan	41° 04′	84° 59′	916	4.7	3.1	1.1	5.3	Inland Lake
Xinjiang	Salt Lake	47° 41′	87° 34′	484	4.9	1.5	1	5.1	Inland Playas
Xinjiang	Han Shuiquan Lake	44° 51′	92° 41′	479	–	–	–	5	Inland Playas
Xinjiang	Pier bass he birds Lake	44° 47′	91° 15′	700	–	–	–	5	Inland Playas
Xinjiang	Hala Lake	46° 21′	83° 16′	60	–	–	–	5	Inland Playas
Xinjiang	Barker Brooks Lake	41° 21′	90° 40′	1,000	–	–	–	5	Inland Playas
Xinjiang	Yanshan Lake	41° 30′	81° 22′	1,300	–	–	–	5	Inland Playas
Xinjiang	Qiao Kule Lake	36° 06′	82° 41′	4,474	–	–	–	5	Inland Playas
Xinjiang	Lake Burak	35° 43′	79° 22′	5,745	–	–	–	5	Inland Playas
Xinjiang	Dead Haizi	47° 18′	87° 32′	486	3.4	1.9	1.4	4.6	Inland Lake
Xinjiang	Unknown Lake	39° 08′	76° 32′	1,231	2.3	2.1	2	4.5	Inland Lake
Xinjiang	Kubase Nuoer	47° 36′	87° 31′	486	–	–	–	4.5	Inland Playas
Xinjiang	Red mountain Lake	35° 28′	78° 56′	4,835	2.6	1.9	1.7	4.3	Inland Playas
Xinjiang	Pierbarrele	38° 28′	77° 24′	3.6	–	1.8	1.1	4.1	Inland Lake
Xinjiang	Unknown Lake	40° 42′	80° 44′	2.6	–	1.8	1.4	4	Inland Lake

(continued)

Table 12.1 (continued)

Province	Name of Lake	Location		Altitude/m	Length/km	Width/km		Area/km^2	Remarks
		N	E			Max	Mean		
Xinjiang	Niujuan Lake	44° 31′	93° 50′	680	–	–	–	4	Inland Playas
Xinjiang	Lake Daboo Sun	45° 50′	90° 45′	890	–	–	–	4	Inland Playas
Xinjiang	Schaalsee Lake	42° 36′	92° 18′	76	–	–	–	4	Inland Playas
Xinjiang	Beilike Lake	36° 43′	89° 03′	4,740	–	–	–	4	Inland Playas
Xinjiang	Xiaoerkule Lake	36° 07′	82° 40′	4,474	4.4	1.4	0.8	3.7	Inland Lake
Xinjiang	Jiangbatawule Lake	41° 31′	90° 42′	735	–	–	–	3.6	Inland Playa
Xinjiang	Little haizi	40° 52′	86° 48′	880	6.8	1.8	0.5	3.6	Inland Lake
Xinjiang	Aggie kenkule	39° 43′	78° 21′	1,124	4.8	1.6	0.8	3.6	Inland Lake
Xinjiang	Unknown Lake	46° 32′	87° 53′	594	4.8	1.1	0.7	3.5	Inland Lake
Xinjiang	Niaolimaiqi Lake	39° 28′	76° 33′	1,219	2.5	2	1.4	3.4	Inland Lake
Xinjiang	Unknown Lake	40° 44′	80° 03′	1,050	2.9	2	1.2	3.4	Inland Lake
Xinjiang	Unknown Lake	41° 12′	85° 59′	891	2.9	1.2	1.1	3.4	Inland Lake
Xinjiang	Unknown Lake	40° 26′	87° 42′	830	3.5	2	0.9	3.2	Inland Lake
Xinjiang	Wulamool Lake	35° 17′	81° 38′	5,390	7	0.8	0.4	3.1	Inland Lake
Xinjiang	Bellekkule Lake	37° 16′	83° 56′	2,874	10	0.8	0.3	3	Inland Lake
Xinjiang	Yongfeng Lake	36° 07′	85° 56′	4,950	2.3	1.6	1.3	3	Inland Lake
Xinjiang	Nasl culler	41° 02′	85° 31′	905	4.3	1.8	0.7	3	Inland Lake
Xinjiang	Rabbit Lake	36° 43′	87° 14′	4,685	4.2	1.6	0.8	3	Inland Playa
Xinjiang	Unknown	40° 28′	80° 34′	1,033	3.3	1.6	0.9	3	Inland Lake
Xinjiang	Carpakkukuke	44° 52′	92° 04′	700	–	–	–	3	Inland Playa
Xinjiang	Top mountain I Lake	46° 52′	87° 40′	285	–	–	–	3	Inland Playa
Xinjiang	Bird of the salt Lake	44° 37′	85° 48′	428	–	–	–	3	Inland Playa

(continued)

Table 12.1 (continued)

Province	Name of Lake	Location		Altitude/m	Length/km	Width/km		Area/km^2	Remarks
		N	E			Max	Mean		
Xinjiang	Salt Lake in north sand	44° 37′	87° 36′	428	–	–	–	3	Inland Playa
Xinjiang	Top mountain II Lake	46° 21′	87° 45′	388	–	–	–	3	Inland Playa
Xinjiang	Yong ji Lake	45° 15′	82° 08′	200	–	–	–	3	Inland Playa
Xinjiang	Taiping tai salt Lake	41° 22′	91° 58′	985	–	–	–	3	Inland Playa
Xinjiang	Salt Lake in winpadai	41° 52′	82° 06′	1,372	–	–	–	3	Inland Playas
Xinjiang	Wuzunqiao Lake	38° 28′	89° 57′	2,500	–	–	–	3	Inland Playa
Xinjiang	Akekule Lake	47° 18′	87° 02′	488	2.4	1.8	1.2	2.9	Inland Playa
Xinjiang	Waterlogging boom swinging	38° 41′	77° 28′	1,190	3.2	1.4	0.8	2.9	Inland Lake
Xinjiang	Stone diffuse Lake	36° 15′	86° 17′	5,014	3.4	1.3	0.8	2.8	Inland Lake
Xinjiang	Reed Lake	47° 40′	88° 02′	654	3	1.5	0.9	2.7	Inland Lake
Xinjiang	Abkharich	39° 26′	76° 39′	1,215	2.5	1	1	2.5	Inland Lake
Xinjiang	Shbarkule	46° 46′	90° 54′	2,722	2.5	1.5	1	2.5	Inland Lake
Xinjiang	Unknown	36° 25′	90° 51′	4,845	3	2.1	0.8	2.5	Inland Lake
Xinjiang	In pond water Lake	36° 11′	86° 10′	5,040	3.9	1.5	0.6	2.4	Inland Lake
Xinjiang	Bill ikzi	39° 15′	77° 55′	1,155	4.6	0.9	–	2.4	Inland Lake
Xinjiang	Unknown	43° 27′	88° 03′	1,085	–	1.3	1	–	
Xinjiang	Tang balak Lake	47° 38′	88° 13′	659	4.8	0.9	0.5	2.4	Inland Lake
Xinjiang	Huangcao Lake	35° 52′	85° 17′	4,930	3.6	0.9	0.6	2.3	Inland Lake
Xinjiang	Tomenotrol	38° 26′	77° 23′	1,205	2	1.4	1.1	2.2	Inland Lake
Xinjiang	Dadaikikul	41° 02′	85° 46′	900	5	0.6	0.4	2.2	Inland Lake
Xinjiang	Togak armubar	39° 48′	76° 14′	1,402	2.8	1.4	0.8	2.2	Inland Lake
Xinjiang	Unknown	40° 27′	87° 21′	868	3.2	1.6	0.7	2.2	Inland Lake

(continued)

Table 12.1 (continued)

Province	Name of Lake	Location		Altitude/m	Length/km	Width/km		Area/km^2	Remarks
		N	E			Max	Mean		
Xinjiang	Unknown	41° 30′	81° 20′	1,255	2.9	1.2	0.7	2.1	Inland Lake
Xinjiang	Unknown	39° 27′	88° 19′	809	2.1	1.8	1	2.1	Inland Lake
Xinjiang	Gammaegele	47° 19′	87° 32′	486	2	1	1	2	Inland Lake
Xinjiang	Unknown	46° 23′	87° 43′	597	2.7	1.2	0.7	2	Inland Lake
Xinjiang	Unknown	45° 45′	85° 30′	268	2.5	1.5	0.8	2	Inland Lake
Xinjiang	Unknown	41° 13′	85° 57′	890	3.7	0.7	0.5	2	Inland Lake
Xinjiang	Naomao IILake	43° 57′	94° 45′	321	–	–	–	2	Inland Playas
Xinjiang	Lake Cameroon	44° 28′	91° 18′	1,000	–	–	–	2	Inland Playas
Xinjiang	Fortuna's salt Lake	47° 14′	87° 34′	480	–	–	–	2	Inland Playas
Xinjiang	Dengta's salt Lake	46° 52′	87° 35′	479	–	–	–	2	Inland Playas
Xinjiang	Kikukuk Lake	44° 18′	90° 36′	700	–	–	–	2	Inland Playas
Xinjiang	Natsuko Lake	46° 26′	87° 25′	388	–	–	–	2	Inland Playas
Xinjiang	Naomao Lake	43° 50′	95° 06′	479	–	–	–	2	Inland Playas
Xinjiang	QiJiaoJing west Lake	43° 22′	91° 05′	1,200	–	–	–	2	Inland Playas
Xinjiang	Pants Lake	41° 35′	92° 13′	115	–	–	–	2	Inland Playas
Xinjiang	Dabancheng west salt Lake	43° 28′	88° 05′	1,072	–	–	–	2	Inland Playas
Xinjiang	South Lake soda lakes I	42° 33′	93° 51′	83	–	–	–	2	Inland Playas
Xinjiang	Shule Noel	42° 38′	93° 10′	81	–	–	–	2	Inland Playas
Xinjiang	Dental kessala Lake	36° 50′	81° 23′	1,442	–	–	–	2	Inland Playas
Xinjiang	Mollik Lake	36° 40′	79° 50′	2,237	–	–	–	2	Inland Playas
Xinjiang	Chao bohai Lake	36° 34′	85° 48′	4,740	–	–	–	2	Inland Playas
Xinjiang	Open Lake	42° 40′	84° 57′	3,609	3.5	1.3	0.5	1.9	Inland Lake

(continued)

Table 12.1 (continued)

Province	Name of Lake	Location		Altitude/m	Length/km	Width/km		Area/km^2	Remarks
		N	E			Max	Mean		
Xinjiang	Unknown	44° 58′	83° 53′	256	2.9	1.2	0.6	1.8	Inland Lake
Xinjiang	Unknown	39° 12′	76° 49′	1,200	2.1	1.1	0.9	1.8	Inland Lake
Xinjiang	Unknown	37° 10′	79° 32′	1,351	3.4	1	0.5	1.8	Inland Lake
Xinjiang	Unknown	39° 46′	78° 22′	1,120	2.3	1.4	0.8	1.8	Inland Lake
Xinjiang	Unknown	39° 45′	78° 48′	1,119	2.1	1.1	0.8	1.7	Inland Lake
Xinjiang	Unknown	40° 34′	87° 45′	848	4.1	0.9	0.4	1.7	Inland Lake
Xinjiang	Meisimailai Lake	39° 31′	76° 49′	1,202	2.2	1.1	0.7	1.6	Inland Lake
Xinjiang	The bitter Lake	35° 42′	79° 22′	4,770	2.1	1.6	1.1	1.5	Inland Lake
Xinjiang	Chickens, ducks haizi	40° 47′	86° 58′	873	2.9	1.1	0.5	1.5	Inland Lake
Xinjiang	Kaite bought jonkair	38° 06′	86° 00′	1,338	5	0.6	0.3	1.5	Inland Lake
Xinjiang	Alanchuba	37° 18′	79° 32′	1,319	2.2	1.1	0.7	1.5	Inland Lake
Xinjiang	Birds from Lake	36° 14′	86° 05′	5,014	2.5	1.1	0.6	1.5	Inland Lake
Xinjiang	Giant waves Lake	36° 02′	85° 53′	4,860	3.3	1	0.5	1.5	Inland Lake
Xinjiang	Wasteland bazaar	38° 48′	77° 26′	1,187	2.1	0.9	0.7	1.5	Inland Lake
Xinjiang	Spring Lake	37° 05′	87° 41′	4,385	–	–	–	1.5	Inland Playas
Xinjiang	The Eric Lake	39° 05′	77° 44′	1,165	2.1	1.3	0.7	1.4	Inland Lake
Xinjiang	Jin Zhong Lake	36° 26′	88° 31′	5,055	1.4	1.2	1	1.4	Inland Lake
Xinjiang	Unknown	39° 53′	79° 15′	1,089	1.9	1.3	0.7	1.4	Inland Lake
Xinjiang	Unknown	40° 33′	81° 05′	1,020	1.8	1.7	0.8	1.4	Inland Lake
Xinjiang	Unknown	40° 31′	87° 32′	882	2	1.9	0.7	1.4	Inland Lake
Xinjiang	Unknown	35° 32′	79° 35′	4,851	2.3	1.5	0.6	1.3	Inland Lake
Xinjiang	Unknown	35° 30′	79° 33′	5,147	3	0.7	0.4	1.3	Inland Lake

(continued)

Table 12.1 (continued)

Province	Name of Lake	Location		Altitude/m	Length/km	Width/km		Area/km^2	Remarks
		N	E			Max	Mean		
Xinjiang	Unknown	40° 29′	87° 22′	871	2	1	0.6	1.3	Inland Lake
Xinjiang	Unknown	47° 08′	87° 41′	506	2.1	1.4	0.6	1.3	Inland Lake
Xinjiang	Alwintera Lake	41° 22′	82° 35′	995	2.2	0.7	0.6	1.3	Inland Lake
Xinjiang	Unknown	40° 54′	80° 07′	1,066	3.2	0.5	0.4	1.3	Inland Lake
Xinjiang	Caocule	39° 27′	88° 24′	808	1.8	1.1	0.7	1.2	Inland Lake
Xinjiang	Tsagaandunledenkule	47° 06′	90° 30′	2,403	3	0.8	0.4	1.2	Inland Lake
Xinjiang	Alwintera Lake	41° 21′	81° 20′	1,149	2.2	0.7	0.6	1.3	Inland Lake
Xinjiang	Unknown	40° 54′	80° 07′	4,797	3.2	0.5	0.4	1.3	Inland Lake
Xinjiang	Hammer Lake	38° 28′	77° 23′	1,202	2.2	1	0.6	1.2	Inland Lake
Xinjiang	Five spot Lake	36° 36′	86° 14′	4,781	2.8	0.5	0.4	1.2	Inland Lake
Xinjiang	Lotus root Lake	35° 49′	85° 20′	4,935	2.7	0.7	0.4	1.2	Inland Lake
Xinjiang	Aqike Lake	35° 44′	81° 31′	4,685	–	–	–	1.2	Inland Playas
Xinjiang	Erzi Lake	35° 42′	84° 53′	4,900	1.9	0.7	0.6	1.1	Inland Lake
Xinjiang	Shibaerkule	46° 49′	90° 52′	2,652	1.6	1.1	0.7	1.1	Inland Lake
Xinjiang	Spring Lake	37° 05′	87° 39′	4,371	1.4	0.9	0.8	1.1	Inland Lake
Xinjiang	Unknown Lake	39° 06′	76° 38′	1,215	1.3	1.1	0.8	1.1	Inland Lake
Xinjiang	Anggekule	36° 15′	83° 03′	3,820	1.6	0.9	0.7	1.1	Inland Lake
Xinjiang	Spring waves Lake	35° 54′	86° 03′	4,980	1.5	0.9	0.7	1.1	Inland Lake
Xinjiang	Kamileha Lake	39° 22′	77° 44′	1,152	1.6	1	0.7	1.1	Inland Lake
Xinjiang	Yulekenkelen	47° 38′	87° 21′	476	2.5	1	0.4	1.1	Inland Lake
Xinjiang	Alaanba	37° 18′	79° 32′	1,318	2	0.9	0.6	1.1	Inland Lake
Xinjiang	Unknown Lake	40° 52′	80° 08′	1,060	1.6	0.7	0.6	1	Inland Lake

(continued)

Table 12.1 (continued)

Province	Name of Lake	Location		Altitude/m	Length/km	Width/km		Area/km^2	Remarks
		N	E			Max	Mean		
Xinjiang	Unknown Lake	47° 19′	87° 51′	519	1.2	1.2	0.8	1	Inland Lake
Xinjiang	Unknown Lake	47° 12′	87° 56′	534	2.3	0.7	0.4	1	Inland Lake
Xinjiang	Sweet water sea	35° 21′	79° 30′	4,827	2.7	0.6	0.4	1	Inland Playas
Xinjiang	Keqike Kailai	38° 59′	77° 09′	1.9	0.7	0.6	–	1	Inland Lake
Xinjiang	Unknown Lake	41° 36′	78° 06′	2.1	0.6	0.5	–	1	Inland Lake
Xinjiang	Three Tang Lake	44° 14′	93° 20′	1,035	–	–	–	1	Inland Playas
Xinjiang	Wu Dong SuYi Lake	44° 32′	91° 40′	702	–	–	–	1	Inland Playas
Xinjiang	Jiaertasi Lake	46° 34′	89° 44′	840	–	–	–	1	Inland Playas
Xinjiang	Small salt Lake	43° 47′	94° 32′	482	–	–	–	1	Inland Playas
Xinjiang	Black Lake	47° 11′	86° 30′	1,214	–	–	–	1	Inland Playas
Xinjiang	Qian Shan Salt Lake	45° 06′	84° 50′	285	–	–	–	1	Inland Playas
Xinjiang	Yongji Lake	44° 40′	82° 10′	210	–	–	–	1	Inland Playas
Xinjiang	Wutai Glauber's salt Lake	44° 48′	82° 10′	300	–	–	–	1	Inland Playas
Xinjiang	South Lake alkali LakeII	42° 32′	93° 32′	84	–	–	–	1	Inland Playas
Xinjiang	Taha Lake	42° 18′	86° 59′	1,137	–	–	–	1	Inland Playas
Xinjiang	Red willow Lake	38° 03′	89° 29′	3,145	–	–	–	1	Inland Playas
Xinjiang	Wild duck Lake	37° 18′	86° 49′	4,160	–	–	–	1	Inland Playas
Xinjiang	Bird Suxiao Lake	38° 24′	90° 02′	2,936	–	–	–	1	Inland Playas
Xinjiang	Xiaokule Lake	37° 10′	86° 50′	4,478	–	–	–	1.0	Inland Playas
Inner Mongolia	Wuliangsuhai	40° 54′	108° 49′	1,021	35.4	12.7	6.6	233	
Inner Mongolia	Huer Chagan Nuoer	43° 24′	114° 53′	1,016	19.1	5.4	5.6	108	

(continued)

Table 12.1 (continued)

Province	Name of Lake	Location		Altitude/m	Length/km	Width/km		Area/km²	Remarks
		N	E			Max	Mean		
Inner Mongolia	East Juyanhai	42° 18′	101° 16′	903	8.5	5.8	4.1	35	
Inner Mongolia	Alashan Gobi	43° 17′	114° 02′	985	7	2.7	2.4	17	
Inner Mongolia	Unknown Lake	41° 57′	101° 36′	890	7.6	2.4	1.2	9.2	Inland freshwater Lake
Inner Mongolia	East shore Lake	41° 31′	113° 15′	1,405	–	–	–	7.8	Inland Playas
Inner Mongolia	Shalage Nuoer	42° 48′	114° 53′	1,180	–	–	–	7.5	Inland Playas
Inner Mongolia	Hexi new Lake	40° 56′	100° 07′	1,080	5.9	2.1	1.2	7.1	Inland Lake
Inner Mongolia	Unknown Lake	41° 31′	113° 16′	1,404	5.2	2	1.3	6.7	Inland Lake
Inner Mongolia	Chagan Nuoer	43° 16′	114° 48′	1,040	4.3	1.9	1.2	5.3	Inland freshwater Lake
Inner Mongolia	Haer Naoer	42° 21′	109° 52′	1,102.9	–	–	–	4.7	Inland Playas
Inner Mongolia	Shepherd Company Haizi	41° 14′	108° 23′	1,021	4.5	1.9	1	4.4	Inland Lake
Inner Mongolia	Chagan Nuoer	42° 28′	111° 38′	1,169	–	–	–	4.3	Inland Playas
Inner Mongolia	Guerbanzhagan Qaidam Lake	39° 59′	105° 53′	1,020	–	–	–	3.6	Inland Playas

(continued)

Table 12.1 (continued)

Province	Name of Lake	Location		Altitude/m	Length/km	Width/km		Area/km²	Remarks
		N	E			Max	Mean		
Inner Mongolia	Bayin Nike Lake	38° 59′	104° 15′	1,309	–	–	–	3.6	Inland Playas
Inner Mongolia	Unknown Lake	41° 29′	113° 09′	1,403	2.6	1.9	1.3	3.4	Inland Lake
Inner Mongolia	Baxinggaole Lake	39° 59′	104° 08′	1,200	–	–	–	3.4	Inland Playas
Inner Mongolia	Ulan Nuoer	43° 45′	114° 46′	1,300	2.2	2	1.5	3.3	Inland Playas
Inner Mongolia	Alatenggadasi Lake	42° 41′	114° 50′	1,120	–	–	–	3.2	Inland Playas
Inner Mongolia	Tonggulou Nuoer	38° 43′	105° 08′	1,288	–	–	–	3.1	Inland Playas
Inner Mongolia	Xiale Nao	42° 02′	101° 01′	940	2.8	2.1	1.1	3	Inland Lake
Inner Mongolia	Halamugetai Lake	40° 11′	103° 21′	1,486	–	–	–	3	Inland Playas
Inner Mongolia	Unknown Lake	41° 17′	108° 27′	1,021	3	1.6	1	2.9	Inland Lake
Inner Mongolia	Bage Naoer	42° 27′	100° 34′	870	–	–	–	2.8	Inland Playas
Inner Mongolia	Prague Lake	43° 40′	112° 30′	920	–	–	–	2.5	Inland Playas
Inner Mongolia	Shang Matara Lake	43° 05′	114° 03′	1,000	–	–	–	2.5	Inland Playas

(continued)

Table 12.1 (continued)

Province	Name of Lake	Location		Altitude/m	Length/km	Width/km		Area/km^2	Remarks
		N	E			Max	Mean		
Inner Mongolia	Saidabusu Lake	42° 16′	110° 04′	1,140	4.5	1	0.6	2.5	Inland Playas
Inner Mongolia	Er Haizi	40° 22′	106° 58′	1,050	2.3	1.7	1	2.4	Inland freshwater Lake
Inner Mongolia	Brown Chagan Nuoer	43° 13′	114° 27′	1,010	3.1	1.3	0.7	2.3	Inland Playas
Inner Mongolia	LiuTian Haizi	41° 12′	108° 23′	1,022	2.5	1.6	0.9	2.2	Inland Lake
Inner Mongolia	Saidabusu Nuoer	42° 16′	110° 03′	1,120	–	–	–	2.2	Inland Playas
Inner Mongolia	Xinjiang Lake	42° 24′	108° 24′	1,149.6	–	–	–	2	Inland Playas
Inner Mongolia	Chagan Naoer	42° 48′	110° 51′	1,154	2	1.2	1	2	Inland Lake
Inner Mongolia	Red orchid Marr Northeast SeaNorth	40° 32′	103° 27′	1,250	–	–	–	2	Inland Playas
Inner Mongolia	HaDe Lake	38° 46′	105° 07′	1,292	–	–	–	2	Inland Playas
Inner Mongolia	Alkali pool	40° 08′	108° 12′	1,060	–	–	–	2	Inland Playas
Inner Mongolia	HaEr Nuoer	43° 51′	114° 31′	1,020	–	–	–	2	Inland Playas
Inner Mongolia	Wansheng Tai Lake	41° 35′	111° 30′	898	–	–	–	2	Inland Playas

(continued)

Table 12.1 (continued)

Province	Name of Lake	Location		Altitude/m	Length/km	Width/km		Area/km^2	Remarks
		N	E			Max	Mean		
Inner Mongolia	Yingde Er Lake	43° 50′	112° 40′	910	–	–	–	2	Inland Playas
Inner Mongolia	Qingshi Lake	39° 05′	103° 30′	1,400	–	–	–	2	Inland Playas
Inner Mongolia	Unknown Lake	39° 27′	108° 50′	1,376	2.1	0.9	0.6	1.8	Inland Playas
Inner Mongolia	Wenggong Nuoer	42° 35′	114° 09′	1,128	2.2	1.5	0.8	1.8	Inland Playas
Inner Mongolia	Unknown Lake	40° 41′	108° 22′	1,047	1.8	1.3	1	1.8	Inland Lake
Inner Mongolia	Nuoer Tu Lake	39° 46′	102° 27′	1,180	–	–	–	1.7	Inland Playas
Inner Mongolia	Haganula Nuoer	43° 56′	114° 47′	1,049.9	–	–	–	1.6	Inland Playas
Inner Mongolia	Aci Haizi	40° 28′	106° 47′	1,050	1.5	1.2	1.1	1.6	Inland freshwater Lake freshwater Lake
Inner Mongolia	Unknown Lake	41° 06′	107° 46′	1,030	3.1	0.6	0.5	1.5	Inland Lake
Inner Mongolia	ErJieNao Haizi	41° 50′	113° 47′	1,454	3.8	1.6	0.4	1.5	Inland Lake
Inner Mongolia	Aiken eek Lake	38° 44′	105° 05′	1,292	–	–	–	1.5	Inland Playas
Inner Mongolia	Ulan Nuoer	43° 20′	114° 31′	1,020	–	–	–	1.5	Inland Playas

(continued)

Table 12.1 (continued)

Province	Name of Lake	Location		Altitude/m	Length/km	Width/km		Area/km²	Remarks
		N	E			Max	Mean		
Inner Mongolia	Wenduoerchabu Nuoer	43° 20′	114° 43′	1,000	–	–	–	1.5	Inland Playas
Inner Mongolia	Unkown Lake	43° 19′	113° 44′	979	1.9	1.5	0.7	1.4	Inland Playas
Inner Mongolia	Adaduo Monao	42° 27′	101° 19′	920	2.4	1.2	0.6	1.4	Inland Playas
Inner Mongolia	Durham Qaidam Lake	39° 58′	105° 57′	1,020	–	–	–	1.3	Inland Playas
Inner Mongolia	Sumbarun Jilin Lake	39° 47′	102° 25′	1,180	–	–	–	1.2	Inland Playas
Inner Mongolia	Che Hu Lake	39° 53′	102° 15′	1,160	–	–	–	1.2	Inland Playas
Inner Mongolia	Kersin Qaidam Lake	39° 57′	105° 55′	1,020	–	–	–	1.2	Inland Playas
Inner Mongolia	Unkown Lake	41° 07′	107° 49′	1,030	2.5	0.8	0.5	1.2	Inland Lake
Inner Mongolia	Hulaguo Lake	42° 29′	100° 31′	898	2.3	1	0.5	1.2	Inland Playas
Inner Mongolia	Sultenuor Lake	43° 07′	114° 08′	1,040	1.6	1	0.7	1.1	Inland Playas
Inner Mongolia	Hulgi Nao Lake	40° 45′	108° 22′	1,064	–	–	–	1.1	Inland Playas
Inner Mongolia	Terry Del Lake	39° 51′	102° 26′	1,200	–	–	–	1.1	Inland Playas

(continued)

Table 12.1 (continued)

Province	Name of Lake	Location		Altitude/m	Length/km	Width/km		Area/km^2	Remarks
		N	E			Max	Mean		
Inner Mongolia	Quasi-Hunderon Qaidam Lake	39° 57′	106° 02′	1,020	–	–	–	1	Inland Playas
Inner Mongolia	KuhJilin Lake	39° 54′	102° 28′	1,200	–	–	–	1	Inland Playas
Inner Mongolia	Doha Hale Jinda Lake	39° 56′	105° 56′	1,020	–	–	–	1	Inland Playas
Inner Mongolia	Eagle Gide Lake	39° 46′	102° 09′	1,151	–	–	–	1	Inland Playas
Inner Mongolia	Zhangji Nao	40° 41′	108° 20′	1,019	–	–	–	1	Inland Playas
Inner Mongolia	Shangkelu Lake	40° 41′	108° 18′	1,060	–	–	–	1	Inland Playas
Inner Mongolia	Quasi-Gede Lake	39° 51′	102° 19′	1,180	–	–	–	1	Inland Playas
Inner Mongolia	Bagnol	41° 39′	109° 42′	1,553	–	–	–	1	Inland Playas
Inner Mongolia	Bayi Nuoer	39° 50′	101° 28′	1,304	–	–	–	1.0	Inland Playas
Inner Mongolia	Wengeng Nuoer	42° 30′	114° 05′	1,214	–	–	–	1.0	Inland Playas
Inner Mongolia	Jiaolugou Lake	40° 05′	100° 58′	1,277	–	–	–	1.0	Inland Playas
Inner Mongolia	Nandawudengnao	41° 35′	113° 30′	986.0	–	–	–	1.0	Inland Playas

(continued)

Table 12.1 (continued)

Province	Name of Lake	Location		Altitude/m	Length/km	Width/km		Area/km²	Remarks
		N	E			Max	Mean		
Inner Mongolia	Baiyinnao Lake	41° 32′	113° 00′	985.0	–	–	–	1.0	Inland Playas
Inner Mongolia	Han Yi Naoer	41° 30′	112° 58′	1,520.0	–	–	–	1.0	Inland Playas
Inner Mongolia	Wangong Nuoer	42° 37′	114° 10′	984.0	–	–	–	1.0	Inland Playas
Inner Mongolia	Gobainoer	42° 54′	113° 41′	1,020.0	–	–	–	1.0	Inland Playas
Inner Mongolia	Buridenao	43° 08′	112° 40′	1,097	–	–	–	1.0	Inland Playas
Inner Mongolia	Xi Polegel Lake	43° 37′	112° 26′	941	–	–	–	1.0	Inland Playas
Inner Mongolia	Hanbanchi	43° 30′	112° 20′	1,018	–	–	–	1.0	Inland Playas
Inner Mongolia	Wulantaoligai	43° 41′	113° 05′	1,036	–	–	–	1.0	Inland Playas
Inner Mongolia	Hala Tu Po	43° 15′	113° 00′	1,025	–	–	–	1.0	Inland Playas
Inner Mongolia	Ulanore	43° 35′	112° 55′	1,060	–	–	–	1.0	Inland Playas
Inner Mongolia	Polegel Lake	43° 38′	112° 33′	930	–	–	–	1.0	Inland Playas
Inner Mongolia	Pemba Lake	43° 47′	112° 40′	920.0	–	–	–	1.0	Inland Playas

(continued)

Table 12.1 (continued)

Province	Name of Lake	Location		Altitude/m	Length/km	Width/km		Area/km²	Remarks
		N	E			Max	Mean		
Inner Mongolia	Wutong Sea	40° 15′	100° 55′	1,225	–	–	–	1.0	Inland Playas
Inner Mongolia	Dagu Sea	40° 05′	100° 52′	1,297	–	–	–	1.0	Inland Playas
Inner Mongolia	Xia Mata La Lake	43° 12′	114° 18′	1,038.0	–	–	–	1.0	Inland Playas
Inner Mongolia	Hitler Nuoer	43° 22′	114° 37′	1,030.0	–	–	–	1.0	Inland Playas
Inner Mongolia	Ula Nuoer	44° 06′	113° 56′	960.0	–	–	–	1.0	Inland Playas
Inner Mongolia	Baya Nuoer	44° 16′	113° 27′	1,060.0	–	–	–	1.0	Inland Playas
Inner Mongolia	Zhunchagan Gan Nuoer	44° 32′	113° 42′	1,160.0	–	–	–	1.0	Inland Playas
Inner Mongolia	Hoh Nuoer	44° 35′	113° 43′	1,165.0	–	–	–	1.0	Inland Playas
Inner Mongolia	Baya Nuoer	39° 52′	101° 08′	1,346	–	–	–	1.0	Inland Playas
Inner Mongolia	Aleid Lake	39° 47′	102° 22′	1,180.0	–	–	–	1.0	Inland Playas
Inner Mongolia	Delma Tara	43° 12′	114° 01′	991	3.1	1.0	0.3	1.0	Inland Playas
Inner Mongolia	Dundee Mauta	43° 13′	114° 08′	1,020	2.0	1.0	0.5	1.0	Inland Playas

(continued)

Table 12.1 (continued)

Province	Name of Lake	Location		Altitude/m	Length/km	Width/km		Area/km^2	Remarks
		N	E			Max	Mean		
Inner Mongolia	Baya Nuer	44° 16′	113° 27′	1,073	1.4	1.0	0.7	1.0	Inland Playas
Inner Mongolia	Ulan Nur	43° 02′	113° 07′	971	1.9	0.7	0.5	1.0	Inland Playas

Table 12.2 Lakes with a surface area larger than 10 km^2 in the arid land of China

Provine	Name of Lake	Location		Area/ km^2	Volume/ million (m^3)	Remarks
		N	E			
Xinjiang	Lop Nor	90° 15′	40° 20′	3 006	No data	Dried up in 1972
Xinjiang	Lake Aibi	82° 35′–83° 10′	44° 00′	1 070	149 8	
Xinjiang	Bosten Lake	87° 03′	41° 58′	960	773 0	
Xinjiang	Lake Ulung	87° 18′	47° 13′	753	602 4	
Xinjiang	Manas Lake	85° 55′	45° 42′	550	No data	Dried up in 1962
Xinjiang	Lake Gary Dobo	76° 52′	40° 07′	231	No data	
Xinjiang	Jili Lake	87° 26′	46° 55′	174	57 000	
Xinjiang	Aiximan Lake	80° 07′	40° 42′	149.6	No data	Dry up
Xinjiang	Jakesalay Lake	81° 22′	36° 49′	148	No data	
Xinjiang	Aiding Lake	42° 38′	89° 14′	124	9 9.20	
Xinjiang	Qinggelike Lake	84° 28′	38° 02′	120	No data	
Xinjiang	Lake Balik	92° 05′	43° 45′	116	No data	
Xinjiang	Quchusu Lake	83° 29′	37° 38′	90	No data	
Xinjiang	Lake Sauer Creek	83° 03′	37° 38′	90	No data	
Xinjiang	Lake Tecoma	86° 53′	40° 54′	88	No data	Dried up in 1972
Xinjiang	Dabson nore	86° 15′	45° 44′	87	No data	
Xinjiang	Lake Baghdad	81° 04′	40° 47′	70	No data	
Xinjiang	Big salt Lake	87° 26′	41° 46′	65	No data	Dry up
Xinjiang	Erika Lake	85° 47′	45° 54′	44	No data	
Inner Mongolia	East Juyanhai	101° 15′	42° 18′	35.5	No data	
Xinjiang	Dabancheng East Salt Lake	88° 08′	43° 23′	35.1	No data	
Xinjiang	Lake Tahr	86° 53′	42° 11′	30	No data	Dry up
Xinjiang	Chaiwo Fort Lake	87° 54′	43° 03′	30	No data	
Xinjiang	Yihu Lake	94° 12′	43° 02′	29.2	No data	
Inner Mongolia	Tengger Nuoer	110° 41′	42° 26′	28.6	No data	
Inner Mongolia	North Pond	107° 27′	37° 58′	20.3	No data	
Xinjiang	Glemy Lele	86° 54′	40° 46′	11	No data	Dry up
Xinjiang	Yingkuer Haizi	86° 50′	40° 47′	10	No data	Dry up
Xinjiang	Sai Yitekule	86° 53′	40° 54′	10	No data	Dry up

Note ① Manas Lake dried up in 1962, in 2002 the Lake area was 600 km^2; ② Aidynkol Lake elevation of −154.31 m; ③ Taiwan Lake dried up in 2000, Lake area in 2001 was 10 km^2, in 2003 it reached more than 200 km^2; ④ Erik Lake dried up in 2000, in 2003 Lake area was 57 km^2; ⑤ East Lake in 2008 dried up and the Lake area was 45 km^2; ⑥ Balkash lake is located in Kazakhstan, the Lake area is18,200 km^2, Lake water volume about 10.6 million m^3

It is well known that water resources in arid areas (including surface water and groundwater) are the result of mountainous runoff area, affected by the geological structure and hydrogeological condition, recharged by groundwater and formed many springs in the vast region outside of runoff zone; springs converge to a small lake, such as there are about 144 in less than 1–10 km^2 freshwater and saltwater lakes and salt lakes, with the total area of 33.42 km^2.

Influenced by the predatory exploitation of land and its rich, environment change of small lake is prominent in arid areas. Some freshwater lakes have disappeared, such as some Small Haizi of Tarim river basin in Xinjiang. Corksu lake, and Algol lake and other small lakes; Algol lake covers an area of about 70 km^2 (water report of 1990), all dried up. Some lakes have disappeared, such as Eric lake and Manas lake in Karamay (Xinjiang).

Eric lake is the rump lake for poplar river. According to the satellite remote sensing image in 1989–2005 for Eric lake, the lake dried up in 2000, the area of lake was 59.6 km^2 in 2003 (including reed area of water).

Salt lake is another type of lake in arid areas, and it is a type of lake which has an important value. Salt lake (also known as Lake brine) usually refers to the lake of which the salinity is equal to or greater than 50 g/l. Lopnor is the second largest salt lake in China, today, with area of about 5,350 km^2. In historical period, the Lopnor was a big freshwater lake collected by runoff of south of Tian Shan, and the Kunlun and Altun. The lake dried up in 1962 due to interference of human activities and became a vast dry lake (about 600 km^2). There are salt crust with crack shape about 10–15 cm and large dead reeds at the rim of lake in the dried bottom about 450 km^2. After drying Lobnor lake, salt deposition of surface and bottom is rich in inter-crystal brine in and inter-layer brine, with relative density of 1.286, pH of 6.07, the content of KCl of 2.08%, and refers to sulfate magnesium and sulfate subtypes with hydrochemical type (Wang and Dong 2002).

According to the information of scientific expedition to Lobnor in the Qinghai Institute of Salt Lakes Chinese Academy of Sciences, there are utilized of resources, brine, and halide salts. Inter-crystalline brine occurs between the surface rock salt, gypsum and glauberite salt layer, rich brine potassium high mineralization, with the thickness of aquifer of more than 100 m, salinity of inter-layer brine of 372.08 g/l. In saline resources, rock salt owes an area of 700 km^2, thickness of 0.5–2.5 m. The content of NaCl is 60–80%, the highest up to 94%, reserves of about hundreds of one hundred million tons, a very large rock salt deposit. There reserves about 250 million t of potassium magnesium salt deposit, 120,000 ton of potassium nitrate, huge amount of sodium calcium sulfate ore (Guo et al. 2003).

Lake Types. Arid area is vast in China, and lakes are in diverse species (Tables 12.1 and 12.2). According to incomplete statistical classification, there are about nine lake types.

(1) Settling lake. Geological period had a descending basin or lowland in the depression, rivers into the retention of water into the lake. Such as the LopNor lake of the Tarim River, the Junggar Lake in the Junggar Basin, the Heihe River, the Shule river basin in the east, the West Jueyan Sea and the Hokkaido,

and the Lindi Basin in the Turpan Basin, the Aidynkol—154.31 m, the lowlands in China. As well as there are many lakes such as Ake kuku in the southern slope of the Kunlun Mountains and the Altun Mountains, Qiangtang plateau.

(2) Subsiding lake. Such as Bosten Lake, Ulungu Lake, Ebinur Lake, Barkol Lake and the Tian Shi Lake belong to this lake type. This type of lake is mainly distributed in the edge of the basin or mountains basin, with a certain catchment area, from the surface water and groundwater into.

(3) Terminal moraine lake. Ancient glaciers or planing erosion in modern glaciers alpine zone, forming shades, ranging in size after the depression, Ice Age, the glaciers receded, moraines (terminal moraine dam) damming the river to form glacier lake. It is characterized by growing up along the river bar, such as Xinjiang famous Fukang Tianchi and Burjin Kanas Lake Scenic Area.

(4) Glacier blocking lake. China arid areas with modern glaciers 21,568 strips are the world's mountains glaciers latitudes subregion up to the cloth (Shi et al. 2002). There are most extensive mountain of Chinese glaciers–Kunlun Mountains (Glacier area is about 12,566 km^2), accounting for 20.7% of the country's glacier area; China's glacier has the largest distribution of the river—Tarim water (glacier area of 19,889 km^2), accounting for 33.5% of the glacier area; there is the largest basin of Chinese glaciers—Kunlun peak area. The total glacier area is 4346 km^2. There are 15 large glaciers with more than 300 km 2 in area distributed in the low latitudes of the world and four of them are found in the arid region of China. (337.9 km^2 in the territory of China, 567.2 km^2 in the territory of China), the Tommel Glacier (337.9 km^2), the Tiggyrizi glacier (313.7 km^2), and the sound of the Karakoram Mountains Glacier (379.9 km^2) (Hu et al. 2002). As in the process of glacier movement, some support glaciers blocked the main river, some of the main glaciers blocked the tributaries of the river, and the formation of glaciers blocked the lake. At the elevation of 4,900 m a.s.l, the lake is 9.48 km long and has an average width of 638 m. The lake area is 6.1 km^2 and the average water depth is 52.6 m depth of 154.5 m, with water storage capacity of up to 318 million m^3. The glacier lake is often caused by important floods in the lower reaches of the basin.

(5) Interfluve lake. Mainly distributed in the middle and lower reaches of the river basin in the arid region, such as the Lauru Lake in the Tarim river basin, Lake Sutherland, Saite Lake, Dada Mukule (length 2 km, width is 1.1 km, area 2.2 km^2). Yili River, Heihe River, and Kaidu River are also distributed.

(6) Oxbow lake. Rivers into the plains or flat basin area, the ground slope is gentle, the flow rate is reduced, such as the Tarim River stream in the British Bazaar to the Chala section, river bed slope of 1/7 700, slow water, river bend). Yeer Qiang River, Hetian River downstream section also has a small amount of Oxbow lake.

(7) Wind erosion lake. In the lower reaches of the Tarim river basin and the lower reaches of the Shule River, there are many low-lying depressions formed by strong directional wind erosion, forming rivers when floods are flooded. Such as the Tarim River mainstream famous Daxi Haizi and so on.
(8) Submerge lake. Tian Shan Mountains, Qilian Mountains, Kunlun Mountains, Altun Mountains, Altai Mountains are the southern slope of the spill overflow zone, overflowing water retention in the depression to form lakes, such as Lake Aidynkol, mushroom lake (built reservoir) and the northern slope of the Kunlun Mountains, such as Berkeley Kul and so on. There are more than 460 large and small lakes in the Badan Jaran Desert and Tengger Desert, and they belong to this lake type.

Water reservoir. According to incomplete statistics, there are about 500 water reservoirs in arid areas of China in different sizes and functions. They play an important supporting role in the economic development of the basin and has different positive and negative effects on the local environment.

12.2 Ebinur Is the Largest Salt Lake in the Northwest China

Ebinur Lake is located in the western part (43° 38′–45° 52′N; 79° 53′–85° 02′E) of the Junggar basin in Xinjiang in northwest China and is a typical rump lake in an arid region (Abuduwaili et al. 2007; Abuduwaili 2009). Ebinur Lake is a shallow, closed lake basin in arid region of northwest China (Fig. 12.1). The lake has a drainage area of 50,321 km^2, including 24,317 km^2 of mountainous terrain and lakes 542 km^2. The Ebinur Lake is the largest lake in the basin as well as the largest salt lake in Xinjiang in northwest China (Abuduwaili et al. 2014). Ala Mountain borders the lake to the north and northwest, Borotala Valley is to the west, the Jing River pluvial fan is to the south, and sand dunes around the Kuitun River are to the east. The Borotala, Jinghe, and Kuitun Rivers are the main feeders into the lake that come from the west, south, and east, respectively. Due to the arid desert climate, there is little rainfall with the average annual rainfall of the basin around only 100–200 mm and a potential evaporation of up to 1500–2000 mm (Abuduwaili et al. 2014). Mean annual precipitation around the lake is about 95 mm, whereas annual evaporation is 1315 mm (Wu et al. 2009). Maximum precipitation falls in summer. In winter, the snow cover is shallow—10–25 cm—and persists up to the late February or early March. The mean July temperature is +27 °C; the mean January temperature is—17 °C.

The lake has a maximum water depth of 3.5 m and an average depth of 1.2 m. The lake water has 85–124 g L^{-1} of total dissolved solids. The lake water salinity is about 120 g/l (Mahpir and Tursunov 1996). The lake has a sodium chloride–sulfate

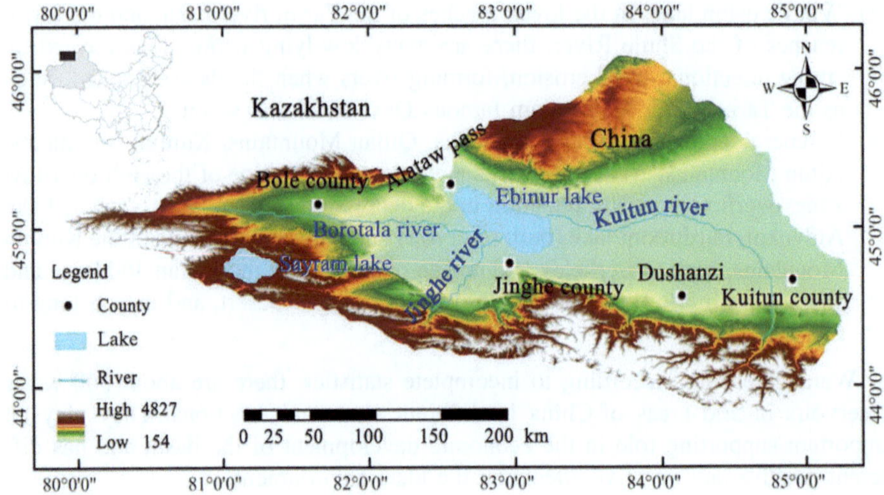

Fig. 12.1 Ebinur lake basin

or sulfate–chloride composition of water (Table 12.3). Based on a field survey conducted in 2009, it was shown that the major ions in Lake Ebinur were chlorine and sodium, and the hydrochemical classification subsequently changed from sulfate-sodium-II type to chloride-sodium-II type (Wu et al. 2014a, b).

Ebinur Lake receives surface water inputs from the Borotala and Jing Rivers. The Ala Mountain pass, northwest of the lake, is a well-known wind corridor, with wind speeds exceeding 20 m s^{-1} on 164 days of the year and maximum wind speeds of up to 55 m s^{-1} (Wu et al. 2009; Ma et al. 2011b) and northwestern winds prevail in the region. Strong winds are frequent in this region.

The Lake basin is considered an important base for grain, cotton, animal husbandry, oil and chemical industries. In Xinjiang in northwest China, it is also an important open channel to the west, known as the "Euro-Asian Continental Bridge" that runs along the Ebinur Lake across the lake basin from the north to the southwest (Zhou and Lei 2005; Mi et al. 2008). Since the 1990s, both the implementation of the "Western Development Policy of China" and the developmental policy by the Xinjiang Uygur Autonomous Region in northwest China have led to prodigious economic development in the Ebinur Lake Basin. In this area, fertilizers and pesticides imprudently used in agricultural processing and large quantities of emissions by the townships have led to pollution of the main feeders, the Jing and Bortala Rivers. This has significantly and negatively influenced the water quality of the main water area of the Lake. This pollution will eventually seep into the sediment and become a threat against the aquatic organisms and the ecological environment (Liu et al. 2011). Over the last 50 years, a combination of the population and large-scale exploit of water and soil of Aibi Lake has led to a dramatic reduction in the amount of water flowing into the lake from rivers (Liu et al. 2011).

Table 12.3 Hydrochemical characteristics of water in the Ebinur lake (Fan and Zhang 1992)

Sampling point	pH	Salinity (g/l)	CO_3^{2-}	HCO_3^-	SO_4^{2-}	Cl^-	Ca^{2+}	Mg^{2+}	$Na^+ + K^+$	Predominant ions
			mmol/l (%)							
Western lakeside	8.0	122.5	Traces	$\frac{8.3}{0.4}$	$\frac{957.2}{59.7}$	$\frac{921.9}{48.8}$	$\frac{32.5}{1.7}$	$\frac{371.5}{19.7}$	$\frac{1483.5}{78.5}$	Cl^-–SO_4^{2-}, Na^+
Southern lakeside	7.9	81.8	None	$\frac{7.1}{0.6}$	$\frac{564.8}{47.3}$	$\frac{622.1}{52.1}$	$\frac{14.1}{1.2}$	$\frac{255.3}{21.4}$	$\frac{924.7}{77.4}$	SO_4^{2-}–Cl^-, Na^+

In the Aibi Lake Basin, the flora is primarily that of Central Asia and Mongolia. There is a total of 385 plant species belonging to 53 families and 191 genera (Qian et al. 2004). The soil types are mainly Piedmont psephitic and Gypsum desert soil; the vegetative cover is mainly H. ammodendron desert and Ephedra desert; and plant growth is Populus euphratica forest, Phragmites australis, and lowland meadow. Since the 1950s, the driving forces of industrial and agricultural development have dramatically decreased the amount of water flowing into the lake, resulting in a reduction in the lake water area from 1000 to 500 km^2. As the lake has dried up, the ecological environment of the lake basin has visibly deteriorated (Liu et al. 2014a, b).

The surface area of this lake was about 1070 km^2 in 1950 and then experienced a rapid contraction. In 1972, the lake area decreased to 589 km^2. In the late 1990s, Ebinur Lake started to expand; however, the lake area shrank sharply from 2004. The variation of Lake Ebinur area is jointly controlled by human activities and climate change. However, the human activity was mainly responsible for Ebinur Lake shrinking quickly over the past half century (Ma et al. 2014).

12.3 Hydrograhical Characteristics of the Sayram Lake and Environmental Issues in the Region

Sayram Lake is located in the western part of the Tian Shan Mountains, northwest China (Fig. 12.2). It is a closed-basin lake in the western Tian Shan. The lake surface area fluctuated between 458.6 and 462.2 km^2 over the decade from 2001 to 2011 (Wu et al. 2014a, b). The water body is surrounded by high mountains and has a catchment area of 1408 km^2 (Wang and Dou 1998). The vegetation of Sayram Lake Basin has relatively simple composition and obviously vertical bands. Above the 3800 m above sea level (asl), the surface is snow-covered region. From 3800 m to 2800 m asl, the vegetation changes from alpine sparse vegetation to alpine meadows.

The surface is covered with subalpine steppe meadows (2800–2400 m asl). Then, the vegetation changes to forest meadows and mountain meadows (2400–2150 m asl). Around the lakeside belts, the vegetation varies to steppe and desert–steppe (Jin 1995). Lake Sayram has a maximum water depth of 99 m and an average water depth of 46.4 m (Wu et al. 2014a, b). The lake is located in the hinterland of the Eurasian continent where Northern Hemisphere westerly winds are characteristic of the semi-arid climate. So the climate of the region is dominated by Northern Hemisphere westerly winds.

Water in Sayram Lake comes mainly from glacial snowmelt and phreatic water (Jin 1995). There are 39 inflows around the lake. Among these, 7 rivers are perennial, 13 ones are spring-fed, and 19 ones are seasonal (Jin 1995). The region receives annual total precipitation of 350 mm, and mean annual temperature is about 0.5 °C (Wang and Dou 1998). Regional grasslands constitute a rich resource

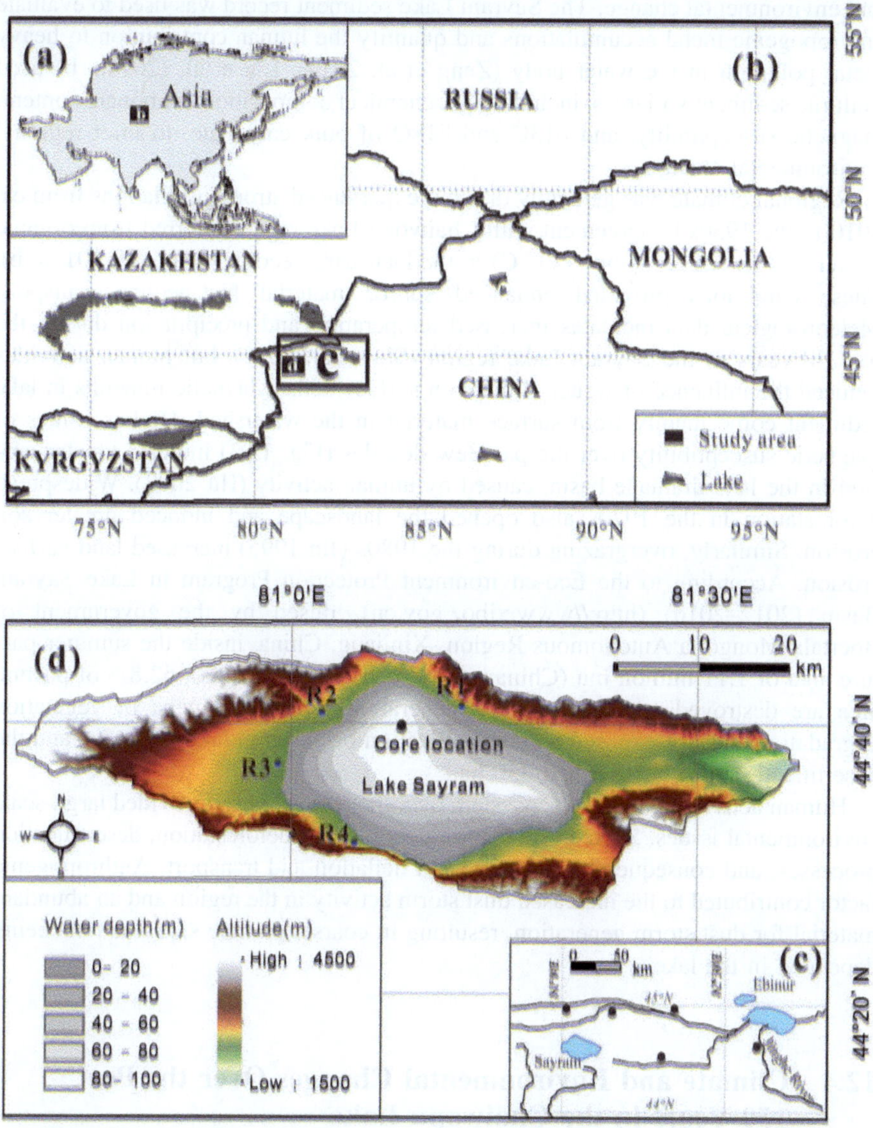

Fig. 12.2 **a, b** Geographic location and bathymetry of Sayram Lake. **c** Sites of dustfall samples. **d** Sites of riverine samples collection (Ma et al. 2015)

that is exploited by traditional Kazak herders. There are no residential villages, and there is no agricultural cultivation around the lake. Nomadic grazing of livestock is the most important activity in the drainage basin.

Given its basin structure and location, Sayram Lake possesses a high-quality sediment archive that can be utilized to increase our understanding of past climate and environmental change. The Sayram Lake sediment record was used to evaluate anthropogenic metal accumulations and quantify the human contribution to heavy metal pollution in the water body (Zeng et al. 2014). Liu et al. (2014a, b) used multiple sediment variables, including geochemical composition, carbonate content, magnetic susceptibility, and δ13C and δ18O of bulk carbonate, to infer regional environmental change.

Regional climate was generally dry, but experienced strong oscillations from ca. 1910 to the 1930s in agreement with Chaiwopu Lake region inferred from organic matter and its stable isotope (13 °C) in the lacustrine records (Ma et al. 2013a, b). These conditions provided enhanced source material for aeolian transport. Meteorological data indicates increased temperature and precipitation during the past 50 years in the Sayram Lake region (Shi et al. 2007), but human activities reduced the influence of wetter climate on surface soils. Magnetic minerals in lake sediment come mainly from surface material in the watershed. Higher values of magnetic susceptibility over the past few decades (Fig. 12.3) indicate greater erosion in the lake drainage basin, caused by human activity (Hu 2002). Widespread deforestation in the 1970s also opened the landscape and induced greater soil erosion. Similarly, overgrazing during the 1980s (Jin 1995) increased land surface erosion. According to the Eco-environment Protection Program in Lake Sayram Basin (2012–2016) (http://www.xjboz.gov.cn), raised by the government of Boertala Mongolia Autonomous Region, Xinjiang, China, inside the summer pasture area of 1.11 million mu (China unit of area; 15 mu = 1 ha), 62.8% of pasture area are destroyed and underwent degeneration. The phenomena of vegetation degradation, surface exposed and desertification occurred in lakeside belts, and the desertification area has reached 250 ha.

Human activity in the lake ecosystem over the past 50 years provided large-scale environmental issues. Such as vegetation degradation, deforestation, desertification processes, and consequently dust and sand deflation and transport. Anthropogenic factor contributed to the increased dust storm activity in the region and an abundant material for dust storm generation, resulting in coarser particle size fractions being deposited in the lake.

12.4 Climate and Environmental Changes Over the Past 150 Years in the Chaiwopu Lake

Chaiwopu Basin is a small, inter-montane structural basin in the central Tian Shan Mountain area, northwest China. Yilianhabierga Mountain lies south of the basin and has an elevation of 4483 m above sea level (a.s.l.), and to the north is Bogeda Mountain (5445 m a.s.l.). Chaiwopu Basin is connected with the Junggar Basin to the west, and the Baiyanggou River links it with the Turpan Basin to the southeast. Chaiwopu Lake is situated in Dabancheng District, approximately 45 km southeast of Urumqi, Xinjiang, northwest China (Fig. 12.4a). The lake is fairly round in

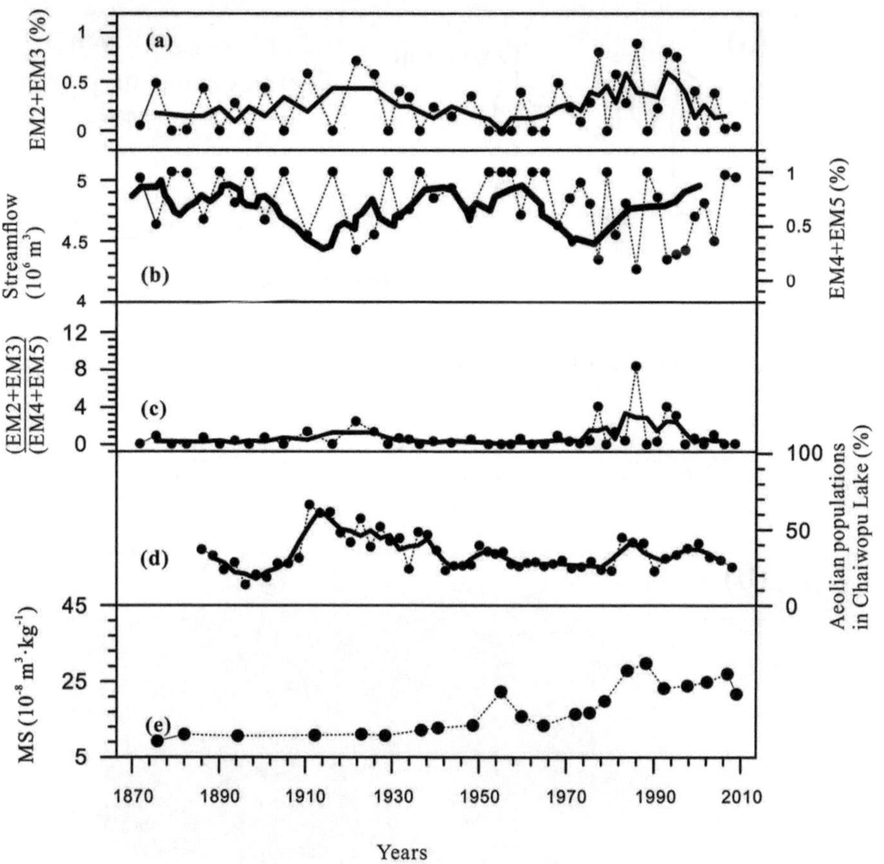

Fig. 12.3 Regional proxies for fluvial and aeolian environments. **a** Relative content of aeolian transported material (EM2 + EM3) in Sayram Lake sediments (dashed line) with the three-point running average (solid line). **b** Relative content of fluvial transported material (EM4 + EM5) in Sayram Lake sediments (dashed line) comparing with annual streamflow of Jinghe River (Shang et al. 2014) (solid line). **c** Ratio of aeolian to fluvial transport fractions in the Sayram Lake core. **d** Content of aeolian particle size populations recovered from the Chaiwopu Lake core (Ma et al. 2013a, b). **e** Magnetic susceptibility (MS) of Sayram Lake sediments

shape and approximately 5–6 km in diameter. The lake water level was 1903.86 m a.s.l. with a lake area of 30 km^2 in 1971, but fell to 1901.66 m a.s.l. with an area of 27 km^2 in 2008. The average water depth is ~2 m with a maximum depth of ~4 m (Fig. 12.4b). The lake is brackish, with a salinity of 6.8 g/l. It has a transparency of 18 cm, a pH of 9.04, and an average conductivity of 0.867 s/m (Ma et al. 2013a, b). Chaiwopu Lake is covered with ice from mid-November to late March or early April. It is a natural cold-water lake that receives water from several streams running from Bogeda Mountain. The lake and its surroundings have been

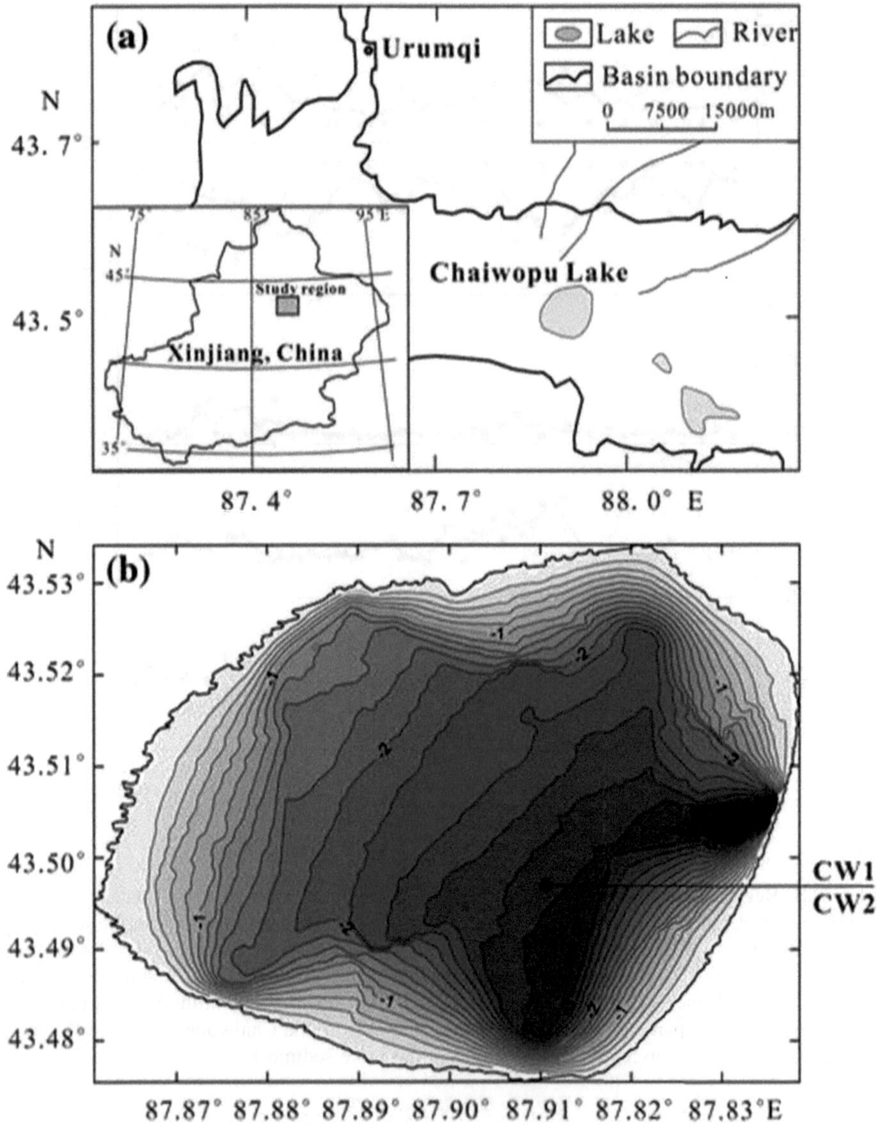

Fig. 12.4 Study site (**a**) and bathymetry of Chaiwopu Lake with core location (**b**)

officially protected since 2009, when Urumqi Chaiwopu Lake National Wetland Park was created by the State Forestry Administration of China.

The meteorological stations of Dabancheng (43.3627°N, 88.3116°E; 918.7 m a.s.l.) and Urumqi (43.8259°N, 87.6174°E; 1105.3 m a.s.l.) have recorded the mean annual temperature and the annual total precipitation during the past 50 years (Fig. 12.5). Mean annual temperature rose gradually since the late 1950s, and stable

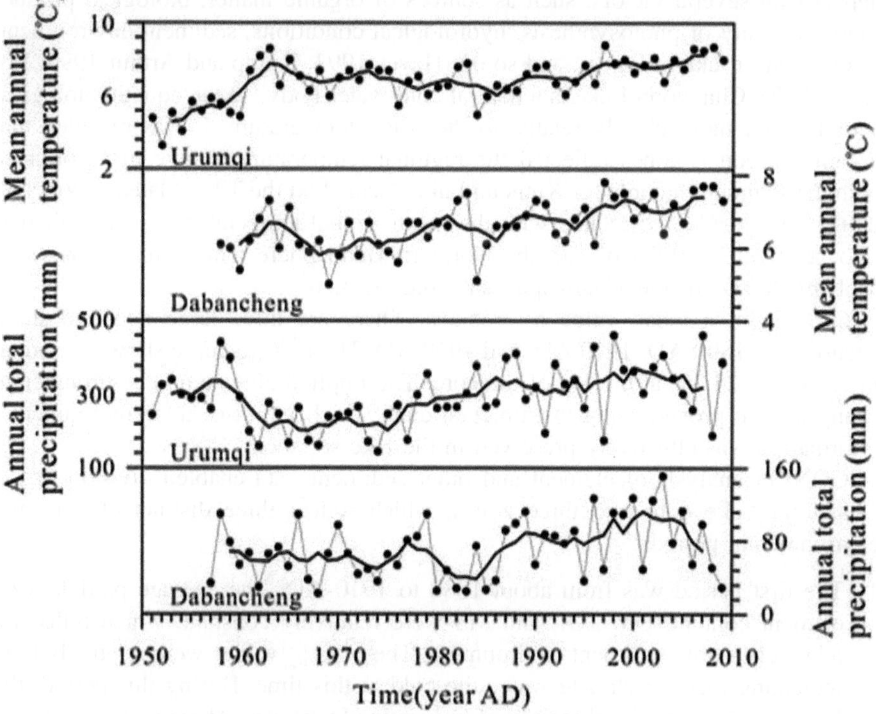

Fig. 12.5 Curves of mean annual temperature and annual total precipitation with 5-year moving average (solid lines) in the region of Chaiwopu Lake (Ma et al. 2013a, b)

higher temperatures have been maintained in recent years. Annual total precipitation increased since 1980 AD, whereas it has decreased over the last 10 years.

Organic matter in the lake sediment is a mix of aquatic and terrestrial plant debris, the latter resulting from watershed erosion (Meyers and Teranes 2001). The organic matter content can thus be used to reconstruct paleoenvironments of lakes and their watersheds and to infer past regional climate change (Brenner et al. 1999). In arid/semi-arid areas, regional moisture is the main factor influencing plant growth (Ma et al. 2011a). In humid climates, plants grow vigorously, leading to a higher content of organic matter in lake sediments, whereas under arid conditions, plant growth is limited and organic matter content in lake sediments is lower. The lower OM-LOI content indicates drier climate period, while the higher reflects wetter climate interval (Liu et al. 2002; Wu et al. 2009; Oldfield et al. 2010; Zhong et al. 2010). The OM-LOI variations are consistent with Palmer Drought Severity Index (PDSI) of the central Tian Shan Mountain area in northwest China (Li et al. 2006). The OM-LOI indirectly reflects humidity variation in Chaiwopu Lake region. OM-LOI analysis showed values between 3.7 and 6.3%. The OM-LOI below 35 cm is relatively low, with an average of 4.3%, whereas above 35 cm, it increases to an average of ~5.4%. The carbon isotope of bulk organic matter depends on several factors, such as sources of organic matter, biological productivity, intensity of photosynthesis, hydrological conditions, sediment environment, reservation of lake sediment, and so on (Hayes 1993; Kump and Arthur 1999; Wu et al. 2007). Chaiwopu Lake is a natural cold-water body, so the aquatic biological activities are more closely related to the water temperature. It was revealed that organic carbon isotopes reflected the regional temperature change in accordance with the sediment records for Xingcuo Lake, located on the Tibet Plateau (Wu et al. 2007). The $\delta^{13}C_{org}$ correlated with the mean annual temperature in Urumqi. It is also positively correlated with the Northern Hemisphere temperature (Jones and Moberg 2003). In the Chaiwopu Lake sediment, $\delta^{13}C_{org}$ ranges from −24.2 to −26.5%, with a mean value of −25.5%. There are three large negative shifts occurring at 1890 AD, 1910 AD, and 1960 AD. The $\delta^{13}C_{org}$ curve shows a gradual increase to −24.2% at the top of the core. The implied climate and environmental changes were proved using instrumental data, which indicates that environmental information was effectively preserved in the lake sediment.

CONISS analysis of element and other sediment data enabled division of the Chaiwopu Lake core into three zones, which reflect three distinct climate and environmental periods:

1. The first period was from about 1880 to 1910 AD. The average particle size, element content, OM-LOI, and MS were relatively constant, which reflects a relatively stable sediment environment. The $\delta^{13}C_{org}$ values were relatively low, indicating that the climate was quite cold at this time. During this period, the lake sediment consisted mainly of fine-grained particles. There is no significant variation in the median grain size, which reflects the stable hydrodynamic conditions. Values of the C index were relatively low and Sr/Ca was relatively high, suggesting that regional climate was dry and lake water salinity was high. The data indicates a relatively stable aquatic environment, with high salinity, under cold and dry climate conditions, after the Little Ice Age.
2. The period from 1910 to 1950 was the second stage. In this period, geochemical indicators OM-LOI and grain size fluctuated significantly. δ^{13} C_{org} increased. In this period, values of the C index were low and Sr/Ca remained high, suggesting that lake water was shallow and the regional climate was unstable, with generally higher temperature and humidity. Contents of mobile elements (Mg, Ca, and Sr) are lower (Fig. 12.7), which suggests that the chemical weathering intensity was still weak.
3. In the last 50 years, which correspond to the upper 25 cm in the core, and the 50-year instrumental data from the Urumqi meteorological station, the MAT displays a warming trend, and the lowest temperature was observed in the 1950s. The $\delta^{13}C_{org}$ also decreased. Variations in annual total precipitation were more complex, and there was a significant increase in ATP since the 1990s. According to satellite images collected at different times, along with respective topographic maps, Chaiwopu Lake has remained stable, with fluctuations in total area of <2 km^2 (Ma et al. 2011b). The grain size remained relatively constant, so the water body was relatively stable. MS is very different from that measured in

previous periods, indicating greater erosion of the Chaiwopu Basin, perhaps caused by human activity. The high value of the C index suggests a high moisture condition in the watershed during this period, a better soil–vegetation environment, and more surface runoff. The low molar Sr/Ca supports this interpretation. Increases in heavy metal and total phosphorus (TP) concentrations were likely influenced by enhanced human activity.

Global warming will have a large impact around the globe, causing differences in precipitation distribution, hydrological cycles, and effective moisture changes. In the Chaiwopu region, the climate became warmer and moister beginning about 1950 AD. Shi et al. (2007) considered that a warm–moist transition may occur in northwest China, based on the instrumental data and glacier and lake changes over the last 50 years.

Global and Northern Hemisphere temperature experienced a significant change about 1910 (Jones and Moberg 2003; Brohan et al. 2006), and in the arid region of northwest China, PDSI also showed the central Tian Shan Mountain area was dry at this time (Li et al. 2006), consistent with the hypothesis that dry climate provides abundant material basis for sandstorms, and explaining how the aeolian sands increased the sediment grain size in Chaiwopu Lake.

References

Abuduwaili J (2009) Effects of wind erosion on the salt emission, transport process of salts: a case study of the area of Lake Ebinur, Xinjiang, China. Arid Land Geogr 32(2):211–217. (In Chinese)

Abuduwaili J (2012) Lakes of Arid Land and Salt-Dust Storms. Environmetal Science Press of China, Beijing 301 p (in Chinese)

Abuduwaili J, Xu JR, Mu GJ, Xu M, Gabchenko MV (2007) Effect of soil dust from Ebinur Lake on soil salts and landscape of surrounding regions. J Glaciol Geocryol 29(6):928–939 (In Chinese)

Abuduwaili J, Zhang Zh, Jiang F (2014) Assessment of the distribution, sources and potential ecological risk of heavy metals in the dry surface sediment of Aibi Lake in Northwest China. PLoS ONE 10(3):e0120001. https://doi.org/10.1371/journal.pone.0120001

Brenner M, Whitmore TJ, Curtis JH, Hodell DA, Schelske CL (1999) Stable isotope ($\delta^{13}C$, $\delta^{15}N$) signatures of sedimented organic matter as indicators of historical lake trophic state. J Paleolimnol 22:205–221

Brohan P, Kennedy JJ, Harris I, Tett SFB, Jones PD (2006) Uncertainty estimates in regional and global observed temperature changes: a new dataset from 1850. J Geophys Res 111:D12106. https://doi.org/10.1029/2005JD006548

Fan Z, Zhang L (1992) Studying the Geochemistry of Xinjiang Lakes. Geogr Arid Regions 3:1–8 (in Chinese)

Guo N, Zhang J, Liang Y (2003) Climate change indicated by the recent change of inland lakes in Northwest China. J Glaciol Geocryol 25:211–214 (in Chinese with English abstract)

Hayes J (1993) Factors controlling ^{13}C contents of sedimentary organic compounds: principles and evidence. Mar Geol 113:111–125

Hu S (2002) Identification of greigite in lake sediments and its magnetic significance. Sci China Ser D 45:81–87

Hu R, Ma H, Fan Z (2002) Climate Trend indicated by lake variation in Xinjiang. Res Environ Arid Regions 16:20–27 (in Chinese with English abstract)

Jin X (1995) Lakes in China: research of their environment 2 (in Chinese). China Ocean Press, Beijing, p 668

Jones PD, Moberg A (2003) Hemispheric and large-scale surface air temperature variations: an extensive revision and an update to 2001. J Climate 16:206–223

Kump LR, Arthur MA (1999) Interpreting carbon-isotope excursions: carbonates and organic matter. Chem Geol 161:181–198

Li J, Gou X, Cook ER, Chen F (2006) Tree-ring based drought reconstruction for the central Tien Shan area in northwest China. Geophys Res Lett 33:L07715. https://doi.org/10.1029/2006GL025803

Liu H, Xu L, Cui H (2002) Holocene history of desertification along the woodland-steppe border in northern China. Quat Res 57:259–270

Liu DW, Jilili A, Lei JQ, Wu GY (2011) Deposition rate and chemical composition of the aeolian dust from a bare saline playa, Ebinur Lake, Xinjiang, China. Water Air Soil Pollut 218(1–4):175–184 (in Chinese) PMID: 21743755

Liu W, Wu J, Ma L, Zeng H (2014a) A 200-year sediment record of environmental change from Lake Sayram, Tianshan Mountains in China. GFF 136:548–555

Liu DW, Jilili A, Wang LX (2014b) Sediment fluxes and material composition of the salt dust in Ebinur Lake region, Xinjiang. J Glaciol Geocryol 36(2):352–359 (in Chinese). https://doi.org/10.1007/s11096-013-9902-9. PMID: 24366678

Ma L, Wu J, Yu H, Zeng H, Abuduwaili J (2011a) The medieval warm period and the Little Ice Age from a sediment record of Lake Ebinur, northwest China. Boreas 40:518–524

Ma L, Wu J, Abuduwaili J (2011b) Area change of Chaiwopu Lake and its environmental effects in the past 30 years. Arid Land Geogr 34:649–653 (in Chinese with English abstract)

Ma L, Wu J, Abuduwaili J (2013a) Climate and environmental changes over the past 150 years inferred from the sediments of Chaiwopu Lake central Tianshan Mountains northwest China. Int J Earth Sci 102:959–967

Ma L, Wu J, Abuduwaili J (2013b) Climate and environmental changes over the past 150 years inferred from the sediments of Chaiwopu Lake, central Tianshan Mountains, northwest China. Int J Earth Sci (Geol Rundsch) 2013(102):959–967. https://doi.org/10.1007/s00531-012-0838-4

Ma L, Wu J, Liu W, Abuduwaili J (2014) Distinguishing between anthropogenic and climatic impacts on lake size: a modeling approach using data from Ebinur Lake in arid northwest China. J Limnol 73(2):148–155

Ma L, Wu J, Abuduwaili J, Liu W (2015) Aeolian particle transport inferred using a ~150-year sediment record from Sayram Lake, arid northwest China. J Limnol 74(3):584–593

Mahpir J, Tursunov AA (1996) Introduction to the hydroecology in Central Asia. Hygiene Science and Technology Press, Urumqi (in Chinese)

Meyers PA, Teranes JL (2001) Sediment organic matter. In: Last WM, Smol JP (eds) Tracking environmental changes using lake sediments—volume II: physical and chemical techniques. Kluwer, Dordrecht, pp 239–269

Mi Y, Chang SL, Shi QD, Gao X, Huang C (2008) Aquatic environmental quality assessment in Ebinur Lake catchment during high flow period. J Lake Sci 21(6):891–894 (in Chinese)

Oldfield F, Battarbee RW, Boyle JF, Cameron NG, Davis B, Evershed RP, McGovern AD, Jones V, Thompson R (2010) Terrestrial and aquatic ecosystem responses to late Holocene climate change recorded in the sediments of Lochan Uaine, Cairngorms, Scotland. Quat Sci Rev 29:1040–1054

Qian WH, Tang X, Quan LS (2004) Regional characteristics of dust storms in China. Atmos Environ 38(29):4895–4907

Shi Y, Shen Y, Hu R (2002) Preliminary study on signal, impact and foreground of climatic shaft from warm-dry to warm-wet in Northwest China. J Glaciol Geocryol 24:219–226 (in Chinese with English abstract)

Shi Y, Shen Y, Kang E, Li D, Ding Y, Zhang G, Hu R (2007) Recent and future climate change in northwest China. Clim Change 80:379–393

Wang S, Dong G (eds) (2002) Environmental characteristics and evolution in west China. In: Qin D (ed) Assessment on the environmental evolution in west China, Science Press, Beijing, 1:49–61 (in Chinese)

Wang S, Dou H (1998) Lakes in China (in Chinese). Science Press, Beijing, p 580

Wu J, Schleser GH, Lucke A, Li S (2007) A stable isotope record from freshwater lake shells of the eastern Tibetan Plateau, China, during the past two centuries. Boreas 36:38–46

Wu J, Yu Z, Zeng H, Wang N (2009) Possible solar forcing of 400-year wet–dry climate cycles in northwestern China. Clim Change 96:473–482

Wu J, Liu W, Zeng H, Ma L, Bai R (2014a) Water quantity and quality of six lakes in the arid Xinjiang Region, NW China. Environ Proc 1:115–125

Wu J, Liu W, Zeng H, Ma L, Bai R (2014b) Water quantity and quality of six lakes in the Arid Xinjiang region, NW China. Environ Proc 1(2):115–125

Xia J, Zuo QT, Shao MCh (2003) Sustainable management of water resources in lake bosten. Chinese Science Press, Beijing, China (in Chinese)

Zeng H, Wu J, Liu W (2014) Two-century sedimentary record of heavy metal pollution from Lake Sayram: a deep mountain lake in central Tianshan, China. Quatern Int 321:125–131

Zhong W, Xue J, Li X, Xu H, Ouyang J (2010) A Holocene climatic record denoted by geochemical indicators from Barkol Lake in the northeastern Xinjiang, NW China. Geochem Int 48:792–800

Zhou CH, Lei JQ (2005) Trend of sand drift disasters in the Ebinur Lake area and their impacts on the Eurasian continental Bridge Arid land. Geography 28(1):98–102 (in Chinese)

Index

C
Central Asia (CA), 1–3, 5–8, 11, 33, 224, 227–230, 246, 247, 249, 266, 298, 301, 319, 320, 329, 330, 336, 338, 361, 370, 371, 400
Commonwealth of Independent States (CIS), 1, 16, 32, 33, 304, 371
Correlation Coefficient (CF), 320, 327, 328, 339

H
Hydroelectric Power Plant (HPP), 16, 17, 19, 23, 255

M
Maximum Allowable Ratios (MAR), 39
Mean Absolute Error (MAE), 339, 349

N
Nash-Sutcliff Efficiency Coefficient (NSE), 339, 345, 349

P
Palmer Drought Severity Index (PDSI), 405, 407

R
Root Mean Square Error (RMSE), 339, 345, 349

S
System Dynamics (SD) model, 331, 332

W
Water-Economic Basins (WEB), 11, 32, 40, 41
Water Pollution Index (WPI), 38
Weather Research and Forecasting (WRF) model, 347
Weather Stations (WSs), 261, 320, 321

J. Abuduwaili et al., *Hydrology and Limnology of Central Asia*, Water Resources Development and Management, https://doi.org/10.1007/978-981-13-0929-8

Index

C
Central Asia (CA), 1–3, 5, 8, 11, 59, 136, 225, 230, 240, 242, 249, 266, 298, 301, 319, 320, 328, 330, 358, 361, 370, 371, 400
Commonwealth of Independent States (CIS), 16, 23, 136, 154, 191
Conditional Temperature (CT), 300, 322, 323, 359

H
Hydroelectric Power Plants (HPP), 10, 11, 14, 158

M
Maximum Allowable Ratios (MAR), 38, 39
Mean Absolute Error (MAE), 338, 349

N
Nash-Sutcliffe Efficiency Coefficient (NSE), 340, 341, 348

P
Palmer Drought Severity Index (PDSI), 405, 407

R
Root Mean Square Error (RMSE), 338, 345, 349

S
System Dynamics (SD) models, 371, 372, …

W
Water economic districts (WED), 71, 81, 83, …
Water Pollution Index (WPI), 38
Weather Research and Forecasting (WRF) model, 415
Weather Stations (WS), 301, 320, 359

L. [illegible] et al., *Hydrology and Limnology of Central Asia*, Water Resources Development and Management, https://doi.org/10.1007/978-981-13-0929-8

MIX
Papier aus verantwortungsvollen Quellen
Paper from responsible sources
FSC® C105338

If you have any concerns about our products,
you can contact us on
ProductSafety@springernature.com

In case Publisher is established outside the EU,
the EU authorized representative is:
Springer Nature Customer Service Center GmbH
Europaplatz 3, 69115 Heidelberg, Germany

Printed by Libri Plureos GmbH
in Hamburg, Germany